**Springer Proceedings in Physics**

Volume 266

Indexed by Scopus

The series Springer Proceedings in Physics, founded in 1984, is devoted to timely reports of state-of-the-art developments in physics and related sciences. Typically based on material presented at conferences, workshops and similar scientific meetings, volumes published in this series will constitute a comprehensive up-to-date source of reference on a field or subfield of relevance in contemporary physics. Proposals must include the following:

- name, place and date of the scientific meeting
- a link to the committees (local organization, international advisors etc.)
- scientific description of the meeting
- list of invited/plenary speakers
- an estimate of the planned proceedings book parameters (number of pages/ articles, requested number of bulk copies, submission deadline).

***Please contact:***

For Americas and Europe: Dr. Zachary Evenson; zachary.evenson@springer.com
For Asia, Australia and New Zealand: Dr. Loyola DSilva; loyola.dsilva@springer.com

More information about this series at http://www.springer.com/series/361

Leonid Bulavin · Nikolai Lebovka
Editors

# Soft Matter Systems for Biomedical Applications

Springer

*Editors*
Leonid Bulavin
Taras Shevchenko National University of Kyiv
Kyiv, Ukraine

Nikolai Lebovka
National Academy of Science of Ukraine
Biocolloidal Chemistry Institute named after F.D. Ovharenko
Kyiv, Ukraine

ISSN 0930-8989 ISSN 1867-4941 (electronic)
Springer Proceedings in Physics
ISBN 978-3-030-80923-2 ISBN 978-3-030-80924-9 (eBook)
https://doi.org/10.1007/978-3-030-80924-9



This Springer imprint is published by the registered company Springer Nature Switzerland AG
The registered company address is: Gewerbestrasse 11, 6330 Cham, Switzerland

# Preface

This volume comprises the peer-reviewed papers and reviews related to biomedical applications of soft matter systems. Initially, these papers were planned to be presented at the International Conference “Physics of Liquid Matter: Modern Problems” (PLMMP-2020) in May 2020 in Kiev, Ukraine. However, due to the COVID-19 pandemic the PLMMP Conference was shifted to May 2022. The collected papers highlight latest research trends and include contributions from global experts in the field. The book contains 15 chapters, and the content is divided into four parts that include considerations of fundamentals of soft matter systems, mechanisms and molecular interactions, magnetic field effects, and biomedical applications of nanosystems and nanomaterials. Particularly, the water contribution to the protein folding, dynamics of biological water, recent data on modeling and experiments of bio-liquids, behavior of lyotropic liquid crystal phases of phospholipids, and colloidal particles in confined and deformed nematic liquid crystals have been reviewed. Book content also includes analysis of intermolecular interactions and effects of magnetic fields in biosciences. Biomedical applications of smart polymer-based multicomponent nanosystems and of Laponite®-based nanomaterials and formulations have been also presented. We hope that this volume will stimulate further research in all these areas.

Leonid Bulavin
Nikolai Lebovka

# Contents

# List of Contributors

**Dina Badreeva** Laboratory of Information Technologies, Joint Institute for Nuclear Research, Dubna, Russia

**Valentino Bianco** Faculty of Chemistry, Chemical Physics Department, Universidad Complutense de Madrid, Madrid, Spain

**Leonid Bulavin** Faculty of Physics, Taras Shevchenko National University, Kyiv, Ukraine

**Tetiana Busko** Faculty of Physics, Taras Shevchenko National University of Kyiv, Kyiv, Ukraine

**Gaia Camisasca** Department of Mathematics and Physics, Roma Tre University, Rome, Italy

**Vasyl Chekhun** R.E. Kavetsky Institute for Experimental Pathology Oncology and Radiobiology, Kyiv, Ukraine

**S. B. Chernyshuk** Institute of Physics, NAS Ukraine, Kyiv, Ukraine

**Ivan Coluzza** Center for Cooperative Research in Biomaterials (CIC biomaGUNE), Basque Research and Technology Alliance (BRTA), Donostia San Sebastián, Spain;
Basque Foundation for Science, IKERBASQUE, Bilbao, Spain

**Oksana Dmytrenko** Faculty of Physics, Taras Shevchenko National University of Kyiv, Kyiv, Ukraine

**Leon de Villiers Engelbrecht** Department of Chemical and Geological Sciences, University of Cagliari, Monserrato, Italy

**Leon Engelbrecht** Department of Chemical and Geological Sciences, University of Cagliari, Monserrato, Italy

**Giancarlo Franzese** Secció de Física Estadística i Interdisciplinària–Departament de Física de la Matèria Condensada, Facultat de Física, Universitat de Barcelona, Barcelona, Spain;
Institute of Nanoscience and Nanotechnology (IN2UB), Universitat de Barcelona, Barcelona, Spain

**Paola Gallo** Department of Mathematics and Physics, Roma Tre University, Rome, Italy

**V. M. Garamus** Helmholtz-Zentrum Hereon, Geesthacht, Germany

**O. Yu. Gorobets** Institute of Magnetism of NAS and MES of Ukraine, Kyiv, Ukraine

**S. V. Gorobets** National Technical University of Ukraine «Igor Sikorsky Kyiv Polytechnic Institute», Kyiv, Ukraine

**Yu. I. Gorobets** Institute of Magnetism of NAS and MES of Ukraine, Kyiv, Ukraine

**Dmytro Gryn** Faculty of Physics, Taras Shevchenko National University, Kiev, Ukraine

**Victor Gubanov** Faculty of Physics, Taras Shevchenko National University, Kiev, Ukraine

**T. R. Heyn** Institute of Human Nutrition and Food Science, Division of Food Technology, Kiel University, Kiel, Germany

**Pavol Hrubovčák** Frank Laboratory of Neutron Physics, Joint Institute for Nuclear Research, Dubna, Russia;
Department of Condensed Matter Physics, University of P. J. Šafárik in Košice, Košice, Slovakia

**Antonio Iorio** Department of Mathematics and Physics, Roma Tre University, Rome, Italy;
CNRS, Université de Paris, UPR9080, Laboratoire de Biochimie Théorique, Institut de Biologie Physico-Chimique-Fondation Edmond de Rothschild, PSL Research University, Paris, France

**Oleksandr Ivankov** Frank Laboratory of Neutron Physics, Joint Institute for Nuclear Research, Dubna, Russia;
Institute for Safety Problems of Nuclear Power Plants NAS of Ukraine, Kyiv, Ukraine

**Mykola Kaniuk** Palladin Institute of Biochemistry of the National Academy of Sciences of Ukraine (NASU), Kyiv, Ukraine

**Natalia A. Kasian** Institute for Scintillation Materials of State Scientific Insitution "Institute for Single Crystals", National Academy of Sciences of Ukraine, Kharkiv, Ukraine

**J. K. Keppler** Food Process Engineering Laboratory, Agrotechnology and Food Sciences Group, Wageningen University, Wageningen, The Netherlands

**Tomáš Kondela** Frank Laboratory of Neutron Physics, Joint Institute for Nuclear Research, Dubna, Russia;
Department of Nuclear Physics and Biophysics, Comenius University in Bratislava, Bratislava, Slovakia

**Olena Korotych** Chemical Engineering Department, The University of Florida, Gainesville, USA;
Biochemistry and Cellular and Molecular Biology Department, The University of Tennessee, Knoxville, USA

**Norbert Kučerka** Frank Laboratory of Neutron Physics, Joint Institute for Nuclear Research, Dubna, Russia;
Department of Physical Chemistry of Drugs, Faculty of Pharmacy, Comenius University in Bratislava, Bratislava, Slovakia

**Alexander Kuklin** Frank Laboratory of Neutron Physics, Joint Institute for Nuclear Research, Dubna, Russia;
Moscow Institute of Physics and Technology, Dolgoprudny, Russia

**Mykola Kulish** Faculty of Physics, Taras Shevchenko National University of Kyiv, Kyiv, Ukraine

**Nataliya Kutsevol** Faculty of Chemistry, Taras Shevchenko National University, Kyiv, Ukraine

**Yuliia Kuziv** Faculty of Chemistry, Taras Shevchenko National University, Kyiv, Ukraine

**Aatto Laaksonen** Division of Physical Chemistry, Department of Materials and Environmental Chemistry, Arrhenius Laboratory, Stockholm University, Stockholm, Sweden;
Laboratory of Materials-Oriented and Chemical Engineering, Centre of Advanced Research in Bionanoconjugates, Nanjing Tech University, Nanjing, P. R. China;
Centre of Advanced Research in Bionanoconjugates and Biopolymers "Petru Poni" Institute of Macromolecular Chemistry, Iasi, Romania

**Nikolai Lebovka** F. D. Ovcharenko Institute of Biocolloidal Chemistry, National Academy of Sciences of Ukraine, Kyiv, Ukraine

**Andrii Lesiuk** Faculty of Physics, Taras Shevchenko National University of Kyiv, Kyiv, Ukraine

**B. I. Lev** Bogolyubov Institute for Theoretical Physics, NAS Ukraine, Kyiv, Ukraine

**Ming-Ru Li** The Key Laboratory of Beam Technology and Material Modification of Ministry of Education, College of Nuclear Science and Technology, Beijing Normal University, Beijing, China;
Beijing Radiation Center, Beijing, China

**Longin N. Lisetski** Institute for Scintillation Materials of State Scientific Insitution "Institute for Single Crystals", National Academy of Sciences of Ukraine, Kharkiv, Ukraine

**Maryna Manilo** F. D. Ovcharenko Institute of Biocolloidal Chemistry, National Academy of Sciences of Ukraine, Kyiv, Ukraine

**Francesca Mocci** Department of Chemical and Geological Sciences, University of Cagliari, Monserrato, Italy

**Andriy Momot** Faculty of Physics, Taras Shevchenko National University of Kyiv, Kyiv, Ukraine

**Tatiana Murugova** Frank Laboratory of Neutron Physics, Joint Institute for Nuclear Research, Dubna, Russia

**Antonina Naumenko** Faculty of Physics, Taras Shevchenko National University, Kiev, Ukraine

**Tymofii Nikolaienko** Faculty of Physics, Taras Shevchenko National University of Kyiv, Kyiv, Ukraine

**Olena Pavlenko** Faculty of Physics, Taras Shevchenko National University of Kyiv, Kyiv, Ukraine

**Sergiy Perepelytsya** Bogolyubov Institute for Theoretical Physics of the National Academy of Sciences of Ukraine, Kyiv, Ukraine

**Kristyna Pospiskova** Department of Nanobiotechnology, Biology Centre, ISB, CAS, Ceske Budejovice, Czech Republic;
Regional Centre of Advanced Technologies and Materials, Palacky University, Olomouc, Czech Republic

**Joan Àguila Rojas** Secció de Física Estadística i Interdisciplinària–Departament de Física de la Matèria Condensada, Facultat de Física, Universitat de Barcelona, Barcelona, Spain;
Institute of Nanoscience and Nanotechnology (IN2UB), Universitat de Barcelona, Barcelona, Spain

**Ivo Safarik** Department of Nanobiotechnology, Biology Centre, ISB, CAS, Ceske Budejovice, Czech Republic;
Regional Centre of Advanced Technologies and Materials, Palacky University, Olomouc, Czech Republic;
Department of Magnetism, Institute of Experimental Physics, SAS, Kosice, Slovakia

**Yurii Samchenko** F. D. Ovcharenko Institute of Biocolloidal Chemistry, National Academy of Sciences of Ukraine, Kyiv, Ukraine

**Olena Samoylenko** Kavetsky Institute of Experimental Pathology, Oncology and Radiobiology, National Academy of Sciences of Ukraine, Kyiv, Ukraine

**K. Schwarz** Institute of Human Nutrition and Food Science, Division of Food Technology, Kiel University, Kiel, Germany

**Volodymyr Shlyakhovenko** Kavetsky Institute of Experimental Pathology, Oncology and Radiobiology, National Academy of Sciences of Ukraine, Kyiv, Ukraine

**Vadim Skoi** Frank Laboratory of Neutron Physics, Joint Institute for Nuclear Research, Dubna, Russia;
Moscow Institute of Physics and Technology, Dolgoprudny, Russia

**Dmitry Soloviov** Frank Laboratory of Neutron Physics, Joint Institute for Nuclear Research, Dubna, Russia;
Moscow Institute of Physics and Technology, Dolgoprudny, Russia;
Institute for Safety Problems of Nuclear Power Plants NAS of Ukraine, Kyiv, Ukraine

**Anja Steffen-Heins** Institute of Human Nutrition and Food Science, Division of Food Technology, Kiel University, Kiel, Germany

**Liliia V. Sviechnikova** Institute for Scintillation Materials of State Scientific Insitution "Institute for Single Crystals", National Academy of Sciences of Ukraine, Kharkiv, Ukraine

**Lorenzo Tenuzzo** Department of Mathematics and Physics, Roma Tre University, Rome, Italy

**O. M. Tovkach** Department of Physics, Georgetown University, Washington, DC, USA

**Olga V. Vashchenko** Institute for Scintillation Materials of State Scientific Insitution "Institute for Single Crystals", National Academy of Sciences of Ukraine, Kharkiv, Ukraine

**Tudor Vasiliu** Centre of Advanced Research in Bionanoconjugates and Biopolymers "Petru Poni" Institute of Macromolecular Chemistry, Iasi, Romania

**Yonglei Wang** Division of Physical Chemistry, Department of Materials and Environmental Chemistry, Arrhenius Laboratory, Stockholm University, Stockholm, Sweden

**Hui-Ting Xu** The Key Laboratory of Beam Technology and Material Modification of Ministry of Education, College of Nuclear Science and Technology, Beijing Normal University, Beijing, China;
Beijing Radiation Center, Beijing, China

**Feng-Shou Zhang** The Key Laboratory of Beam Technology and Material Modification of Ministry of Education, College of Nuclear Science and Technology, Beijing Normal University, Beijing, China;
Beijing Radiation Center, Beijing, China;
Center of Theoretical Nuclear Physics, National Laboratory of Heavy Ion Accelerator of Lanzhou, Lanzhou, China

**Nan Zhang** The Key Laboratory of Beam Technology and Material Modification of Ministry of Education, College of Nuclear Science and Technology, Beijing Normal University, Beijing, China;
Beijing Radiation Center, Beijing, China

# Part I Fundamentals of Soft Matter Systems

# Chapter 1
# Water Contribution to the Protein Folding and Its Relevance in Protein Design and Protein Aggregation

**Giancarlo Franzese, Joan Àguila Rojas, Valentino Bianco, and Ivan Coluzza**

**Abstract** Water plays a fundamental role in protein stability. However, the effect of the properties of water on the behaviour of proteins is only partially understood. Several theories have been proposed to give insight into the mechanisms of cold and pressure denaturation, or the limits of temperature and pressure above which no protein has a stable, functional state, or how unfolding and aggregation are related. Here we review our results based on a theoretical approach that can rationalize the water contribution to protein solutions' free energy. We show, using Monte Carlo simulations, how we can explain experimental data with our recent results. We discuss how our findings can help in developing new strategies for the design of novel synthetic biopolymers or new possible approaches for mitigating neurodegenerative pathologies.

G. Franzese (✉) · J. Àguila Rojas
Secció de Física Estadística i Interdisciplinària–Departament de Física de la Matèria Condensada, Facultat de Física, Universitat de Barcelona, Martí i Franquès 1, 08028 Barcelona, Spain
e-mail: gfranzese@ub.edu

Institute of Nanoscience and Nanotechnology (IN2UB), Universitat de Barcelona, Martí i Franquès 1, 08028 Barcelona, Spain

V. Bianco
Faculty of Chemistry, Chemical Physics Department, Universidad Complutense de Madrid, Plaza de las Ciencias, Ciudad Universitaria, 28040 Madrid, Spain

I. Coluzza
Center for Cooperative Research in Biomaterials (CIC biomaGUNE), Basque Research and Technology Alliance (BRTA), Paseo de Miramon 182, 20014 Donostia San Sebastián, Spain
e-mail: icoluzza@cicbiomagune.es

Basque Foundation for Science, IKERBASQUE, 48013 Bilbao, Spain

V. Bianco
e-mail: vabianco@ucm.edu



L. Bulavin and N. Lebovka (eds.), *Soft Matter Systems for Biomedical Applications*, Springer Proceedings in Physics 266, https://doi.org/10.1007/978-3-030-80924-9_1

# List of Abbreviations and Symbols

## *Abbreviations*

| | |
|---|---|
| 2D | two-dimensional |
| 3D | three-dimensional |
| b | bulk |
| f | folding |
| GPU | graphics processing unit |
| h | hydration |
| HB | hydrogen bond |
| IFP | inverse folding problem |
| IU | internal units |
| MD | molecular dynamics |
| R | residue |
| SR | stability region |
| u | unfolding |
| w | water |

## *Symbols*

| | |
|---|---|
| $C_p$ | isobaric heat capacity |
| $C_p$ | isobaric heat capacity |
| $F$ | effective free energy |
| $H$ | enthalpy |
| $k$ | factor of compressibility |
| $k_B$ | Boltzmann constant |
| $N$ | number of cells |
| $P$ | pressure |
| $r$ | distance |
| $S$ | entropy |
| $T$ | temperature |
| $V$ | volume |
| $v$ | proper volume |
| $\alpha$ | thermal expansivity factor |
| $b$ | isothermal compressibility factor |
| $\epsilon$ | depth of the potential well |
| $\Delta x$ | difference in magnitude x between final and initial state |
| $\sigma$ | bonding index |
| $\Phi$ | hydrophobic amino acid |
| $\zeta$ | hydrophilic amino acid |
| $\chi$ | mixed amino acids |

## 1.1 Introduction

Proteins are complex molecules consisting of one or more amino acids chains and are essential for all living organisms. They can be part of structural components, as in muscles and collagen, or can develop fundamental functions, like antibodies and enzymes. To perform their biological activity, proteins in many cases need to fold into unique conformations (native states) and be soluble. Protein misfolding can lead to their aggregation with fibrillar morphology (amyloids), which, as a consequence, gives rise to pathological conditions. These conditions include neurodegenerative disorders, as Alzheimer's or Parkinson's disease, and others amyloidosis, i.e., pathological states associated with the formation of extracellular amyloid deposits [1]. Among them, two pathologies of high interest are organ-limited amyloidosis, e.g., the senile cardiac amyloidosis that causes heart failure, and systemic amyloidoses, e.g., the AL amyloidoses affecting kidneys, heart, peripheral nervous system, gastrointestinal tract, blood, lungs, and skin, among other organs.

A folded protein undergoes a conformational change (denaturation) and unfolds when its temperature $T$ is increased above some limiting value. The unfolding by heating stops the protein's biological function and changes its properties, including optical qualities and solubility. A typical example occurs when we cook an egg. Heat makes the transparent and liquid albumen, a protein solution made of about 90% water and 10% proteins, opaque-white and gelatinous.

The thermal denaturation is well understood as a consequence of the increase of the solution's kinetic energy, which induces disorder into the ordered native state. Hence, hot unfolding is a process somehow similar to the melting of a crystal. The transition from the high-$T$ unfolded state to the lower-$T$ folded configuration is understood as a consequence of the protein chain collapse due to the effective attraction among the amino acid residues [2].

However, proteins unfold also when they are pressurized [3]. Unlike the case of high temperature, the effect of pressure $P$ on the stability of the native state is more debated, with some emphasis on the roles of cavities in the folded protein [4].

More intriguing is that proteins can also unfold by decreasing $T$ [5]. Cold denaturation is counterintuitive if we assume that folding is similar to melting, and the protein stability is controlled by an effective attraction that increases when $T$ decreases [6, 7]. It reveals that protein stability results from a balance among different water-mediated forces, including, but not limited to, effective hydrophobic interaction.

To rationalise all the experimental data, Hawley proposed a theory that, based on free energy relations, predicts for any protein with a native state a close stability region (SR) with, in general, an elliptic shape in the $T$-$P$ plane. Depending on the specific protein and its measurements, the elliptic SR can also extend toward negative pressures. Hence, the theory predicts that a protein would denature under tension, consistent with experimental extrapolations [8] and direct observations [9].

Cold and $P$ unfolding can be rationalised, assuming an enthalpic gain of the solvent upon the denaturation process without clarifying this gain's origin and its relationship with molecular interactions [10]. Alternative theories have hypothesised a $P$-dependent hydrophobic collapse [11] or enthalpic gain and entropic cost [12]. Further analytic studies have shown that indirect accounting for the $P$-dependent water-water and water-protein interactions are enough to explain why in the experimental is more difficult to observe the high-pressure reentrant helix–coil transition in DNA than in protein folding [13]. However, none of these approaches can reproduce the entire elliptic SR as Hawley predicted and supported by experiments. As we discuss in this short review, cold and $P$ unfolding can be better understood as a competition between different free energy contributions coming from water, one from hydration water and another from bulk water, reproducing the whole elliptic SR.

Also, within our approach, we rationalise why proteins, stable at ambient conditions, do not necessarily work at extreme $T$-$P$ conditions and why there are $T$-$P$ above which no protein is stable in its native state. By explicitly including water, we can show its evolutionary driving force toward thermophilic proteins with higher surface hydrophilicity at high $T$ and ice-binding proteins with lower surface hydrophilicity at low $T$. On the other hand, the proteins have smaller segregation between the hydrophilic surface and the hydrophobic core if they are stable at high $P$. As we discuss in the following, our theory allows us to introduce a new protein design protocol useful for engineering proteins and drugs working far from ambient conditions.

Finally, we discuss the protein interface effect on the water fluctuations in the protein hydration shell and their relevance in the protein–protein interaction. We study the folding and aggregation of proteins as a function of their concentration. The mechanisms leading to the folding process's failure and the formation of potentially dangerous protein aggregates are a matter of considerable scientific debate [14]. We show that the propensity to aggregate is not strictly related to the surface hydrophobicity of the protein. The increase of the concentration of individual protein species can induce a partial unfolding of the native conformation without the occurrence of aggregates. The concentration at which the proteins aggregate is, indeed, higher than that at which they unfold. Hence, we discuss two-steps smooth transition between the folded, unfolded, and aggregated states of proteins and how this process could be affected by a nearby hydrophobic interface, such as that of a nanoparticle.

This chapter does not pretend to represent an exhaustive review of the vast literature about protein folding, aggregation, and design generated in many decades by the community of renowned biophysicists. It is just a summary of a part of our contributions to the field. It can be considered as a structured introduction to the topic for beginners in this field. It assumes the knowledge of basic notions of Statistical Physics and Thermodynamics and provides a minimal list of references where the interested reader can find further details and more comprehensive reviews.

## 1.2 Hawley's Theory

In 1971, Hawley proposed a theory based on the assumption that the folding (f) unfolding (u) transition can be modelled as a reversible first-order phase transition and that equilibrium thermodynamics hold during the denaturation [15]. The theory is based on a Taylor expansion of the difference in Gibbs' free energy $\Delta G$ between the folded and the unfolded state truncated at the second order:

$$\Delta G(P,T) = \frac{\Delta\beta}{2}(P-P_0)^2 + 2\Delta\alpha(P-P_0)(T-T_0) - \frac{\Delta C_P}{2T_0}(T-T_0)^2 + \Delta V_0(P-P_0) - \Delta S_0(T-T_0) + \Delta G_0 \quad (1.1)$$

where $T_0$ and $P_0$ are the temperature and pressure of the ambient conditions, $\Delta V_0$ and $\Delta S_0$ are the volume and entropy variation along the folded-to-unfolded (f-u) transition, $\Delta\alpha$, $\Delta C_P$ and $\Delta\beta$ are the differences along the f-u transition in the thermal expansivity factor, the isobaric heat capacity and the isothermal compressibility factor, respectively. This equation is constrained by $\Delta\alpha > \Delta C_p \Delta\beta / T_0$ resulting in the SR having an elliptic shape (Fig. 1.1). Even if it is a phenomenological model, the ability of Eq. (1.1) to describe the experimental data makes it a useful tool when studying the f-u transformation.

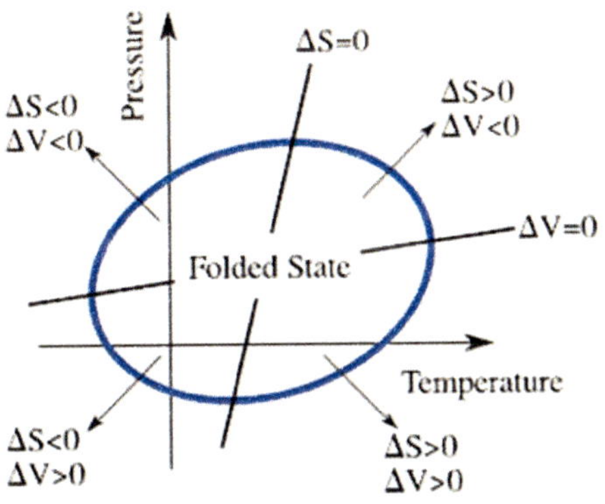


**Fig. 1.1** Stability Region of a protein according to the Hawley theory in a temperature–pressure *T-P* representation. The elliptic line separates the native (folded) from the denatured state. The folded-to-unfolded transition occurs by varying either *T* or *P*, or both, and implies changes in the volume *V* or the entropy *S* as illustrated along the arrows. Loci with $\Delta V = 0$ and $\Delta S = 0$ cross the ellipsis where its tangent has infinite or zero slope, respectively

## 1.3 The Model for the Hydrated Protein

We consider a single protein, represented by a flexible chain of coarse-grained residues ($R$), suspended in water ($w$) and described by the enthalpy

$$H \equiv H_{R,R} + H_{R,w} + H^{(h)}_{w,w} + H^{(b)}_{w,w} \qquad (1.2)$$

where the first term accounts for the residue-residue contribution, the second for the residue-water contribution at the protein interface, the third for the hydration ($h$) water contribution, and the fourth for the bulk ($b$) water contribution away from the protein interface. Before describing in detail each of these terms in the following Eqs. (1.3), (1.8), (1.13), we observe that this minimal model comprises:

1. The covalent (peptide) bonds between the amino-terminal and the carboxyl-terminal of consecutive monomers along the chain, which constrains the possible configurational changes of the protein and the total conformational entropy.
2. The interactions between non-consecutive residues, including a repulsive and an attractive part, represented as a pairwise term, sum of all the $R$-$R$ (van der Waals, hydrogen bond, and electrostatic) contributions. This term is one of the driving forces leading to the native state.
3. The $R$-$w$ van der Waals and hydrogen bonding enthalpy at the protein surface, influencing the protein secondary structure, exposing the hydrophilic residues to water, and burying the hydrophobic residues into the folded protein core.
4. The $w$-$w$ van der Waals and hydrogen bond enthalpy in the hydration layer, influenced by the nearby protein residues.
5. The $w$-$w$ van der Waals and hydrogen bond enthalpy in the bulk, not affected by the protein residues.

By changing the pressure $P$ and the temperature $T$, as we discuss in the following, the enthalpy and the Gibbs free energy of the system change and drive the protein toward folded or unfolded states.

### *1.3.1 The Franzese-Stanley Coarse-Grained Water Model*

In a diluted protein solution, water represents more than 90% of the system. It is the thermal bath that continuously interchanges energy, entropy, and enthalpy with the protein. This interchange allows the protein to overcome the potential energy barriers, separating metastable states from the native state, within the range of $P$ and $T$ at which the folded protein is stable. Fully atomistic models can explicitly account for the water contribution to the free energy but are limited in size and time by their large computational cost. All atoms molecular dynamics (MD) simulations with explicit water are very expensive due to the large number of water molecules necessary to solvate even small proteins. For example, MD can simulate proteins up to 80 amino acid residues in ~ 11,000 water molecules for a cumulative simulation time of 643 μs on a specialized Anton supercomputer [16]. On more affordable CPU or GPU hardware all atoms MD calculations with specialized algorithms can simulate, e.g., 42-residues-long proteins in ~5,000 water molecules for a cumulative simulation time of 52 μs [17] or 200 μs [18], or a 29-residue-long region of a disordered protein in ~8,200 water molecules for an aggregated simulation time of ~1.4 ms [19].

If, instead, the model has an implicit solvent with effective $T$ and $P$-dependent force-fields, larger proteins, up to 92 amino acids, can be simulated for milliseconds. However, the drawback, in this case, is that the water effect on the residues interactions is included only in an effective way, losing the transferability of the model to others $T$ and $P$ and changing the dynamics. Possible alternatives combine MD with experimental constraints reaching predictions of the structures of proteins up to 326 amino acids [20] and competing with structural bioinformatics methods [21]. However, also in these cases, it is hard to tell if the protein-folding kinetics resembles the experiments. More details can be found in recent reviews [22, 23].

Another common option is coarse-grain groups of atoms into larger rigid units and to combine this grouping with implicit water or pseudo-explicit water (charged Lennard–Jones particles replacing groups of water molecules) [24]. These models gain speed in simulations by reducing the resolution of the description. Hence, their use is limited to those cases in which no atomic resolution is needed, and the principal aim is, in general, a qualitative understanding of a specific process. However, these approaches, although useful in several studies, are unable to account for the most specific contribution of water, the hydrogen bonds.

To explicitly include the water into a coarse-grain protein representation, we developed in these years a water model in which we retain the description of the water hydrogen bonds at the molecular level but lose the molecular resolution of the water position, replacing it with a density field. Furthermore, we account in the model for the many-body contribution of the hydrogen bonds [25–30]. We have shown that this contribution is relevant to better describe the water anomalous properties at least on a qualitative level [31–40]. Furthermore, the simple formulation of the model allows us to perform approximated analytic calculations that facilitate the understanding of the relevant mechanisms underlying the properties of the proteins [41–45].

The large reduction of water degrees of freedom that we attain with our specific coarse graining allows us to study systems as large as 160.000 water molecules [29] or more (512.000 water molecules, unpublished), reaching sizes that are challenging for atomistic models. Furthermore, it allows to reach simulation timescales as long as 100 s [46], comparable to the experimental times.

In particular, we focused on the water under confinement or at hydrophobic interfaces [47–54] or the interface with proteins [46, 55], see also [56–60]. In the following, we will define the model focusing on the protein folding problem [61–65], the protein design [66, 67], and the protein aggregation [68, 69].

In this approach, we partition the bulk into $N$ cells, one per molecule, and consider that, at high $T$ and ambient $P$, when there are no hydrogen bonds, the molecules are homogeneously distributed in volume $V$, each with a proper volume $v = V/N$. We consider the system at constant *N, P, T* with a fluctuating volume $V_{TOT}^{(b)}$ and an enthalpy [29, 30]

$$H_{w,w}^{(b)} \equiv \sum_{ij} U(r_{ij}) - JN_{HB}^{(b)} - J_\sigma N_{coop} + PV_{TOT}^{(b)} \tag{1.3}$$

where the first term

$$\sum_{ij} U(r_{ij}) \equiv 4\sum_{ij}\left[\left(\frac{r_0}{r_{ij}}\right)^{12} - \left(\frac{r_0}{r_{ij}}\right)^{6}\right] \tag{1.4}$$

is the Lennard–Jones (LJ) potential energy (isotropic term), being $\epsilon = \frac{5.8\,\mathrm{kJ}}{\mathrm{mol}}$ the depth of the potential well, $r_0 = 2.9$Å the distance from the particle to the well (molecule's hard-core), and $r_{ij}$ the O–O distance. A cut-off of the LJ potential was established at $r = 6r_0$ to accelerate the calculations.

The second term accounts for the directional contribution of the hydrogen bond (HB) interaction, with $J = 1.2\epsilon$ being the energetic gain for each HB. Here,

$$N_{HB}^{(b)} \equiv \sum_{\langle ij\rangle} n_i n_j \delta_{\sigma_{ij}\sigma_{ji}} \tag{1.5}$$

is the number of HBs in the bulk, depending on the local density index $n_i$ and the bonding index $\sigma_{ij}$, with $\delta_{\sigma_{ij}\sigma_{ji}} = 1$ if molecules *i* and *j* have the same bonding index, $\delta_{\sigma_{ij}\sigma_{ji}} = 0$ otherwise. The first index is $n_i = 1$ if the proper volume of the molecule *i* has a characteristic size at most ~3.7 Å, the distance at which a HB breaks [28, 29], $n_i = 0$ otherwise. The bonding index $\sigma_{ij} = 1, \ldots, q$ describes the relative orientation of molecules *i* to its neighbour *j*. We choose the parameter *q* by selecting 30° as the maximum deviation from a linear bond (i.e., $q \equiv 180°/30° = 6$) [29, 30]. The sum in (1.5) runs over neighbouring molecules and, to simplify the model, we assume that each molecule can form at most four HBs.

The third term of Eq. (1.3) takes into account the quantum many-body interaction caused when a new HB is formed, reinforcing all the other HBs formed by the same molecule [28, 29]. It is

$$N_{coop} \equiv \sum_i n_i \sum_{(l,k)_i} \delta_{\sigma_{ik}\sigma_{il}} \tag{1.6}$$

and it mimics the cooperativity of the HBs between the possible pairs of the $\sigma_{ij}$ indices of the molecule *i*. The choice $J_\sigma = 0.2\epsilon$, an order of magnitude smaller than *J*, ensures that the cooperative term is relevant only below the temperature at which the HBs are formed.

In Eq. (1.3) the total volume is [29, 30]

$$V_{TOT}^{(b)} \equiv V + N_{HB}^{(b)} v_{HB}^{(b)} \tag{1.7}$$

where $cv_{HB}^{(b)}/v_0 = 0.5$, the average volume increase associated with the formation of a HB, is given by the average volume increase between high-density ices VI and VIII and low-density (tetrahedral) ice Ih, and $v_0$ is the proper volume of a water molecule without the HB. The volume term $V$ is free to fluctuate as a function of the LJ interaction and depends weekly on $T$ and $P$ near ambient conditions. The local formation of HBs gives rise to the local density fluctuations responsible for the density and compressibility anomaly of water [31–40].

### 1.3.2 The Coarse-Grained Hydration Water Model

As we describe in the following, we coarse-grained the proteins as self-avoiding heteropolymers with residue that can occupy only one of the cells of the system. Hence, when we include proteins in the system, we replace water cells with coarse-grained residues, generating an interface. Each residue can be a neighbour of another residue or can be hydrated. The water can form HBs within the hydration shell. The energy of the hydration-shell HBs depends on the nature of the nearby amino acids. For water molecules, forming a HB, near two hydrophobic ($\Phi$) amino acids, two hydrophilic ($\zeta$) amino acids, or one of each kind (mixed, $\chi$), the enthalpy is [63–65]

$$H_{w,w}^{(h)} \equiv H_{w,w}^{(\Phi)} + H_{w,w}^{(\zeta)} + H_{w,w}^{(\chi)} \tag{1.8}$$

where

$$H_{w,w}^{(\Phi)} \equiv \sum_{ij} U\left(r_{ij}\right) - J^{(\Phi)} N_{HB}^{(\Phi)} - J_{\sigma}^{(\Phi)} N_{coop}^{(\Phi)} + P V^{(\Phi)} \tag{1.9}$$

with

$$V^{(\Phi)} \equiv N_{HB}^{(\Phi)} v_{HB}^{(\Phi)} \tag{1.10}$$

and

$$v_{HB}^{(\Phi)} / v_{HB,0}^{(\Phi)} \equiv 1 - k_1^{(\Phi)} P \tag{1.11}$$

where $v_{HB,0}^{(\Phi)}$ is the volume increase for $P = 0$, and $k_1^{(\Phi)} = \frac{v_0}{4\epsilon}$ is a factor accounting for the compressibility of the hydrophobic hydration shell [64–66]. Based on numerical and experimental observations that point to a higher correlation in the HBs between water molecules near hydrophobic residues, Bianco et al. [64–66] adopted $J^{(\Phi)} = 1.83J$, while $J_{\sigma}^{(\Phi)} = J_{\sigma}$.

Expressions similar to Eq. (1.9)-(1.11) hold for the hydrophilic ($\zeta$) and the mixed ($\chi$) cases in Eq. (1.8). Parameters in the mixed case are set equal to the average of the $\Phi$ and the $\zeta$ case. To simplify the model, the parameters for the $\zeta$ case are set equal to the bulk case and $k^{(\zeta)} = 0$ because the compressibility effect is due mainly to the hydrophobic interface [64–66].

The total volume occupied by the water is

$$V_{TOT} \equiv V_{TOT}^{(b)} + V^{(\Phi)} + V^{(\zeta)} + V^{(\chi)} \tag{1.12}$$

and fluctuates with $T$ and $P$. The volume occupied by the proteins is constant.

### 1.3.3 The Coarse-Grained Protein-Water Model

The protein's enthalpy, including the interaction with the hydration shell, is modelled as [64–66]

$$H_p \equiv H_{R,R} + H_{R,w} \equiv \sum_i^{N_c} \left[ \sum_{i \neq j}^{N_c} C_{ij} S_{ij} + \sum_{j'}^{N_w} C_{ij'} S_i^W \right] \tag{1.13}$$

where the first sum runs over the $N_c$ residues indices, the second over the $N_w$ hydration water molecules, $C$ is a contact matrix with $C_{mn} = 1$ if $m$ and $n$ are first neighbours and 0 otherwise, $S$ is the Miyazawa-Jernigan matrix [70, 71] where $S_{ij}$ accounts for the interaction between amino acids $i$ and $j$, and $S_i^W$ is the interaction energy the residue $i$ with the hydration water molecule and depends on residue's hydropathy, with $S_i^W = -\epsilon^{(\Phi)}$ or $-\epsilon^{(\zeta)}$ if the residue is hydrophobic or hydrophilic, respectively.

With this model, Bianco and Franzese [64] identify the different mechanisms with which water participates in the cold and pressure denaturation. Furthermore, Bianco et al. in Ref. [65] analyse how changes in i) the specific protein residue-residue interactions in the native state of the amino acids sequence, and ii) the water properties at the hydration interface, affect the protein stability region. They show that the solvent properties are essential to rationalize the stability region shape at low $T$ and high $P$, independent of the model's parameters and for both proteins with a native state and disordered proteins. These results open the perspective to develop advanced computational design tools for protein engineering.

## 1.4 Protein Design

Protein design has as main objective to understand how a protein sequence encodes specific structural and functional properties [72]. It consists of searching for the ensemble of sequences that fold into a target protein backbone structure. However, it remains one of the major challenges across the disciplines of biology, physics, and chemistry [73, 74].

Protein design is often referred to as the inverse folding problem (IFP). IFP is still one of the toughest major challenges in biophysics, and the most common approach is to search for the sequence that minimizes the energy or a scoring function into the target protein [72, 75–77]. When considering extreme conditions, it is important to observe that protein sequences that are robust against mutations tend to have larger thermal stability [74, 77]. To this end, previous studies tried to identify how intra-protein interactions affect the thermal stability of the native structure [78], or to optimize specific interactions, such as the charge, to improve the stability of the proteins [79].

However, until the work by Bianco et al. [66], the way the design changes at extreme conditions was not disentangled. In Ref. [66] the authors develop a strategy for protein design, following the previous approach introduced by Coluzza et al. [75, 76], and focusing on the relationship between the sequence and the thermodynamic conditions at which the target structure is stable. They adopt the coarse-grained model for hydrated proteins described in the previous section to perform the design of heteropolymers on a lattice.

According to previous design studies on the lattice a solution to the folding problem can be found by looking for the sequence with the lowest energy under the constraint that the composition of the amino acids maximizes the number of letter permutations $N_P = N!/(n_A!n_B!..n_Y!)$, where $N$ is the sequence length and $n_x$ is the number of residues corresponding to the letter $x$.

In the presence of explicit water, the solvent interaction must be taken into account during the energy minimization. However, a simple downhill search does not work because it would be too sensitive to the large noise produced by the water configurations. Hence, the best strategy is to average the energy of each sequence on the ensemble of water configurations, getting an effective free energy per sequence $F_S$. Such an approach is currently only feasible in a water model on a lattice, where a large number of solvent configurations per sequence can be scanned efficiently, allowing the design protocol to explore many solutions. The design process consists of finding the sequence that minimizes $F_S$ under the constraint of maximizing $N_P$. As a consequence of this approach, the ensemble of the design solutions is influenced by the environmental parameters, such as temperature and pressure, via their effect on the solvent properties. An interesting extension would be to include also the effect of pH changes on the solvent.

Finally, it is important to stress that we considered only fully compact structures. The presence of internal pockets could be used to design stable proteins with internal water molecules. Following the pioneering work of Sterpone et al. [80–82] we can extrapolate that internal water could have an additional stabilization role, a prospect that will definitely deserve future investigation.

### 1.4.1 Design Protocols

Bianco et al. [66] follow two protocols to design a stable protein at a given $T$ and $P$. Both protocols are based on the average enthalpy associated with the hydrated protein. The first (MIN ENTHALPY) consists of minimizing the average enthalpy in the protein folded state. The second (MAX GAP) maximizes the enthalpy gap between the folded and the unfolded protein conformations.

In the following, we adopt the first protocol. We start from an initial random sequence for the protein already folded in the target structure and we sample the space of sequences. At each sampling step, we replace an amino acid of the polymer either by exchanging it with another of the 20 possible monomers or by swapping the position of two amino acids in the sequence. We perform 100 replacements for each swap.

For each move, we check (i) if it does not lead to a more homogeneous sequence, that would be unable to fold, and (ii) if it decreases the protein's enthalpy. If both conditions are met, we accept the protein move, otherwise we accept it with probability [66]

$$p_{acc} \equiv min\{1, exp(-\Delta/T_0\}min\{1, (\Pi_n/\Pi_o)^{\omega}\} \quad (1.14)$$

where $\Delta$ is the change in enthalpy, expressed in internal units, between the new and the old sequence, $T_0 = 0.05$ is an optimization temperature, $\Pi_n$ and $\Pi_o$are the numbers of permutations of the new and old sequence, respectively, $\omega = 14$ is a weighting parameter. The interested reader can find further details and references in Ref. [66]. Next, we perform a series of Monte Carlo steps to equilibrate the water with cluster and single variable moves [28, 29]. Once the water is equilibrated, we compute the protein's average enthalpy and check that it is smaller than the previous one. Hence, the new sequence is a better candidate for the protein to fold into the target structure at the given pressure and temperature.

For each thermodynamic state point, we sample more than $10^8$ independent sequences and consider, for characterization and stability analysis, only the best 5 or 15 sequences. For these selected sequences we check their validity (ability to fold into the target structure) with isobaric, isothermal Monte Carlo simulations, starting from a stretched protein conformation. We finally select the valid sequence with the smallest enthalpy in the native state [66].

### 1.4.2 Design Results in Two Dimensions

In Ref. [66], Bianco et al. identify 5–15 optimized sequences for each design pressure and temperature, for a total of more than $1.5 \times 10^3$ optimizations, a number far beyond the capability of any fully atomistic protein model. For each designed sequence, they test the stability in $T$–$P$ by checking if the protein folds into the target structure.

Consistent with natural proteins, they show that sequences that are stable at ambient conditions do not necessarily work at extreme conditions of $T$ and $P$ and that the range of stability in $T$ and $P$ increases with the design temperature $T_d$ at ambient pressure (Fig. 1.2). Also, they find that there are limits of $T$ and $P$ above which no protein has a stable functional state [83–85] and no stable sequence can be found. Furthermore, small proteins undergo cold denaturation for temperatures that are lower than those for longer proteins, with a higher content of hydrophobic residues [66]. They demonstrate that these limits, and the selection mechanisms for proteins, depend on how the properties of the surrounding water change with $T$ and $P$.

They study how the design temperature and pressure, $T_d$ and $P_d$, affect the segregation of a hydrophilic surface and a hydrophobic core in the selected proteins. They find that the segregation is never extreme but is larger at high $T_d$ and low $P_d$ (Fig. 1.3). This result is consistent with experimental trends for thermophilic, mesophilic [86], and ice-binding proteins [66].

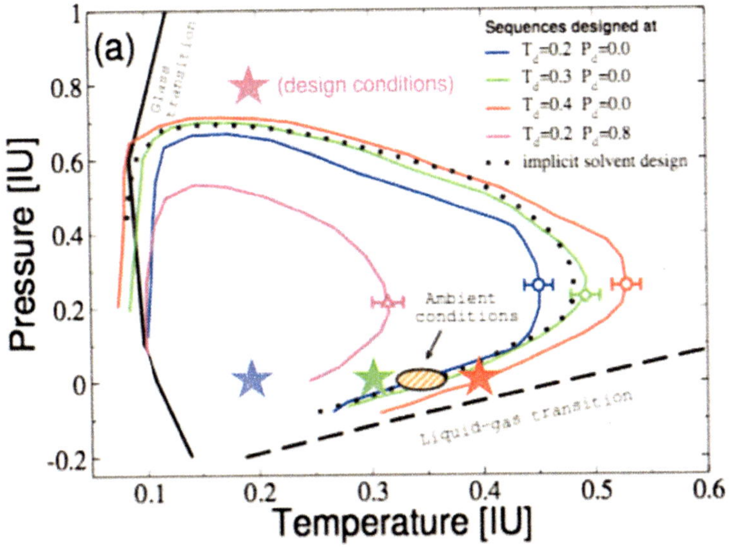


**Fig. 1.2** *T-P* stability region (SR) for proteins designed to fold in a given native structure, with the explicit water model described in the text. The coloured continuous lines enclose the SRs for sequences designed at the temperature and pressure indicated in the legend and marked by corresponding-coloured stars. The dotted line marks the SR found with implicit water. The continuous black line locates the glass transition below which the relaxation times toward equilibrium exceed our observation times (here the Monte Carlo simulation times) as in real experiments. The dashed line traces the liquid–gas spinodal. The orange-shaded ellipse evidences the near-ambient conditions. The symbols with error bars represent the typical error over the estimate of the SRs. Pressure and temperature are expressed in internal units (IU), corresponding to $4\varepsilon/v_0$ and $4\varepsilon/k_B$, respectively, where $k_B$ is the Boltzmann constant. Adapted from Ref. [66]

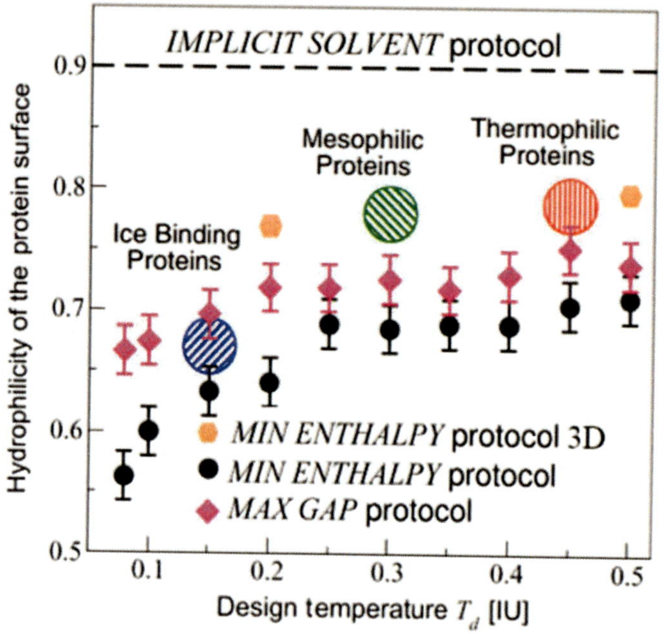


**Fig. 1.3** Hydrophilicity of the surface of the designed protein for a range of design temperatures $T_d$ at $P_d = 0$ calculated with explicit water. Sequences selected with different protocols in 2D (black dots and purple rhombuses) qualitatively follow the trend observed for the average hydropathy of real thermophilic (large red circle), mesophilic (large green circle), and ice-binding proteins (large blue circle). Differences between the two protocols are discussed in Ref. [65]. Preliminary results in 3D and two different design temperatures (orange hexagons) show a consistent trend. The discontinuous line corresponds to the calculations with implicit water. The design temperature $T_d$ is expressed in internal units (IU), as in Fig. 1.2, and the hydrophilicity as a dimensionless number normalized to 1, corresponding to the fraction of hydrophilic surface residues. Adapted from Ref. [66]

Surprisingly, they observe that larger segregation reduces the stability range in $T$ and $P$. The more segregated sequences, selected at high T, are highly resistant (*superstable*) to both cold and pressure denaturation. Superstable proteins have an average stability region that encompasses the average stability region of proteins designed at ambient conditions. These results are potentially useful for engineering proteins and drugs working far from ambient conditions and offer an alternative rationale to the evolutionary action exerted by the environment in extreme conditions [66].

### 1.4.3 *Design Results in Three Dimensions*

Here we describe our preliminary results for the extension of the design approach, described above, to the 3D case. We aim to assess how the dimensionality affects them when compared with those achieved in 2D.

The change of dimensionality implies some differences in the relative weight of the enthalpy and energy terms described in Sect. 1.3. In particular, if we consider a partition of the 3D volume into cubic cells, each molecule (or residue) has six neighbours, instead of four as in the 2D partition into squares. On the one hand, each water molecule can form only up to four HBs, hence the entropy associated with the HB formation largely increases in 3D compared to 2D. On the other hand, the six neighbours in 3D give rise to more terms in the energy contributions to the total enthalpy, partially balancing the increased entropy.

To keep our preliminary calculations affordable on cheap GPUs, we design a 48-residue-long heteropolymer with a compact target structure (Fig. 1.4). We hydrate the protein with 681 water molecules within a cubic volume partitioned into $9^3$ cells with periodic boundary conditions. The target structure has a core of four residues and a surface of 44 amino acids on three different layers of the partition. We optimize the sequence using the MIN ENTHALPY protocol at three thermodynamic conditions, listed in Table 1.1. For each thermodynamic condition, we observe that the design protocol leads to sequences with converging enthalpies within $10^8$ iterations (Fig. 1.5).

First, we observe that in 3D (Fig. 1.4) 38, i.e., 80%, of the 44 surface residues are hydrophilic. This value is 30% higher than the surface hydropathy found in 2D at the same $P$ and $T$. However, the 2D result is for a heteropolymer with a smaller number of surface residues, due to its shorter length (30 residues) and to the lower dimensionality. We find that the trend in the difference in surface hydropathy between 2 and 3D design is independent of P and $T$, being always higher in 3D (Table 1.2).

On the contrary, the fraction of hydrophobic residues in the core in 3D is less than that observed in 2D. Here, only one of the four core residues is hydrophobic, while in 2D approximately 40% of the core was hydrophobic. However, due to the small size of the present core (four residues), any hydropathy smaller than 50% (corresponding to two residues), would be discretized to one single residue.

We rationalize these differences in surface and core hydropathy as due to the different sizes of surface and core sequences. In particular, these results suggest that the smaller the sequence, the smaller the fraction of surface hydrophilic residues or core hydrophobic amino acids. Longer proteins, with larger cores in 3D and larger surfaces in 2D, should be investigated to clarify this point.

Despite these differences between 2D [66] and 3D preliminary results, we find that the $T_d$-dependence of the surface hydropathy in 3D is consistent with that found in 2D and in real thermophilic, mesophilic, and ice-binding proteins (Table 1.2 and Fig. 1.3). In particular, the residue segregation increases by increasing the design temperature $T_d$. More systematic studies, beyond the scope of this minireview, are needed in this direction to support these preliminary results.

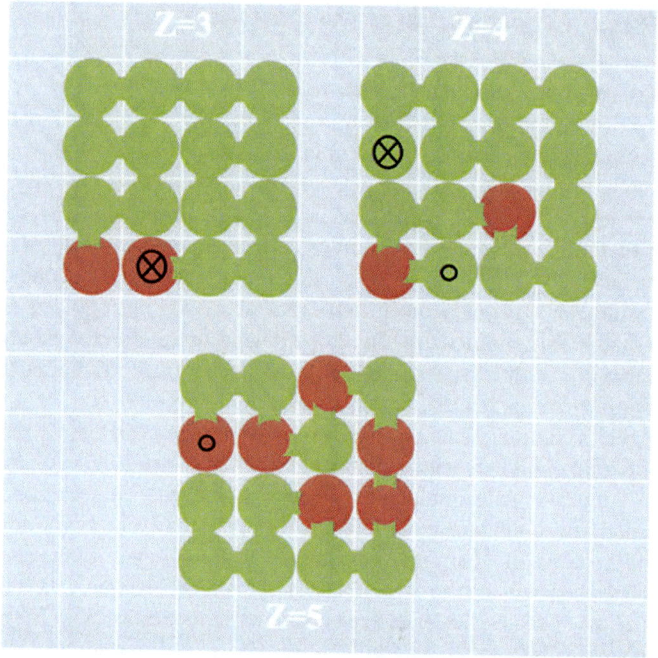


**Fig. 1.4** Schematic representation of the sequence selected by our design protocol at $P = 0.8\,(4\varepsilon/v_0)$ and $T = 0.2\ (4\varepsilon/k_B)$ for a 48-residue-long heteropolymer with the compact target structure as in the figure. The structure occupies three planes in the cubic partition of the volume, corresponding to the $z$-coordinate $z = 3, 4, 5$, with $1 \leq z \leq 9$. Each plane, and the corresponding $z$ value, are represented separately for clarity. We mark the chain structure by green segments within the plane and by a crossed or an empty circle—depending on if the chain moves up or down, respectively––between the $z$-planes. We represent the hydrophilic and the hydrophobic residues in green and brown, respectively, according to the Kyte-Doolittle hydropathy scale [87]. All the residues, but the (central) core four at $z = 4$, are exposed to water

**Table 1.1** Design pressure and temperature at which we optimize the protein sequence with the target structure in Fig. 1.4. Initial and final enthalpy refers to the calculated values at the beginning and the end of the optimization protocol. All the quantities are expressed in internal units (IU): $4\varepsilon/v_0$, $4\varepsilon/k_B$, $4\varepsilon$ for pressure, temperature, and enthalpy, respectively

| Pressure [IU] | Temperature [IU] | Initial enthalpy [IU] | Final enthalpy [IU] |
|---|---|---|---|
| 0.2 | 0.2 | −0.2 | −12 |
| 0.2 | 0.5 | −0.2 | 14.12 |
| 0.8 | 0.2 | −0.2 | −33.71 |

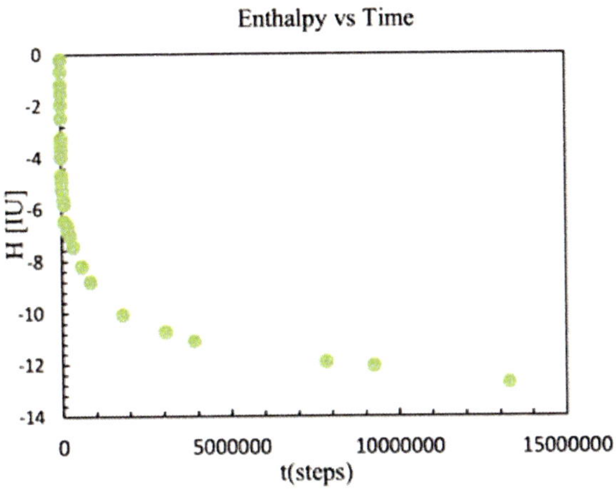


**Fig. 1.5** Enthalpy changes as a function of the number of iterations (time steps) for our design protocol at $P = 0.2\ (4\varepsilon/v_0)$ and $T = 0.2\ (4\varepsilon/k_B)$ for a 48-residue-long heteropolymer with the compact target structure as in Fig. 1.4. Symbols mark the iterations that lead to sequences with lower enthalpy. Enthalpy is measured in internal units (IU), $4\varepsilon$

**Table 1.2** Surface hydropathy (fraction of hydrophilic residues on the protein surface) for the sequences selected by the MIN ENTHALPY protocol in 2D and 3D calculations at different design pressures $P_d$ and temperatures $T_d$. Values in 2D are from Ref. [66]. All quantities are expressed in internal units (IU)

| $P_d$ [IU] | $T_d$ [IU] | 2D Surface hydropathy | 3D Surface hydropathy |
|---|---|---|---|
| 0.2 | 0.2 | 0.65 | 0.77 |
| 0.2 | 0.5 | 0.70 | 0.80 |
| 0.8 | 0.2 | 0.55 | 0.80 |

## 1.5 Protein Unfolding as a Precursor of Protein Aggregation

Protein aggregation is a mechanism related to a large number of diseases, such as Spongiform encephalopathies, Parkinson's, or Alzheimer's disease [1, 88–90], and is a relevant issue in biopharmaceutical production [91]. It can be related to the variation of external factors [92] or the milieu composition, as, e.g., ions concentration [93] and it is mostly inevitable when protein concentrations exceed the natural values [68]. Nevertheless, the mechanisms leading to the aggregation are still debated [94] and the objective of extended research. In particular, coarse-grained models represent valid tools to investigate protein solutions at large concentrations in explicit water [68, 95, 96].

In a recent publication, Bianco et al. [68] use the coarse-grain model described in the previous sections to study by Monte Carlo simulations how the increase of concentration of individual protein species affects their folding and aggregation in an aqueous solution. They follow the design protocol described in this chapter to select eight proteins with native structures with maximally compact conformations, composed of 36 or 49 amino acids. The selected proteins have different compositions, with a range of surface and core hydropathy combinations. The fraction of hydrophilic surface amino acids varies from 0.5 to 0.7, while that of hydrophobic residues in the protein core is between 0.25 and 0.45.

They calculate the free energy profile of these proteins as a function of the native contacts and inter-protein contacts, for different protein concentrations $c$, defined as the volume fraction of the system occupied by the protein and expressed in percentage. In all the cases they observe that for low concentrations, $c \lesssim 5\%$, all the proteins reach their native folded state and, on average, are not in contact with each other. They show that the increase of $c$ can induce a partial unfolding of the native conformation without inducing aggregation [68].

At very high (sequence-dependent) concentrations, e.g., $c \geq 20\%$ in some of the studied cases, Bianco et al. find that the proteins lose completely their folded structures and aggregate. Comparing proteins with different compositions, they show that, as long as the design explicitly accounts for the water environment, the propensity to aggregate is not strictly related to the hydrophobic content of the protein's surfaces. Also, proteins with the same native structure, but different sequences, have thresholds for the unfolding and the aggregation that are different in each case [68].

By approaching two isolated proteins of the same species, they show that their unfolding is water-mediated and starts before the residues can interact directly (Fig. 1.6). This occurs at a distance close to the average separation between proteins at the unfolding concentration. Furthermore, by switching off the water terms

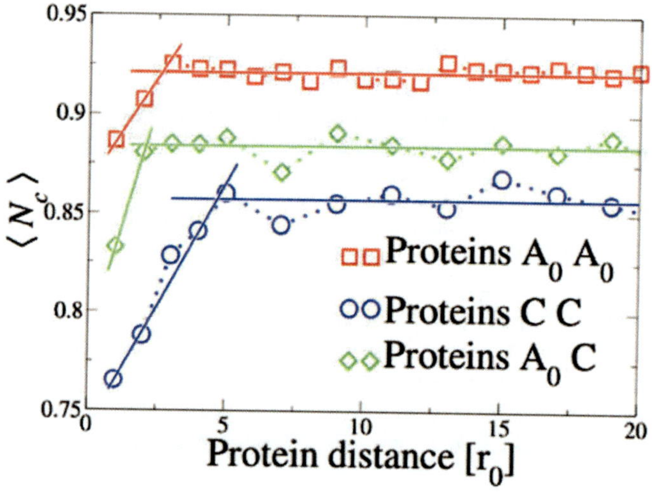


**Fig. 1.6** Percentage of Native contacts $< N_C >$ as a function of the minimum distance between two appr proteins (of the same type, $A_0$ or C, or of different types, $A_0$ and C). As they approach, the two proteins partially unfold even before touching

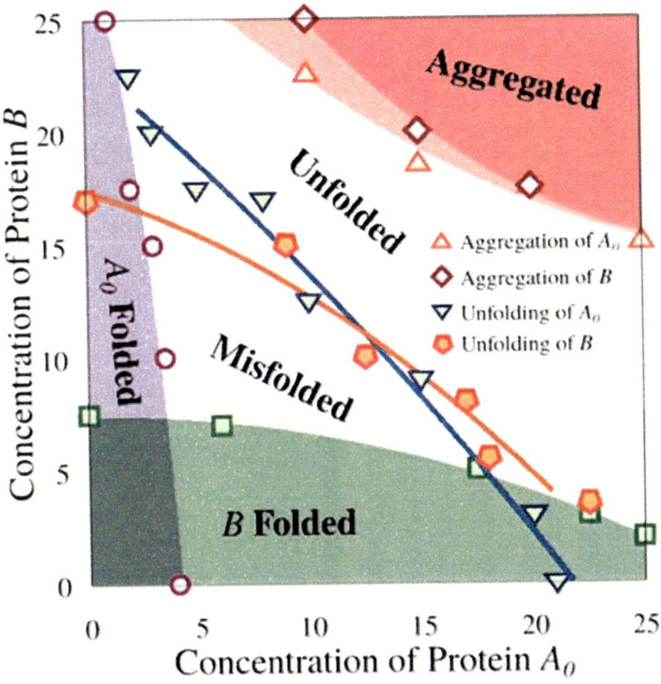


**Fig. 1.7** Phase diagram of the folding and aggregation of two designed proteins $A_0$ and B. The folded regions of each protein are orthogonal to each other, indicating that each folded protein remains rather insensitive to the increase in the concentration of the other. Volume concentrations are expressed in percentage

between the two approaching proteins, they show that the proteins aggregate without unfolding. Hence, the water causes a free-energy barrier against aggregation [68].

Bianco et al. suggest that the distance at which the two approaching proteins start to unfold can be considered as the water-mediated interaction radius of a protein. To support this conclusion, they perform a percolation analysis of the cluster sizes of statistically correlated water molecules between the two proteins when they are folded or unfolded.

This size is a proxy of the correlation length in water affected by the nearby proteins. Bianco et al. find that this size increases when the proteins unfold. Hence, the water-mediated protein–protein interaction increases in the unfolded state. This role of the water in inducing the unfolded regime, which is a precursor of the fully aggregated state, is an unexpected and not previously observed prediction of their simulations [68].

The water-mediated protein–protein interactions are essential also when different proteins are mixed at different concentrations. Bianco et al. [97] show that each component of a protein mixture is capable of maintaining its folded state at densities greater than the one at which they would precipitate in single-species solutions

(Fig. 1.7). They demonstrate that the free energy of each protein in the mixture remains unaffected by the presence of the other protein species and depends mainly on their individual concentrations.

Moreover, by increasing the protein concentrations, they find that the protein aggregates mostly between proteins of the same species. The aggregation propensity is essentially regulated by similarities in protein sequence more than in protein structure [97].

They show the generality of their observation over many different proteins, adopting the coarse-grained model of hydrated proteins described here and based on the Franzese-Stanley water model. Furthermore, they confirm their simulation results by performing dynamic light scattering experiments to evaluate the protein aggregation in a binary mixture of bovine serum albumin and consensus tetratricopeptide repeat, clarifying critical aspects of the cellular mechanisms regulating the expression and aggregation of the proteins [97].

## 1.6 Conclusions

The results summarized in this chapter show that water plays a complex role in the enthalpy balance of protein solutions. In particular, protein (i) cold denaturation, (ii) unfolding by pressurization, and (iii) unfolding by depressurization, are energy, density, and enthalpy driven by water, respectively, in addition to other suggested mechanisms [4].

The understanding of these mechanisms and the water contribution to these processes have opened the perspective to develop advanced computational design tools for protein engineering. In particular, including water contribution into the analysis of the selection of stable proteins at ambient and extreme thermodynamic conditions, allows us to elucidate why there are limits of temperature and pressure above which no protein has a stable functional state. Also, it clarifies why the large segregation of hydrophilic amino acids on the surface and hydrophobic residues into the protein core reduces the thermodynamic stability range of a protein.

We presented a design protocol that explains the hydropathy profile of proteins as a consequence of a selection process influenced by water. Proteins selected to be stable at high temperatures are stable also at extremely low temperatures and high pressures and are characterized by non-extreme segregation of a hydrophilic surface and a hydrophobic core. These results shed light on how environmental extreme conditions could have exerted an evolutionary action regulated by the water-mediated interactions.

In particular, the water fluctuations in the hydration shell are essential to understand the protein–protein interactions. The partial unfolding can occur for increasing protein concentration without aggregation. The hydration water is responsible for the enthalpy barrier that stabilizes misfolded proteins when their concentration further increases. As a consequence, the water contribution pushes toward high values the concentration thresholds above which the proteins aggregate, possibly representing an important regulatory mechanism against neurodegenerative aggregation-related diseases.

In a recent work Koga et al. [98] show that an extensive mutation of large hydrophobic residues to smaller ones for a de novo designed ideal protein, with an extremely high thermal stability, does not affect significantly the stability of the protein at 100 °C. Hence, by taking into account explicit water in the design of highly stable protein, we could also shed light on the observation that hydrophobic tight cores might not be key for the stability of a protein at high temperatures.

**Acknowledgements** G.F. acknowledges the support of the Spanish Grant N. PGC2018-099277-B-C22 (MCIU/AEI/ERDF) and the support by ICREA Foundation (ICREA Academia prize). I.C. acknowledges the support of the Maria de Maeztu Units of Excellence Programme – Grant No. MDM-2017-0720 Spanish Ministry of Science, Innovation and Universities, the Austrian Science Fund (FWF) project 26253-N27, the Spanish Ministerio de Economià y Competitividad (MINECO) Grant. N. FIS2017-89471-R, the Programa Red Guipuzcoana de Ciencia, Tecnología y Informacion SN 2019-CIEN-000051-01, the BIKAINTEK program (grant No. 008- B1/2020), the COST Action CA17139.

## References

1. Chiti F, Dobson C (2006) Protein misfolding, functional amyloid, and human disease. Annu Rev Biochem 75:333–366. https://doi.org/10.1146/annurev.biochem.75.101304.123901
2. Lau K, Dill K (1989) A lattice statistical mechanics model of the conformational and sequence spaces of proteins. Macromolecules 22(10):3986–3997. https://doi.org/10.1021/ma00200a030
3. Meersman F, Dobson C, Heremansa K (2006) Protein unfolding, amyloid fibril formation and configurational energy landscapes under high-pressure conditions. Chem Soc Rev 35:908–917. https://doi.org/10.1039/B517761H
4. Roche J, Caro J, Norberto D, Barthe P, Roumestand C, Schlessman J, Garcia A, Royer C (2012) Cavities determine the pressure unfolding of proteins. Proc Natl Acad Sci 109:6945–6950. https://doi.org/10.1073/pnas.1200915109
5. Privalov P (1990) Cold denaturation of proteins. Crit Rev Biochem Mol Biol 25:281–306. https://doi.org/10.3109/10409239009090612
6. Bruscolini P, Casetti L (2000) Lattice model for cold and warm swelling of polymers in water. Phys Rev E 61:R2208. https://doi.org/10.1103/PhysRevE.61.R2208
7. Bastolla U, Grassberger P (2001) Exactness of the annealed and the replica symmetric approximations for random heteropolymers. Phys Rev E 63:031901. https://doi.org/10.1103/PhysRevE.63.031901
8. Lesch H, Stadlbauer H, Friedrich J, Vanderkooi J (2002) Stability diagram and unfolding of a modified cytochrome c: what happens in the transformation regime? Biophys J 82:1644–1653. https://doi.org/10.1016/S0006-3495(02)75515-X
9. Larios E, Gruebele M (2010) Protein stability at negative pressure. Methods 52:51–56. https://doi.org/10.1016/j.ymeth.2010.04.010
10. Muller N (1990) Search for a realistic view of hydrophobic effects. Acc Chem Res 23:23–28. https://doi.org/10.1021/ar00169a005
11. Marqués M, Borreguero J, Stanley HE, Dokholyan N (2003) Possible mechanism for cold denaturation of proteins at high pressure. Phys Rev Lett 91:138103. https://doi.org/10.1103/PhysRevLett.91.138103

12. Patel B, Debenedetti P, Stillinger F, Rossky P (2007) A water-explicit lattice model of heat-, cold-, and pressure-induced protein unfolding. J Biophys J 93:4116–4127. https://doi.org/10.1529/biophysj.107.108530
13. Badasyan AV, Tonoyan ShA, Mamasakhlisov YSh, Giacometti A, Benight AS, Morozov VF (2011) Competition for hydrogen-bond formation in the helix-coil transition and protein folding. Phys Rev E 83:051903. https://doi.org/10.1103/physreve.83.051903
14. Dobson C (2004) Principles of protein folding, misfolding and aggregation. Semin Cell Dev Biol 15:3–16. https://doi.org/10.1016/j.semcdb.2003.12.008
15. Hawley SA (1971) Reversible pressure–temperature denaturation of chymotrypsinogen. Biochemistry 10:2436–2442. https://doi.org/10.1021/bi00789a002
16. Lindorff-Larsen K, Piana S, Dror R, Shaw D (2011) How fast-folding proteins fold. Science 334:517. https://doi.org/10.1126/science.1208351
17. Rosenman D, Connors C, Chen W, Wang C, García A (2013) Aβ Monomers transiently sample oligomer and fibril-like configurations: ensemble characterization using a combined MD/NMR approach. J Mol Biol 425:3338. https://doi.org/10.1016/j.jmb.2013.06.021
18. Lin Y, Bowman G, Beauchamp K, Pande V (2012) Investigating how peptide length and a pathogenic mutation modify the structural ensemble of amyloid beta monomer. Biophys J 102:315–324. https://doi.org/10.1016/j.bpj.2011.12.002
19. Herrera-Nieto P, Pérez A, De Fabritiis G (2020) Characterization of partially ordered states in the intrinsically disordered N-terminal domain of p53 using millisecond molecular dynamics simulations. Sci Rep 10:12402. https://doi.org/10.1038/s41598-020-69322-2
20. Robertson J, Nassar R, Liu C, Brini E, Dill K, Perez A (2019) NMR-assisted protein structure prediction with MELDxM. Proteins 87:1333–1340. https://doi.org/10.1002/prot.25788
21. Brini E, Simmerling C, Dill K (2020) Protein storytelling through physics. Science 370:3041. https://doi.org/10.1126/science.aaz3041
22. Tajkhorshid E, Aksimentiev A, Balabin I, Gao M, Isralewitz B, Phillips J, Zhu F, Schulten K (2003) Large scale simulation of protein mechanics and function. Adv Protein Chem 66:195–247. https://doi.org/10.1016/S0065-3233(03)66006-7
23. Nowak W (2017) Applications of computational methods to simulations of proteins dynamics. In: Leszczynski J, Kaczmarek-Kedziera A, Puzyn T, Papadopoulos M, Reis H, Shukla M (eds) Handbook of Computational Chemistry. Springer, Cham, pp 1–43. https://doi.org/10.1007/978-3-319-27282-5_31
24. Marrink SJ, Risselada HJ, Yefimov S, Tieleman DT, de Vries AH (2016) Coarse-grained protein models and their applications. Chem Rev 116:7898–7936. https://doi.org/10.1021/acs.chemrev.6b00163
25. Franzese G, Stanley HE (2002) Liquid-liquid critical point in a Hamiltonian model for water: analytic solution. J Phys: Condens Matter 14:2201. https://doi.org/10.1088/0953-8984/14/9/309
26. Franzese G, Stanley HE (2002) A theory for discriminating the mechanism responsible for the water density anomaly. Phys A 314:508. https://doi.org/10.1016/S0378-4371(02)01186-X
27. Franzese G, Marqués M, Stanley HE (2003) Intramolecular coupling as a mechanism for a liquid-liquid phase transition. Phys Rev E 67:011103. https://doi.org/10.1103/PhysRevE.67.011103
28. Mazza M, Stokely K, Strekalova EG, Stanley HE, Franzese G (2009) Cluster Monte Carlo and numerical mean-field analysis for the water liquid-liquid phase transition. Comput Phys Commun 180:497–502. https://doi.org/10.1016/j.cpc.2009.01.018
29. Bianco V, Franzese G (2019) Hydrogen bond correlated percolation in a supercooled water monolayer as a hallmark of the critical region. J Mol Liq 285:727–739. https://doi.org/10.1016/j.molliq.2019.04.090
30. Coronas L, Vilanova O, Bianco V, Santos F, Franzese G (2020) The Franzese-Stanely coarse grained model for hydration water. arXiv:2004.03646
31. Franzese G, Yamada M, Stanley HE (2000) Hydrogen-bonded liquids: Effects of correlations of orientational degrees of freedom. In: Tokuyama M, Stanley HE (eds) AIP Conference Proceedings, CP519, Statistical Physics, pp 281–287. https://doi.org/10.1063/1.1291569

32. Franzese G, Stanley HE (2005) The metastable liquid-liquid phase transition: From water to colloids and liquid metals. In: Beck C, Benedek G, Rapisarda A, Tsallis C (eds) Complexity, Metastability And Nonextensivity. World Scientific, Singapore, pp 210–214. https://doi.org/10.1142/9789812701558_0024
33. Stanley HE, Kumar P, Xu L, Franzese G, Yan Z, Mazza M, Buldyrev SV, Chen SH, Mallamace F (2008) Liquid polyamorphism: possible relation to the anomalous behavior of water. Eur Phys J Spec Top 161:1–17. https://doi.org/10.1140/epjst/e2008-00746-3
34. Stanley HE, Buldyrev S V, Chen S-H, Franzese G, Han S, Kumar P, Mallamace F, Mazza MG, Xu L, Yan Z (2009) Liquid polyamorphism and the anomalous behavior of water. In: Haug R (ed) Advances in Solid State Physics. Springer, Heidelberg, pp 249–266. https://doi.org/10.1007/978-3-540-85859-1_20
35. De los Santos F, Franzese G (2009) Influence of intramolecular couplings in a model for hydrogen-bonded liquids. In: Marro J, Garrido PL, Hurtado PI (eds) AIP Conference Proceedings. CP1091 Modelling and Simulation of New Materials. Tenth Granada Lectures, Granada, Spain, 15–19 September 2008. pp 185–197. https://doi.org/10.1063/1.3082279
36. Stokely K, Mazza MG, Stanley HE, Franzese G (2010) Metastable water under pressure. In: Rzoska SJ, Drozd-Rzoska A, Mazur V A (eds) Metastable Systems under Pressure. NATO Science for Peace and Security Series A: Chemistry and Biology. Proceedings of the NATO Advanced Research Workshop on Metastable Systems under Pressure: Platform for New Technologies and Environmental Applications, Odessa, Ukraine, 4–8 October 2008. Springer, Netherlands, pp 197–216. https://doi.org/10.1007/978-90-481-3408-3_14
37. Franzese G, Stanley HE (2006) Understanding the unusual properties of water. In: Lynden-Bell RM, Conway S, Barrow JD, Finney JL, Harper C (eds) Water and Life. The Unique Properties of. CRC Press, Taylor & Francis Group, Boca Raton, FL, USA, pp 101–118. https://doi.org/10.1201/EBK1439803561-c7
38. Bianco V, Ruberto R, Ancherbak S, Franzese G (2012) Water and anomalous liquids. In: Mallamace F, Stanley HE (eds) Complex Materials in Physics and Biology. Proceedings of the International School of Physics "Enrico Fermi." IOS Press, The Netherlands, pp 113–128. https://doi.org/10.3254/978-1-61499-071-0-113
39. Bianco V, Vilanova O, Franzese G (2013) Going deep into supercooled water. In: Fernández D, Frob M B, Latella I, Dector A (eds) Proceedings of the Barcelona Postgrad Encounters on Fundamental Physics. Barcelona: Facultat de Física. Universitat de Barcelona; Charleston, Spain, pp 103–126. https://cataleg.ub.edu/record=b2082772~S1*cat
40. Bianco V, Vilanova O, Franzese G (2013) Polyamorphism and polymorphism of a confined water monolayer: liquid-liquid critical point, liquid-crystal and crystal-crystal phase transitions. In: Viswanathan GM, Raposo EP, da Luz MGE (eds) Perspectives and Challenges in Statistical Physics and Complex Systems for the Next Decade. World Scientific, Singapore, pp 126–149. https://doi.org/10.1142/9789814590143_0008
41. Franzese G, Stanley HE (2007) The widom line of supercooled water. J Phys Condens Matter 19:205126. https://doi.org/10.1088/0953-8984/19/20/205126
42. Kumar P, Franzese G, Stanley HE (2008) Predictions of dynamic behavior under pressure for two scenarios to explain water anomalies. Phys Rev Lett 100:105701. https://doi.org/10.1103/PhysRevLett.100.105701
43. Kumar P, Franzese G, Stanley HE (2008) Dynamics and thermodynamics of water. J Phys Condens Matter 20:244114. https://doi.org/10.1088/0953-8984/20/24/244114
44. Stokely K, Mazza MG, Stanley HE, Franzese G (2010) Effect of hydrogen bond cooperativity on the behavior of water. Proc Natl Acad Sci 107:1301–1306. https://doi.org/10.1073/pnas.0912756107
45. Mazza MG, Stokely K, Stanley HE, Franzese G (2012) Effect of pressure on the anomalous response functions of a confined water monolayer at low temperature. J Chem Phys 137:204502. https://doi.org/10.1063/1.4767355
46. Mazza MG, Stokely K, Pagnotta SE, Bruni F, Stanley HE, Franzese G (2011) More than one dynamic crossover in protein hydration water. Proc Natl Acad Sci USA 108:19873. https://doi.org/10.1073/pnas.1104299108

47. Franzese G, De los Santos F (2009) Dynamically slow processes in supercooled water confined between hydrophobic plates. J Phys Condens Matter 21:504107. https://doi.org/10.1088/0953-8984/21/50/504107
48. Franzese G, Hernando-Martinez A, Kumar P, Mazza MG, Stokely K, Strekalova EG, De los Santos F, Stanley HE (2010) Phase transitions and dynamics in bulk and interfacial water. J Phys Condens Matter 22:284103. https://doi.org/10.1088/0953-8984/22/28/284103
49. Strekalova EG, Mazza MG, Stanley HE, Franzese G (2011) Large decrease of fluctuations for supercooled water in hydrophobic nanoconfinement. Phys Rev Lett 106:145701. https://doi.org/10.1103/PhysRevLett.106.145701
50. De los Santos F, Franzese G (2011) Understanding diffusion and density anomaly in a coarse-grained model for water confined between hydrophobic walls. J Phys Chem B 115:14311. https://doi.org/10.1021/jp206197t
51. Strekalova EG, Corradini D, Mazza MG, Buldyrev SV, Gallo P, Franzese G, Stanley HE (2012) Effect of hydrophobic environments on the hypothesized liquid-liquid critical point of water. J Biol Phys 38:97. https://doi.org/10.1007/s10867-011-9241-9
52. Strekalova EG, Mazza MG, Stanley HE, Franzese G (2012) Hydrophobic nanoconfinement suppresses fluctuations in supercooled water. J Phys: Condens Matter 24:064111. https://doi.org/10.1088/0953-8984/24/6/064111
53. De los Santos F, Franzese G (2012) Relations between the diffusion anomaly and cooperative rearranging regions in a hydrophobically nanoconfined water monolayer. Rapid Commun Phys Rev E 85:010602. https://doi.org/10.1103/PhysRevE.85.010602
54. Bianco V, Franzese G (2014) Critical behavior of a water monolayer under hydrophobic confinement. Sci Rep (Nature Publishing Group) 4:4440. https://doi.org/10.1038/srep04440
55. Franzese G, Stokely K, Chu XQ, Kumar P, Mazza MG, Chen SH, Stanley HE (2008) Pressure effects in supercooled water: comparison between a 2D model of water and experiments for surface water on a protein. J Phys Condens Matter 20:494210. https://doi.org/10.1088/0953-8984/20/49/494210
56. Stanley H E, Kumar P, Xu L, Franzese G, Yan Z, Mazza MG, Chen S H, Mallamace F, Buldyrev SV (2007) Liquid polyamorphism: some unsolved puzzles of water in bulk, nanoconfined, and biological environments. In: Tokuyama M, Oppenheim I, Nishiyama H (eds) International Conference on Flow Dynamics, AIP Conference Proceedings, vol 982, pp 251–271. https://doi.org/10.1063/1.2897798
57. Kumar P, Franzese G, Buldyrev S, Stanley H (2009) Dynamics of water at low temperatures and implications for biomolecules. In: Franzese G, Rubi M (eds) Aspects of Physical Biology. Lecture Notes in Physics, vol. 752, Springer, Heidelberg, pp 3–22. https://doi.org/10.1007/978-3-540-78765-5_1
58. Stanley HE, Kumar P, Han S, Mazza MG, Stokely K, Buldyrev SV, Franzese G, Mallamace F, Xu L (2009) Heterogeneities in confined water and protein hydration water. J Phys Condens Matter 21(50):504105. https://doi.org/10.1088/0953-8984/21/50/504105
59. Stanley HE, Buldyrev SV, Franzese G, Kumar P, Mallamace F, Mazza MG, Stokely K, Xu L (2010) Liquid polymorphism: water in nanoconfined and biological environments. J Phys Condens Matter 22:284101. https://doi.org/10.1088/0953-8984/22/28/284101
60. Stanley HE, Buldyrev SV, Kumar P, Mallamace F, Mazza MG, Stokely K, Xu L, Franzese G (2011) Water in nanoconfined and biological environments. J Non-Cryst Solids 357:629. https://doi.org/10.1016/j.jnoncrysol.2010.07.029
61. Franzese G, Bianco V, Iskrov S (2011) Water at interface with proteins. Food Biophys 6:186. https://doi.org/10.1007/s11483-010-9198-4
62. Bianco V, Iskrov S, Franzese G (2012) Effect of hydrogen bonds on protein stability. J Biol Phys 38:27
63. Franzese G, Bianco V (2013) Water at biological and inorganic interfaces. Food Biophys 8:153. https://doi.org/10.1007/s11483-013-9310-7
64. Bianco V, Franzese G (2015) Contribution of water to pressure and cold denaturation of proteins. Phys Rev Lett 115:108101. https://doi.org/10.1103/PhysRevLett.115.108101

65. Bianco V, Pagès N, Coluzza I, Franzese G (2017) How the stability of a folded protein depends on interfacial water properties and residue-residue interactions. J Mol Liq 245:129. https://doi.org/10.1016/j.molliq.2017.08.026
66. Bianco V, Franzese G, Dellago C, Coluzza I (2017) Role of water in the selection of stable proteins at ambient and extreme thermodynamic conditions. Phys Rev X 7:021047. https://doi.org/10.1103/PhysRevX.7.021047
67. Vilanova O, Bianco V, Franzese G (2017) Multi-scale approach for self-Assembly and protein folding. In: Coluzza I (ed) Design of Self-Assembling Materials. Springer, Cham, pp 107–128. https://doi.org/10.1007/978-3-319-71578-0_5
68. Bianco V, Franzese G, Coluzza I (2019) Cover feature: In silico evidence that protein unfolding is a precursor of protein aggregation. Chem Phys Chem 21(5):377–384. https://doi.org/10.1002/cphc.201900904
69. March D, Bianco V, Franzese G (2021) Protein unfolding and aggregation near a hydrophobic interface, polymers 13:1-156. https://doi.org/10.3390/polym13010156
70. Miyazawa S, Jernigan R (1985) Estimation of effective interresidue contact energies from protein crystal structures: Quasi-chemical approximation, Macromolecules 18:3:534–552. https://doi.org/10.1021/ma00145a039Coluzza
71. (2017) Computational protein design: a review. J Phys Condens. Matter 29:143001. https://doi.org/10.1088/1361-648X/aa5c76
72. Fung HK, Welsh WJ, Floudas CA (2008) Computational de novo peptide and protein design: rigid templates versus flexible templates Ind. Eng Chem Res 47:993–1001. https://doi.org/10.1021/ie071286k
73. Samish I, Macdermaid C, Perez-Aguilar J, Saven J (2011) Theoretical and computational protein design. Ann Rev Phys Chem 62:129–149. https://doi.org/10.1146/annurev-physchem-032210-103509
74. Coluzza I, Muller HG, Frenkel D (2003) Designing refoldable model molecules. Phys Rev E 68:046703. https://doi.org/10.1103/PhysRevE.68.046703
75. Coluzza I (2011) A coarse-grained approach to protein design: learning from design to understand folding. PLoS ONE 6(7):e20853. https://doi.org/10.1371/journal.pone.0020853
76. Das R, Baker D (2008) Macromolecular modeling with Rosetta Ann. Rev Biochem 77:363–382. https://doi.org/10.1146/annurev.biochem.77.062906.171838
77. Vendruscolo M (1998) Protein stability and foldability-designability complementarity. Phys A 249(1–4):576–580. https://doi.org/10.1016/S0378-4371(97)00518-9
78. Pucci F, Rooman M (2017) Physical and molecular bases of protein thermal stability and cold adaptation. Curr Opin Struct Biol 42:117–128. https://doi.org/10.1016/j.sbi.2016.12.007
79. Tzul FO, Schweiker KL, Makhatadze GI (2015) Modulation of folding energy landscape by charge-charge interactions: linking experiments with computational modeling. Proc Natl Acad Sci U S A 112(3):E259–E266. https://doi.org/10.1073/pnas.1410424112
80. Katava M, Marchi M, Madern D, Sztucki M, Maccarini M, Sterpone F (2020) Temperature unmasks allosteric propensity in a thermophilic malate dehydrogenase via dewetting and collapse. J Phys Chem B 124(6):1001–1008. https://doi.org/10.1021/acs.jpcb.9b10776
81. Chakraborty D, Taly A, Sterpone F (2015) Stay wet, stay stable? how internal water helps the stability of thermophilic proteins. J Phys Chem B 119(40):12760–12770. https://doi.org/10.1021/acs.jpcb.5b05791
82. Rahaman O, Kalimeri M, Melchionna S, Hénin J, Sterpone F (2015) Role of internal water on protein thermal stability: the case of homologous G domains. J Phys Chem B 119(29):8939–8949. https://doi.org/10.1021/jp507571u
83. Gross M, Jaenicke R (1994) Proteins under pressure. The influence of high hydrostatic pressure on structure, function and assembly of proteins and protein complexes. Eur J Biochem 221:617. https://doi.org/10.1111/j.1432-1033.1994.tb18774.x
84. Floriano WB, Nascimento MA, Domont GB, Goddard WA (1998) Effects of pressure on the structure of metmyoglobin: molecular dynamics predictions for pressure unfolding through a molten globule intermediate. Protein Sci 7(11):2301–2313. https://doi.org/10.1002/pro.5560071107

85. Smeller L (2002) Pressure-temperature phase diagrams of biomolecules. Biochim Biophys Acta 1595:11. https://doi.org/10.1016/s0167-4838(01)00332-6
86. Fukuchi S, Nishikawa K (2001) Protein surface amino acid compositions distinctively differ between thermophilic and mesophilic bacteria. J Mol Biol 309(4):835–843. https://doi.org/10.1006/jmbi.2001.4718
87. Kyte J, Doolittle RF (1982) A simple method for displaying the hydropathic character of a protein. J Mol Biol 157(1):105–132. https://doi.org/10.1016/0022-2836(82)90515-0
88. Ross CA, Poirier MA (2005) What is the role of protein aggregation in neurodegeneration? Nat Rev Mol Cell Biol 6:891–898. https://doi.org/10.1038/nrm1742
89. Aguzzi A, O'Connor T (2010) Protein aggregation diseases: pathogenicity and therapeutic perspectives. Nat Rev Drug Discov 9(3):237–248. https://doi.org/10.1038/nrd3050
90. Knowles TPJ, Vendruscolo M, Dobson CM (2014) The amyloid state and its association with protein misfolding diseases. Nat Rev Mol Cell Biol 15:384–396. https://doi.org/10.1038/nrm3810
91. Chi EY, Krishnan S, Randolph TW, Carpenter JF (2003) Physical stability of proteins in aqueous solution: mechanism and driving forces in nonnative protein aggregation. Pharm Res 20:1325–1336. https://doi.org/10.1023/A:1025771421906
92. Tartaglia GG, Pechmann S, Dobson CM, Vendruscolo M (2007) Life on the edge: a link between gene expression levels and aggregation rates of human proteins. Trends Biochem Sci 32:204–206. https://doi.org/10.1016/j.tibs.2007.03.005
93. Song J (2009) Insight into "insoluble proteins" with pure water. FEBS Lett 583:953–959. https://doi.org/10.1016/j.febslet.2009.02.022
94. Dobson CM (2004) Principles of protein folding, misfolding and aggregation. Semin Cell Dev Biol 15(1):3–16. https://doi.org/10.1016/j.semcdb.2003.12.008
95. Cellmer T, Bratko D, Prausnitz JM, Blanch HW (2007) Protein aggregation in silico. Trends Biotechnol 25(6):254–261. https://doi.org/10.1016/j.tibtech.2007.03.011
96. Nasica-Labouze J, Nguyen PH, Sterpone F, Berthoumieu O, Buchete N-V, Coté S, De SA, Doig AJ, Faller P, Garcia A, Laio A, Li MS, Melchionna S, Mousseau N, Mu Y, Paravastu A, Pasquali S, Rosenman DJ, Strodel B, Tarus B, Viles JH, Zhang T, Wang C, Derreumaux P (2015) Amyloid beta protein and Alzheimer's disease: when computer simulations complement experimental studies. Chem Rev 115:3518–3563. https://doi.org/10.1021/cr500638n
97. Bianco V, Alonso-Navarro M, Di Silvio D, Moya S, Cortajarena AL, Coluzza I (2019) Proteins are solitary! Pathways of protein folding and aggregation in protein mixtures. J Phys Chem Lett 10:4800–5480. https://doi.org/10.1021/acs.jpclett.9b01753
98. Koga R, Yamamoto M, Kosugi T, Kobayashi N, Sugiki T, Fujiwara T, Koga N (2020) Robust folding of a de novo designed ideal protein even with most of the core mutated to valine. Proc Natl Acad Sci U S A 117(49):31149–31156. https://doi.org/10.1073/pnas.2002120117

# Chapter 2
# Slow Dynamics of Biological Water

**Gaia Camisasca, Antonio Iorio, Lorenzo Tenuzzo, and Paola Gallo**

**Abstract** Water hydrating biomolecules shows a more complex dynamical behavior when compared to the bulk. Its translational slow dynamics can be described by two mechanisms characterized by two well distinct time scales. One mechanism is the $\alpha$-relaxation typical of supercooled bulk water and other glass forming liquids. Upon cooling, this relaxation shows a fragile-to-strong crossover due to the activation of hopping phenomena which permits to the water molecules in the hydration layer to escape from nearest neighbors cage. The second mechanism is a much slower relaxation that is present only in hydration water and it is coupled with the biomolecule dynamics. This long-relaxation shows upon cooling a strong-to-strong crossover in coincidence with the well-known Protein Dynamical Transition. Structural rearrangements of biomolecules can trap hydration water molecules over length-scale larger than nearest neighbors distances. This causes a new hopping regime specific only of hydration water and already active at high temperature.

G. Camisasca · A. Iorio · L. Tenuzzo · P. Gallo (✉)
Department of Mathematics and Physics, Roma Tre University, Via della Vasca Navale 84, 00146 Rome, Italy
e-mail: paola.gallo@uniroma3.it

G. Camisasca
e-mail: gaia.camisasca@uniroma3.it

A. Iorio
e-mail: iorio@ibpc.fr

L. Tenuzzo
e-mail: lorenzo.tenuzzo@uniroma3.it

A. Iorio
CNRS, Université de Paris, UPR9080, Laboratoire de Biochimie Théorique, Institut de Biologie Physico-Chimique-Fondation Edmond de Rothschild, PSL Research University, 13 rue Pierre et Marie Curie, 75005 Paris, France



L. Bulavin and N. Lebovka (eds.), *Soft Matter Systems for Biomedical Applications*, Springer Proceedings in Physics 266, https://doi.org/10.1007/978-3-030-80924-9_2

## List of Abbreviations

| | |
|---|---|
| FSC | Fragile-to-Strong Crossover |
| MCT | Mode Coupling Theory |
| MSD | Mean Square Displacement |
| MSF | Mean Square Fluctuation |
| PDT | Protein Dynamical Transition |
| RDF | Radial Distribution Function |
| SISF | Self Intermediate Scattering Function |
| SSC | Strong-to-Strong Crossover |

## 2.1 Introduction

Biological water refers to the water present in biological environment. In bio-systems water plays a role of paramount importance and it is essential for life [1]. Water is present in biology at different levels: for instance, water participates in many chemical reactions and bio-mechanisms such as protein folding, from a molecular point of view few water molecules are typically structurally bound to proteins in order to contribute to their stability. Since water is the natural ambient for living organisms, it facilitates the transport of molecules and drugs inside the body in order to reach theirs targets and can be used safely as a solvent for many biomedical applications, such as the storage of biomaterials without the hazard of toxicity.

But water is also one of the most peculiar liquid on Earth, due to the behavior often opposite to the behavior of most of the other simple liquids: for example it is usual for simple liquids to contract upon decreasing temperature, while water reaches a density maximum which at ambient pressure is observed at 4 °C. Starting from the striking ability of a single water molecule to form up to four hydrogen bonds, which is at the base of its cooperative dynamics and particular tetrahedral structure, water anomalies can be actually encountered all over its phase diagram and have great impact also in the dynamics. For these reasons, many different scenarios have been proposed in order to frame water anomalies [2], including the most fascinating presence of a liquid–liquid critical point inside the no-man's land [3] and the singularity-free scenario [4]. Both these scenarios predict a line of finite maxima in the different thermodynamic response functions, which in the first scenario it is known as Widom line [5–8]. The Widom line is the line of correlation length maxima that emanates from a second order critical point in the one-phase region. Close to this line and close to the critical point thermodynamic response functions, like specific heat and isothermal compressibility, show lines of maxima that converge to the Widom line upon approaching to the critical point.

Modifications in the thermodynamics [9–17], in the structure [18–26] and especially in the dynamics [18–20, 25, 27–46] of the water contained in biological environment are due to the interaction of water with the biomolecules and this changes profoundly the situation. Besides the fundamental knowledge, the study of biological water dynamics and structure is essential for advancements.

In this chapter we will focus on the behavior of biological water in the supercooled regime. The study of water at low temperatures has several important practical aspects, especially in connection to biomedical applications. For instance, we can consider the case of the biological role of water in the low-temperature activation of proteins. In fact, protein functions are inhibited if the protein is not minimally hydrated. If hydrated, a protein can undergo the so-called protein dynamical transition [47] (PDT) which takes place in the supercooled regime of water and which is essential to the activation of conformational changes necessary to carry out their biological functions [48–51]. Connected to the supercooled regime, there are the cryostorage of biomaterials and the problem of maintaining biological functions. For example, stabilization of membranes and proteins at low temperature can be improved by employing sugar-water solutions which protect them from freezing stresses, but the mechanism of protection is still today mostly debated.

How does the biomolecule dynamically couple with its hydration water? Does this coupling modify the behavior of hydration water with respect to the behavior of bulk water? Is it possible to predict the bioprotection ability of a solute by studying its interaction with water? In order to answer these questions a profound study of the relaxation mechanisms in biological water is mandatory and fundamental to improve biomedical protocols. Here we review some results obtained by molecular dynamics simulations on two water solutions of biological interest. The first system consists of the lysozyme protein immersed in pure water and it permits to investigate the relaxation mechanisms of the water in contact with the protein and their coupling. The second system consists of a trehalose-water solution employed in cryoprotocols and it permits to investigate how the relaxation mechanism may change when the size and the nature of the solute change. We investigated these systems upon supercooling, therefore we also characterized the slow dynamics of hydration water as a function of the temperature, which is an important parameter in biology.

## 2.2 Methods

The results shown in this chapter come from two series of simulations. Two different systems have been simulated by means of classical molecular dynamics.

The lysozyme-water system is composed by 1 lysozyme protein immersed in 13,982 water molecules. 8 $Cl^-$ ions are also inserted in this solution in order to neutralize the charges on the lysozyme residues. A snapshot from the simulation box is shown in Fig. 2.1 (left). Water molecules were modeled with the SPC/E potential [52] and the lysozyme by the CHARMM force-field for proteins [53, 54]. Chlorine

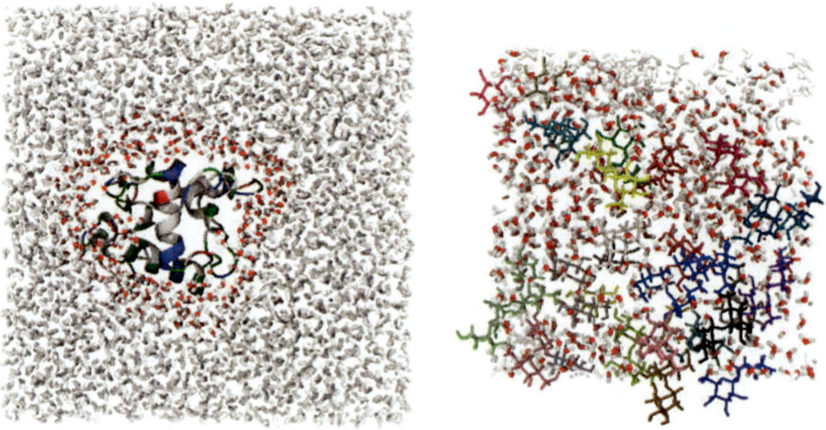

**Fig. 2.1** Left: Cubic simulation box containing one lysozyme protein and water molecules. The linear dimension of the box is 75.6 Å and the temperature is 280 K. The protein is depicted according to its secondary structure. Residues are colored according to the following criterion: blue indicates basic residues, red indicates acid residues, green indicates polar residues and white indicates non-polar residues. Water molecules are shown in white and the hydration water molecules in red (oxygen) and white (hydrogens). In order to optimize the visualization, only a slice of 15 Å of water molecules in the direction perpendicular to the book page is rendered. Lysozyme is shown entire. Right: Cubic simulation box containing a solution of trehalose and water molecules. The linear dimension of the box is 40.0 Å and the temperature is 290 K. Trehalose molecules are shown in different colors with respect to each other, to better identify trehalose clusters. Water molecules outside trehalose hydration shells are colored white, while hydration water molecules are colored in red (oxygen) and white (oxygens). In order to optimize the visualization, only a slice of 15 Å box along the direction perpendicular to the book page is shown here.

force-field was also taken from CHARMM. The system was simulated at constant pressure for temperature in the range from 320 to 200 K every 10 K. Production runs used to calculate dynamic quantities last 20 ns at high temperature and 30 ns for the lowest temperatures in refs. [23, 41, 55–57]. These runs were recently extended to 100 ns at each temperature in ref. [58].

The trehalose-water system is composed by 54 $\alpha-\alpha$ trehalose molecules and 1498 water molecules, corresponding to a weight percentage $\phi = 40.65\%$ wt in trehalose. A snapshot from the simulation box is shown in Fig. 2.1 (right). Water molecules were modeled with the SPC/E potential [52] and trehalose with the CHARMM force-field for sugars [59, 60]. The system was simulated at constant pressure for temperatures in the range from 320 to 200 K every 10 K. Production runs varied from 30 ns at the highest temperature to 95 ns at the lowest temperature [25, 44, 56, 57].

Classical molecular dynamics simulations were run with GROMACS simulation package [61]. Details on the simulation procedures and preparation of the systems can be found in the cited literature.

## 2.3 Hydration Water

Here we focus on the dynamical behavior of the water moving closer to the lysozyme or the trehalose bio-molecules, which we call hydration water. In the simulations we adopted a geometrical criterion to define hydration water: we classified a water molecule as a hydration water molecule if the distance between its oxygen atom and the closest biomolecule atom is less than or equal to 6 Å. We therefore analyze the structure and the dynamics of water molecules which lie and move within a 6 Å-shell around the lysozyme or a 6 Å-shell around the trehalose molecules, depending on the system. Hydration water molecules have been colored differently from the molecules outside the hydration shell for both systems, see Fig. 2.1. In the case of lysozyme, most of the water molecules contained in the box are bulk-like and around 1200 molecules over the more 13 thousand are classified as lysozyme hydration water molecules. In the case of the trehalose system, the solution is concentrated, therefore we observe that most of the water is hydration water. Trehalose molecules also cluster together upon cooling and consequently hydration water is shared among several trehalose molecules.

We decided to take a 6 Å shell for hydration water based on the analysis of the radial distribution function (RDF) of water around the lysozyme protein and adopted the same definition around trehaloses.

In Fig. 2.2 the oxygen–oxygen RDF of pure water is shown in blue dashed line. Water organizes into defined shells: the first peak of the function corresponds to the first nearest neighbors, while in the second shell the neighbor molecules organize in almost tetrahedral arrangement. In the same figure we report the RDF of water around two particular sites of the lysozyme protein, these sites are shown in the inset. The orange site is buried inside the protein, i.e., it is not exposed directly to the water. The RDF calculated between this site and the water oxygens in the solution corresponds to the orange curve. We see that there are on average no water molecules within 6 Å, and consequently, we have labelled the site "not-hydrated". Very different behavior is observed for the green site that is directly exposed outward the water. The RDF, see the green line, it is similar to the bulk-water RDF: it does have the typical oscillations of the blue line, approximately at similar distances. The water contained in the solution, therefore, organizes into defined shells also around the protein sites which are exposed and hydrophilic and consequently we label them "hydrated site". With the adopted criterion for the definition of hydration water, we see now from these RDFs that our lysozyme hydration water corresponds to about two hydration shells around the exposed protein sites. In the case of trehalose hydration water, the RDFs between hydration water and the oxygen atoms of the hydroxyl groups of trehaloses are similar to the RDF shown in Fig. 2.2 for the lysozyme hydrated site, and also in this case (not shown) the second shells ends at about 6 Å [25].

In Fig. 2.3 we compare the hydrogen bonding properties of pure water (left) with the properties of lysozyme hydration water (right). We report on these plots the percentage of water molecules involved in $n$ hydrogen bonds at different temperatures starting from 300 K down to 200 K. In the case of bulk water, we see that most of the

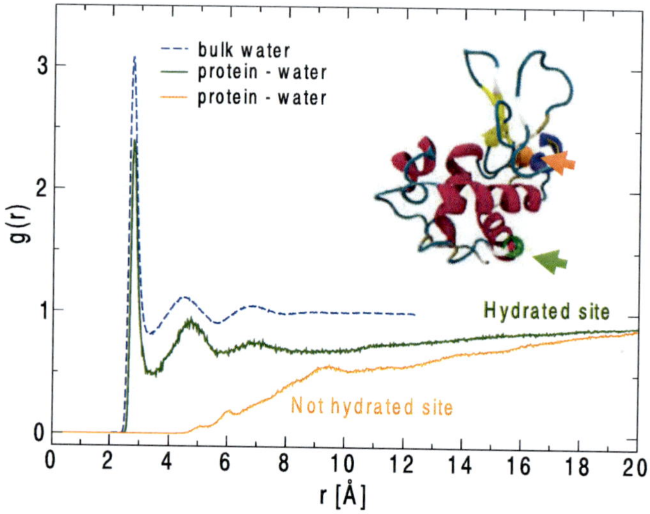


**Fig. 2.2** Radial Distribution Functions (RDF) between water oxygens and two selected lysozyme sites: an exposed/hydrated site (green curve) and an inner/non-hydrated site (orange curve). For comparison also the oxygen–oxygen RDF of bulk water is reported (blue curve). The temperature for all curves is 300 K. In the inset, we show the location of the two selected sites of the lysozyme protein highlighted as balls superimposed to its secondary structure. The color of the arrow coincides with the color of the corresponding RDF (reprinted with permission from [23])

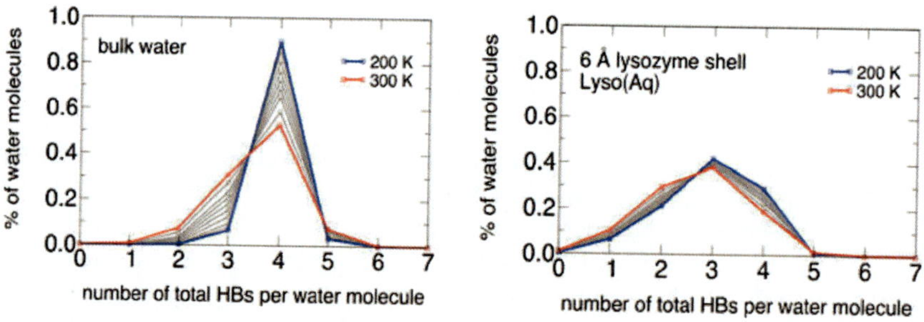


**Fig. 2.3** Distributions of the population of water molecules engaging $n$ hydrogen bonds (HB) with another water molecule in bulk water (left panel) and in the hydration layer around the lysozyme protein (right panel) at different temperatures

population forms four hydrogen bonds already at high temperature and upon cooling almost all the molecules are sharing four hydrogen bonds. This indicates that the structure of water becomes more and more coordinated into tetrahedral configuration as we cool down the system. In the case of lysozyme hydration water, we observe that a hydration water molecule forms with another hydration molecule more likely 2 or 3 hydrogen bonds at high temperature. As the temperature goes down, the number of molecules involved in 4 hydrogen bonds increases at the expense of the population involved in 2 hydrogen bonds, but the most likely value remains 3 hydrogen bonds. The fourth hydrogen bond of hydration water can in fact be directly formed with the atoms of lysozyme.

The hydrogen bonds distribution of trehalose hydration water is practically similar to that of lysozyme hydration water (not shown) [25].

In the next sections, we will discuss results about the dynamics of the water which moves close to these two biomolecules.

## 2.4 Slow Dynamics of Hydration Water Probed with Self Intermediate Scattering Functions

The dynamics of liquids can be studied by means of the Self Intermediate Scattering Function (SISF), $F_{SELF}(Q, t)$, that is defined as:

$$F_{SELF}(Q,t) = \frac{1}{N}\left\langle \sum_{i=1}^{N} e^{i\vec{Q}\cdot(\vec{r}_i(t)-\vec{r}_i(0))} \right\rangle, \tag{2.1}$$

where $N$ is the number of water molecules, $\vec{r}_i(t)$ is the position of the oxygen atom of the $i$-th water molecule at time $t$, and $\vec{Q}$ is the transferred wave-vector. This correlation function can be directly computed from the atomistic trajectories stored from molecular dynamics simulations and it is able to track through the single particle dynamics the relaxations of the system, and how they evolve upon supercooling. We will, here in the following, concentrate on this correlation function calculated for the oxygens. Being the oxygen located very close to the center of mass of the water molecules, this function will have information only relative to the translational dynamics of the water molecules.

The Self Intermediate Scattering Function of water was computed in pioneering works on the supercooled regime [62, 63], where it was modeled according to the Mode Coupling Theory (MCT) [64] with the following formula:

$$F_{SELF}(Q,t) = \left(1 - f_Q\right)e^{-\left(\frac{t}{\tau_s}\right)^2} + f_Q e^{-\left(\frac{t}{\tau_\alpha}\right)^\beta}. \tag{2.2}$$

In this model, the correlation function decays at early times according to a Gaussian dependence on time characterized by a short relaxation time $\tau_s$ to take into account the initial ballistic motion of the tagged particle. In this regime, the water molecule moves as a free particle and its motion does not depend greatly on the temperature. When approaching the neighbors, the water molecule starts to feel the environment that in the context of the Mode Coupling Theory corresponds to the cage formed by the neighbors. Inside the cage, the water molecules are characterized by a rattling motion. The cage regime is strongly temperature dependent, and upon increasing the supercooling it becomes more and more effective in trapping transiently the molecules. $f_Q$ is the Lamb-Mössbauer factor which is the dynamic signature of the cage effect in the self-intermediate scattering function and it quantifies the rattling motion of the water molecule and consequently it can be used to estimate the typical cage radius [57, 62, 63]. Once the cage melts, the water molecules can explore the space outside the cage. This happens at longer time and the liquid structural relaxation is modeled with the stretched exponential function, with a characteristic time $\tau_\alpha$, called structural $\alpha$-relaxation time and $\beta$ is the stretching parameter of the $\alpha$ -relaxation. The $\alpha$-relaxation is strongly temperature dependent.

The Self Intermediate Scattering Functions of lysozyme and trehalose hydration water were calculated by considering only the water molecules belonging to a shell of 6 Å around the protein or around the sugars in a series of papers [23, 25, 31, 41, 44, 55, 56]. A selection of results is presented in Fig. 2.4 for lysozyme hydration. Bulk water curves are also reported for comparison. Similar results are obtained for trehalose hydration water, see ref. [44]. All the curves were calculated at the $Q_{max} = 22.5$ Å$^{-1}$, which corresponds to the position of the first sharp peak of the oxygen–oxygen structure factor of water. Here, the peculiar features of the MCT caging due to supercooling are enhanced [65], because this $Q$ value corresponds to the typical interparticle nearest neighbors distance.

The correlation functions of hydration water show the typical behavior of bulk water and other glass-forming liquids upon supercooling described by the Mode

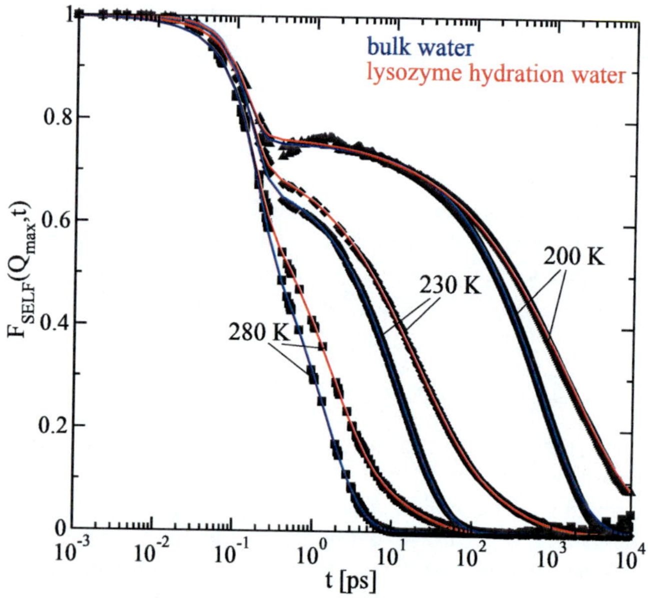


**Fig. 2.4** Self-intermediate scattering functions (SISFs) of the oxygen atoms of bulk water and lysozyme hydration water at three selected temperatures (symbols). SISFs are calculated at the peak of the oxygen–oxygen structure factor $Q = 2.25$ Å$^{-1}$. Continuous lines are fits to the model described by Eq. 2.2 for water (blue lines), and by Eq. 2.3 for lysozyme hydration water (red lines)

Coupling Theory. Upon cooling the so-called two step relaxation, the signature of the presence of two well-defined time scales of relaxation, is clearly detectable.

At high temperatures the correlators, after the first ballistic regime, decay quickly to zero. Upon decreasing temperature, the time scale of the ballistic regime approximately remains the same and the curves nearly coincide up to $\approx 0.2$ ps. At lower temperature correlators develop a plateau region, where transiently the particles are trapped by the neighbors cage. Eventually, the correlators decay to zero when the cages melt, and the liquid can structurally rearrange.

By comparing the curves, we see that at a given temperature hydration water correlators decay to zero slower than the correlators of bulk water, meaning that the translational dynamics of hydration water is slower due to its interaction with the biosurface, in particular both lysozyme and trehalose hydration water correlators show upon cooling a very stretched tail.

The Self Intermediate Scattering functions of hydration water of both lysozyme and trehalose cannot in fact be described by the model of bulk water which misses to describe a very-long time decay that is the visible tail in the curves of Fig. 2.4. These very stretched tails which extend correlations in the self-motion of particles over times, are not compatible with a single stretched relaxation mechanism, such as the model of Eq. 2.2. For this reason, the model must be modified by adding a second structural long-relaxation. The new model, firstly proposed in ref. [31] and then adopted to described the Self Intermediate Scattering functions of hydration water, is given by:

$$\begin{aligned} F_{SELF}(Q,t) =& (1 - f_\alpha - f_{long})e^{-\left(\frac{t}{\tau_s}\right)^2} \\ &+ f_\alpha e^{-\left(\frac{t}{\tau_\alpha}\right)^\beta} + f_{long} e^{-\left(\frac{t}{\tau_{long}}\right)^{\beta_{long}}} \end{aligned} \tag{2.3}$$

where $(f_\alpha, \tau_\alpha, \beta_\alpha)$ and $(f_{long}, \tau_{long}, \beta_{long})$ are the amplitude, relaxation time and stretching parameters of the α-relaxation and the long-relaxation respectively.

The second stretched exponential, which described the long-relaxation absent in bulk water, can describe the curves very well, as we can see from fit reported in Fig. 2.4 on top of the curves. To appreciate the quality of the fit with this model with respect to the model of bulk water, we refer the reader to Fig. 1 of ref. [56].

From the fits of the Self Intermediate Scattering Functions, we can obtain the two important slow time scales of the single-particle dynamics of hydration water. One is the alpha relaxation also present in the bulk water and the other is the long-relaxation characteristic only of hydration water.

The presence of two different relaxations of hydration water is supported by several experimental studies, including depolarized light scattering experiments [35, 45, 66, 67], femtosecond resolved fluorescence experiments [68–71] and neutron scattering experiments [29].

We will discuss these two timescales in details in the following.

### 2.4.1 *The α-relaxation of Hydration Water*

The $\alpha$-relaxation time of the lysozyme hydration water and of the trehalose hydration water extracted from the fits via Eq. 2.3 are shown in Fig. 2.5. As the temperature decreases, the structural relaxation becomes slower and slower. The time scale of this relaxation varies from the order of 1 picosecond at temperature T = 300 K to the order of 1 ns at the lowest investigated temperature T = 200 K in both cases. The $\alpha$-relaxation of bulk water lies on the same time scale of that of the biological hydration water in the same range of temperatures. Direct comparison with the $\alpha$-relaxation of bulk water was made in the original references, refs. [41, 44, 55].

The $\alpha$-relaxation of bulk water follows the predictions of the Mode Coupling Theory upon cooling, and the temperature behavior of the relaxation time can be described by a power law of the form:

$$\tau_\alpha = \tau_0(T - T_c)^{-\gamma}, \tag{2.4}$$

from high temperature down to the mild supercooling regime. $T_c$ and $\gamma$ are the Mode Coupling temperature and exponent respectively and $\tau_0$ is a time constant. At pressure of 1 bar, SPC/E water shows approximately at $T_{FSC} = 210$ K [41, 62, 72] a crossover from the power law regime to an Arrhenius law regime, where the relaxation time is described by:

$$\tau_\alpha = \tau_0' e^{-\frac{E_a}{k_B T}}, \tag{2.5}$$

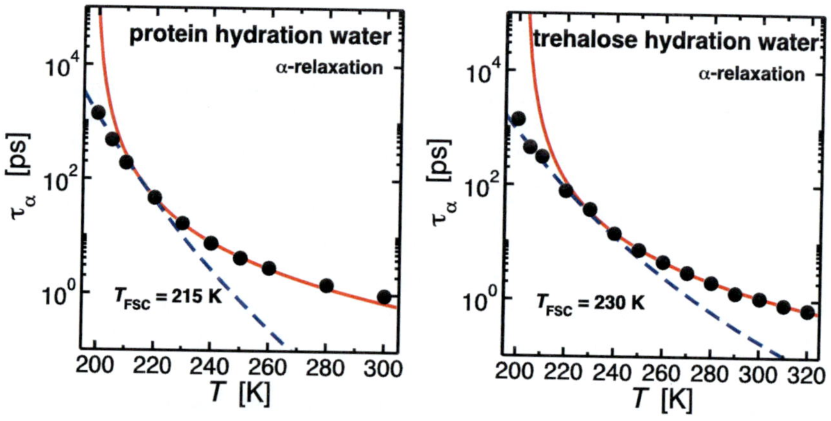


**Fig. 2.5** Left: α-relaxation times of lysozyme hydration water as a function of the temperature. Right: α-relaxation times of trehalose hydration water as a function of the temperature. The points fit the MCT power law, Eq. 2.4, at high temperatures (continuous red lines) and the Arrhenius law, Eq. 2.5, at low temperatures (dashed blue lines). $T_{FSC}$ indicates the temperature of the fragile-to-strong crossover of $\tau_\alpha$ (left panel is reprinted with permission from [55])

upon further supercooling. $E_a$ is the activation energy for the hopping mechanisms, $\tau_0'$ is a time constant and $k_B$ is the Boltzmann constant.

Glass-forming liquids are conventionally classified into two broad classes, strong or fragile liquids, depending on the temperature behavior of their structural relaxation times [73]. When the $\alpha$-relaxation time follows the Arrhenius law, Eq. 2.5, the liquid is termed strong. Fragile liquids show a relaxation time which follows a super-Arrhenius behavior, that can be phenomenologically described with the MCT power law, Eq. 2.4, or with the Vogel – Fulcher – Tamman law [2]. Because of this classification, the crossover of $\tau_\alpha$ takes the name of fragile-to-strong (FSC) crossover. The fragile-to-strong crossover of the $\alpha$-relaxation time along a thermodynamic path, here along the 1 bar isobar, is connected to the crossing of the Widom line, the line of maxima in the thermodynamic response functions that is connected to the characteristics of water of having both a high density structure and a low density structure [74, 75], and consequently to the phase-diagram of water [5]. Different models employed in Molecular Dynamics simulations of bulk water show the fragile-to-strong crossover and its connection to the Widom Line [72, 76–80]. The FSC and the Widom line were also observed in simulations of water in diluted solution with electrolytes [81] and confined water [8, 82, 83].

The $\alpha$-relaxation of both lysozyme hydration water and trehalose hydration water has the same phenomenology of the $\alpha$-relaxation of bulk water. As we can see from Fig. 2.5, both the water hydrating the lysozyme and the trehalose is described at high temperature by the fragile MCT power law, and crosses to the strong Arrhenius law at low temperature. The temperature of fragile-to-strong crossover is $T_{FSC} = 215$ K and $T_{FSC} = 230$ K in the case of lysozyme hydration water and trehalose hydration water respectively. The increase of the crossover temperature indicates a higher glass transition temperature. The presence of the fragile-to-strong transition in hydration water could be connected to the Widom Line and consequently to the presence of a liquid–liquid transition and of a liquid–liquid critical point in the phase diagram of hydration water as it has been discussed for bulk water. Such hypothesis has been discussed in literature, see refs. [9, 15, 17, 84].

The MCT makes precise predictions for example, about the scaling behavior of the Self Intermediate Scattering Functions in temperature and time, the time temperature superposition principle, and on the numerical values of the scaling exponents. Tests of the MCT theory in the fragile region were conducted in ref. [56] in the case of lysozyme and trehalose hydration water. Results were fully consistent with the predictions of the theory.

Finally, the $\alpha$-relaxation of lysozyme hydration water was also probed at different length scales in ref. [57]. The Self Intermediate Scattering functions were calculated at different $Q$-values from 6 $nm^{-1}$ to 30 $nm^{-1}$ for several temperatures. From the curves it was then obtained $\tau_\alpha(Q, T)$. The inverse of the $\alpha$-relaxation time as a function of $Q$ is reported in Fig. 2.6. $\tau_\alpha$ is an estimate of the time a water molecule needs to diffuse over a distance $Q^{-1}$. Therefore, it decreases with increasing $Q$ at every temperature. Besides, it was observed for the lysozyme hydration water a quadratic dependence $\tau_\alpha^{-1} \sim Q^2$, for all the investigated temperature regime. A

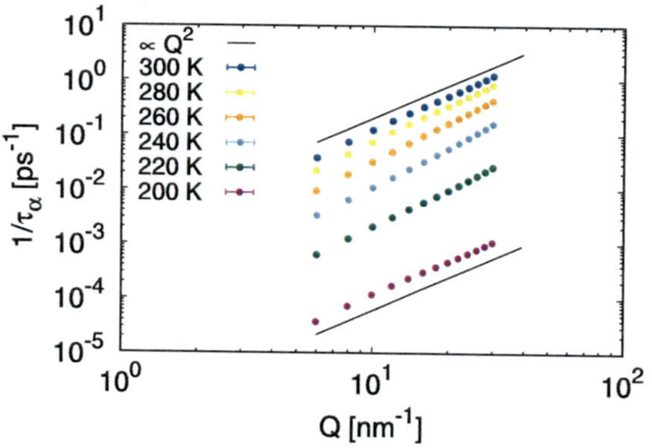


**Fig. 2.6** Inverse of the α-structural relaxation times of lysozyme hydration water extracted from the self Intermediate Scattering Functions calculated at several different values of the wave vector $Q$ and for different temperatures. Black lines reporting quadratic behavior are guides for the eyes (reprinted with permission from [57])

similar result was observed in the case of confined water [85], while in the case of bulk water deviations from the quadratic dependence are found at high $Q$ and low temperatures [63].

In conclusion, the $\alpha$-relaxation of lysozyme hydration water and trehalose hydration water is the analogous $\alpha$-relaxation of bulk water and other glass-formers. The interaction of hydration water with the biomolecule preserves the characteristics of this relaxation: the time scale and the temperature behavior, including the FSC, of the $\alpha$-relaxation is the same of the one of bulk water. Only the value of $T_{FSC}$ appears to depend on the details of the biomolecules.

### 2.4.2 *The Long-Relaxation of Hydration Water and Its Connections to the Biomolecule Dynamics*

In this section we discuss the temperature behavior of the second longer time scale extracted from the density correlators of hydration water of lysozyme and trehalose. This second slower relaxation is absent in bulk water and arises in hydration water from the dynamic coupling between the hydration water and the solute. In order to investigate the coupling, we also show in this section the connections with the biomolecule dynamics. We start from the lysozyme hydration water.

The long-relaxation time extracted from the fit of the SISFs of lysozyme hydration water is shown in Fig. 2.7 (left panel). The time-scale spans from several picoseconds at higher temperatures to several nanoseconds at low temperatures, about 5 times slower than the $\alpha$-relaxation process at the highest temperature and about 2.5 slower at the lowest temperature. Similar slowing factors at high temperature were evaluated from experiments of dynamic light scattering [35].

The long-relaxation is not only slower than the $\alpha$-relaxation at each temperature, but it shows a different temperature behavior. $\tau_{long}$ is described by two Arrhenius laws at high and low temperature, with two different values of activation energy. The crossing between the two strong regimes, a strong-to-strong crossover (SSC),

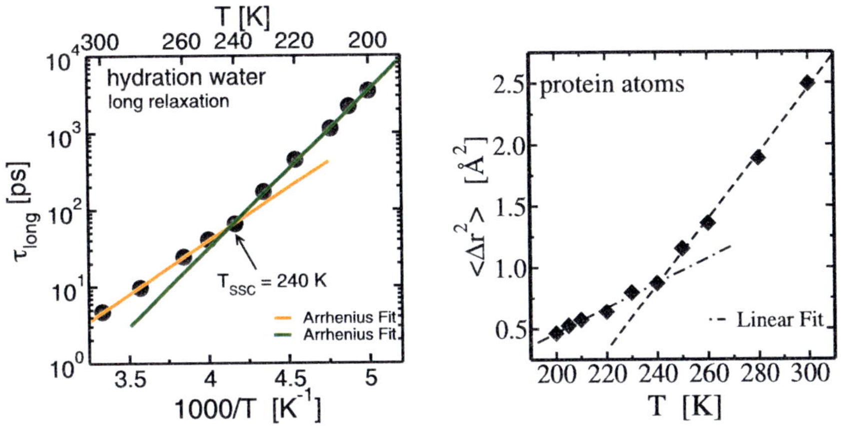


**Fig. 2.7** Left: Plot of the long-relaxation time $\tau_{\rm long}$ of lysozyme hydration water. The points fit Arrhenius laws with different activation energies at high and low temperature. The strong-to-strong crossover (SSC) temperature is $T_{ssc} = 240$ K. Right: Mean Square Fluctuations (MSF), $\Delta r^2$, of the structure of lysozyme as a function of temperature. Dashed lines are best linear fits to determine the protein dynamical transition, the crossover take places at $T_{PDT} = 240$ K. Note the coincidence of the SSC of $\tau_{\rm long}$ and the protein dynamical transition at 240 K (reprinted with permission from [41])

takes place at $T_{SSC} = 240$ K. As said at the beginning of this section, the long-process arises from the dynamic coupling between the protein and its hydration water layer. As a supporting evidence of this point, in ref. [41] it was explicitly investigated the protein internal motion. The mean square fluctuations (MSF) of the lysozyme structure were evaluated at each temperature, by a procedure of structure alignment of the protein along the Molecular Dynamics trajectory which removes rigid translational and rotational motion of the protein during the simulations and permits therefore to evaluate fluctuations of its structure. The obtained MSF are shown as a function of the temperature in Fig. 2.7 (right panel). The fluctuations of the protein structure increase linearly with temperature, but two different slopes are observed at low and high temperature. The MSF of proteins consists of two contributions, one due to vibrations and a second one due to conformational changes [86]. At low temperatures the MSF consists of the vibrational contribution only and it increases linearly with the temperature. The conformational contribution is activated above a specific temperature only if the protein is minimally hydrated. If this is the case, when the conformational contribution sets on, a change in the slopes of the MSF of proteins can be detected. This transition between a regime where the protein is essentially structurally hindered and a regime where it can change conformations and it exerts its different functions is called protein dynamical transition (PDT) [47]. From the calculated MSF, the protein dynamical transition of the lysozyme happens at $T_{PDT} = 240$ K, in coincidence therefore with the strong-to-strong crossover of $\tau_{long}$ at $T_{SSC}$ in the dynamics of hydration water. The coincidence of the crossover temperatures of two quantities, one solely of the hydration water and one solely of the protein, strongly evidences the dynamic coupling between the protein internal dynamics and its hydration water.

In the case of the trehalose hydration water surprisingly similar results were obtained in ref [44]. It was observed in the study that trehalose molecules aggregate consistently below 280 K in the investigated solution. In particular, the size of the biggest trehalose cluster of the solution was monitored in the simulations as a function of the temperature, and it was found that the biggest cluster is composed by more than 50% of the total trehalose molecules of the solution below 280 K and increases to almost 90% at 200 K. The presence of an extended cluster of trehalose molecules, with a total number of atoms comparable to the lysozyme but with a larger surface exposed to the solution, gives rise to a collective surface dynamic of the cluster which behaves as a flexible macromolecule, similarly to a protein.

The long-relaxation times extracted from the fits of the Self Intermediate Scattering Functions calculated using the positions of the oxygen atoms of water molecules moving inside the trehalose hydration shell [44] are plotted in Fig. 2.8 (left panel).

The time scale of the long-relaxation of trehalose hydration water spans from several picoseconds at higher temperatures to hundreds of nanoseconds at low temperature.

The $\tau_{long}$ shows a first crossover upon cooling at 280 K, in correspondence of the formation of the extended cluster of trehalose molecules. Once the cluster is formed, by decreasing further the temperature we observe a second strong-to-strong crossover around 250 K. In order to test whether a dynamic coupling between the hydration water and the trehalose clusters holds also in this system, the mean square displacements of the hydroxyl groups of the trehalose molecules were evaluated at each temperature. Its value at 500 ps is plotted in Fig. 2.8, right panel. We observe

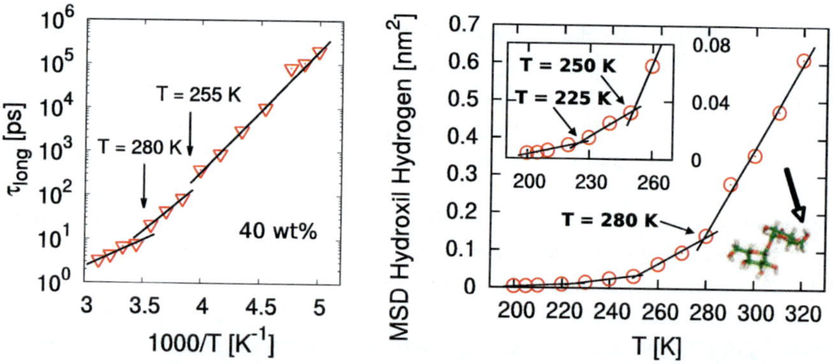


**Fig. 2.8** Left: Plot of temperature behavior of the long-relaxation time $\tau_{long}$ of trehalose hydration water. The points fit Arrhenius laws with different activation energies. Right: Mean square displacement (MSD) of the hydroxyl groups of trehalose as a function of the temperature. Lines are best linear fit lines done to determine the crossover locations. Both the $\tau_{long}$ and the MSD exhibit crossovers at T = 280 and at about 250 K. The MSD bears information also about the $\alpha$-relaxation exhibiting an extra crossover at T = 225 K (see the inset) (reprinted with permission from [44])

that the formation of the trehalose cluster impacts also this quantity as we find again the signature of the cluster formation at 280 K as a first crossover upon cooling. A second change of slope happens approximately at 250 K, therefore in correspondence of the (second) strong-to-strong crossover of the $\tau_{long}$. Since at this temperature the cluster is formed, by analogy with the lysozyme this transition was called trehalose-dynamical transition. This crossover was also observed experimentally in ref. [87]. The presence of a transition in the dynamics of hydrated solute could be general feature of water closer to macromolecules [88], and in fact it has been recently also observed in hydrated PNIPAM microgels [89, 90]. Finally, since the mean square displacement probes translational dynamics, it is also sensitive to the $\alpha$-relaxation. In the inset, the mean square displacement is magnified in the low temperature region where we observe a third, low temperature crossover around 230 K which corresponds to the FSC of the $\tau_{\alpha}$.

With a follow-up study [25], it was unambiguously proved that the $\alpha$-relaxation is related to only water, while the long-relaxation arises from the coupling with the macromolecules. This was possible by calculating the correlation function of hydrogen bonds between hydration water molecules and the correlations function of hydrogen bonds between hydration water and the trehaloses. The first correlations functions probes therefore only water properties, and it shows a unique crossover in correspondence of the FSC at 230 K. The second correlation functions, which directly probes the coupling between hydration water and trehalose, shows a unique crossover in correspondence of the strong-to-strong crossover at 250 K.

In conclusion, the long-relaxation of lysozyme hydration water and trehalose hydration water is a new relaxation, absent in bulk water, which arises from the dynamical coupling between the biomolecules and its hydration water. The time scale is much slower with respect to the $\alpha$-relaxation and the temperature behavior is strong. A crossover between two different strong regimes occurs simultaneously with the protein or the trehalose dynamical transition.

## 2.5 Slow Dynamics of Hydration Water Probed with Self van Hove Correlation Functions

Single particle dynamics can be probed in the real space through the Fourier transform of the Self Intermediate Scattering Function, the so-called van Hove Self Correlation Function. This function can be written as:

$$G_{SELF}(r,t) = \frac{1}{N}\left\langle \sum_{i=1}^{N} (r - |\vec{r}_i(t) - \vec{r}_i(0)|) \right\rangle, \tag{2.6}$$

and it can be calculated directly from the trajectories stored in molecular dynamics simulations, without performing a Fourier transform operation on the Self Intermediate Scattering Function, thus avoiding truncation problems.

For isotropic liquids the radial part, $4\pi r^2 G_{SELF}(r, t)$, is particularly relevant because it gives directly the probability density that a particle of the liquid performs a displacement $r$ in a time interval $t$. van Hove Self Correlation Functions have been studied for many liquids, including glass-formers [91–93] and bulk water [4, 63, 75, 94–96].

In the case of supercooled water, it was shown how this particular correlation function can probe the relaxation mechanism which permits to the liquid to relax structurally below the FSC avoiding the divergence of the $\alpha$-relaxation time below the Mode Coupling Temperature $T_C$: few degrees above $T_C$, in fact, new temperature-activated phenomena, called hopping processes, set on. These mechanisms smear out of the divergence of the $\tau_\alpha$ when approaching $T_C$ from above, and correspondently $\tau_\alpha$ crosses from the MCT regime to the Arrhenius regime. Without hopping phenomena, below $T_C$ water molecules would remain trapped inside the frozen neighbors cage, because thermal fluctuations would be too small to relax the cage. In this case, the liquid would transform into a glass.

By means of the van Hove Self Correlation Functions, the activation of hopping phenomena can be observed by monitoring their shape. Besides, since they are correlators in real space, they give a direct estimate of the length-scale of the hopping distances involved in the relaxation.

In the fragile regime, the van Hove Self Correlation Functions of bulk water plotted as a function of the distance $r$ are unimodal curves at every time $t$, with a single maximum that evolves at longer $r$ as $t$ increases. In the strong regime, below the FSC, the van Hove Self Correlation Functions at long times develop multiple peaks [63, 95] and these peaks coincide with the coordination shells of the radial distribution functions, because water molecules can escape the neighbors cage by jumping toward outer coordination shells.

In the case of hydration water, we observe these hopping phenomena below the FSC. Moreover, we find a new hopping mechanism which is characteristic of hydration water only and that appears at high temperatures and long times.

### 2.5.1 *α-relaxation of Hydration Water and Cage-Escaping Hopping*

van Hove Self Correlation Functions were calculated for the oxygen atoms of hydration water of lysozyme at 300 K in ref. [23]. The study was extended to the supercooled regime in ref. [58].

We show in Fig. 2.9 an example of a curve calculated for a long time and a low temperature, where we can observe the presence of hopping phenomena related to the neighbors cage-escaping mechanism related to the $\alpha$-relaxation and therefore similar to bulk water [72, 95]. The curve is calculated at $t = 50$ ns for T = 200 K, therefore below the FSC of the lysozyme hydration water. The van Hove Self Correlation Functions of lysozyme hydration water shows a main peak around 0.18 nm and

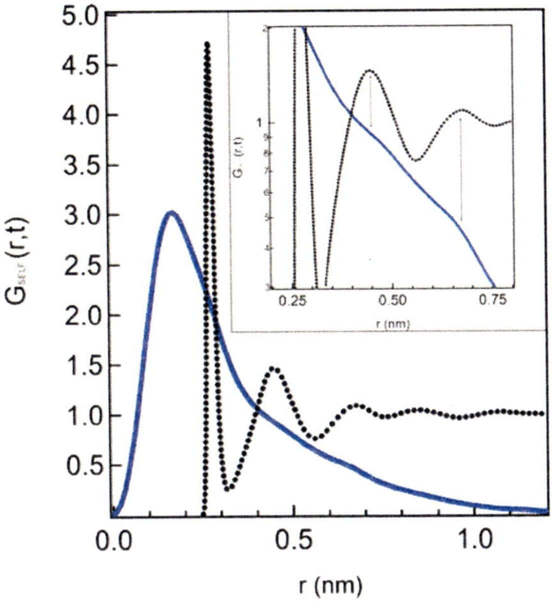


**Fig. 2.9** van Hove Self Correlation Function of lysozyme hydration water at t = 50 ns and T = 200 K (blue continuous line). The radial distribution function is also reported (dotted black line). The shoulders at large $r$ of the van Hove Self Correlation Function evidence the presence of hopping phenomena, as they are aligned with the shells of the RDF. The inset is a blowup to best evidentiate the coincidence between the van Hove Self Correlation Function peaks and the RDF peaks (reprinted with permission from [58]).

several shoulders at larger distance. The main peak is well within the first peak of the radial distribution function (also shown in the same figure) which means that it corresponds to a fraction of particles that are still localized inside a very small region of space centered at the position of the particle at $t = 0$ after 50 ns. Nonetheless, we find also finite fractions of particles at other preferred positions (the shoulders) that correspond to the RDF shells, mainly the second and the third ones. This point is particularly visible by looking at the inset of Fig. 2.9, where the region of the shoulders of the van Hove Self Correlation Functions is magnified. The shoulders are due to molecules which successfully escape from the neighbors cage where they were trapped, by jumping toward outer water coordination shells and effectively bypassing the molecular cage.

### *2.5.2 Long-Relaxation of Hydration Water and Protein-Related Hopping*

In addition to the cage-escaping mechanism, new hopping phenomena arise in lysozyme hydration water already at high temperature. These mechanisms, absent in bulk water, happen on a longer length scale with respect to the low-temperature hopping due to the MCT cage-effect.

In order to show this mechanism, we plot in Fig. 2.10 the van Hove Self Correlation Functions of hydration water at several different times from $t = 240$ fs to $t = 15$ ns, at

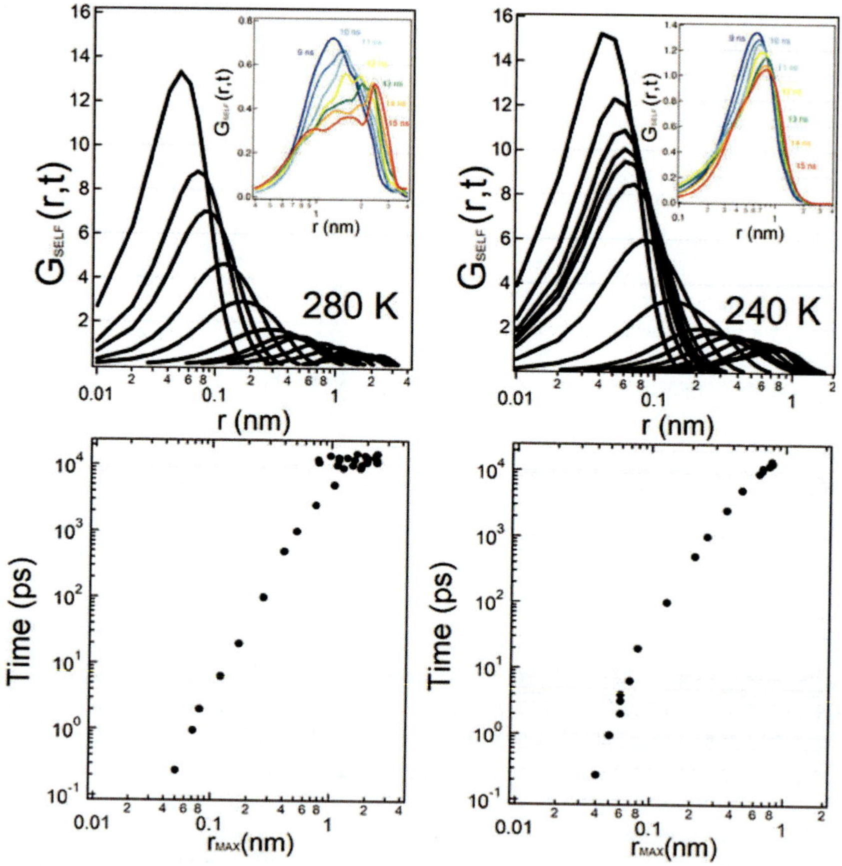


**Fig. 2.10** Hydration water oxygens van Hove Self Correlation Functions in logarithmic scale at $T = 280$ K (panel A) and $T = 240$ K (panel B) and the position of the peaks of the curves as a function of the time at $T = 280$ K (panel C) and $T = 240$ K (panel D). At $T = 280$ K we clearly observe the presence of multiple peaks concentrated between 9 and 15 ns (reprinted with permission from [58])

two selected temperature, $T = 280$ K (panel A) and at the protein dynamical transition temperature, $T = 240$ K (panel B). As the time increases, the curves broaden, and the main peaks move toward large distances because of the translational motion of the water molecules that will move away from the origin of their motion. In the insets, a magnification of the long-time curves in logarithmic scale is reported, which allows to identify clearly this new hopping-phenomena at $T = 280$ K. We note that the hopping peaks are very well defined, and we identify them as the channels for the long-relaxation. A closer inspection of the curves at $T = 280$ K reveals that the curves develop several local maxima and secondary shoulders for times longer than $t = 9$ ns. The long-relaxation persists at all temperatures for our Self Intermediate Scattering

Functions, but the time scale stretches very much and we are not able to follow the related hopping processes in the van Hove Self Correlation Function for the low temperatures as they are outside of the temporal window that we can investigate in the direct space. We in fact see from our Figure that at $T = 240$ K the hopping processes are not visible any longer. We report the position of the single maximum or the positions of the several local maxima as a function of the time in the bottom panels, at $T = 280$ K (panel C) and $T = 240$ K (panel D). The presence of multiple peaks is the signal of hopping phenomena. The shoulders in the van Hove Self Correlation Functions develop at very large distances, 0.8–3 nm, compared to the shoulders in the self van Hove functions of Fig. 2.9 due to the hopping of MCT where the hopping distances are those of the first shells of the RDF. Moreover, at this high temperature, there is no clear cage effect (no plateau in the Self Intermediate Scattering Functions of Fig. 2.4) and finally the time scale here is very long, 9–15 ns, when compared to the $\alpha$-relaxation time that is circa 1 picosecond at T = 280 K. Based on these observations, we conclude that the present hopping is different from the hopping related to the $\alpha$-relaxation and its cage-effect. The undergoing mechanism here is due instead to water molecules with very long residence time inside the lysozyme hydration shell, trapped in protein domains that can rearrange on a time scale that we can probe above the protein dynamical transition. In this respect, these hopping phenomena were observed at high temperature mainly and down to T = 250 K. Below the protein dynamical transition, structural fluctuations of the lysozyme protein are strongly suppressed. This turns in a substantial increase of the time scales required for observing the present hopping mechanism.

## 2.6 Conclusions

In this chapter we presented results on the translational dynamics of the hydration water of a lysozyme protein and of trehalose molecules. Translational dynamics was first studied by calculating the self Intermediate Scattering Functions which have been analyzed under the Mode Coupling theory framework. The dynamics of hydration water can be described by two slow translational relaxations. One process is the analogous $\alpha$-relaxation of bulk water and other glass-forming liquids. In bulk water, lysozyme hydration water and trehalose hydration water the time scale of the $\alpha$-relaxation is similar. A fragile-to-strong crossover is detected in all these three systems upon decreasing temperature. The second, slower relaxation, the long-relaxation, is absent in bulk water and arises from the interactions of the biomolecules with its hydration water. The long-relaxation happens on a time scale that is at every temperature well separated and much slower than the $\alpha$-relaxation time. The long-relaxation times exhibits upon cooling a strong-to-strong crossover and this happens in coincidence with the protein/trehalose dynamical transitions, which are transitions that regard the biomolecules dynamics.

A complementary study of the translational dynamics of lysozyme hydration water was conducted by calculating the van Hove Self Correlation Functions. This study proved that below the fragile-to-strong crossover, the structural relaxation of hydration water is possible thanks to hopping phenomena which permit the molecules to escape from the confinement exerted by the nearest neighbors, the so-called cage-effect. These hopping phenomena are the same mechanisms which permit the relaxation of bulk water below the fragile-to-strong crossover. For their nature, they are characterized by supercooled temperatures, distances of the order of few coordination shells and long timescales of the order of tenths of nanoseconds. The same study also revealed a new trapping-regime and hopping phenomena very different from the mechanism discussed above. These new hopping phenomena start to take place at high temperatures, well above the protein dynamical transition and at high temperatures the length scale is approximately 10 times larger than length scale of the neighbors cage-related hopping and the time scale is long (nanoseconds). We related these new mechanisms to the rearrangements of protein domains activated above the protein dynamical transition, that permit to detect translational motions of that group of hydration water molecules influenced by the lysozyme and with very long residence time through the second, long-relaxation mechanism. Very recent results on the hydration water of lysozyme immersed in a trehalose-water solution [97] show that the addiction of the trehalose is very effective in slowing down the dynamics of the lysozyme hydration water, in particular trehalose tremendously slows down the long-relaxation process while leaving unaltered the $\alpha$-relaxation time. The long-relaxation of hydration water seems therefore to play an important role in the context of bioprotection.

## References

1. Franks F (2000) Water: A Matrix of Life, 2nd edn. The Royal Society of Chemistry, Cambridge
2. Gallo P, Amann-Winkel K, Angell CA, Anisimov MA, Caupin F, Chakravarty C, Lascaris E, Loerting T, Panagiotopoulos AZ, Russo J, Sellberg JA, Stanley HE, Tanaka H, Vega C, Xu L, Pettersson LGM (2016) Water: a tale of two liquids. Chem Rev 116:7463–7500
3. Poole PH, Sciortino F, Essmann U, Stanley HE (1992) Phase behaviour of metastable water. Nature 360:324–328
4. Sastry S, Debenedetti PG, Sciortino F, Stanley HE (1996) Singularity-free interpretation of the thermodynamics of supercooled water. Phys Rev E 53:6144–6154
5. Xu L, Kumar P, Buldyrev SV, Chen S-H, Poole PH, Sciortino F, Stanley HE (2005) Relation between the Widom line and the dynamic crossover in systems with a liquid-liquid phase transition. Proc Natl Acad Sci U S A 102:16558–16562
6. Franzese G, Stanley HE (2007) The Widom line of supercooled water. J Phys Condens Matter 19:205126
7. Gallo P, Stanley HE (2017) Supercooled water reveals its secrets. Science 358:1543–1544
8. De Marzio M, Camisasca G, Conde MM, Rovere M, Gallo P (2017) Structural properties and fragile to strong transition in confined water. J Chem Phys 146:084505
9. Kumar P, Yan Z, Xu L, Mazza MG, Buldyrev SV, Chen SH, Sastry S, Stanley HE (2006) Glass transition in biomolecules and the liquid-liquid critical point of water. Phys Rev Lett 97:177802
10. Kumar P, Franzese G, Buldyrev SV, Stanley HE (2008) Dynamics of water at low temperatures and implications for biomolecules. Lect Notes Phys 752:3–22

11. Stanley HE, Buldyrev SV, Franzese G, Kumar P, Mallamace F, Mazza MG, Stokely K, Xu L (2010) Liquid polymorphism: water in nanoconfined and biological environments. J Phys Condens Matter 22:284101
12. Franzese G, Hernando-Martínez A, Kumar P, Mazza MG, Stokely K, Strekalova EG, De Los Santos F, Stanley HE (2010) hase transitions and dynamics of bulk and interfacial water. Phys Condens Matter 22:284103
13. Mallamace F, Corsaro C, Mallamace D, Baglioni P, Stanley HE, Chen SH (2011) A possible role of water in the protein folding process. J Phys Chem B 115:14280–14294
14. Bianco V, Iskrov S, Franzese G (2012) Understanding the role of hydrogen bonds in water dynamics and protein stability. J Biol Phys 38:27–48
15. Franzese G, Bianco V (2013) Water at Biological and Inorganic Interfaces. Food Biophys 8:153–169
16. Bianco V, Franzese G (2015) Contribution of water to pressure and cold denaturation of proteins. Phys Rev Lett 115:108101
17. Mallamace F, Corsaro C, Mallamace D, Vasi S, Vasi C, Stanley HE, Chen SH (2015) Some thermodynamical aspects of protein hydration water. J Chem Phys 142:215103
18. Roberts CJ, Debenedetti PG (1999) Structure and dynamics in concentrated, amorphous carbohydrate-water systems by molecular dynamics simulation. J Phys Chem B 103:7308–7318
19. Bizzarri AR, Cannistraro S (2002) Molecular dynamics of water at the protein-solvent interface. J Phys Chem B 106:6617–6633
20. Lee SL, Debenedetti PG, Errington JR (2005) A computational study of hydration, solution structure, and dynamics in dilute carbohydrate solutions. J Chem Phys 122:204511
21. Giuffrida S, Panzica M, Giordano FM, Longo A (2011) SAXS study on myoglobin embedded in amorphous saccharide matrices. Eur Phys J E 34:87
22. de Barbosa R, C, Barbosa MC, (2015) Hydration shell of the TS-Kappa protein: higher density than bulk water. Phys A 439:48–58
23. Camisasca G, Iorio A, De Marzio M, Gallo P (2018) Structure and slow dynamics of protein hydration water. J Mol Liq 268:903–910
24. Martelli F, Ko HY, Borallo CC, Franzese G (2018) Structural properties of water confined by phospholipid membranes. Front Phys 13:136801
25. Iorio A, Camisasca G, Rovere M, Gallo P (2019) Characterization of hydration water in supercooled water-trehalose solutions: the role of the hydrogen bonds network. J Chem Phys 151:044507
26. Martelli F, Crain J, Franzese G (2020) Network topology in water nanoconfined between phospholipid membranes. ACS Nano 14:8616–8623
27. Rocchi C, Bizzarri AR, Cannistraro S (1998) Water dynamical anomalies evidenced by molecular-dynamics simulations at the solvent-protein interface. Phys Rev E 57:3315–3325
28. Cordone L, Cottone G, Giuffrida S, Palazzo G, Venturoli G, Viappiani C (2005) Internal dynamics and protein–matrix coupling in trehalose-coated proteins. Biochem Biophys Acta - Proteins Proteomics 1749:252–281
29. Paciaroni A, Cornicchi E, Marconi M, Orecchini A, Petrillo C, Haertlein M, Moulin M, Sacchetti F (2009) Coupled relaxations at the protein-water interface in the picosecond time scale. J R Soc Interface 6(Suppl 5):S635–S640
30. Paolantoni M, Comez L, Gallina ME, Sassi P, Scarponi F, Fioretto D, Morresi A (2009) Light scattering spectra of water in trehalose aqueous solutions: evidence for two different solvent relaxation processes. J Phys Chem B 113:7874–7878
31. Magno A, Gallo P (2011) Understanding the mechanisms of bioprotection: a comparative study of aqueous solutions of trehalose and maltose upon supercooling. J Phys Chem Lett 2:977–982
32. Corradini D, Strekalova EG, Stanley HE, Gallo P (2013) Microscopic mechanism of protein cryopreservation in an aqueous solution with trehalose. Sci Rep 3:1218
33. Cordone L, Cottone G, Cupane A, Emanuele A, Giuffrida S, Levantino M (2015) Proteins in saccharides matrices and the trehalose peculiarity: biochemical and biophysical properties. Curr Org Chem 19:1684–1706

34. Franzese G, Bianco V, Iskrov S (2011) Water at Interface with Proteins. Food Biophys 6:186–198
35. Comez L, Lupi L, Morresi A, Paolantoni M, Sassi P, Fioretto D (2013) More is different: experimental results on the effect of biomolecules on the dynamics of hydration water. J Phys Chem Lett 4:1188–1192
36. Giuffrida S, Cottone G, Bellavia G, Cordone L (2013) Proteins in amorphous saccharide matrices: structural and dynamical insights on bioprotection. Eur Phys J E Soft Matter 36:79
37. Mallamace F, Corsaro C, Mallamace D, Stanley HE, Chen SH (2013) Water and biological macromolecules. Adv Chem Phys 152:263–308
38. Schirò G, Fomina M, Cupane A (2013) Communication: protein dynamical transition vs liquid-liquid phase transition in protein hydration water. J Chem Phys 139:121102
39. Mallamace F, Baglioni P, Corsaro C, Chen SH, Mallamace D, Vasi C, Stanley HE (2014) The influence of water on protein properties. J Chem Phys 141:165104
40. Mallamace F, Corsaro C, Mallamace D, Cicero N, Vasi S, Dugo G, Stanley HE (2015) Dynamical changes in hydration water accompanying lysozyme thermal denaturation. Front Phys 10:106104
41. Camisasca G, De Marzio M, Corradini D, Gallo P (2016) Two structural relaxations in protein hydration water and their dynamic crossovers. J Chem Phys 145:044503
42. Comez L, Paolantoni M, Sassi P, Corezzi S, Morresi A, Fioretto D (2016) Molecular properties of aqueous solutions: a focus on the collective dynamics of hydration water. Soft Matter 12:5501–5514
43. Köhler MH, de Barbosa RC, da Silva LB, Barbosa MC (2017) Role of the hydrophobic and hydrophilic sites in the dynamic crossover of the protein-hydration water. Phys A Stat Mech Appl 468:733–739
44. Iorio A, Camisasca G, Gallo P (2019) Slow dynamics of hydration water and the trehalose dynamical transition. J Mol Liq 282:617–625
45. Corezzi S, Paolantoni M, Sassi P, Morresi A, Fioretto D, Comez L (2019) Trehalose-induced slowdown of lysozyme hydration dynamics probed by EDLS spectroscopy. J Chem Phys 151:015101
46. Dos Santos MAF, Habitzreuter MA, Schwade MH, Borrasca R, Antonacci M, Gonzatti GK, Netz PA, Barbosa MC (2019) Dynamical aspects of supercooled TIP3P-water in the grooves of DNA. J Chem Phys 150:235101
47. Doster W, Cusack S, Petry W (1990) Dynamic instability of liquidlike motions in a globular protein observed by inelastic neutron scattering. Phys Rev Lett 65:1080–1083
48. Combet S, Zanotti J-M (2012) Further evidence that interfacial water is the main "driving force" of protein dynamics: a neutron scattering study on perdeuterated C-phycocyanin. Phys Chem 14:4927–4934
49. Schirò G, Fichou Y, Gallat F-X, Wood K, Gabel F, Moulin M, Härtlein M, Heyden M, Colletier J-P, Orecchini A, Paciaroni A, Wuttke J, Tobias DJ, Weik M (2015) Translational diffusion of hydration water correlates with functional motions in folded and intrinsically disordered proteins. Nat Commun 6:6490
50. Schirò G, Weik M (2019) Role of hydration water in the onset of protein structural dynamics. J Phys Condens Matter 31:463002
51. Chen S-H, Lagi M, Chu X, Zhang Y, Kim C, Faraone A, Fratini E, Baglioni P (2010) Dynamics of a globular protein and its hydration water studied by neutron scattering and MD simulations. Spectrosc Int J 24:1–24
52. Berendsen HJC, Grigera JR, Straatsma TP (1987) The missing term in effective pair potentials. J Phys Chem 91:6269–6271
53. MacKerell AD, Bashford D, Bellott M, Dunbrack RL, Evanseck JD, Field MJ, Fischer S, Gao J, Guo H, Ha S, Joseph-McCarthy D, Kuchnir L, Kuczera K, Lau FT, Mattos C, Michnick S, Ngo T, Nguyen DT, Prodhom B, Reiher WE, Roux B, Schlenkrich M, Smith JC, Stote R, Straub J, Watanabe M, Wiórkiewicz-Kuczera J, Yin D, Karplus M (1998) All-atom empirical potential for molecular modeling and dynamics studies of proteins. J Phys Chem B 102:3586–3616

54. MacKerell AD, Feig M, Brooks CL (2004) Extending the treatment of backbone energetics in protein force fields: limitations of gas-phase quantum mechanics in reproducing protein conformational distributions in molecular dynamics simulations. J Comput Chem 25:1400–1415
55. Camisasca G, De Marzio M, Rovere M, Gallo P (2017) Slow dynamics and structure of supercooled water in confinement. Entropy 19:185
56. Iorio A, Camisasca G, Gallo P (2019) Glassy dynamics of water at interface with biomolecules: a Mode Coupling Theory test. Sci China Phys Mech Astron 62:107011
57. Iorio A, Minozzi M, Camisasca G, Rovere M, Gallo P (2020) Slow dynamics of supercooled hydration water in contact with lysozyme: examining the cage effect at different length scales. Philos Mag 100:2582–2595
58. Tenuzzo L, Camisasca G, Gallo P (2020) Protein-water and water-water long-time relaxations in protein hydration water upon cooling—a close look through density correlation functions. Molecules 25:4570
59. Guvench O, Greene SN, Kamath G, Brady JW, Venable RM, Pastor RW, MacKerell AD (2008) Additive empirical force field for hexopyranose monosaccharides. J Comput Chem 29:2543–2564
60. Guvench O, Hatcher ER, Venable RM, Pastor RW, MacKerell AD (2009) CHARMM additive all-atom force field for glycosidic linkages between hexopyranoses. J Chem Theory Comput 5:2353–2370
61. Pronk S, Páll S, Schulz R, Larsson P, Bjelkmar P, Apostolov R, Shirts MR, Smith JC, Kasson PM, van der Spoel D, Hess B, Lindahl E (2013) GROMACS 4.5: a high-throughput and highly parallel open source molecular simulation toolkit. Bioinformatics 29:845–854
62. Gallo P, Sciortino F, Tartaglia P, Chen S-H (1996) Slow dynamics of water molecules in supercooled states. Phys Rev Lett 76:2730–2733
63. Sciortino F, Gallo P, Tartaglia P, Chen S-H (1996) Supercooled water and the kinetic glass transition. Phys Rev E 54:6331–6343
64. Götze W, Sjogren L (1992) Relaxation processes in supercooled liquids. Reports Prog Phys 55:241–376
65. Götze W (2009) Complex Dynamics of Glass-Forming Liquids. Oxford University Press, New York
66. Perticaroli S, Comez L, Sassi P, Paolantoni M, Corezzi S, Caponi S, Morresi A, Fioretto D (2015) Hydration and aggregation of lysozyme by extended frequency range depolarized light scattering. J Non Cryst Solids 407:472–477
67. Corezzi S, Sassi P, Paolantoni M, Comez L, Morresi A, Fioretto D (2014) Hydration and rotational diffusion of levoglucosan in aqueous solutions. J Chem Phys 140:184505
68. Pal SK, Peon J, Bagchi B, Zewail AH (2002) Biological water: femtosecond dynamics of macromolecular hydration. J Phys Chem B 106:12376–12395
69. Pal SK, Peon J, Zewail AH (2002) Biological water at the protein surface: dynamical solvation probed directly with femtosecond resolution. Proc Natl Acad Sci 99:1763–1768
70. Zhang L, Wang L, Kao Y-T, Qiu W, Yang Y, Okobiah O, Zhong D (2007) Mapping hydration dynamics around a protein surface. Proc Natl Acad Sci U S A 104:18461–18466
71. Li T, Hassanali AA, Kao Y-T, Zhong D, Singer SJ (2007) Hydration dynamics and time scales of coupled water-protein fluctuations. J Am Chem Soc 129:3376–3382
72. Starr FW, Sciortino F, Stanley HE (1999) Dynamics of simulated water under pressure. Phys Rev E 60:6757–6768
73. Angell CA (2002) Liquid fragility and the glass transition in water and aqueous solutions. Chem Rev 102:2627–2650
74. Camisasca G, Schlesinger D, Zhovtobriukh I, Pitsevich G, Pettersson LGM (2019) A proposal for the structure of high- and low-density fluctuations in liquid water. J Chem Phys 151:034508
75. Camisasca G, Galamba N, Wikfeldt KT, Pettersson LGM (2019) Translational and rotational dynamics of high and low density TIP4P/2005 water. J Chem Phys 150:224507
76. Gallo P, Rovere M (2012) Mode coupling and fragile to strong transition in supercooled TIP4P water. J Chem Phys 137:164503

77. De Marzio M, Camisasca G, Rovere M, Gallo P (2016) Fragile-to-strong crossover in supercooled water : A comparison between TIP4P and TIP4P/2005 models. IL NUOVO CIMENTO C 39:1–8
78. De Marzio M, Camisasca G, Rovere M, Gallo P (2016) Mode coupling theory and fragile to strong transition in supercooled TIP4P/2005 water. J Chem Phys 144:074503
79. De Marzio M, Camisasca G, Rovere M, Gallo P (2018) Fragile to strong crossover and Widom line in supercooled water: A comparative study. Front Phys 13:136103
80. Stanley HE, Kumar P, Han S, Mazza MG, Stokely K, Buldyrev SV, Franzese G, Mallamace F, Xu L (2009) Heterogeneities in confined water and protein hydration water. J Phys Condens Matter 21:504105
81. Gallo P, Corradini D, Rovere M (2013) Fragile to strong crossover at the Widom line in supercooled aqueous solutions of NaCl. J Chem Phys 139:204503
82. Gallo P, Rovere M, Chen S-H (2010) Dynamic crossover in supercooled confined water: understanding bulk properties through confinement. J Phys Chem Lett 1:729–733
83. Gallo P, Rovere M, Chen S-H (2012) Water confined in MCM-41: a mode coupling theory analysis. J Phys Condens Matter 24:064109
84. Coronas LE, Vilanova O, Bianco V, Santos F de los, Franzese G (2020) The Franzese-Stanley Coarse Grained Model for Hydration Water, arXiv:2004.03646v2
85. Gallo P, Rovere M, Spohr E (2000) Glass transition and layering effects in confined water: a computer simulation study. J Chem Phys 113:11324–11335
86. Fenimore PW, Frauenfelder H, McMahon BH, Young RD (2004) Bulk-solvent and hydration-shell fluctuations, similar to alpha- and beta-fluctuations in glasses, control protein motions and functions. Proc Natl Acad Sci U S A 101:14408–14413
87. Magazù S, Maisano G, Migliardo F, Mondelli C, Romeo G (2004) An elastic neutron scattering on dynamical transition in hydrogen-bonded systems. J Mol Struct 700:225–227
88. He Y, Ku PI, Knab JR, Chen JY, Markelz AG (2008) Protein dynamical transition does not require protein structure. Phys Rev Lett 101:178103
89. Zanatta M, Tavagnacco L, Buratti E, Bertoldo M, Natali F, Chiessi E, Orecchini A, Zaccarelli E (2018) Evidence of a low-temperature dynamical transition in concentrated microgels. Sci Adv 4:eaat5895
90. Tavagnacco L, Chiessi E, Zanatta M, Orecchini A, Zaccarelli E (2019) Water-polymer coupling induces a dynamical transition in microgels. J Phys Chem Lett 10:870–876
91. Kob W, Andersen HC (1995) Testing mode-coupling theory for a supercooled binary Lennard-Jones mixture. II. Intermediate scattering function and dynamic susceptibility. Phys Rev E 52:4134–4153
92. Kob W, Andersen HC (1995) Testing mode-coupling theory for a supercooled binary Lennard-Jones mixture I: the van Hove correlation function. Phys Rev E 51:4626–4641
93. Sastry S, Debenedetti PG, Stillinger FH (1998) Signatures of distinct dynamical regimes in the energy landscape of a glass-forming liquid. Nature 393:554–557
94. Svishchev IM, Zassetsky AY (2000) Three-dimensional picture of dynamical structure in liquid water. J Chem Phys 112:1367–1372
95. De Marzio M, Camisasca G, Rovere M, Gallo P (2017) Microscopic origin of the fragile to strong crossover in supercooled water: The role of activated processes. J Chem Phys 146:084502
96. Iwashita T, Wu B, Chen W-R, Tsutsui S, Baron AQR, Egami T (2017) Seeing real-space dynamics of liquid water through inelastic x-ray scattering. Sci Adv 3:e1603079
97. Camisasca G, De Marzio M, Gallo P (2020) Effect of Trehalose on Protein Cryoprotection: Insights on the Mechanism of Slowing Down of Hydration. J Chem Phys 153:224503

# Chapter 3
# Molecular Perspective on Solutions and Liquid Mixtures from Modelling and Experiment

**Leon de Villiers Engelbrecht, Francesca Mocci, Yonglei Wang, Sergiy Perepelytsya, Tudor Vasiliu, and Aatto Laaksonen**

**Abstract** Liquid solutions and mixtures are part of our everyday lives and also important for their chemical and industrial applications. While considered fairly unattractive substances when kept in bottles and containers, their behavior as molecules can be completely the opposite, continuously attracting scientists to explain it better. Very strong repulsive and attractive interactions between the molecules can create most intriguing local structures, aggregates and complexes, whose spatial organization is often difficult to rationalize. Also, the same mixture can behave completely differently depending on the composition ratio, affecting strongly its macroscopic properties. To gain insight into the complex world of binary liquid mixtures, deep eutectic solvents and ionic liquid systems, combined theoretical and experimental studies are necessary. In this chapter we introduce the methodology of computer simulations and illustrate with several examples of the often-unexpected behavior of many liquid mixtures.

L. de V. Engelbrecht · F. Mocci (✉)
Department of Chemical and Geological Sciences, University of Cagliari, 09042 Monserrato, Italy
e-mail: fmocci@unica.it

Y. Wang · A. Laaksonen (✉)
Division of Physical Chemistry, Department of Materials and Environmental Chemistry, Arrhenius Laboratory, Stockholm University, 10691 Stockholm, Sweden
e-mail: aatto.laaksonen@mmk.su.se

S. Perepelytsya
Bogolyubov Institute for Theoretical Physics of the National Academy of Sciences of Ukraine, Kyiv 03143, Ukraine
e-mail: perepelytsya@bitp.kiev.ua

T. Vasiliu · A. Laaksonen
Centre of Advanced Research in Bionanoconjugates and Biopolymers "Petru Poni" Institute of Macromolecular Chemistry, 700487 Iasi, Romania
e-mail: vasiliu.tudor@icmpp.ro

A. Laaksonen
Laboratory of Materials-Oriented and Chemical Engineering, Centre of Advanced Research in Bionanoconjugates and Biopolymers, Nanjing Tech University, Nanjing 210009, People's Republic of China



L. Bulavin and N. Lebovka (eds.), *Soft Matter Systems for Biomedical Applications*, Springer Proceedings in Physics 266, https://doi.org/10.1007/978-3-030-80924-9_3

# List of Abbreviations, Symbols, and Chemical substances

## *Abbreviations*

| | |
|---|---|
| BBGKY | Bogolyubov-Born-Green-Kirkwood-Yvon |
| CG | Coarse-grained |
| FF | Force field |
| GAFF | Amber General Force Field |
| ILs | Ionic liquids |
| LqE | Low-*q* excess |
| KFV | Key feature vectors |
| MC | Monte Carlo |
| MD | Molecular Dynamics |
| MM | Molecular Mechanics |
| NMR | Nuclear Magnetic Resonance |
| NPT | Isothermal–isobaric ensemble |
| PGM | Platinum group metal |
| RDF | Radial distribution function |
| SAXS | X-ray scattering |
| SDF | Spatial distribution functions |
| THF | Tetrahydrofuran |
| VP | Voronoi polyhedron |

## *Symbols*

| | |
|---|---|
| $H^{\mathrm{E}}$ | Excess molar enthalpies |
| $R_G$ | Radius of gyration |
| $R_H$ | Hydrodynamic radius |
| $V^{\mathrm{E}}$ | Excess molar volumes |

## *Chemical substances*

| | |
|---|---|
| BMB | bis(mandelato)-borate |
| BOB | bis(oxalato)-borate |
| DBE | di-*n*-butyl ether |
| DME | 1,2-Dimethoxyethane |
| DMSO | Dimethyl sulfoxide |
| EAN | Ethylammonium nitrate |
| NMP | N-methyl-2-pyrrolidone |
| P6,6,6,14 | trihexyl(tetradecyl)phosphonium |

## 3.1 Introduction

Mixtures, solutions and solubility affect our daily life in many different ways, from the preparation of the food we eat to many industrial and chemical applications. It is highly important to understand the solvent properties from the points of view of solubilities, either to develop separation techniques, or for the use in spectroscopy, or to speed up chemical reactions. Beside their practical importance, these systems are of great relevance for very fundamental research in the study of intermolecular interactions, because solutions and mixtures are where they manifest themselves best. While several aspects of such interactions are understood reasonably well, many phenomena occurring in the dissolving and mixing processes are still the subject of intensive studies. Of particular interest are mixed solvents, ionic liquids, deep eutectic solvents and their mixtures. These systems offer by far the richest possibilities for molecules to like or dislike each other and form the most interesting local structures with short- and long-range correlations.

Experimentally, solutions and mixtures are studied using thermodynamic and spectroscopic techniques; or measuring physical properties such as the density, the viscosity, or transition temperature. However, to explain the obtained results, especially when the behaviour deviates considerably from the ideal one, the experiments should be supported by modelling techniques. This combination opens a window into the molecular world, where different interactions compete with each other until a dynamic equilibrium among a wide variety of optimal arrangements is reached. In molecular simulations, it is possible to follow the decisions molecules make to interact and give space to the co-solvent successively invading more and more of the territory. There are many compromises to be made, while sometimes there are smart solutions, and the mixing becomes smooth. Molecular simulations can be used as a virtual microscope to investigate the interactions at the atomic scale. However, understanding the molecular organization from the atomic coordinates is not straightforward, and different types of analysis methods can be employed depending on the type of information required.

As experiments need theory for in-depth understanding, simulations need experiments for validation: atomistic computer simulations are based on many approximations, and the results obtained by the simulations should be always combined with experiments to be verified.

In this book chapter we illustrate some possible combinations of analysis methods. By choosing several interesting cases from our own research, we illustrate how molecular simulations can be analysed to understand the solvent organization at different levels and how the macroscopic properties arise from the microscopic organization.

This chapter is organized as follows. In Sect. 3.2 we briefly introduce the basic concepts of the classical molecular modelling techniques mostly used to study solvents, ionic liquid and mixtures. In Sect. 3.3, we discuss how the simulation results can be analysed to obtain information on the solvent structure. In Sect. 3.4, we discuss the physical observables that can be calculated from the simulations. In Sect. 3.5 we present selected examples from our own recent research on how to combine the parameters discussed in Sects. 3.3 and 3.4 to explain the solvation phenomena. Finally, in Sect. 3.6, we outline the main conclusions.

## 3.2 Classical Molecular Simulations

The most common classical particle-based simulation methods are Molecular Dynamics (MD) and Monte Carlo (MC) [1, 2]. MD is a fully deterministic method based on numerically solving the Newton's equations of motion for a large number of particles. The simulated particles interact with each other, feeling a force directed on them from the neighbouring particles, making them to move this way, mimicking real molecules in condensed phases.

In MC simulations, the particles are moved stochastically by randomly choosing one particle at a time and moving it to a random direction. The move is accepted if the energy goes down ($\Delta E \leq 0$) or fulfils the Boltzmann distribution (exp(−(1/kT)ΔE), otherwise it is rejected.

In both MD and MC, the inter-particle interactions are calculated from the force field (FF), which is a mathematical expression of the total potential energy of the system partitioned among different contributions, typically bonded and non-bonded terms. Each of these terms is defined by a functional form, which may vary across different force fields, and by a set of parameters (see Sect. 3.2.1).

In MD the forces come from negative spatial derivatives of the potential function(s), while in MC the energies are used directly. While both methods can be employed in studying solvation phenomena [3–6], the investigations we will discuss in this book chapter make use of MD, and therefore we will describe this methods in some more detail. Simulations can be carried out at different conditions corresponding to constant volume, temperature, pressure or chemical potential, based on statistical mechanical microcanonical, canonical, isobaric, etc. ensembles. In MD, the microcanonical (E, N, V) ensemble corresponds to pure Newtonian dynamics, while for canonical (N, V, T) ensemble a thermostat is needed to keep temperature constant, and for isobaric (N, P, T) ensemble also a barostat to have a constant pressure. While until some decades ago it was almost common practice to develop or adapt an own simulation code, implementing the chosen method, there is now a wide variety of freely available software packages that can be used. We normally choose the one that we are already familiar with. MD simulations normally are the best choice for particle-based simulations. MC simulations may be more efficient in simulations of polymer solutions. While both static and time-dependent properties can be accessed using MD, only static properties can be calculated in MC simulations.

### *3.2.1 Force Fields*

The force field (FF) is the most important input to molecular simulations, since it is behind the molecular conformations that will be obtained for the studied molecules, their internal flexibility and their interactions [7]. FF consists of conceptually simple molecular mechanics (MM) mathematical functions where different contributions are assumed to be additive, incorporating non-bonded and bonded interactions. Non-bonded interactions consist of short-range Van der Waals (Lennard–Jones) and long-range Coulombic terms and bonded interactions describe bond-stretching, angle-bending and rotation around covalent bonds. Additional terms can be added to keep molecules planar, describe better specific H-bonds or to include information from nuclear magnetic resonance (NMR) and crystallography as restraints. Only the non-bonded terms give interaction energies while the other terms are penalty functions, most often harmonic wells. The non-bonded interactions are given by pair-potentials.

FFs are parameterized based on experiments or quantum chemical calculations. Common FFs, such as CHARMM [8, 9], Amber [10, 11], OPLS [12, 13], and many others, are continuously improved by refining the parameters [14]. Lately, machine learning techniques are becoming common in constructing and parameterizing more accurate FFs [15]. The purpose of FFs is to empirically try to mimic the molecular forces, which can be described accurately only by first-principles of quantum mechanics. While MD first-principles simulation methods do exist, like the Car-Parrinello and other Born–Oppenheimer MD methods, they are computationally too expensive when studying large systems, and/or for long simulation times, and have limited application for liquid mixtures.

Existing empirical MM FFs differs in the functional form and/or in how they were parameterized, and thus the simulation results are FF-dependent. The choice of an appropriate FF is often one of the most critical points in the simulation setup, albeit often overlooked. In general, it is suggested to test at least two or three FFs or different set of parameters, especially when the investigation concern a new system where the molecules are not of a common type, as is often the case with pharmaceutical compounds. In studies of solvents with co-solvents over wider concentration ranges as described in this book chapter, it is highly important to verify the FF and results, and sometimes it is necessary to tune some parameters. In fact, partial atomic charges, in particular, often have to be adjusted depending on the surrounding media. As an example, increasing the polarity of a mixture, the atomic charges of the molecular component might need to be recalculated to reproduce the polarization effect of the surrounding media [16].

## 3.3 Structure of Molecular Liquids and Liquid Mixtures

Before getting into the details of the mathematical formulas behind the calculation of some structural descriptors from the sets of particle coordinates that constitute a simulation trajectory, we want to underline the importance of a proper visualization of the trajectory. Many freely available codes exist such as VMD [17] or gOpenMol [18].

To describe the structure of a liquid, averaged density distributions are used as a standard method [19]. The most complete description of a liquid structure could be obtained from a full $N$-body distribution function $g^{(N)}$. This is a function of all $N$ particle positions $\boldsymbol{r}$ and their orientations $\boldsymbol{\omega}$ in space. However, this function is far too complex for any practical purpose. To describe the properties of the system, the $N$-body distribution function may be reduced to a set of distribution functions that satisfy the chain of linked equations by Bogolyubov-Born-Green-Kirkwood-Yvon (BBGKY chain). More often, pair distribution function $g^{(2)}$ which gives the probability to find any two particles in a liquid at certain positions with given orientations in space is used. As it is still too complicated to compute, it can be further reduced to find the probability to find the particles at a certain mutual distance and orientation. Assuming further an isotropic monatomic liquid, we end up in radial distribution functions with only the mutual distance as variable which can be used for site–site interactions in arbitrary molecular systems and displayed as a two-dimensional graph $g_{AB}(\mathrm{r})$ (see Figs. 3.1 and 3.2).

**Fig. 3.1** Reduction of N-body distribution function $g^{(N)}$ to pair distribution function $g^{(2)}$ and to radial distribution function $g_{AB}$ between to sites A and B (see the text)

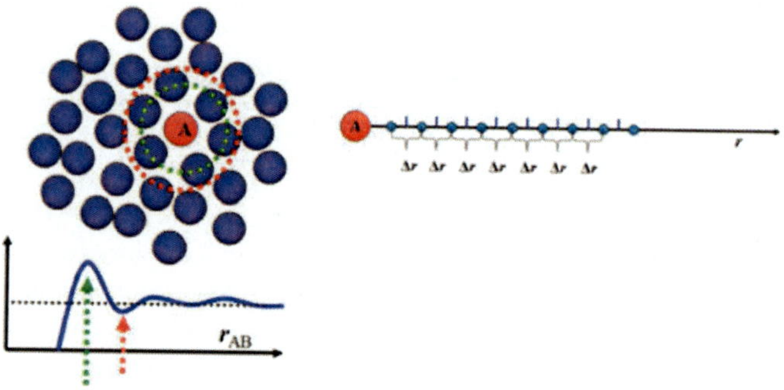


**Fig. 3.2** Left: (top) 2D projection of a model liquid and (bottom) the probability density to find atoms $B$ at a given distance from atom $A$. Right: Discretization of the A-B distance to be used to calculate the histogram to be used to compute the RDF according to Eq. (3.3)

### 3.3.1 Radial Distribution Functions

The radial distribution function $g_{AB}(r)$ gives a probability to find atom $B$ at a distance $r$ from atom $A$ and can be expressed as radial density–density correlation function:

$$g_{AB}(r) = \frac{\rho_A(r|r_A = 0)}{\rho_B}, \quad (3.1)$$

where the microscopic density $\rho_A(\mathbf{r})$ is defined as:

$$\rho_A(\boldsymbol{r}) = \langle \sum_A \delta(\boldsymbol{r} - \boldsymbol{r}_A) \rangle, \quad (3.2)$$

$\rho_B$ is the bulk density of $B$. Radial distribution functions are calculated in nearly all MD and MC simulation studies of liquids and solutions to obtain information of the average liquid structure. Pictorially, it means that we use as reference each of the atoms at a time. From this atom, which we call "A", we calculate the distances to all neighbouring atoms within a cut-off radius in the simulation configuration (Fig. 3.3). This is done for each saved frame from the simulation and averaged over all saved trajectory data. In practice, this means calculation of a histogram by discretizing the $r_{\mathrm{AB}}$ distance to a number of bins with a width of $\Delta r$.

Defining $N_{\mathrm{B}}(\boldsymbol{r}, \Delta\boldsymbol{r})$ as the average number of $B$ particles found within the interval $\boldsymbol{r} - ½\Delta\boldsymbol{r}$ and $\boldsymbol{r} + ½\Delta\boldsymbol{r}$ from the central particle $A$, and $V_{\mathrm{B}}(\boldsymbol{r}, \Delta\boldsymbol{r})$ as the volume of the spherical slice between interval $\boldsymbol{r} - ½\Delta\boldsymbol{r}$ and $\boldsymbol{r} + ½\Delta\boldsymbol{r}$ from the central particle $A$, the radial distribution function can be calculated as:

$$g_{AB}(r) = \left(\frac{V}{N_B}\right) \frac{N_B(r, \Delta r)}{V_B(r, \Delta r)} = \left(\frac{V}{N_B}\right) \frac{N_B(r, \Delta r)}{4\pi r^2 \Delta r}. \quad (3.3)$$

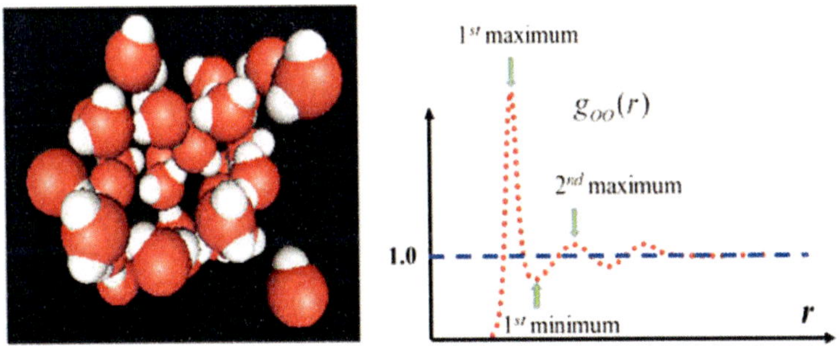


**Fig. 3.3** Left: configuration of liquid water from MD simulations. Right: $g_{OO}(r)$ radial distribution function where the first maximum is shown together with the first minimum and second maximum. More details are in the text

As an example, we calculate $g_{OO}(r)$ from simulations of liquid water, shown on the right in Fig. 3.3.

The first maximum corresponds to the donor–acceptor hydrogen bond distance for water. The blue dashed line corresponds to the bulk density of water which is a result of the normalization done in Eq. (3.3) where the distance dependent number density is normalized with the bulk number density $N_B/V$ of particles $B$. In Fig. 3.3 on right both $A$ and $B$ are oxygens of water. Note that value of 1.0 of the bulk number density of water, is only coincidentally equal to the bulk density of water expressed in g/cm$^3$. RDF in Fig. 3.3 oscillates showing three maxima which represent the three hydration shells around each oxygen atom. Water is very structured due to strong H-bonds. In fact, many structural features are averaged out in calculating the RDFs. We will show a tool later which can catch details which appear when the orientation is taken into account. This tool is the spatial distribution function (SDF).

A very useful quantity, the coordination number, can be obtained by integrating the value of the RDF until the first minimum. This value, depending on the case and system, is also called solvation number or hydration number $n$. It can be obtained from

$$n = 4\pi \frac{N}{V} \int_{0}^{1^{st}\text{minimum}} r^2 g(r) dr. \tag{3.4}$$

Here, $n$ tells us for example how many molecules are in the first coordination shell. This number is normally not an integer, as it represents the average of different coordination configurations, also comprising those capturing the exchange of molecules between the considered coordination shell and the next outer shell or the bulk. Calculation of a running integral of the RDF allows to count the number of neighbours as a function of the distance from the reference atom.

### 3.3.2 *Spatial Distribution Functions*

Spatial distribution functions (SDF) were invented by Peter Kusalik and Igor Svishchev in early 1990's [20, 21]. SDFs can be calculated as:

$$g_{AB}(\vec{r}) = \frac{\rho_B \langle \vec{r} | \vec{r}_A = 0 \rangle}{\rho_B}. \tag{3.5}$$

The Eq. (3.5) is seemingly similar to that for the RDFs in Eq. (3.3). However, there is one important difference: we do not have plain distances in Eq. (3.5) but rather vectors. As vectors have both magnitude and direction these distribution functions depend on the orientation of molecules in the liquid or solution. To describe a vector, we need three dimensions. 3D space can conveniently be spanned by either Cartesian or spherical polar coordinates. We use here Cartesian coordinates to define a cubic 3D grid as a reference system for the SDFs. The SDF $g_{AB}$(i, j, k) can then be given as:

$$g_{AB}(i, j, k) = \frac{\left\langle \frac{1}{N_A} \sum_{n_A=1}^{N_A} \sum_{n_B=1}^{N_B} I_{ijk}\left[R_{n_A}^{A}\left(r_{n_B}^{B} - r_{n_A}^{A}\right)\right] \right\rangle}{\rho_B l^3}, \tag{3.6}$$

where $i, j$ and $k$ are integers representing the small cubic grids in the mesh to build a 3D space for density probabilities. $N_A$ and $N_B$ are numbers of atoms of type $A$ and $B$, respectively. $\rho_B$ is the bulk density of $B$ and $l$ is cubic box length. $I_{ijk}$ is a 3D grid and has the same function as the estimator for the histogram in calculations of RDFs but the histogram here is in three dimensions and cannot be visualized. $R$ is a rotation matrix to transform the vectors from laboratory frame to local molecular frame of atom $n_A$ of type $A$.

The RDFs do not need any spatial reference frame as they are functions of the distance which is a scalar. SDFs need to be fixed in a local frame which can be the principal coordinate system of the molecule whose neighbourhood we are studying. However, it can be defined differently from case to case and all we need is three atoms to create a local Cartesian coordinate system. The latter is required if the molecule is large or flexible as we cannot accurately calculate the density distribution around the whole molecule, as the whole structure may fluctuate too much to give well-localized densities in the laboratory coordinate system, even if the translational and rotational motion of the reference molecule are removed.

SDFs are straight-forward to calculate but there is one principal problem to visualize them. They are four-dimensional quantities because they are calculated as functions of x, y, z coordinates and intensity. Therefore, to visualize the SDF, we need to freeze one of the dimensions. To demonstrate this issue, we use liquid water from our own simulations.

In Fig. 3.4 on the left, there is a representation of the water oxygen–oxygen SDF where one of the three Cartesian dimensions is omitted so that only two dimensions

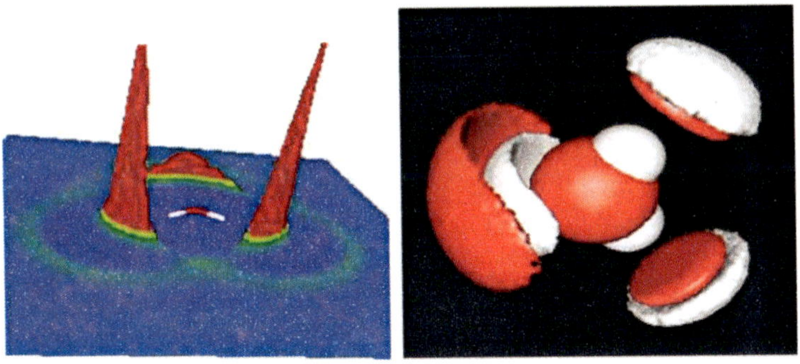

**Fig. 3.4** Two different strategies to display the 4D spatial functions. Left: One Cartesian dimension is not displayed to allow the full intensity included. Right: the intensity is frozen to a threshold value to give an iso-density (intensity) surface together with three Cartesian dimensions

**Fig. 3.5** Water oxygen–oxygen SDFs with the thresholds chosen: from left to right 4.0, 3.0, 2.5, 2.0, 1.8 (adapted with permission from [22])

of the reference coordinate system are displayed. We can see that the water molecule is on the x–y plane. The third dimension is the intensity. The two highest intensities arise from the oxygen atoms of the water molecules in the first hydration shell, accepting the hydrogens from the water on which the local coordinate system is built.

On the right in Fig. 3.4 the same SDF is represented, but instead of showing how the intensity varies on a given plane, as in the picture on the left, the intensity is fixed to an arbitrary threshold value. Note that like RDFs these functions are averages collected during the whole simulation. It is like to take a photo on nanoscale with the shutter open during the duration of the entire simulation. The moving atoms coordinating and hydrogen bonding to the water fixed in the local coordinate frame centre leave traces successively building the SDF. It is obvious that the image on the right gives a better overview of the four-coordinated water H-bond structure of the first hydration sphere, which could hardly be inferred from the representation on the left. With the representation on the right, also called isosurface representation, we can see that the two waters hydrogen bonded to the oxygen lone-pairs are much more mobile than those bound to the H atoms and their representations overlap. But if we increase the threshold for the isodensity they become separated. Nearly all the SDFs in existing literature use the iso-density representation. In Fig. 3.5 we demonstrate how to choose a suitable iso-density value. The threshold values are chosen with respect to bulk density, that as discussed for the RDF is by definition $= 1.0$, see Eq. 3.5.

It is interesting to see how the structures of the hydration shells become visible by varying the isodensity value. At 2.5, 2.0 and 1.8 we can observe also the second hydration shell. When we go down to threshold 2.5, we can see already features of the third hydration shell. It is not possible to go much lower than 1.5 because approaching the bulk density the details gradually disappear.

SDFs are very useful in studies of solutions and liquid mixtures. We give below two examples of using SDFs to study an equimolar mixture of water and acetonitrile. Both components are polar so they should mix easily as we know from our Chemistry classes that “like dissolves like”. This is sometimes not completely true, and the mixtures of polar molecules do not need to be homogeneous on molecular level but may exhibit so called micro-heterogeneity. This is the case with the water-acetonitrile mixture [23].

The SDFs in Fig. 3.6 show regions of high density of water oxygen (red), nitrogen of acetonitrile (blue), and the methyl of acetonitrile (green) around water molecules. It is worth noting the competition between oxygen and nitrogen to accept the hydrogens from the reference water. Also, interestingly there is a high probability to find the methyl on both sides of the water plane. The same water structure as in Fig. 3.4 (right) appears clearly in Fig. 3.6, indicating that water largely retains its characteristic structure in the mixture. A closer analysis of simulation configurations shows small local water clusters together with linear, branched and cyclic H-bonded polymers of water. SDFs can be modified and tailored for specific systems to provide more information, as shown in the following two examples.

DMSO-water mixture, a well-known cryo-solvent in the molar ratio 1:3, freezes at −62 °C, a much lower temperature than that of the respective neat components. It was long-hypothesized that, in this mixture, two water molecules strongly coordinate the sulfoxide oxygen, with an additional water molecule forming an H-bonded bridge between the other two. This stable complex structure was assumed to prevent freezing of the mixture at normal (expected) freezing temperatures.

In simulations we did find [24] two water molecules strongly coordinating to the DMSO sulfoxide oxygen, which is seen in the violet SDF shell near the oxygen atom in Fig. 3.7, left, obtained with a reference frame fixed on the DMSO atoms. No evidence of the hypothesized coordination mode described above can be seen. In Fig. 3.7, right, the reference frame is built on the oxygen atoms of DMSO and the

**Fig. 3.6** Spatial distribution function showing water and acetonitrile around water in an equimolar mixture (reprinted with permission from [37])

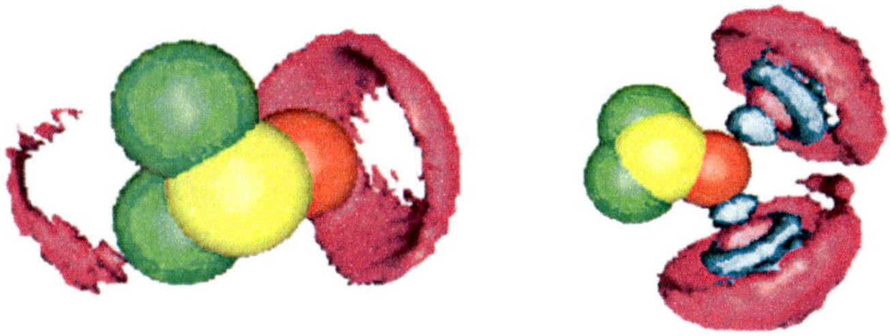

**Fig. 3.7** SDFs from an equimolar mixture of DMSO-water at room temperature. Left: water density around DMSO as a reference. Right: Water density around DMSO and to its sulfoxide oxygen two tightly bound water molecules. The reference frame is built on the oxygens from DMSO and the two water molecules coordinating to it by a strong H-bond (adapted with permission from [24])

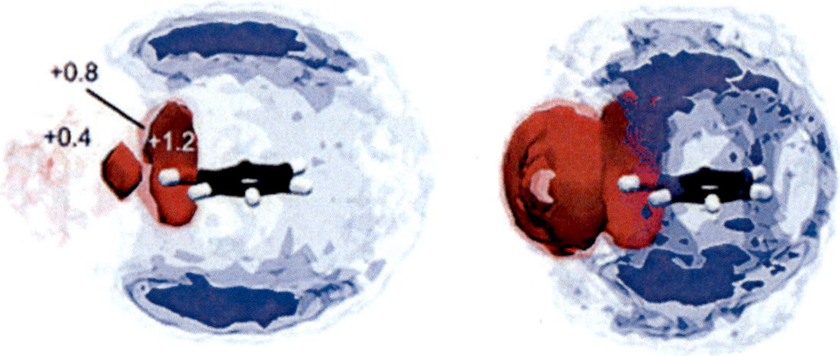


**Fig. 3.8** ΔSDFs (Eq. 3.7) showing excess water oxygen densities in red, excess co-solvent densities in blue (ethanol oxygen or acetonitrile nitrogen). Left: water–ethanol mixture. Right: water-acetonitrile mixtures. SDFs were calculated using a reference frame fixed on the phenol molecules (adapted with permission from [25])

two water molecules coordinating to it by strong H-bonds; also, this SDF shows only a weak probability for a third water molecule bridging those in the first solvation shell. However, this latter representation allows explaining the very low freezing temperature as most likely due to fast "ball-bearing"-like dynamics. Indeed, we can see how the two water molecules coordinating to sulfoxide oxygen rotate while H-bonded to DMSO, and how surrounding water H-bonds to them.

Another quantity that can be obtained from two SDFs is the ΔSDFs defined as:

$$\Delta\text{SDF: } g_{\text{water-cosolvent}}(x, y, z) = g_{\text{water}}(x, y, z) - g_{\text{cosolvent}}(x, y, z). \quad (3.7)$$

The ΔSDF can be very useful to study the differences in the solvation of a given molecule, as shown in Fig. 3.8, where this function is used to highlight differences in the solvation of phenol in two solvent mixtures: water–ethanol and water-acetonitrile [25].

Our groups have been widely using the SDF to understand solvation phenomena, and besides the examples given here and in Sect. 3.5, we refer the interested reader to ref. [26]. All the SDFs represented in this book chapter were produced using gOpenMol [18, 22]

### 3.3.3 *Cluster Analysis and H-Bonded Structures*

The RDFs and SDFs presented in the previous sections are routinely used to describe the local solution structure, i.e., the average molecular composition and arrangements around molecules, in computer simulations of liquid systems. However, the structures of H-bonded liquids, such as water or alcohols, and their mixtures with other liquids are often characterised by the presence of H-bonded aggregates ("clusters") or even extended networks which continually break up and recombine, shedding and adding

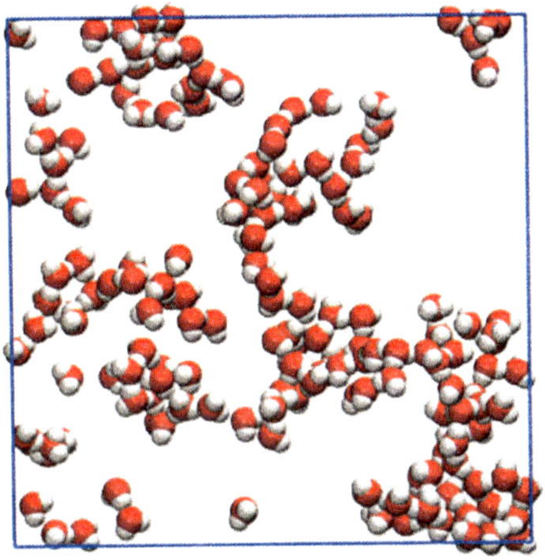

**Fig. 3.9** Distribution of water molecules in a simulated equimolar mixture of water and 1,2-dimethoxyethane (DME) [27]. The representation shows only water molecules within a 15 Å-thick slice of the simulation configurations, with DME molecules removed, in the interest of clarity. The simulation box boundaries are shown in blue

H-bonded members (Fig. 3.9). Depending on the system studied, molecular clusters may consist of tens to hundreds of members, such that RDFs and SDFs do not provide detailed information on their average sizes, size distributions, shapes (topologies) or dynamic properties.

The binary mixtures of water and acetonitrile, introduced in Sect. 3.3.2, for example, have been shown to have micro-heterogeneous liquid structures, in which water molecules tend to self-associate via strong H-bonding interactions [23, 28]. Similar molecular clustering is found in other binary mixtures of H-bonding or polar liquids, and detailed knowledge of the nature of this phenomenon may help to explain their anomalous physical properties, e.g. higher/lower than expected densities, [29] and have important implications for their solvent characteristics [27].

In computer simulations of aqueous systems, the formation of H-bonded water clusters may be studied by choosing an appropriate H-bond definition [30, 31], and iteratively applying this definition to a given water molecule, its newly-found H-bonded neighbours, then their neighbours, and so on, combining this information in order to identify all unique H-bonded water clusters in the particular simulation configuration (Fig. 3.10).

A very simple H-bond definition, for example, may consider two water molecules H-bonded if their oxygen atoms are separated by a distance $r_{OO} \leq r_{min}$, where $r_{min}$ is the position of the first minimum of the water-water $g_{OO}(r)$; additional geometric or energetic H-bond criteria are typically implemented [32]. Once established, a cluster identification algorithm may be applied to other simulation frames in order to determine the average size of H-bonded water clusters, their size distribution, etc.

Ionic liquids (ILs), a diverse family of ionic substances that are liquid at low temperatures, typically below 100 °C, have also been found to have micro-heterogeneous liquid structures, often consisting of percolating ionic (polar) and non-polar regions, or "domains" [33]. Computer simulations have been instrumental in the characterisation of IL domain structures, where a Voronoi tessellation-based

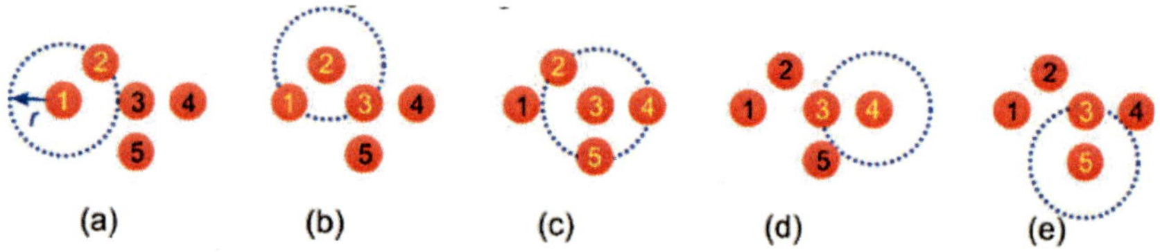


**Fig. 3.10** Schematic representation of H-bonded water cluster identification using a simple O…O distance criterion, i.e., two water molecules are considered H-bonded neighbours if their oxygen atoms are separated by a distance $r \leq r_{\text{min}}$. Starting from an arbitrarily chosen water molecule 1 in panel (a), only water 2 may be seen to satisfy the neighbour criterion (inside the dotted blue circle with radius *r*, numbers in yellow), thus water molecules 1 and 2 are considered H-bonded and part of the same H-bonded water cluster. Considering water molecule 2 in panel (b), and applying the same H-bond criterion, it may be seen that one *new* H-bonded neighbour, 3, is found and added to the cluster list, which in turn finds *two* new H-bonded neighbours, 4 and 5, in panel (c). As shown in panels (d) and (e), no new H-bonded neighbours are found for either water molecules 4 or 5, thus concluding the H-bonded cluster member list (1–5). Practical implementations of the simplistic workflow described here are more complex in order to effectively identify branched and cyclic H-bonded structures

cluster identification method has proven highly successful [34], as it allows for the construction of neighbour list without prior knowledge of the interaction type or characteristic parameters (distances, angles etc.), i.e. it can be applied to study atomic contacts in a more general sense.

This method is illustrated in Fig. 3.11. Essentially, a unique Voronoi polyhedron (VP) is assigned to each atom of the simulated system according to established rules [35], thus partitioning the entire simulation configuration into such polyhedra; any two atoms sharing a polyhedron face are now considered Voronoi neighbours. All atoms, in turn, are assigned to IL domains, e.g., atoms forming part of highly charged groups are assigned to the "polar/ionic" domain, whereas those forming aliphatic chains are assigned to the "non-polar" domain. The unique Voronoi neighbour lists of all atoms may now be used in order to determine all "polar" and "non-polar" domains, such that the number of domains, their sizes and shapes may be determined. As for the H-bonded cluster analysis described above, the VP procedure may be repeated for the remaining simulation trajectory frames in order to obtain time-averaged properties or their time evolution. Finally, the Voronoi tessellation-approach has also been used to study the structures of molecular liquids, e.g. water under different conditions [36–38].

A method for analysing clustering of molecular pairs and applied on ionic liquids is suggested by Pei and Laaksonen [39] based on key feature vectors (KFV) and shown to be very useful for liquids and solutions with strong intermolecular interactions, effectively locating local molecular pair structures.

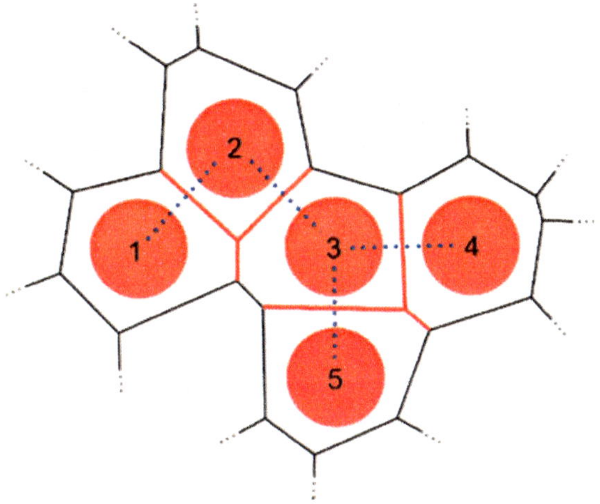


**Fig. 3.11** Schematic drawing illustrating the essential Voronoi tessellation-based cluster identification concept in two dimensions. The water molecule oxygen atom arrangement in Fig. 3.10 is shown, surrounded by atoms of another solvent (grey circles) as part of a larger simulation configuration, with the water-water H-bond connectivity established before (O...O distance criterion) shown using blue dotted lines. Following Voronoi tessellation of the entire simulation configuration according to established rules, [35] each atom is assigned (located inside) a unique Voronoi polyhedron (VP); all pairs of atoms sharing VP faces (shown as edges in this two-dimensional drawing) are considered neighbours. Considering water molecule 1, note that it shares two of its VP faces with other water molecules, 2 and 3 (red edges), identifying water molecules 1–3 as part of the same water cluster, or "domain". Note that the VP connectivity differs from that established using an O...O distance criterion (blue dotted lines), but that the cluster assignment is the same. The VP face shared by 1 and 2 is larger than that shared by the more distant 1 and 3, illustrating a common and useful feature of the method: close neighbours share larger VP faces, allowing for filtering of neighbours based on the shared VP face size (or area). The water VP neighbours of 2 and 3, may now be similarly identified and the procedure iterated in order to identify the entire water cluster

## 3.4 Physical Observables

Studies of liquid mixtures often require a large number of simulations at different concentrations and corresponding analysis of trajectories to compute characteristic observables, such as excess molar volumes or enthalpies, viscosities or x-ray diffraction patterns. Here below, we briefly describe some of the most commonly calculated properties in studies of solutions and mixture. Many other experimental techniques are used in combination with simulations, such as spectroscopic measurements; for NMR chemical shift or relaxation parameters we refer the interested reader to refs [40, 41], where we describe the combination of NMR methods with simulations. The analysis also involves visual inspection of the images from simulations, i.e., snapshots and animations.

### 3.4.1 Density

By far the easiest property to calculate from computer simulations is the average density of the liquid or solution. Densities can be obtained from isothermal–isobaric (NPT) simulations at the corresponding experimental conditions, normally 25 °C and 1 atm. In NPT simulations the volume of the simulation cell is not constant but is periodically adjusted based on the desired pressure and temperature, thus fluctuating around these values. Knowing the number of molecules of different types and their masses gives us the density when divided by the equilibrated average system volume.

Comparison of calculated and measured density is a routine test to check the validity of the model used in the simulations. Discrepancies are common, and very often when studying mixtures, it is of importance to ensure that the models catch the trends in the variation of the density with the composition, rather than the absolute values of the density itself. Furthermore, it should always be taken into account that the simulations do not contain any contaminants and impurities that are occasionally contained in the real samples, and that can sometimes have unexpectedly large effects on the experimentally measured density and other properties.

### 3.4.2 Excess Molar Volume of Mixing

Excess molar volume of mixing is calculated as:

$$V_m^E = \frac{x_1 M_1 + x_2 M_2}{\rho_{mix}} - \frac{x_1 M_1}{\rho_1} - \frac{x_2 M_2}{\rho_2}, \tag{3.8}$$

where $x_i$ is the mole fraction and $M_i$ is the molar mass of component $i$ and $\rho_{mix}$ and $\rho_i$ are the mixture density and the density of component $i$, respectively.

### 3.4.3 Excess Molar Enthalpy of Mixing

Excess enthalpy of mixing is calculated as:

$$\Delta H^{excess} = H_{mix} - x_1 H_1 - x_2 H_2, \tag{3.9}$$

where $H_{mix}$ is the enthalpy of the mixture $x_i$, and $H_i$ are the molar fraction and enthalpy of component $i$. $H_{mix}$ can be calculated from intermolecular energies

$$\Delta U^{excess} = U_{mix} - x_1 U_1 - x_2 U_2, \tag{3.10}$$

using $\Delta H = \Delta U - p\Delta V$. However, since $\Delta V$ is negligible for liquids, $\Delta H \approx \Delta U$.

### 3.4.4 Viscosity

Viscosity $\eta$ is a quantity expressing the resistance of a fluid to flow and perceived as its thickness. It results from intermolecular interactions and can be seen as a friction between molecules, and molecules and a wall. The viscosity can be calculated in equilibrium simulations by Green–Kubo integrals from shear stress correlation functions

$$\eta = \frac{V}{k_B T} \int_0^\infty dt \langle \sigma_{zx}(t) \sigma_{zx}(0) \rangle_{eq}, \tag{3.11}$$

where $\sigma_{zx}$ are zx-components of the shear stress tensor.

### 3.4.5 Translational Diffusion

Diffusion is a spontaneous motion of molecular particles in a liquid or solution from high concentrations to less-crowded environment driven by concentration gradients. The diffusion coefficient, $D$, is an important dynamic quantity, characterized by intermolecular interactions. It can be measured experimentally, for example by NMR. It can be calculated either from velocity autocorrelation functions (Eq. (3.12)) or from mean square displacement by the Einstein relationship (Eq. (3.13))-

$$\begin{aligned} D = \frac{1}{3} \int_0^\infty \langle \vec{v}(t) \cdot \vec{v}(0) \rangle \mathrm{dt} &= \frac{k_B T}{m} \int_0^\infty \frac{\langle \vec{v}(t) \cdot \vec{v}(0) \rangle}{\langle \vec{v}(0) \cdot \vec{v}(0) \rangle} \mathrm{dt} \\ &= \frac{k_B T}{m} \int_0^\infty \widehat{\mathrm{C}}_v(t) \mathrm{dt} = \frac{k_B T}{m} \tau_v \end{aligned}, \tag{3.12}$$

$$6Dt = \left\langle |\vec{r}(t) - \vec{r}(0)|^2 \right\rangle. \tag{3.13}$$

Both methods, using the normalized velocity autocorrelation functions and the means square displacement, should give identical results. This comparison is a good way to verify that the simulation is long enough.

Viscosity and diffusion are inversely proportional to each other, which can be seen in the Stokes–Einstein relationship:

$$D = \frac{k_B T}{6\pi \eta R_H}, \tag{3.14}$$

where $R_H$ is the "hydrodynamic radius", the effective radius in solution. In simulations we can easily calculate the radius of gyration of a molecule ($R_G$), as the root-mean-square distance of all particles from the centre of mass. The relationship between $R_G$ and $R_H$ is not unique and depend on the shape and size of the molecule and should be evaluated properly when the value of the viscosity is to be estimated from the diffusion, or *vice versa.*

### 3.4.6 Structure Factors

RDFs from computer simulations, calculated in real space, can be compared with experimental quantities called structure factors, obtained in reciprocal space, if these are available. The reason why there are experimental RDFs is that the total structure factors obtained from scattering studies are decomposed into partial structure factors, which can be Fourier transformed from k-space to real space to provide RDFs. Also, the RDFs can be transformed to partial structure factors and added together, with weighting factors, to obtain total structure factors to compare directly with experimental scattering functions. As the standard method to compute the Coulomb interactions is Ewald summation, and in Ewald the short-range interactions are calculated in real space and long-range in reciprocal space, we calculate the reciprocal functions needed to compute structure factors. Fourier transform of the liquid density given in Eq. 3.15 is

$$\rho(\boldsymbol{k}) = \sum_{i}^{N} e^{\mathrm{i}\,\mathbf{k}\boldsymbol{r}_i}. \tag{3.15}$$

In Ewald summation, these functions are calculated at each time step, and can be used for calculation of structure factors. Partial structure factor $S_{AB}(k)$ is defined as:

$$S_{AB}(k) = \frac{1}{N}\langle \rho(k)\rho(-k)\rangle = \frac{1}{N}\left\langle \sum_{i\in A}^{N} e^{-\mathrm{i}\mathbf{k}\boldsymbol{r}_i} \sum_{j\in B}^{N} e^{\mathrm{i}\mathbf{k}\boldsymbol{r}_j} \right\rangle. \tag{3.16}$$

Applying periodic boundary conditions, k-vectors can be given in a grid:

$$\boldsymbol{k} = \left(\frac{2\pi}{\mathrm{BOXL}}\right)(k_x, k_y, k_z), \tag{3.17}$$

where BOXL is the cubic box length and $k_x$, $k_y$ and $k_z$ are integers.

For isotropic liquids the RDFs and structure factors are related as:

$$S_{AB}(k) = \frac{N_A}{N}\left\{\delta_{AB} + \frac{N_A}{V}\frac{4\pi}{k}\int_0^{\infty}[g_{AB}(r) - 1]\sin(kr)r\,dr\right\}. \tag{3.18}$$

Scattering studies are very useful for liquid systems with long-range correlations, for example ionic liquids and deep eutectics, as we will see later in this chapter.

## 3.5 Case Studies

### *3.5.1 Unusual Density Variation*

In most applications of liquid mixtures, these consist of macroscopically fully miscible liquids, and the variation of the physical–chemical properties with mixture composition can be roughly predicted knowing the properties of the individual components. In general, one might expect that when mixing two solvents with different density, the density of the resulting mixture should be in-between those of the two pure components.

However, in the case of the mixtures of the widely used solvent N-methyl-2-pyrrolidone (NMP), this is not the case: when water (less dense) is added to NMP (more dense), the resulting mixture is, over a large composition range, more dense than pure NMP itself and, as a consequence, the density curve as a function of the molar fraction NMP presents a maximum at the NMP: water molar ratio 2:1. To understand the structural organization of water and NMP in their mixtures, and the origin of the maximum in the density curve, and in those of other properties [42] at varying water/NMP compositions, we performed several MD simulations [16, 42].

In Fig. 3.12 (left) are plotted the experimental density values as a function of the mole fraction of NMP, $x_1$, together with the average density values obtained from three sets of MD simulations at 298 K, displaying a more-or-less pronounced maximum at an NMP:WAT ratio of about 2:1 ($x_1 \sim 0.3$).

To reproduce the experimental density trend of this binary system over the whole composition range, it is important to consider that the variation in the solvation around NMP affects the charge polarization. Indeed, the force field parameters that well reproduce the behaviour of neat NMP [43] (Set 1) or those that reproduce well

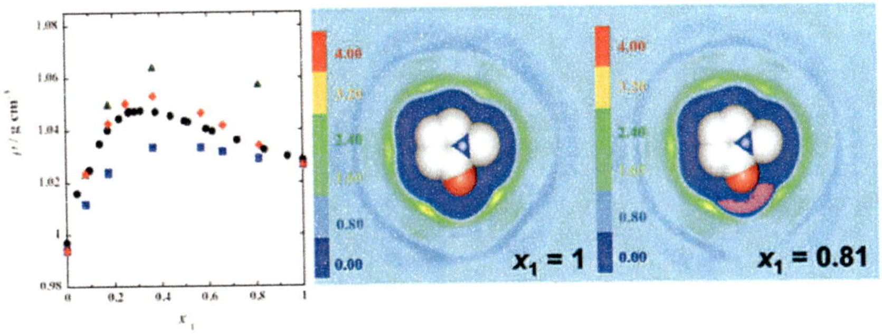


**Fig. 3.12** Left: Experimental density values (black circles) as a function of the mole fraction of NMP, $x_1$, and the average density values obtained from three sets of MD simulations at 298 K: Set 1 (blue squares), Set 2 (green triangle), Set 3 (red diamond). Center: Cross-section representation of the SDFs of heavy atoms of NMP in neat NMP ($x_1 = 1$). Right: same as in Centre for the mixture with water ($x_1 = 0.81$). Isovalues: 3 and 13 respectively for NMP and water atoms (adapted with the permission from [16])

the behaviour at low NMP concentration in water solutions (Set 2) (blue squares or green triangle, respectively, in Fig. 3.12) reproduce only roughly the experimental trend. Notably, Set 3 with the charges obtained by a linear fitting of charges in Set 1 and Set 2 for each value of $x_1$, reproduces very well the experimental trend and has absolute values quite close to the experimental ones over the whole concentration range.

As described in ref [16], inspection of the RDFs and coordination numbers alone does not allow explaining the presence of the density maximum, and SDF are extremely useful in this case. Visual inspection of the SDFs, represented with the software gOpenMol [18, 22], explains the density trend: Water molecules interact with the NMP carbonyl oxygen, accessing a "*cavity*" not accessible to the atoms of solvating NMP molecules, without altering significantly the NMP structural organization observed in the neat solvent. The latter consideration is supported also by the experimental X-ray structure function, which shows that that addition of water to NMP up to $x_1 = 0.37$ does not affect the relative balance of structural correlations and the resulting X-ray structure functions. SDFs show that two water molecules can occupy the cavity of the NMP network, which is empty in neat NMP, thus explaining why addition of water to NMP leads to an increase in density up to the molecular ratio of 2:1. At smaller NMP mole fractions, the new intermolecular interactions originated by the water molecules that cannot be accommodated inside the structure of NMP dominate, and the density decreases.

### 3.5.2 *Excess Molar Enthalpies and Volumes of Alcohol Ether Mixtures and Their Dependence on the Clustering*

Alcohol + ether binary liquid mixtures have attracted considerable fundamental interest as model systems containing one strongly H-bonding self-associated component, the alcohol, and one that is unable to do so, the ether, yet can accept H-bonds from the former. These mixtures are also of significant practical interest as biofuels or additives for internal combustion engines [44, 45]. The excess molar enthalpies ($H^E$) and volumes ($V^E$) of binary mixtures of butanol isomers + di-*n*-butyl ether (DBE) under ambient conditions have been extensively measured, showing a clear dependence on the degree of butanol isomer alkyl group branching [46–48]. Both the $H^E$ and $V^E$ of the mixtures in this series increase in the order 1-butanol < *iso*-butanol < 2-butanol < *tert*-butanol, with their $V^E$ exhibiting a particularly intriguing sign change from negative (1- and *iso*-butanol) to positive (2- and *tert*-butanol); these observations have been largely explained in terms of the expected same-order increasing steric hinderance on butanol-DBE cross-species H-bonding. Interestingly, explanations of the strikingly similar $H^E$ and $V^E$ variations of the analogous series of butanol isomer + tetrahydrofuran (THF) or 2-methyl THF binary mixtures prominently feature differences in butanol isomer self-association via H-bonding [49, 50], which has not been studied experimentally nor by computer simulation methods for any of these mixtures.

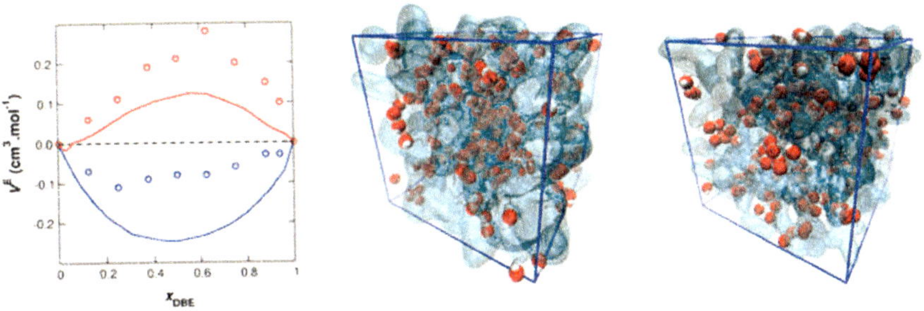


**Fig. 3.13** Left: Experimental and simulated excess molar volumes ($V^E$) of 1- (blue) and 2-butanol (red) + DBE binary liquid mixtures at 298 K, 0.1 MPa. In each case, solid lines connect the experimental data, with simulated data shown by circle markers; the horizontal dashed black line indicates the position of $V^E = 0$ cm$^3$.mol$^{-1}$. Centre: Representative configuration from the simulation of an equimolar 1-butanol + DBE mixture, showing clustering of 1-butanol molecules. 1-butanol -OH groups are shown by a space-filling representation, and alkyl groups by black lines. DBE molecular details have been removed in the interest of clarity and are shown by a transparent cyan surface. Right: equimolar 2-butanol + DBE simulation, same representation as in Centre. Comparing the figure in the centre with that on the right, it can be seen that the latter has smaller H-bonded clusters and 4-membered cyclic configurations

In this context, classical MD simulations were performed for two representative mixture series, 1- and 2-butanol + DBE, using the Amber General Force Field (GAFF) [51]. The simulations, which reproduce the essential experimental $H^E$ and $V^E$ differences of these mixtures, (Fig. 3.13) reveal marked differences in butanol isomer self-association, with 1-butanol more extensively H-bonded, but also suggest that differences in the detailed H-bonded cluster *topologies* may contribute significantly to the anomalous volumetric property differences. In particular, 2-butanol shows a strong tendency to form 4-membered cyclic H-bonded clusters ("rings") in its mixtures with DBE. The formation of 4-membered 2-butanol H-bonded rings was found to correlate with the simulated system volume, suggesting that these play an important role in the higher $V^E$ of these mixtures compared to those of 1-butanol.

### 3.5.3 Is the $[PtCl_6]^{2-}$ Ion Preferentially Solvated by the Organic Solvent in a Water + Organic Solvent Solution?

Binary aqueous solvent mixtures are widely used as solvents in chromatographic separation of platinum group metal (PGM) complexes [52]. The solvent characteristics of such mixtures depend on their composition, which also affects their physical properties and microscopic liquid structures. To shed light on the solvation of an industrially important platinum (IV) complex, the octahedral $[PtCl_6]^{2-}$ anion, in binary mixtures of water with methanol or 1,2-dimethoxyethane (DME), we used a combination of $^{195}$Pt NMR spectroscopy and MD computer simulations [27].

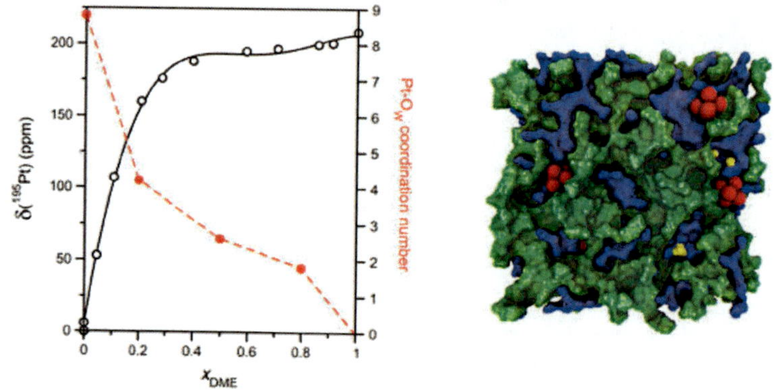


**Fig. 3.14** Left: Experimental $^{195}$Pt NMR chemical shift for the $[PtCl_6]^{2-}$ complex anion in water ($D_2O$) + 1,2-dimethoxyethane (DME) binary mixtures (128 MHz, 303 K) plotted as a function of mole fraction DME (open black circle markers, solid black line); average $[PtCl_6]^{2-}$ hydration numbers from MD computer simulation (solid red circle markers, dashed red line). Right: Space-filling representation of a simulation configuration showing the solvation of $[PtCl_6]^{2-}$ complexes (red) at the boundaries of water-rich regions in a water (blue) + DME (green) binary mixture. Hydronium counter-ions are shown in yellow (adapted with permission from [27])

The $^{195}$Pt NMR chemical shift, $\delta(^{195}\text{Pt})$, of $[PtCl_6]^{2-}$ is thought to be dependent on the local solution environment of the platinum complex [53, 54], and in our experimental measurements $\delta(^{195}\text{Pt})$ showed a non-linear dependence on the mixture composition for both methanol and DME mixtures, Fig. 3.14 (left), suggesting preferential solvation of the complex by these organic solvents, especially DME [55, 56]. MD computer simulations showed that the methanol and DME + water binary mixtures studied exhibit structural microheterogeneity, in which water molecules tend to self-associate, or cluster, by strong H-bonding interactions. The simulations revealed that in DME + water mixtures, the platinum complex tends to locate at the surface of water-rich regions, i.e. an interfacial solvation preference, such that it is partly solvated by water and partly DME, as shown in Fig. 3.14 (right).

These interesting MD simulation results show that the inherent mixed solvent microheterogeneity plays an important role in the apparent preferential solvation of such metal complexes.

### 3.5.4 *The Low-Q Excess*

Ethylammonium nitrate (EAN) is a prototypical protic ionic liquid (IL), having been discovered more than a century ago. More recently, neutron scattering experiments have revealed the liquids structure of EAN to be characterized by nanometer-scale heterogeneity (microheterogeneity), consisting of polar and non-polar regions, resulting in a sponge-like percolating structure [57]. This microheterogeneous structure is also responsible for a particular feature, the low-$q$ peak, in the X-ray scattering (SAXS) pattern of EAN [58].

Interestingly, addition of certain molecular solvents, e.g. $n$-alcohols [59, 60], to EAN results in a significant increase in X-ray scattering intensity at much lower $q$, in the so-called extreme low-$q$ region, indicative of the formation of large-scale molecular aggregates (Fig. 3.15, left). This phenomenon, referred to as the low-$q$ excess (L$q$E), has since also been observed for EAN mixtures containing other molecular cosolvents; in fact, certain EAN + acetonitrile binary mixtures show the most intense L$q$E scattering reported to date (at low EAN content, e.g. mole fraction EAN $x_{\mathrm{EAN}} = 0.1$) [61]. The particular L$q$E scattering patterns of these EAN + acetonitrile mixtures, shown in Fig. 3.15, suggest the presence of large molecular aggregates of effectively cylindrical shape at low EAN concentrations, such that a structure characterized by worm-like EAN self-associates surrounded by acetonitrile has been proposed.

MD computer simulations were performed to study the structure of one such EAN + acetonitrile mixture, $x_{\mathrm{EAN}} = 0.1$ (i.e. EAN:acetonitrile 1:9), known to show a very intense experimental L$q$E. Initially, an all-atom model was simulated using the Amber GAFF parameter set [62]. The simulation revealed that EAN and acetonitrile form separate nanoscale phases, with geometry essentially consistent with the proposed worm-like EAN aggregates, see Fig. 3.15 (right).

Nevertheless, full recovery of the experimental L$q$E from computer simulation trajectories requires a model system with dimensions of more than 100 nm, a prohibitive computational cost using current atomic-resolution models, e.g. GAFF. To this end, a simplified coarse-grained (CG) computational model was developed

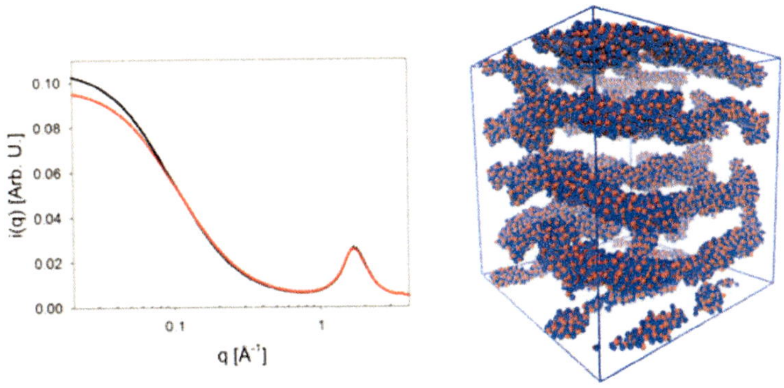


**Fig. 3.15** Left: SAXS patterns of an EAN + acetonitrile binary mixture ($x_{\mathrm{DBE}} = 0.1$) at 298 K (black) and 313 K (red), showing the low-$q$ peak ($q \approx 1$ Å$^{-1}$), and LqE at very low $q$. Adapted with permission from reference [61]. Copyright © 2017 American Chemical Society. Right: Coarse-grained MD simulation configuration showing only EAN (acetonitrile omitted in the interest of clarity), with ethylammonium cations in blue and nitrate anions in red; note the cylindrical worm-like EAN aggregate structures

using the original atomic-resolution MD trajectory as reference, in which certain atomic groups (e.g. methyl groups) are represented by a single interaction site [63, 64]; the CG model consisted of ca. 179,000 CG particles, representing nearly 600,000 atoms. The CG model reasonably reproduces the experimental SAXS pattern in the L*q*E region, though not at higher *q* due to its lack of atomic resolution, and shows that EAN does indeed form extended worm-like aggregates as previously proposed [61].

### 3.5.5 *Hydration of DNA Counterions*

The DNA double helix is a polyanionic macromolecule that in water solutions is neutralized by physiological cations, many of which are metal ions. These neutralizing ions are known as counterions. The ability of the ions to stabilize the water network around them is known as the positive hydration effect, while the effect of water disordering is also observed for some ions and known as the effect of negative hydration. The positively hydrated ions have increased potential barrier separating the hydration shell from the bulk water, and *vice versa* in the case of negatively hydrated ones (Fig. 3.16). See the review [65] for details on the hydration of alkali metal ions. Both effects are important for the stabilization of the DNA double helix and the recognition of specific sequence motifs of nucleotides.

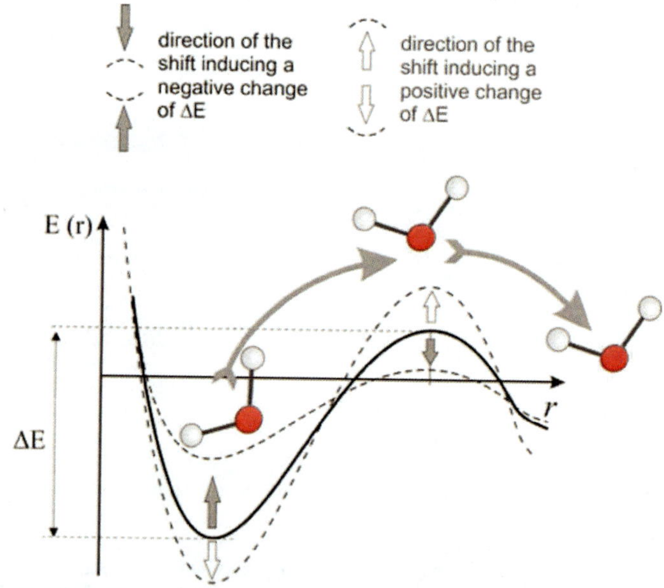


**Fig. 3.16** Character of ions hydration. Positive and negative hydration is related to the increased and decreased values of the potential barriers

The hydration effects of DNA counterions have been extensively studied by MD simulation [66–70]. The results for alkali metal ions have shown that the dynamics of the hydration shell of such counterions depend on the particular region of the DNA double helix where the interaction occurs: minor groove, major groove, or backbone [69]. The longest average residence time has been observed for water molecules in contact with counterions, localized in the minor groove of the double helix. In the case of positively hydrated $Na^+$ counterions, it is about 50 ps, while in the case of negatively hydrated $K^+$ and $Cs^+$ counterions, it is lower than 10 ps. The analysis shows that water molecules in the hydration shell of the ion are strongly restrained, and the effect of negative hydration for $K^+$ and $Cs^+$ counterions has not been observed [70]. The reason for this unexpected behavior is in the model parameters of ions and the water model that were used in the simulation.

### 3.5.6 *Is It Worth Trying to Calculate Viscosity for Ionic Liquids from Computer Simulations When the Experiments Fail?*

Viscosity is a property strongly dependent on temperature and, for the non-Newtonian fluids, also on the shear stress. Non-Newtonian fluids can be shear-thinning or shear-thickening. Blood is an example of a shear-thinning non-Newtonian fluid. Some ionic liquids show similar shear-thinning behaviour [33], as in the case of the trihexyl(tetradecyl)phosphonium bis(mandelato)-borate [P6,6,6,14] [BMB] ionic liquid, the viscosity of which starts to decrease after a certain shear rate [71], as is shown in Fig. 3.17.

Viscosity can be measured using many experimental techniques, from NMR spectroscopy and oscillating disks to simple capillary and falling cylinder methods. Normally, the measurement of viscosity is routine work, however different protocols are practised in different laboratories, from empirical to rigorous. Ionic liquids seem to be a category of liquids for which it is difficult to obtain a reliable value of the

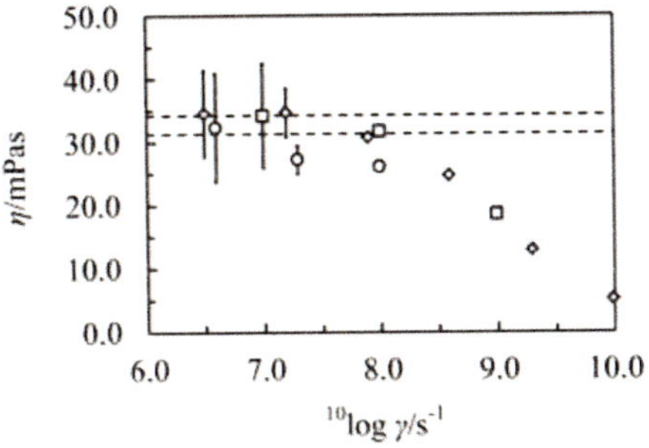


**Fig. 3.17** Viscosity, $\eta$, for [P6,6,6,14] [BMB] at 423 K as a function of shear rate, $\gamma$, for 96 (squares), 735 (diamonds) and 6000 (circles) ion pairs from MD simulations (reprinted with permission from [71])

viscosity. As can be seen from Fig. 3.18 adapted from [72] and showing the temperature dependence of viscosity of [P6,6,6,14] [Cl] ionic liquid from three independent measurements and also calculated in computer simulations, there are large differences among the published experimental results for the same system. Diogo et al. [73] write "*How can it be that in the twenty-first century the claims of uncertainty for viscosity data are greatly exceeded by the differences between the results, obtained with essentially the same sample?*".

If it is more difficult to measure a correct viscosity for ionic liquids than for other substances, it is because of the peculiar properties of ILs. Many of them are very hygroscopic, meaning that it is difficult to keep them dry as they easily absorb moisture from the air.

Figure 3.19 from ref. [77] illustrates well the problem. With a tiny amount of water as an impurity (open circles) the viscosity is already significantly reduced with respect to the pure sample. In the water saturated samples (solid triangles) the

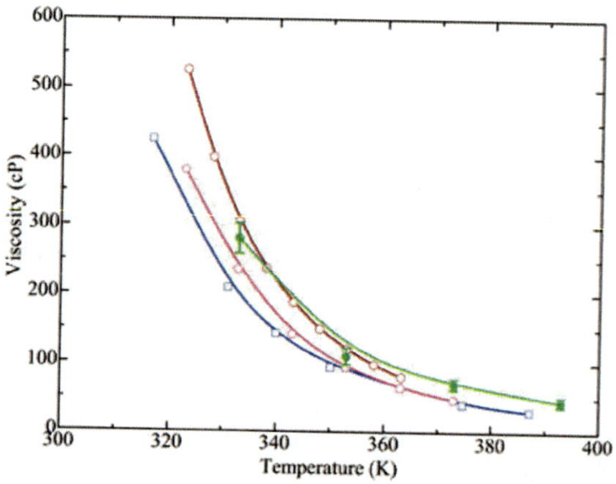


**Fig. 3.18** Temperature dependence of the viscosity of [P6,6,6,14] [Cl] measured in three different laboratories and from MD simulations. Representation code: green, values from MD; red, violet and blue represent experimental data taken from ref. [74] (red), ref [75] (violet), ref [76] (blue) (adapted with permission from [72])

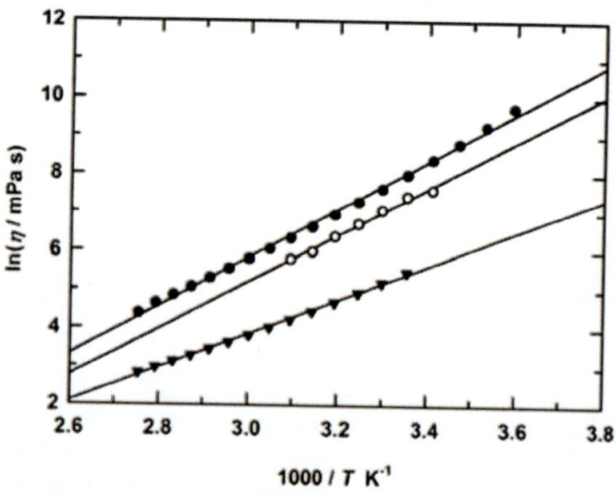


**Fig. 3.19** Viscosity data for pure [P6,6,6,14] [Cl] (solid circles) and water-saturated (solid triangles), taken from ref. [74] and with a small impurity (open circles) as measured by McAtee and Heitz [77] (reprinted with permission from [77])

difference with respect to the neat IL can be two orders of magnitude. Note also the temperature dependence is the same for the pure sample and that with a trace amount of water while it is different for the water-saturated sample.

Differently from what it happens in the "wet laboratory", in computer simulations the pure ionic liquid can really be pure and free from water and other contaminants occasionally left from the synthesis or after drying. Comparisons between computed and experimental data are to be carefully done, always keeping this difference in mind.

It is interesting to note that having (a known amount of) water in ionic liquids is not necessarily a disadvantage. It can be a strategy. Here applies "if you cannot beat them, join them!". Indeed, ionic liquids are highly viscous substances, making them problematic in many applications. As discussed above, by adding already a tiny amount of water they become very much less viscous. Furthermore, ILs are generally expensive, and adding water can compensate some of that cost. However, it may not be as simple as just mixing ionic liquids with water. The aqueous ionic liquids may sometimes show very peculiar behavior, as discussed in our recent review [78]. Also, ionic liquids and the closely related deep eutectic solvents themselves are extremely challenging molecular liquids to characterize. They show very complicated modes of micro-heterogeneities, both structurally and dynamically, with very long-ranging correlations, as discussed in previous examples and also summarized in our very recent review [33].

As discussed in Sect. 3.5.4, viscosity and diffusion are closely related properties, and both should be characterized when studying lubricants. As an example of the study of diffusion, we present here the results from a study that we performed on trihexyl(tetradecyl)phosphonium bis(oxalato) borate ([P6,6,6,14] [BOB]), a promising lubricant. To use charged ionic liquids (molten salts) as lubricants may seem counter-intuitive because of the strong interaction between oppositely charged ions; however, under harsh conditions (high temperature and high pressure) some ILs are characterized by very low friction [79]. As simulations can be performed at any conditions, they are expected to be useful to explain such phenomena when experiments are not possible.

We performed MD simulations on aqueous solution of [P6,6,6,14] [BOB] [80], and selected results, concerning the diffusion coefficient from MD simulations and experiments, are displayed in Fig. 3.20, left. The diffusion of water becomes faster with increasing water content, and faster diffusion corresponds to lower viscosity. However, the slope of the diffusion coefficient of the IL plotted as a function of water mole fraction ($X_w$) displays a much more variable trend. We have identified four distinct diffusion regions with increasing water content: $0 < X_w < 0.5$, $0.5 < X_w < 0.8$, $0.8 < X_w < 0.95$, and $X_w > 0.95$.

The variation of the diffusion trend, depending on the composition, is found to be ultimately linked to the structural organization in the mixture. In particular, in the region $0.5 < X_w < 0.8$, water aggregates bridge the ionic species among the polar domains (see Fig. 3.20, right), and induce an increase in their spatial correlations, thus reducing their mobility. As discussed in ref. [78], spectacular phenomena can occur when water is added to ionic liquids. Much of this is not understood at all, so

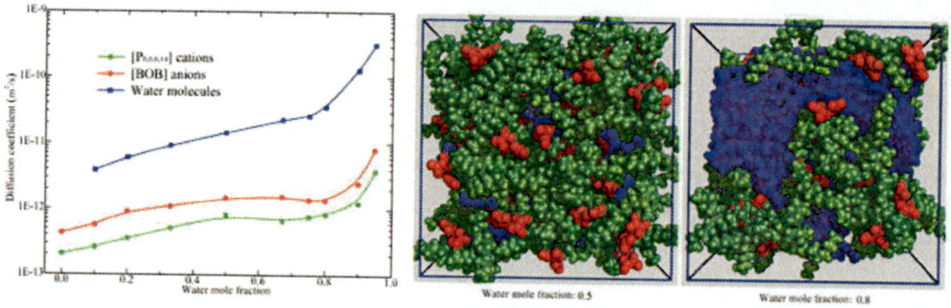


**Fig. 3.20** Left: Translational diffusion coefficients of [P6,6,6,14] cations, [BOB] anions and water molecules in [P6,6,6,14] [BOB] IL–water mixtures at 333 K from MD simulations. Right: snapshots of [P6,6,6,14] [BOB] IL–water at Xwater 0.5 or 0.8. The [P6,6,6,14] cations, [BOB] anions and water molecules in these mixtures are green, red and blue beads, respectively (adapted with permission from [80])

it is an area where the molecular modelling methods introduced in this chapter can be advantageously applied.

## 3.6 Conclusions

Computational molecular modeling offers a unique insight to understand solvation phenomena, and the properties of liquids, neat or in mixtures, neutral or charged, as in ionic liquids. The aim of this chapter was to give an overall orientation on how to study liquids and liquid mixtures to obtain a 3D molecular picture of how different interactions compete with each other to eventually find an equilibrium at the studied concentration of the components. While molecular simulations provide this picture, its reliability has to be always verified by calculating key properties for comparison with experimental results. Systematically carrying out the analysis of the simulation data, and properly averaging it to provide statistically reliable values, can be the most powerful method to study different phenomena in solutions. We have introduced the toolbox and how to use it. We have presented a number of case studies from our own research in recent years to illustrate what computer simulations can offer. We hope to inspire some of our more experimental colleagues with molecular questions to consider in silico studies as a way to obtain answers. Maybe also get questions and hypotheses to carry out new experiments?

**Acknowledgements** S. Perepelytsya acknowledges the support from the National Academy of Sciences of Ukraine (project 0117U000240) F. Mocci thanks the Regione Sardegna for support, grant number RASSR 81788 and MIUR, Project PRIN2017 "CANDL2" (Grant 2017W75RAE). A. Laaksonen also thanks the Swedish Research Council for support (Project No. 2019-03865) and for a partial support by a grant from the Ministry of Research and Innovation of Romania (CNCS-UEFISCDI, Project No. PN-III-P4- P4-ID-PCCF-2016-0050, within PNCDI III. Many simulations described were performed using resources provided by the Swedish National Infrastructure for Computing (SNIC) at PDC, HPC2N and NSC.

## References


1. Allen MP, Tildesley DJ (1987) Computer simulations of liquids. Clarendon Press, Oxford
2. Frenkel D, Smit B (1997) Understanding molecular simulations: from algorithms to applications. Academic Press, Cambridge
3. Lyubartsev AP, Laaksonen A, Vorontsov-Velyaminov PN (1994) Free energy calculations for lennard-jones systems and water using the expanded ensemble method a monte carlo and molecular dynamics simulation study. Mol Phys 82:455–471. https://doi.org/10.1080/00268979400100344
4. Luhmer M, Van Belle D, Reisse J, Odelius M, Kowalewski J, Laaksonen A (1993) Magnetic relaxation of xenon-131 dissolved in benzene. A study by molecular dynamics and Monte Carlo simulations. J Chem Phys 98:1566–1578. https://doi.org/10.1063/1.464273
5. Laaksonen A, Kowalewski J, Jönsson B (1982) Intermolecular nuclear spin-spin coupling and scalar Relaxation, a quantum-mechanical and statistical-mechanical study for the aqueous fluoride ion. Chem Phys Lett 89:412–417
6. Hamani AWS, Bazile JP, Hoang H, Luc HT, Daridon JL, Galliero G (2020) Thermophysical properties of simple molecular liquid mixtures: on the limitations of some force fields. J Mol Liq 303:112663. https://doi.org/10.1016/j.molliq.2020.112663
7. Leach AR (2001) Molecular modelling: principles and applications, 2nd edn. Prentice Hall, Harlow
8. Brooks BR, Bruccoleri RE, Olafson BD, States DJ, Swaminathan S, Karplus M (1983) {CHARMM}: a program for macromolecular energy, minimization, and dynamics calculations. J Comp Chem 4:187–217
9. MacKerell AD, Bashford D, Bellott M, Dunbrack RL, Evanseck JD, Field MJ, Fischer S, Gao J, Guo H, Ha S, Joseph-McCarthy D, Kuchnir L, Kuczera K, Lau FTK, Mattos C, Michnick S, Ngo T, Nguyen DT, Prodhom B, Reiher WE, Roux B, Schlenkrich M, Smith JC, Stote R, Straub J, Watanabe M, Wiorkiewicz-Kuczera J, Yin D, Karplus M (1998) All-atom empirical potential for molecular modeling and dynamics studies of proteins. J Phys Chem B 102:3586–3616
10. Weiner PW, Kollman PA (1981) AMBER: assisted model building with energy refinement. J Comput Chem 2:287–303
11. Cheatham TE, Miller JL, Fox T, Darden TA, Kollman PA (1995) Molecular dynamics simulations on solvated biomolecular systems: the particle mesh Ewald method leads to stable trajectories of DNA, RNA, and proteins. J Am Chem Soc 117:4193–4194
12. Jorgensen W, Swenson C (1985) Optimized intermolecular potential functions for amides and peptides. Structure and properties of liquid amides. J Am Chem Soc 107:569–578
13. Jorgensen WL, Tirado-Rives J (1988) The OPLS potential functions for proteins. Energy minimizations for crystals of cyclic peptides and crambin. J Am Chem Soc 110:1657–1666
14. Dauber-Osguthorpe P, Hagler AT (2019) Biomolecular force fields: where have we been, where are we now, where do we need to go and how do we get there? J Comput Aided Mol Des 33:133–203
15. Gkeka P, Stoltz G, Farimani AB, Belkacemi Z, Ceriotti M, Chodera J, Dinner AR, Ferguson A, Maillet J-B, Minoux H, Peter C, Pietrucci F, Silveira A, Tkatchenko A, Trstanova Z, Wiewiora R, Leliévre T (2020) Machine learning force fields and coarse-grained variables in molecular dynamics: application to materials and biological systems. J Chem Theory Comput 16:4757–4775
16. Usula M, Mocci F, Cesare Marincola F, Porcedda S, Gontrani L, Caminiti R (2014) The structural organization of N -methyl-2-pyrrolidone + water mixtures: a densitometry, X-ray diffraction, and molecular dynamics study. J Chem Phys 140:124503
17. Humprey W, Dalke A, Schulten K (1996) VMD - visual molecular dynamics. J Mol Graph 14:33–38
18. Laaksonen L (1992) A graphics program for the analysis and display of molecular dynamics trajectories. J Mol Graph 10:33–34
19. Barker JA, Henderson D (1972) Theories of liquids. Annu Rev Phys Chem 23:439–484

20. Kusalik PG, Svishchev IM (1994) The spatial structure in liquid water. Science 265:1219–1221
21. Svishchev IM, Kusalik PG (1993) Structure in liquid water: a study of spatial distribution functions. J Chem Phys 99:3049–3058
22. Bergman DL, Laaksonen L, Laaksonen A (1997) Visualization of solvation structures in liquid mixtures. J Mol Graph Model 15(301–6):328–333
23. Kovacs H, Laaksonen A (1991) Molecular dynamics simulation and NMR study of water-acetonitrile mixtures. J Am Chem Soc 113:5596–5605
24. Vishnyakov A, Lyubartsev AP, Laaksonen A (2001) Molecular dynamics simulations of dimethyl sulfoxide and dimethyl sulfoxide - water mixture. J Phys Chem A 105:1702–1710
25. Dahlberg M, Laaksonen A (2006) Preferential solvation of phenol in binary solvent mixtures. A molecular dynamics study. J Phys Chem A 110:2253–2258. https://doi.org/10.1021/jp056463e
26. Laaksonen A, Lyubartsev A, Mocci F (2012) M.DynaMix studies of solvation, solubility and permeability. In: Wang PL (ed) Molecular dynamics - studies of synthetic and biological macromolecules. IntechOpen, pp 85–103
27. Engelbrecht L, Mocci F, Laaksonen A, Koch KR (2018) 195Pt NMR and molecular dynamics simulation study of the solvation of [PtCl6]2- in water-methanol and water-dimethoxyethane binary mixtures. Inorg Chem 57:12025–12037. https://doi.org/10.1021/acs.inorgchem.8b01554
28. Bergman D, Laaksonen A (1998) Topological and spatial structure in the liquid-water–acetonitrile mixture. Phys Rev E 58:4706–4715
29. de Villiers Engelbrecht L, Farris R, Vasiliu T, Demurtas M, Piras A, Cesare Marincola F, Laaksonen A, Porcedda S, Mocci F (2021) Theoretical and experimental study of the excess thermo-dynamic properties of highly non-ideal liquid mixtures of butanol isomers + DBE. J Phys Chem B 125:587-600
30. Stillinger FH (1980) Water revised. Science 209:451–457
31. Luzar A, Chandler D (1996) Effect of environment on hydrogen bond dynamics in liquid water. Phys Rev Lett 76:928–931
32. Kumar R, Schmidt JR, Skinner JL (2007) Hydrogen bonding definitions and dynamics in liquid water. J Chem Phys 126:4107
33. Wang YL, Li B, Sarman S, Mocci F, Lu ZY, Yuan J, Laaksonen A, Fayer MD (2020) Microstructural and dynamical heterogeneities in ionic liquids. Chem Rev 120:5798–5877
34. Brehm M, Weber H, Thomas M, Hollõczki O, Kirchner B (2015) Domain analysis in nanostructured liquids: a post-molecular dynamics study at the example of ionic liquids. ChemPhysChem 16:3271–3277
35. Gellatly BJ, Finney JL (1982) Calculation of protein volumes: An alternative to the Voronoi procedure. J Mol Biol 161:305–322
36. Ruocco G, Sampoli M, Vallauri R (1991) Molecular dynamics simulations of liquid water: Voronoi polyhedra and network topology. J Mol Struct 250:259–270
37. Shih JP, Sheu SY, Mou CY (1994) A Voronoi polyhedra analysis of structures of liquid water. J Chem Phys 100:2202–2212
38. Rapaport DC (1983) Density fluctuations and hydrogen bonding in supercooled water. Mol Phys 48:23–31
39. Pei HW, Laaksonen A (2019) Feature vector clustering molecular pairs in computer simulations. J Comput Chem 40:2539–2549
40. Mocci F, Laaksonen A, Wang Y, Saba G, Lai A, Cesare Marincola F (2014) CompChem and NMR probing ionic liquids. In: Caminiti R, Gontrani L (eds) The structure of ionic liquids, soft and biological matter series. Springer International Publishing, Switzerland, pp 97–126
41. Mocci F, Laaksonen A (2015) Combining MD simulations and NMR spectroscopy for molecular insight and methodological synergy: the integrated MD-NMR method. In: Nuclear magnetic resonance Volume 44, Kamieńska-Trela K (Ed), The Royal Society of Chemistry, Cambridge, pp 592–616
42. Usula M, Porcedda S, Mocci F, Gontrani L, Caminiti R, Marincola FC, Caminiti R, Cesare Marincola F (2014) NMR, calorimetry, and computational studies of aqueous solutions of N-Methyl-2-pyrrolidone. J Phys Chem B 118:10493–10502

43. Gontrani L, Caminiti R (2012) The structure of liquid N-methyl pyrrolidone probed by x-ray scattering and molecular simulations. J Chem Phys 136:074505
44. Rezanova EN, Kammerer K, Lichtenthaler RN (1999) Excess properties of binary alkanol + diisopropyl ether (DIPE) or + dibutyl ether (DBE) mixtures and the application of the extended real associated solution model. J Chem Eng Data 44:1235–1239
45. Alaoui F, Montero E, Bazile JP, Comuñas MJP, Galliero G, Boned C (2011) Liquid density of 1-butanol at pressures up to 140MPa and from 293.15K to 403.15K. Fluid Phase Equilib 301:131–136
46. Patil KR, Pathak G, Pradhan SD (1989) Excess volumes of isomeric butanols with di-n-butyl ether. Proc Indian Acad Sci - Chem Sci 101:443–447
47. Kammerer K, Lichtenthaler RN (1998) Excess properties of binary alkanol-ether mixtures and the application of the ERAS model. Thermochim Acta 310:61–67
48. Bernazzani L, Carosi MR, Duce C, Gianni P, Mollica V (2006) Volumetric properties of binary mixtures of isomeric butanols and C8 solvents at 298.15 K. J Solution Chem 35:1567–1585
49. Valén A, López MC, Urieta JS, Royo FM, Lafuente C (2002) Thermodynamic study of mixtures containing oxygenated compounds. J Mol Liq 95:157–165
50. Giner B, Artigas H, Carrión A, Lafuente C, Royo FM (2003) Excess thermodynamic properties of isomeric butanols with 2-methyl-tetrahydrofuran. J Mol Liq 108(1–3):303–311
51. Wang J, Wang W, Kollman PA, Case DA (2006) Automatic atom type and bond type perception in molecular mechanical calculations. J Mol Graph Model 25:247–260
52. Bernardis FL, Grant RA, Sherrington DC (2005) A review of methods of separation of the platinum-group metals through their chloro-complexes. React Funct Polym 65:205–217
53. Pesek JJ, Mason WR (1977) Platinum-195 magnetic resonance spectra of some platinum(II) and platinum(IV) complexes. J Magn Reson 25:519–529
54. Freeman W, Pregosin PS, Sze SN, Venanzi LM (1976) Platinum-195 NMR using Fourier transform techniques. The PtCl42- ion. J Magn Reson 22:473–478
55. Frankel LS, Stengle TR, Langford CH (1965) A study of preferential solvation utilizing nuclear magnetic resonance. Chem Commun 17:393–394
56. Frankel LS, Langford CH, Stengle TR (1970) Nuclear magnetic resonance techniques for the study of preferential solvation and the thermodynamics of preferential solvation. J Phys Chem 74:1376–1381
57. Atkin R, Warr GG (2008) The smallest amphiphiles: nanostructure in protic room-temperature ionic liquids with short alkyl groups. J Phys Chem B 112:4164–4166
58. Mariani A, Caminiti R, Campetella M, Gontrani L (2016) Pressure-induced mesoscopic disorder in protic ionic liquids: first computational study. Phys Chem Chem Phys 18:2297–2302
59. Russina O, Sferrazza A, Caminiti R, Triolo A (2014) Amphiphile meets amphiphile: beyond the polar-apolar dualism in ionic liquid/alcohol mixtures. J Phys Chem Lett 5:1738–1742
60. Jiang HJ, Fitzgerald PA, Dolan A, Atkin R, Warr GG (2014) Amphiphilic self-assembly of alkanols in protic ionic liquids. J Phys Chem B 118:9983–9990
61. Mariani A, Caminiti R, Ramondo F, Salvitti G, Mocci F, Gontrani L (2017) Inhomogeneity in ethylammonium nitrate-acetonitrile binary mixtures: the highest "low q excess" reported to date. J Phys Chem Lett 8:3512–3522
62. Wang J, Wolf RM, Caldwell JW, Kollman PA, Case DA (2004) Development and testing of a general Amber force field. J Comput Chem 25:1157–1174
63. Mirzoev A, Lyubartsev AP (2013) MagiC: software package for multiscale modeling. J Chem Theory Comput 9:1512–1520
64. Wang YL, Lyubartsev A, Lu ZY, Laaksonen A (2013) Multiscale coarse-grained simulations of ionic liquids: comparison of three approaches to derive effective potentials. Phys Chem Chem Phys 15:7701–7712
65. Smirnov PR, Trostin VN (2007) Structures of the nearest surroundings of the K+, Rb +, and Cs+ ions in aqueous solutions of their salts. Russ J Gen Chem 77:2101–2107
66. Mocci F, Saba G (2003) Molecular dynamics simulations of A. T-rich oligomers: sequence-specific binding of NA+ in the minor groove of B-DNA. Biopolymers 68:471–485. https://doi.org/10.1002/bip.10334

67. Mocci F, Laaksonen A, Lyubartsev A, Saba G, Chimiche S, Cagliari U, Universitaria C (2004) Molecular dynamics investigation of 23 Na NMR relaxation in oligomeric DNA aqueous solution. J Phys Chem B 108:16295–16302
68. Lyubartsev AP, Laaksonen A (1998) Molecular dynamics simulations of dna in solution with different counter-ions. J Biomol Struct Dyn 16:579–592
69. Perepelytsya S (2018) Hydration of counterions interacting with DNA double helix: a molecular dynamics study. J Mol Model 24:171
70. Perepelytsya SM (2020) Positively and negatively hydrated counterions in molecular dynamics simulations of DNA double helix. Ukr J Phys 65:510
71. Sarman S, Wang YL, Rohlmann P, Glavatskih S, Laaksonen A (2018) Rheology of phosphonium ionic liquids: a molecular dynamics and experimental study. Phys Chem Chem Phys 20:10193–10203. https://doi.org/10.1039/c7cp08349a
72. Wang Y, Sarman S, Glavatskih S, Antzutkin ON, Rutland MW, Laaksonen A (2015) Atomistic insight into Tetraalkylphosphonium-Bis(oxalato)borate 2 Ionic Liquid/Water Mixtures. I. Local Microscopic Structure. J Phys Chem B 119:5251-5264
73. Diogo JCF, Caetano FJP, Fareleira JMNA, Wakeham WA (2014) Viscosity measurements on ionic liquids: a cautionary tale. Int J Thermophys 35:1615–1635
74. Neves CMSS, Carvalho PJ, Freire MG, Coutinho JAP (2011) Thermophysical properties of pure and water-saturated tetradecyltrihexylphosphonium-based ionic liquids. J Chem Thermodyn 43:948–957. https://doi.org/10.1016/j.jct.2011.01.016
75. Fraser KJ, Izgorodina EI, Forsyth M, Scott JL, MacFarlane DR (2007) Liquids intermediate between "molecular" and "ionic" liquids: liquid ion pairs? Chem Commun 3817–3819.
76. Vaughan JW, Dreisinger D, Haggins J (2006) Density, viscosity, and conductivity of tetraalkyl phosphonium ionic liquids. ECS Trans 2:381
77. McAtee ZP, Heitz MP (2016) Density, viscosity and excess properties in the trihexyltetradecylphosphonium chloride ionic liquid/methanol cosolvent system. J Chem Thermodyn 93:34–44. https://doi.org/10.1016/j.jct.2015.09.030
78. Ma C, Laaksonen A, Ji X, Lu X (2018) The peculiar effect of water on ionic liquids and deep eutectic solvents. Chem Soc Rev 47:8685–8720
79. Shah FU, Glavatskih S, MacFarlane DR, Somers A, Forsyth M, Antzutkin ON (2011) Novel halogen-free chelated orthoborate-phosphonium ionic liquids: synthesis and tribophysical properties. Phys Chem Chem Phys 13:12865–12873. https://doi.org/10.1039/c1cp21139k
80. Wang Y-L, Sarman S, Golets M, Mocci F, Lu Z-Y, Laaksonen A (2019) 4. Multigranular modeling of ionic liquids. In: Fehrmann R, Santini C (eds) Ionic liquids. Walter De Gruyter GmbH, Berlin, pp 55–100

# Chapter 4
# Lyotropic Liquid Crystal Phases of Phospholipids as Model Tools in Molecular Biophysics and Pharmacology

**Longin N. Lisetski, Olga V. Vashchenko, Natalia A. Kasian, and Liliia V. Sviechnikova**

**Abstract** Model phospholipid membranes are unique objects that naturally bridge the gap between rather distant fields of science. In fact, they are typical lyotropic liquid crystals where ordered molecular arrangement can easily change due to variation of temperature and the presence of dopants, giving rise to different readily characterizable phase states. On the other hand, they are generally recognized as convenient biomimetic objects for modeling various structural and functional features of cell membranes. After a general description of structural and thermodynamic features of model membranes as objects of biophysics, effects of drug substances introduced into the membranes as dopants are reviewed, with various effects of drug-membrane interactions manifested in changes of physico-chemical properties. Special attention is paid to joint action of different drugs, as well as to the process of vitamin D formation in cell membranes induced by UV irradiation.

## List of Abbreviations, Symbols, and Chemical substances

### *Abbreviations*

| | |
|---|---|
| CLC | Cholesteric liquid crystal |
| Ch | Cholesteric liquid crystal phase |
| DSC | Differential scanning calorimetry |
| $L_{\alpha'}$ | Gel phase of phospholipid membrane |

L. N. Lisetski (✉) · O. V. Vashchenko (✉) · N. A. Kasian · L. V. Sviechnikova
Institute for Scintillation Materials of State Scientific Insitution "Institute for Single Crystals", National Academy of Sciences of Ukraine, 60, Nauky Avenue, Kharkiv 61072, Ukraine
e-mail: lisetski@isma.kharkov.ua

O. V. Vashchenko
e-mail: olga_v@isma.kharkov.ua

N. A. Kasian
e-mail: kasian@isma.kharkov.ua



L. Bulavin and N. Lebovka (eds.), *Soft Matter Systems for Biomedical Applications*, Springer Proceedings in Physics 266, https://doi.org/10.1007/978-3-030-80924-9_4

| | |
|---|---|
| $L_{\beta'}$ | Liquid crystalline phase of phospholipid membrane |
| LC | Liquid crystal |
| $P_{\beta'}$ | Ripple phase of phospholipid membrane |
| SmA | Smectic-A liquid crystal phase |
| UV | Ultraviolet |

## Symbols

| | |
|---|---|
| $CN$ | Cooperative number |
| $\Delta H_{\mathrm{m}}$ | Enthalpy |
| $\Delta H_{\mathrm{VH}}$ | Van't Hoff enthalpy |
| $\lambda_{\max}$ | Wavelength of maximum selective reflection |
| $\log D$ | Distribution coefficient |
| $\log P$ | Lipophilicity coefficient |
| $p$ | helical pitch of cholesteric liquid crystal |
| $T_{\mathrm{m}}$ | Main phase transition temperature |
| $T_{\mathrm{p}}$ | Pretransition temperature |
| $\Delta T_{\mathrm{m}}$ | Shift of main phase transition temperature |
| $\Delta T_{1/2}$ | Half-width of the main transition peak |

## Chemical substances

| | |
|---|---|
| 5CB | 4′-Pentyl-4-cyanobiphenyl, $CH_3(CH_2)_7C_6H_4C_6H_4CN$ |
| CTAB | Hexadecyltrimethylammonium bromide, $C_{19}H_{42}NBr$ |
| DMSO | Dimethylsulfoxide, $C_2H_6OS$ |
| DPPC | L-α-dipalmitoylphosphatidycholine, $C_{40}H_{80}NO_8P$ |
| ProD | Provitamin $D_2$, $C_{28}H_{44}O$; provitamin $D_3$, $C_{27}H_{44}O$ |
| PreD | Previtamin $D_2$, $C_{28}H_{44}O$; previtamin $D_3$, $C_{27}H_{44}O$ |
| VitD | Vitamin $D_2$, $C_{28}H_{44}O$; vitamin $D_3$, $C_{27}H_{44}O$ |

## 4.1 Introduction

A generally followed pattern in the studies of nature is the development of certain simplified models that allow the use of the means, approaches and facilities common to physical sciences. These models are expected to retain all essential features of the relevant objects and phenomena, reducing the complicated real picture to a set of accountable features. An example of such approach are the so-called "model phospholipid membranes", or "hydrated phospholipids", which are, in fact, lyotropic liquid crystal (LC) phases formed by phospholipid molecules in water dispersions [1–3]. The natural cell membranes can be basically considered as phospholipid bilayers comprising sterols, proteins and other constituting components, as well as water in various degrees of binding. Thus, lyotropic LC phases of phospholipids can really be considered as a model for studies of cell membranes. Moreover, they can serve as a certain model (less adequate, but still feasible) of other biological tissues where large anisotropic organic molecules possess some degree of long- or short-range ordering.

In this respect, several ideas of implementing physical approaches to biomedical studies arise. One of these ideas is to check all the already known, developed or envisaged drug substances (both active ingredients and excipients) for their "membranotropic action", i.e., their effects on the LC phase transitions in model phospholipid membranes. This would present information that, in some cases, could be essential for establishing the mechanisms of pharmacological action of some drugs. There have been multiple works on this subject, which, despite many excellent reviews [4–9], still lack systematization and generalization.

Our own efforts and contributions in these aspects are presented in Sect. 4.4 of this chapter. The most challenging point in this field is the question of joint action of several drug substances administered simultaneously. Both synergic and antagonistic effects can be expected, modifying their therapeutical action. Mutual compatibility of different drugs is an important point in pharmacology. So, finding out how membranotropic action of different substances is affected by their joint introduction into a model membrane outlines a promising area of biophysical research.

Another aspect of our interest is the ability of model membranes to yield a physically measurable response to various external factors acting on the human organism. Thus, the effects of such harmful factors as ionizing radiation or such ambiguous factors as UV radiation could presumably be studied using certain model systems – and you guess that we suggested the use of model phospholipid membranes for this purpose, which is the subject of Sect. 4.5. In fact, both chapters are closely interrelated, because the direct physico-chemical factor affecting the model membrane is the change in its chemical composition emerging under the applied irradiation, which is basically (in the sense of the description models used) similar to introducing a drug or a bio-active substance.

## 4.2 General Properties of Phospholipid Membranes

We start with the notion of a standard phospholipid DPPC (L-α-dipalmitoylphosphatidycholine) as a certain analog of Drosophyla fruit fly that had served its fundamental role in genetics (to not mention the liquid crystal 5CB that since its synthesis in 1973 by Prof. George Gray has been playing a similar role in LC science).

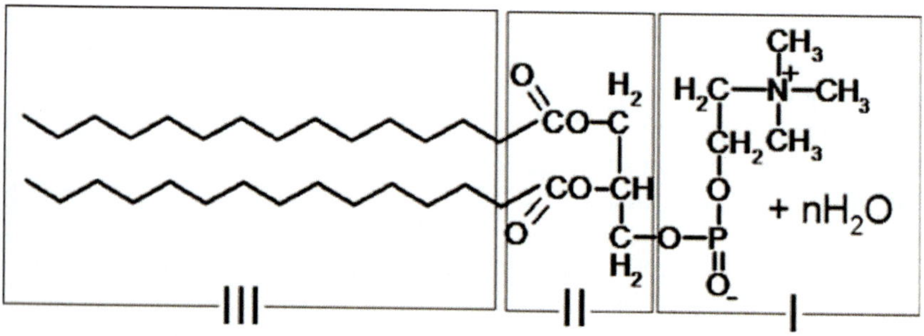


**Fig. 4.1** The chemical structure of DPPC (L-α-dipalmitoylphosphatidycholine). I – polar moiety; II – hydrophilic-hydrophobic interface; III – non-polar moiety

### 4.2.1 *Chemical Structure of Phospholipid Molecules*

The chemical structure of DPPC is presented in Fig. 4.1. From the physico-chemical viewpoint, all basic features of different phospholipids noted in multiple papers on this subject are presented in this scheme. One can see the polar head (I), which, in this particular case, bears both the positive and negative charge, characterizing this phospholipid as zwitterionic.

Also, we see the non-polar hydrophobic moiety (III) with two alkyl chains, which can be of different length (detailed characteristics of DPPC homologues are presented in [10, 11]), one of these chains (or both) can contain double bonds, mostly in the middle of the chain (analogs of stearoyl-oleoyl or dioleyl type are rather common in the literature [12–14]). A peculiar feature of such type of molecules is the presence of the "intermediary" part (II) between the hydrophilic and hydrophobic parts of the molecule, acting as a sort of hydrophilic-hydrophobic interface. The importance of singling out these parts of phospholipid molecules becomes clear when we consider lipid bilayers containing the introduced "alien" molecules. Depending on their chemical structure, these "dopant" molecules tend to interact specifically with one or another part of the "host" phospholipid molecule.

In general, the structure of the phospholipid molecule is, in fact, relatively straightforward, with its constituent parts quite suitable for physical modeling. At the same time, it is clearly more complex than the molecular structure of standard surfactants (e.g., CTAB [15, 16]).

### 4.2.2 *Phase Transitions of Phospholipid Membranes*

Phospholipids, due to their ability to form various readily characterizable LC phases in broad concentration and temperature ranges, are, from the viewpoint of molecular physics, more suited for theoretical and experimental studies. In Fig. 4.2, the arrangement of phospholipid molecules in lyotropic LC phases is shown, together

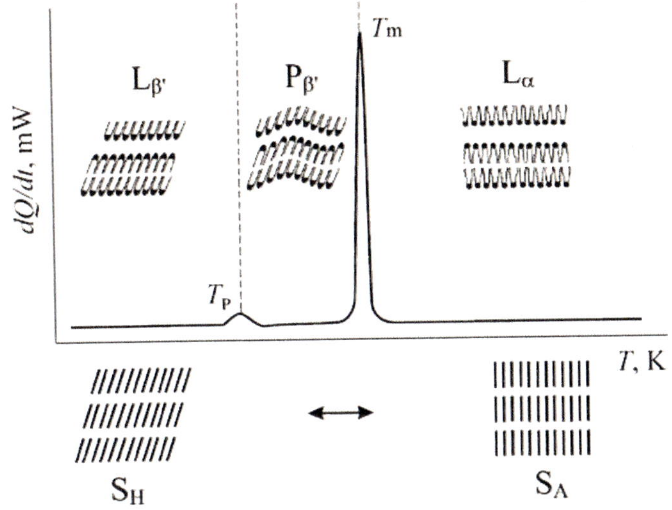


**Fig. 4.2** DSC thermograms of LC phases in hydrated DPPC in physiological temperature range. Here, *dQ/dt* is the heat flow and *T* is the temperature; $T_m$ and $T_p$ are the temperatures of main phase transition and pre-transition, correspondingly

with a scheme of the LC phase transitions (as they are measured by a differential scanning calorimeter).

The hydrated DPPC (L-α-dipalmitoylphosphatidylcholine) can be set as a characteristic example, as it is a predominant lipid in mammalian cells, and it undergoes clear and well-reproducible LC phase transitions in physiological temperature range. The temperatures of these phase transitions are practically unchanged in a broad range of water content (45 to 95% or even wider) and pH values. The sequence of LC phases in hydrated DPPC is $L_{\beta'} \rightarrow P_{\beta'} \rightarrow L_{\alpha,}$ (often described as "gel → ripple phase → LC phase") with mesomorphic phase transitions at ~36 °C (pre-transition, $T_p$) and ~42 °C (main phase transition, $T_m$) [10, 17, 18]. A typical picture of differential scanning calorimetry (DSC) thermograms of such transitions is shown in Fig. 4.2. As one can see, the structures of lyotropic DPPC phases are essentially similar to thermotropic smectic A ($L_\alpha$) or smectic H ($L_{\beta'}$) phases, which bridges the alleged gap between thermotropic and lyotropic LCs.

An alien component can appear in a membrane by different ways. This may be due to (patho)-physiological activity, drug administration or some kind of external physical influence. In any case, it is incorporated into the existing LC structure. This process is governed either by specific interactions with certain moieties of the phospholipid molecule, or by general long-range interactions with the LC ordering. With artificially created model membranes, we can imitate this process by introducing these biologically active substances as dopants in specified concentrations, as they are could be introduced into other lyotropic or thermotropic LC systems. The response of the model membrane, measured by DSC, optical spectroscopy or other physicochemical techniques, would be a certain reflection of the corresponding processes in a real cell membrane under the same conditions.

This assumption outlines the next chapters of our review, i.e., response of model membranes to administration of drug substances (with individual or joint application) and external factors, such as, e.g., UV or ionizing radiation.

## 4.3 Membranotropic Effects of Individual Drug Substances

### *4.3.1 Effects of Drug-Membrane Interactions*

The modern concepts on drug-membrane interactions have been formed for nearly a century [19]. At present, the interaction features of drug substances with lipid membranes are often considered in relationship to their bioavailability and other pharmacokinetic properties. Such studies have been performed for various types of substances, including tranquilizers, anesthetics, antidepressants, myorelaxant agents, steroids, surfactants, narcotic substances, sedative drugs, antihystamines, etc. [20–23].

The transport of drug substances across cell membranes is one of the most important stages of their action, since it largely stimulates the distribution of these substances in biological fluids and tissues of the organism. In most cases, the drug molecules permeate through the lipid bilayer by means of passive diffusion [24–26], with non-specific binding of a drug substance to the membrane being one of the aspects of its pharmacological action. Small amphiphilic molecules can easily overcome not only a cell membrane, but the blood–brain barrier as well [24]. From the data of [27] one can see a clear correlation between increased permeability (i.e., more intense passive diffusion) and lowered values of $T_m$ and resistivity of lipid membranes.

The effects of drug-membrane interactions are not limited by the process of overcoming the hydrophilic-hydrophobic barrier, but are reflected in many properties of the membrane (and the cell as a whole). For many pharmacological groups of substances, the pharmacological action was shown to be dependent on their interaction with lipid membranes:

- **Anesthetics.** The molecular structure of most anesthetics is not specific, suggesting that their therapeutic action is targeted upon membranes rather than proteins [28, 29]. This assumption is also supported by the data of [4], where the anesthetic molecule in the active form showed a more deep penetration into the membrane as compared with the inactive form, favoring stronger depression of $T_m$ and $T_p$. Also, anestetics were shown to modify the distribution of free volume in the bilayer [30], changing the lateral pressure profile and affecting the conformation of membrane protein, thus realizing one of the mechanisms of therapeutic action. Anesthetics were reported to affect phase state, curvature, thickness and hydration of membranes, as well as photosynthesis, active transport, passive diffusion, oxidative metabolism etc. [19, 31, 32].

- **Antitubercular drugs.** Membrane destabilization and significant perturbation in their molecular ordering was observed under antitubercular drugs insertion [33, 34].
- **Antidepressants.** Changes in the contribution of the lipid bilayer to the energy of conformational transformations of membrane lipids due to binding of the drug substance to the membrane was noted as one of the action mechanisms of antidepressants [35].
- **Anti-inflammatory drugs.** Substantial changes in properties of the cell barriers were noted as the main action mechanism of non-steroid anti-inflammatory drugs [31, 36]. Thus, introduction of licophelone led to higher local ordering of alkyl chains of lipids in the membrane, which was accompanied by lower density of their packing [36].
- **Analgesics.** Interaction of analgesics (e.g., paracetamol) with lipid bilayers can affect the electrostatic potential of the membrane, which is important for functioning of the ion channels [37].
- **Antimicrobial drugs.** Numerous experimental data reveal changes in membrane properties (in partuclular, singnificant elevation of membrane permeability) as contributions to the action mechanism of many antimicrobial drugs [37, 38].
- **Antibiotics.** Interactions of antibiotics with cell membranes is critically important for their bioavailability and non-toxicity [33, 34, 39, 40].

Generally, for many drug substances, increased membrane fluidity was noted, which was interpreted as a decrease in lipid ordering due to binding effects [41]. Amphiphilic substances of various pharmacological groups (anesthetics, antibiotics, antidepressants etc.) give rise to multiple effects due to their interactions with membranes. The specific mechanisms include changes in lipid organization, membrane desintegration, formation of channels, non-lamellar structures or interdigitated phases, induction of flip-flop transition, etc. [42].

### 4.3.2 Molecular Aspects of Drug-Membrane Interactions

In studies of membranotropic action of drug substances, an important aspect is outlining the molecular mechanisms of the related effects [21–23, 43]. Thus, it is generally accepted that the membranotropic activity of a drug substance is largely determined by its lipophilicity, log*P*, and distribution coefficient, log*D* (the effective lipophilicity at a given pH value); however, there are many exceptions [24]. The value of log*P* reflects not only the general distribution of the introduced substance in the membrane, but also its more detailed location. Thus, molecular dynamics calculations have shown that lanosterol (log*P* 8,7) penetrates into the membrane more deeply than cholesterol (log*P* 8,5). A similar effect was noted in [44] for merocyanine derivatives with different hydrocarbon chain lengths. The membranotropic activity is also affected by pH-dependent ionization of the drug substances which largely determines its distribution, diffusion, etc. [34, 45, 46].

It should be noted that there were certain attempts to classify the types of interactions between various organic substances and lipid membranes [19, 47–49], but they are predominantly empirical. The authors of [19] by analyzing DSC data on more than 500 different substances, characterized four different types of their interactions with DPPC-based model phospholipid membranes, which were distinguished by location of the molecules in the bilayers. According to this classification, the substances of type A are located in the region of hydrocarbon chains $C_1$-$C_8$, smearing DSC peaks of phase transitions and decreasing the size of cooperative domains [50]. The substances of type B are located in the bilayer interface region, inducing smaller cooperative domain sizes and phase separation. The substances of C type are located in the range of hydrocarbon chains $C_{10}$-$C_{16}$, they lower the phase transition temperatures. Finally, the type D substances are bound to the membrane surface, inducing phase separation of lipids without decreasing the cooperative domain size. When concentration of substances in the membrane is changed, the interaction type can also be modified, e.g., C → A, D → B or B → A.

The role of location of the drug substance molecules is clearly demonstrated in [51], where phenolic and methylated derivatives of an antiviral drug substance resveratrol were studied in DPPC bilayers. Results of DSC and molecular dynamics studies have shown that phenolic groups, unlike methyl groups, contain hydroxyl radicals, which are capable of forming hydrogen bonds with the membrane surface. This leads to changes in the main phase transition (i.e., transition to $L_\alpha$-phase) and especially in the pre-transition (Fig. 4.2), and can be related to higher anti-oxidant activity of such substances.

The location of steroids in the lipid bilayer is determined by the hydrophilic-hydrophobic balance, electrostatic potential and hydration of the membrane surface [52]. Polarization of the π-system and hydroxyl groups, as well as number and direction of hydrogen bonds, are of great importance for membranotropic activity of steroids, since it controls their orientation in the membrane and their interaction with surrounding lipid molecules [53].

Another important factor is the isomerism of drug substance molecules. Thus, for enantiomers of anesthetic bupivacaine, there is an evident difference in membrane localization, area per lipid molecule and average inter-lipid distance [53]. Higher fluidity of the membranes under introduction of R(+)-enantiomer can be related with its ability to block the activity of sodium channels and membrane receptors.

## 4.4 Membranotropic Effects of Drugs Combinations

In the previous paragraphs, we made an attempt to describe and systematize numerous works on interaction of individual drugs with phospholipid membranes. At the same time, the available information on joint action of different drug substances simultaneously introduced into lipid membranes is, however, scarce and fragmentary. Joint effects of different dopants could be, at least in some cases, qualitatively different from the straightforward additivity. Various changes in thermodynamic parameters of lipid membranes occurring under joint application of two or several drug substances were noted [54–57]. In case of direct molecular interactions, such effects can be studied, in addition to DSC, by other methods, e.g., mass spectroscopy [58–60].

The effects of joint application of different drug substances upon the cell membranes can be estimated from the effects of their joint administration (i.e., joint introduction to the model phospholipid membrane as dopant components) upon DSC peaks of phase transitions in these systems. A certain starting point in such studies was marked in [61, 62], where introduction of two drug substances (decamethoxin and acetylsalycic acid, as well as their analogs) showed clear "antagonistic" effect. The temperatures of the main phase transition $T_m$ were decreased by individual action of each of these substances, but they were increased when both membranotropic agents were added simultaneously. In other studies, various non-trivial effects of joint drug action in model phospholipid membranes were also noted [63–66]. Some particular cases have been found [63, 66] when the resulting effect was "synergic", i.e., joint application of two drug substances enhanced the effects of their individual action.

### 4.4.1 Kosmotrope and Chaotrope Substances

Here, we endeavor a further step in this direction, i.e., we aimed at characterizing individual dopants of different chemical nature as "kosmotropic" or "chaotropic" (using the common classification of [67–69]) and assess their individual and joint effect on the model membranes. Nowadays, «kosmotropicity/chaotropicity» conception is expanded to a wide group of substances, but biological significance of the phenomenon has still been under consideration [67, 70–73].

Lipid bilayers are also influenced by Hoffmeister effects [70, 71, 74–76], i.e., effects of quantitative and qualitative differences in hydration parameters of water soluble substances. They manifest themselves in a variety of biological and physico-chemical systems (e.g. as changes in water viscosity and proteins solubility, as well as in shifts of membrane phase transitions). Indeed, the surface of lipid bilayers is abundant in highly hydrated phosphate groups, and hydrophobic interactions inside the layer are strong. A number of important regularities concerning membranotropic action of kosmotropes and chaotropes were noted in [67]. For instance, kosmotropic dopants favor formation of the high-temperature inverted hexagonal phase ($H_{II}$) as well as the low-temperature lamellar crystalline ($L_c$) and gel ($L_\beta$ or $L_{\beta'}$) phases at the expense of the lamellar liquid-crystalline phase ($L_\alpha$). The effect is accompanied by certain decrease in the half-width of the melting peak as well as increased hysteresis [74]. On the contrary, chaotropic dopants expand the existence range of $L_\alpha$ phase not only in monolipid membranes but also in lipid mixtures [77, 78]. The magnitude of the solute effect is proportional to the hydration difference of the adjacent lipid phases and inversely proportional to the transition latent heat [72]. Linear correlation between the shifts of $L_{\beta'} \rightarrow L_{\alpha}$ phase transition temperatures ($\Delta T_{\rm m}$) and the values of cationic radii was shown in our previous work [79]. It was noted that $\Delta T_{\rm m}$ changed the sign (from positive to negative) at a certain critical ionic radius which divides kosmotropic and chaotropic cations of IA subgroup. At the same time, for IB subgroup cations greater effects were observed under similar values of ionic radii, which was probably due to higher electron density [6, 80].

### 4.4.2 *Kinetics of Joint Drug-Membrane Interactions*

It would be interesting to consider the effects of kosmotrope and chaotrope substances on the phase transitions in model phospholipid membranes, choosing as typical examples dimethylsulfoxide (DMSO) and tilorone, respectively. Such studies using DSC were reported in [81, 82]. DMSO is known as a water-soluble membranotropic agent with concentration-dependent effects on the membrane, e.g. [83–90]. In our studies, we used relatively small concentration of DMSO to avoid any significant influence on thermotropic properties of DPPC membrane. Tilorone is an interferon-inducing antiviral and immunomodulatory agent [91–94], but, to our knowledge, its membranotropic effects have not been studied. We found that tilorone substantially decreases the membrane melting temperature $T_m$ when added to DPPC membrane, with the pre-transition peak disappearing.

Upon addition to the DPPC membrane at 2.5% wt/wt, tilorone and DMSO induce opposite effects on the melting temperature ($T_m$), which is decreased in the presence of tilorone and increased with DMSO. The effects of joint introduction of tilorone and DMSO on $T_m$ and $T_p$ (when measured as function of their mass ratio in the fixed total amount of the dopants) showed non-additivity. The extra decrease of the transition temperatures could be caused by either interaction of these substances in the membrane or by certain preference in sorption of tilorone on DPPC membrane as compared to DMSO. So, for better understanding of the mechanisms involved, kinetic studies of this system were carried out.

Several different schemes were used for joint introduction of DMSO and tilorone:

I. Sequential introduction, first tilorone, then DSC scans until constant peak profiles marking the equilibrium, then DMSO, and further scans;
II. The same, but first DMSO, then tilorone;
III. Simultaneous introduction of both substances.

The observed changes in the DSC profiles were, in fact, due to a number of processes, such as drug diffusion through the water phase, its interaction with and sorption onto lipid membranes, diffusion through the membrane, partitioning to interbilayer water phase, sorption on the next layers, etc.

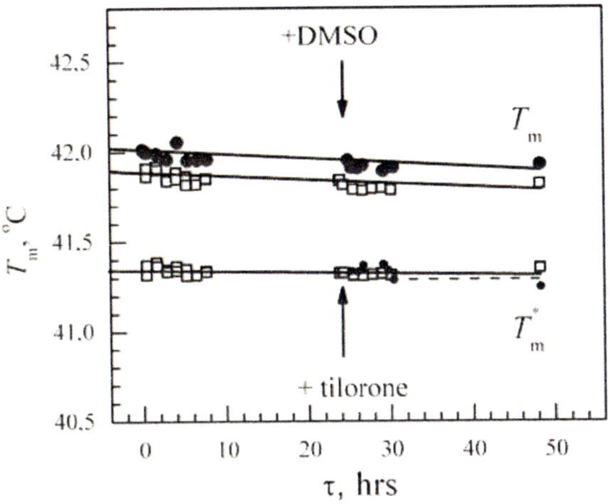


**Fig. 4.3** The temperatures of the deconvoluted peaks of DSC profiles of DPPC membranes doped with tilorone and DMSO according to Scheme I (open symbols) and Scheme II (solid symbols). Here, $T_m$ is the temperature of main phase transition peak of neat DPPC membrane, $T^*_m$ is the temperature of main phase transition peak of DPPC membrane saturated with tilorone, τ is exposition time

In Scheme I, the shape of the main transition peak was distorted after addition of tilorone, and Gaussian deconvolution reveals the appearance of low-temperature peak corresponding to DPPC membrane saturated with tilorone [81]. The temperatures of the deconvoluted peaks remained constant in time (Fig. 4.3).

For further analysis, we considered tilorone-saturated fraction of DPPC membrane, which was determined as the area of the low-temperature peak with respect to the total peak area. In Scheme I (Fig. 4.4, a), addition of DMSO after tilorone caused further increase in the tilorone-saturated fraction, indicating increased amount of lipids interacting with tilorone. In Scheme II (Fig. 4.4, b), DMSO addition before tilorone did not cause any noticeable changes of DSC peaks during 24 h, but subsequent tilorone introduction led to well-defined splitting of the melting peak in consecutive scans indicating that larger and larger amount of tilorone entered the membrane. The tilorone-saturated fraction increased during ~9 h, gradually reaching the equilibrium. Finally, the simultaneous introduction of both substances in Scheme III (Fig. 4.4, c) caused rather rapid increase in the tilorone-saturated fraction, but somewhat slower than in Scheme II.

Thus, kinetic studies allow discerning between sorption of drug molecules to the membrane and their diffusion through stacks of lipid bilayers and water interlayers, which is not possible under equilibrium conditions. It was shown that the joint action of tilorone and DMSO in lipid membranes is different when they are administered in different order. This effect results from the superposition of two mechanisms, namely, deceleration of DMSO sorption on the membrane due to the tilorone advantage in competitive sorption, and acceleration of trans-membrane diffusion of tilorone due to generation of inhomogeneity caused by DMSO. In other words, the effects of joint introduction of these two substances are not only non-additive. They appear also to be non-commutative (the order of introduction is important) and non-associative (introduction one-by-one is different from simultaneous introduction of both substances).

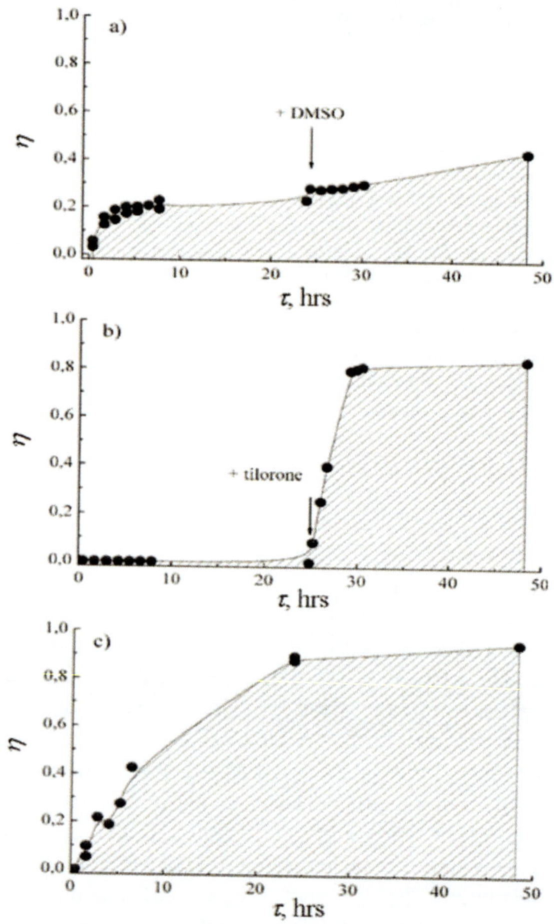


**Fig. 4.4** Variation of the fraction of tilorone-saturated lipid membrane, $\eta$, for different schemes of tilorone and DMSO introduction into DPPC model membrane: Scheme I (a), Scheme II (b), Scheme III (c). Here, $\tau$ is exposition time

The physical reason is the structural change induced in the lipid bilayer by the first introduced substance. Hence, the effects from the second substance are substantially modified. Basing on such approach, possible mechanisms can be considered describing how sorption of a dopant onto lipid membrane could affect the membrane permeability for other dopants. One of possible models was proposed in [82] in terms of dopant-induced membrane non-homogeneity.

## 4.5 Lipid Membranes Affected by External Factors: From Membranotropic Effects to Biomimetic Sensor Materials

### *4.5.1 Effects of Provitamin D Photo-Transformations on Thermodynamic Parameters of Model Lipid Membranes*

It is well known that in the organisms of humans and animals vitamin D (VitD) is formed from its natural precursor provitamin D (ProD) in the epidermal cell membranes. Under UV radiation ProD isomerises to previtamin D (PreD), which gives a set of photoisomers (tachysterol, lumisterol etc.) or transforms to VitD ("dark" stage) [95–97]. The ProD photoisomerisation process and efficiency of VitD formation are strongly affected by both spectral composition of UV radiation and the reaction medium [98–101]. In biological and model lipid membranes spatial constraints imposed on conformationally flexible PreD favour VitD synthesis, and the rate of "dark" reaction PreD → VitD in skin and liposomes is 10–15 times higher than that in isotropic solvents [102–106]. Therefore, lipid bilayer, serving as a reaction medium, predetermines high efficiency of VitD synthesis in vivo.

The membranotropic activity of VitD has been widely studied [107–114]. At high concentrations VitD increases the fluidity and decreases the order parameter of lipid membranes [108]. But there have been no studies on the impact of the ProD and its isomerisation products on lipid membrane. Studying membranotropic effect of the VitD biosynthesis may deepen our understanding of the mechanism of its impact at the molecular level. Therefore, we aimed to study the effects of ProD, the products of its photoisomerisation and VitD on the thermodynamic parameters of the model lipid membrane [115].

The multilayer structures of hydrated DPPC with VitD and ProD as dopants were prepared by standard procedures as in our previous works [79, 80]. ProD and VitD were used in the form of 7-dehydrocholesterol ($ProD_3$) and cholecalciferol ($VitD_3$) respectively. DSC thermograms of DPPC membranes with 7% wt/wt ProD before and after UV irradiation and with VitD were obtained. Both ProD and VitD decrease the main phase transition temperature ($T_m$), the corresponding peak is widened, and the pre-transition peak vanishes, indicating disordering of the lipid bilayer. In the Table 4.1 the thermodynamic parameters of the main phase transition of the DPPC membranes with dopants are shown.

These are the temperature ($T_m$), enthalpy ($\Delta H_m$), half-width of the melting peak ($\Delta T_{1/2}$), the cooperative number ($CN$), which was calculated according to [116] as

$$CN = \Delta H_{VH}/\Delta Hm \approx 7Tm^2/\Delta T_{1/2} \times \Delta Hm,$$

where $\Delta H_{VH}$ is the Van't Hoff enthalpy; $T_m$, $\Delta H_m$ and $\Delta T_{1/2}$ are the temperature, enthalpy and half-width of the main transition peak, respectively.

**Table 4.1** The effect of ProD, VitD as well as ProD UV irradiation and dark storage on the thermodynamic characteristics of the main phase transition of the DPPC membrane

| Dopant | Heating | | | Cooling | | CN |
|---|---|---|---|---|---|---|
| | $T_m$, °C | $\Delta H_m$, kJ/mol | $\Delta T_{1/2}$, °C | $T_m$, °C | $\Delta H_m$, kJ/mol | |
| no dopants | 41.4 ± 0.1 | 36.7 ± 2.6 | 1.1 ± 0.2 | 40.8 ± 0.1 | 37.4 ± 2.7 | 151 ± 39 |
| ProD | 39.9 ± 0.1 | 22.6 ± 1.9 | 2.5 ± 0.2 | 39.2 ± 0.1 | 40.7 ± 2.0 | 103 ± 17 |
| ProD + UV | 39.4 ± 0.1 | 23.8 ± 2.2 | 3.4 ± 0.2 | 38.8 ± 0.1 | 41.0 ± 2.2 | 71 ± 11 |
| ProD + UV + dark | 39.0 ± 0.1 | 23.5 ± 2.2 | 3.7 ± 0.2 | 38.5 ± 0.1 | 39.7 ± 2.3 | 67 ± 10 |
| VitD | 38.9 ± 0.1 | 27.0 ± 2.3 | 5.5 ± 0.2 | 38.0 ± 0.1 | 39.7 ± 2.4 | 38 ± 5 |

Let us consider in details the differences in the membranotropic effect of the additives (Fig. 4.5). The introduction of ProD leads to the $T_m$ decrease by 1.5 °C relative to the pure DPPC membrane. The UV irradiation of this system results in further $T_m$ decrease (by 0.5 °C more). The dark storage of the irradiated sample for 24 h resulted in further $T_m$ decline (by 0.4 °C more). The most significant shift of the $T_m$ value (as compared to the undoped DPPC membrane) was observed for VitD (value $T_m$ decreased by 2.5 °C). Introduction of ProD, UV irradiation and subsequent dark storage also caused noticeable changes in $T_m$. One should also note a clear $\Delta T_{1/2}$ increase and a marked decrease in *CN*. At the same time, there was no considerable effect on $\Delta H_m$ on cooling, while on heating all studied dopants decreased $\Delta H_m$ by ~30%. (Table 4.1).

Thus, during UV irradiation of the DPPC + ProD system and its further dark storage the $T_m$ shifts towards the values representative for the DPPC + VitD system. Such changes are accompanied by transformation of the UV spectrum of the system (Fig. 4.6).

The increase in the optical density of the DPPC + ProD system after UV irradiation reflects formation of ProD photoisomers including tachysterol [97, 98]. Subsequent

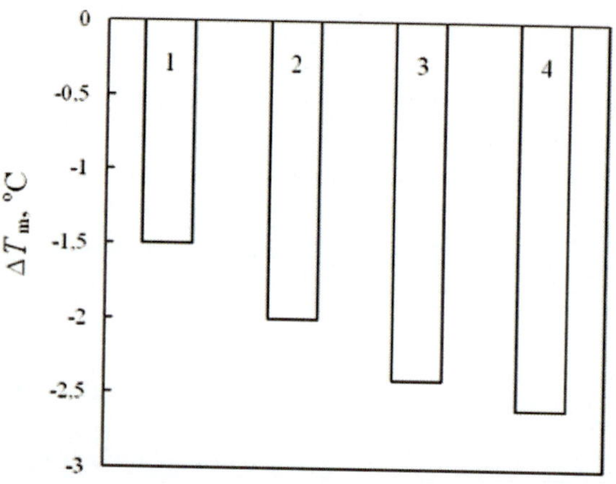


**Fig. 4.5** Shift of the melting temperature ($\Delta T_m$) of DPPC membrane with dopants: 1—ProD, 2—ProD + UV irradiation, 3—ProD + UV irradiation + dark storage, 4—VitD. $\Delta T_m$ was calculated as $\Delta T_m = T_m$ (DPPC + additive) – $T_m$ (DPPC)

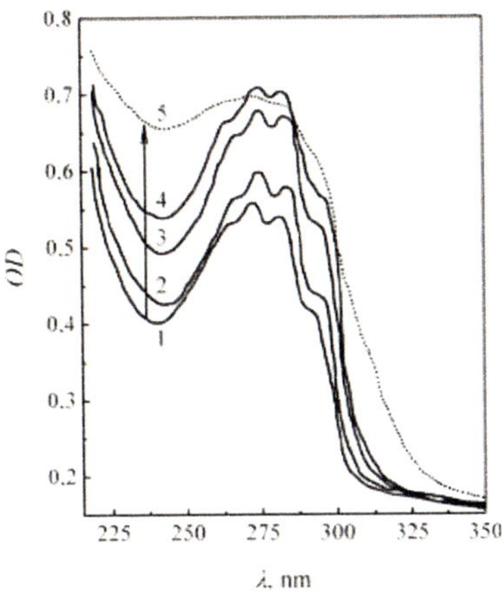


**Fig. 4.6** Changes of ProD absorption spectrum in the DPPC membrane (here, *OD* is the optical density, λ is the wavelength): 1—initial spectrum; 2—5 min UV irradiation; 3—10 min UV irradiation; 4—15 min UV irradiation; 5—subsequent dark storage for 24 h

dark storage of the irradiated sample for 24 h leads to spectrum broadening and an increase in the optical density at 265 nm (Fig. 4.6, curve 5), indicating VitD formation [98].

Thus, it was noted the regular decrease of the $T_m$ values of DPPC membranes with dopants in the sequence: ProD < ProD + UV < ProD + UV + dark storage < VitD. A possible explanation of this phenomenon is related to the formation under UV radiation of a multicomponent mixture of ProD isomers with different degrees of conformational flexibility and anisometry (so called secosteroids – PreD, tachysterol, VitD) [117] instead of one "rigid" ProD. It leads to additional blurring of the transition peak and more notable decrease of the $T_m$, as compared to the effect of ProD.

Thus, it was revealed that the destabilizing effect of VitD on the DPPC membrane is more pronounced than that of ProD and its photoisomers, the probable molecular mechanism is related to different degrees of conformational flexibility and anisometry of these molecules. It means that VitD biosynthesis in vivo leads to the local disordering of lipid membrane which facilitates VitD escape out of the membrane into the intracellular space.

### 4.5.2 *Provitamin D Photo-Isomerisation in Cholesteric Liquid Crystals: Development of Sensor Materials for Bioequivalent Dosimetry of UV Radiation*

In the previous section we demonstrated that phototransformation of UV sensitive molecule of biological origin (ProD) in model lipid membrane (which is an example of a lyotropic liquid crystal) results in the changes of lipid membrane properties. We propose to expand this idea to thermotropic liquid crystals and to study the effects of ProD photoisomerisation in cholesteric liquid crystals (CLC) and to develop a bioequivalent sensor material based on these effects. "Bioequivalent" means that unlike existing UV sensors, in this case the sensor response is based on UV-induced transformation of biologically relevant molecule ProD, which is a natural process in human skin [101–106].

We propose two variants of realization of sensor materials based on two types of effects in CLC:

1. Shifting of the maximum selective reflection wavelength ($\lambda_{max}$) of CLC;
2. Isothermal phase transition from smectic-A to cholesteric.

Both effects are caused by UV induced photoisomerisation of ProD. In addition, we will consider the corresponding effects during "light" and "dark" stages of ProD isomerisation.

It is well-known that CLC possess helical twisting which leads to the selective reflection of light with wavelength $\lambda_{max}$ proportional to helical pitch $p$: $\lambda_{max} = n{\cdot}p$ (where $n$ is the refractive index). It is expected that ProD isomerisation in CLC could be easily monitored instrumentally or in some cases visually by color change.

To make these effects clearly visible, one has to use CLC mixtures ensuring the $\lambda_{max}$ vs. temperature change that is the most suitable for each specific case. In our studies, the following CLC mixtures were used:

- M6: cholesteryl nonanoate 30%, cholesteryl caprinate 10%, cholesteryl caprylate 10%, 4-butylcyclohexanecarboxylic acid 25%, 4-hexylcyclohexanecarboxylic acid 25%;
- M8: cholesteryl formate 18%, cholesteryl butyrate 3%, cholesteryl nonanoate 39%, 4-butylcyclohexanecarboxylic acid 20%, 4-hexylcyclohexanecarboxylic acid 20%;
- M9: cholesteryl nonanoate 43.5%, cholesteryl valerate 27%, cholesteryl caprinate 23.5%, cholesteryl caprylate 6%;
- M15: cholesteryl formate 12%, cholesteryl butyrate 2%, cholesteryl nonanoate 26%, 4-butylcyclohexanecarboxylic acid 60%.

Figure 4.7 presents the effects of ProD, VitD (both 5% wt/wt) and UV irradiation on CLC (M8). It should be noted that ProD and VitD shifts $\lambda_{max}$ of CLC matrix (based on cholesterol esters) in the opposite directions because of the opposite sign on helical twisting power of the isomers [118–120].

UV irradiation of the ProD doped systems leads to $\lambda_{max}$ decreasing towards the values obtained as the effect of VitD introduction. During UV irradiation the $\lambda_{max}$ shift reaches ~100 nm, which can be visually observed as color change from orange to blue.

The main advantage of the sensor material based on ProD photoreaction is its selective sensitivity to biologically active UV radiation (UV-C and UV-B range), with no response to UV-A and visible light (Fig. 4.8) [121, 122]. Such selectivity is predetermined by ProD absorption spectrum [96–100].

Another variant of sensor material is based on different influence of ProD and VitD on the smectic A to cholesteric (SmA → Ch) phase transition temperature.

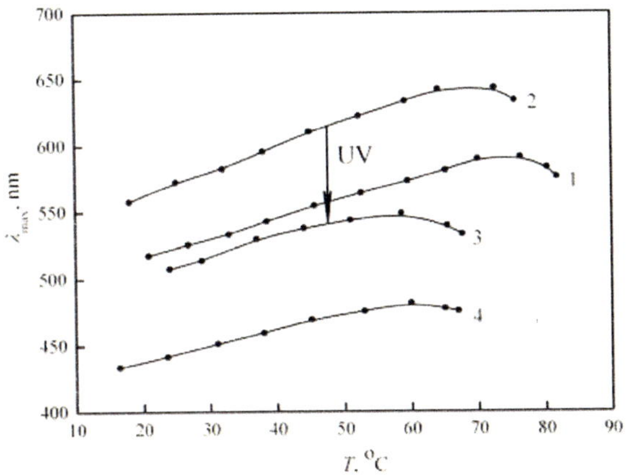


**Fig. 4.7** Effect of ProD, VitD and UV irradiation on the wavelength of maximum selective reflection ($\lambda_{max}$) of the CLC at different temperatures (*T*): 1—M8, 2—M8 + 5% ProD, 3—M8 + 5% ProD + UV irradiation, 4—M8 + 5% VitD

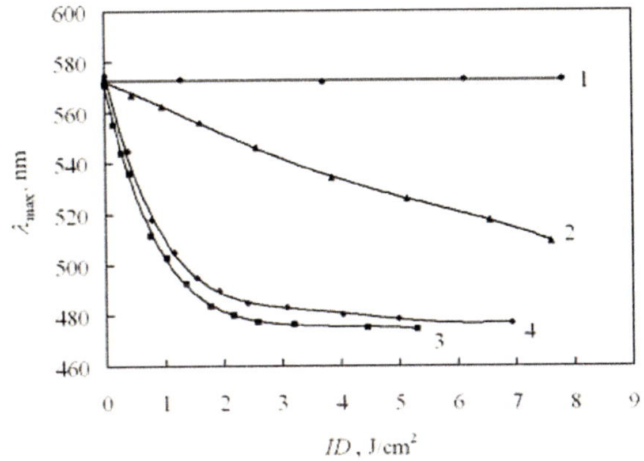


**Fig. 4.8** Dependence of the wavelength of maximum selective reflection ($\lambda_{max}$) of CLC M8 + 10% ProD on the irradiation dose (*ID*) and UV spectral composition controlled by the use of appropriate filters: 1—$\lambda >$ 320 nm (UV-A + visible); 2—$\lambda >$ 280 nm (UV-B); 3—260 $< \lambda <$ 380 nm (UV-C + UV-B); 4—$\lambda >$ 240 nm (without filter)

As described above, decreasing of the lipid membrane melting temperature is more pronounced in the case of VitD (and ProD photoisomers), rather than with ProD.

Qualitatively the same effects were shown in the thermotropic liquid crystals based on cholesterol esters for the cholesteric to isotropic and SmA → Ch phase transitions [123]. These are demonstrated in Fig. 4.9, where the temperature region of sharp unwinding of the cholesteric helix (which is a marker of approaching the Ch → SmA phase transition in cooling mode) for CLC + ProD system is dramatically decreased after UV irradiation. Optimizing the LC + ProD composition, a prospective sensor material was developed, with colorless SmA phase at room temperature (~20–25 °C) turning into Ch phase under UV irradiation, which leads to the appearance of visible colors (Fig. 4.9). It should be noted that at a fixed temperature, UV induced SmA → Ch phase transition will be realized after absorption of a certain minimum UV radiation dose. So, the proposed detection method of UV radiation has threshold character.

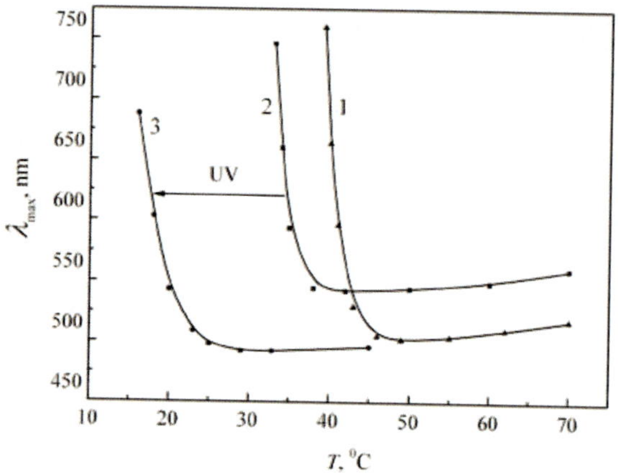


**Fig. 4.9** Color-temperature characteristics for CLC + 6% ProD (here, $\lambda_{max}$ is the wavelength of maximum selective reflection, $T$ is the temperature): 1—initial M6 (mixture based on cholesterol esters); 2—M6 + ProD; 3—M6 + ProD, after UV irradiation (0.68 J/cm$^2$)

The described above effects in LC (Figs. 4.7, 4.8, 4.9) are conditioned by "light" stage of ProD isomerisation process, i.e. ProD photoisomers formation during UV irradiation. The "dark" stage of the process, leading to VitD formation, could be also visualized by monitoring corresponding changes in LC matrix properties.

In Fig. 4.10, the UV induced $\lambda_{max}$ shifts after UV irradiation and dark storage are presented for CLC + ProD. The "dark" process constitutes ~25% relatively total $\lambda_{max}$ shift. For comparison, $\lambda_{max}$ changes during continuous UV irradiation are shown as dotted lines (curves 1, 3 in Fig. 4.10).

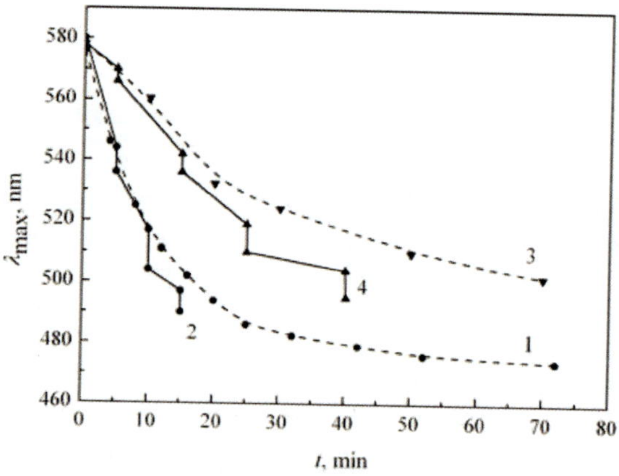


**Fig. 4.10.** Shifting of the wavelength of maximum selective reflection ($\lambda_{max}$) for CLC M15 (based on cholesterol esters) with 10% ProD under UV irradiation and dark storage (here, $t$ is irradiation time). UV dose rate was 0.88 mW/cm$^2$ (curves 1, 2) and 0.45 mW/cm$^2$ (curves 3, 4). Vertical segments on curves 2 and 4 correspond to $\lambda_{max}$ change after 1 day of dark storage. Dotted curves (1, 3) correspond to continuous UV irradiation

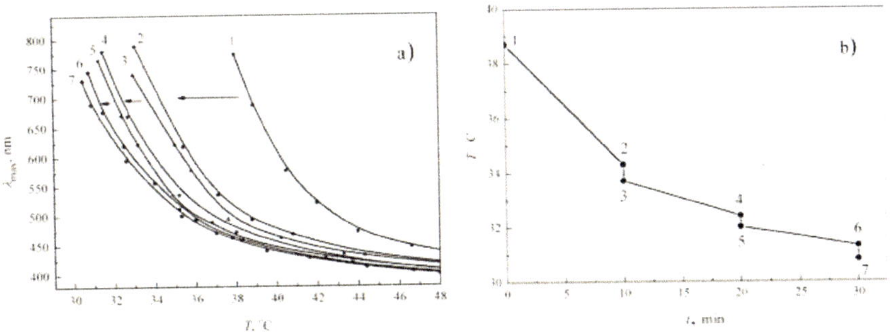


**Fig. 4.11.** Color-temperature characteristics of CLC M9 (based on cholesterol esters) with 5% ProD (here, $\lambda_{max}$ is the wavelength of maximum selective reflection, $T$ is the temperature) (a) and Ch → SmA phase transition temperature ($T$) (b): before UV irradiation (1), 10 min. UV irradiation (0.45 mW/cm$^2$) (2), +1 day dark storage (3), +10 min irradiation (4), +1 day dark storage (5), + 10 min irradiation (6), +1 day dark storage (7)

In Fig. 4.11a a color-temperature characteristics are shown for CLC + ProD at consecutive cycles of UV irradiation and dark storage. Corresponding gradual decrease of the temperature of SmA → Ch phase transition is shown in Fig. 4.11b. More pronounced changes of the phase transition temperature are observed after UV irradiation rather than dark storage, which gives about ~20% of total temperature decrease.

So, for both types of the LC systems (based the on the UV induced $\lambda_{max}$ change or on the SmA → Ch temperature shift) a possibility was shown to use them for detection of the accumulated dose of UV radiation in consecutive cycles of irradiation and dark storage. Their characterization as sensor materials for bioequivalent (or biomimetic) detectors is based on the use of UV-sensitive substance naturally present in the organism (human skin), and on similarity of the physical mechanisms of the response in thermotropic liquid crystals and model phospholipid membranes.

## 4.6 Final Remarks and Perspectives

In this review, we have considered model phospholipid membranes under two different aspects—as examples of lyotropic liquid crystals that can undergo phase transformations due to external factors, and as convenient tools for biophysical and pharmacological studies. The following conclusions can be made on the basis of our analysis:

1. Model phospholipid membranes can be recommended as adequate media at initial stages of pre-clinical testing of drug substances—both newly developed and the already known.
2. The effects of joint administration of different drugs, as well as combinations of active ingredients with different excipients, being of great importance in

pharmacology, can be preliminarily estimated in relatively simple experiments with the tested drug substances and their combinations used as dopants to the model membranes.

3. Since basic phase states and dopant-induced transitions between them are, from the physical viewpoint, essentially similar for different types of membrane lipids, this approach can also provide simulation of various natural cell membranes by appropriate choice of the model membrane composition.
4. Exposition of the membrane to the influence of certain external factors (e.g., ionizing or UV radiation) can result to chemical transformations of constituent molecules, which can formally be treated as introduction of new dopant substances.
5. In particular, formation of vitamin D in human organism occurs in epidermal cell membranes under UV irradiation. This process can be studied in model membrane systems by monitoring the emerging effects; this can be used in the development of bioequivalent (biomimetic) detectors of biologically active UV irradiation.

## References


1. Coster HGL (2003) The physics of cell membranes. J Biol Phys 29:363–399. https://doi.org/10.1023/A:1027362704125
2. Cullis PR, De Kruijff B (1979) Lipid polymorphism and the functional roles of lipids in biological membranes. BBA - Rev Biomembr 559:399–420. https://doi.org/10.1016/0304-4157(79)90012-1
3. Israelachvili JN, Marcelja S, Horn RG, Israelachvili JN (1980) Physical principles of membrane organization. Q Rev Biophys 13:121–200. https://doi.org/10.1017/S0033583500001645
4. Mavromoustakos TM (2007) The use of differential scanning calorimetry to study drug-membrane interactions. In: Dopico AM (ed) Methods in Membrane Lipids. Methods in Molecular Biology™, vol 400. Humana Press Inc., Totowa, pp 587–600
5. Avdeef A (2005) Physicochemical profiling (solubility, permeability and charge state). Curr Top Med Chem 1:277–351. https://doi.org/10.2174/1568026013395100
6. Lisetski LN, Vashchenko OV, Kasian NA, Krasnikova AO (2015) Liquid crystal ordering and nanostructuring in model lipid membranes. In: Karachevtsev VA (ed) Nanobiophysics: fundamentals and applications. Pan Stanford Publishing Singapore, pp 163–191
7. Bilginer R, Arslan Yildiz A (2019) Biomimetic model membranes as drug screening platform. In: Kök F, Arslan YA, Inci F (eds) Biomimetic lipid membranes: fundamentals, applications, and commercialization. Springer, Cham, pp 225–247
8. Mouritsen OG, Bagatolli LA (2016) Life - As a Matter of Fat. Lipids in a Membrane Biophysics Perspective. Springer, Heidelberg
9. Seydel JK, Wiese M (2003) Drug-Membrane Interactions: Analysis, Drug Distribution, Modeling. Wiley-VCH, Weinheim, Germany
10. Wack DC, Webb WW (1989) Synchrotron x-ray study of the modulated lamellar phase P in the lecithin-water system. Phys Rev A 40:2712–2730. https://doi.org/10.1103/PhysRevA.40.2712
11. Koynova R, Tenchov B (2013) Transitions between lamellar and nonlamellar phases in membrane lipids and their physiological roles. OA Biochem 1:1–9. https://doi.org/10.13172/2052-9651-1-1-602

12. Koynova R, Caffrey M (1998) Phases and phase transitions of the phosphatidylcholines. Biochim Biophys Acta (BBA)-Reviews Biomembr 1376:91–145
13. Quinn PJ, Joo F, Vigh L (1989) The role of unsaturated lipids in membrane structure and stability. Prog Biophys Mol Biol 53:71–103. https://doi.org/10.1016/0079-6107(89)90015-1
14. Róg T, Murzyn K, Gurbiel R, Takaoka Y, Kusumi A, Pasenkiewicz-Gierula M (2004) Effects of phospholipid unsaturation on the bilayer nonpolar region. J Lipid Res 45:326–336. https://doi.org/10.1194/jlr.M300187-JLR200
15. Iwamoto K, Ohnuki Y, Sawada K, Seno M (1981) Solid-solid phase transitions of long-chain n-alkyltrimethylammonium halides. Mol Cryst Liq Cryst 73:95–103. https://doi.org/10.1080/00268948108076264
16. Alekseev OM, Lazarenko MM, Puchkovs'ka GO, Bezrodnaya TV, Sendzyuk AA, (2010) Peculiarities of the thermal motion in crystals formed by cetyltrimethylammonium bromide molecules. Ukr J Phys 55:973–979
17. Nagle JF, Tristram-Nagle S (2000) Structure of lipid bilayers. Biochim Biophys Acta (BBA)-reviews Biomembr 1469:159–195. https://doi.org/10.1016/S0304-4157(00)00016-2
18. Pikin SA, Kharakoz DP, Tiktopulo LI, Pikina ES (2005) Molecular aspects of the main phase transition in lipid systems as a weak first-order phase transition: 1. Model of thermodynamic behavior of lipid membranes. Crystallogr Reports 50:308–315. https://doi.org/10.1134/1.1887906
19. Jain MK, Wu NM (1977) Effect of small molecules on the dipalmitoyl lecithin liposomal bilayer: III. Phase transition in lipid bilayer. J Membr Biol 34:157–201. https://doi.org/10.1007/BF01870299
20. Flaten GE, Skar M, Luthman K, Brandl M (2007) Drug permeability across a phospholipid vesicle based barrier: 3. Characterization of drug-membrane interactions and the effect of agitation on the barrier integrity and on the permeability. Eur J Pharm Sci 30:324–332. https://doi.org/10.1016/j.ejps.2006.11.017
21. Peetla C, Stine A, Labhasetwar V (2009) Biophysical interactions with model lipid membranes: applications in drug discovery and drug delivery. Mol Pharm 6:1264–1276. https://doi.org/10.1021/mp9000662
22. Lucio M, Lima JLFC, Reis S (2010) Drug-membrane interactions: significance for medicinal chemistry. Curr Med Chem 17:1795–1809. https://doi.org/10.2174/092986710791111233
23. Triggle DJ (2003) Drug−membrane interactions. analysis, drug distribution, modeling. methods and principles in medicinal chemistry, Volume 15 By Seydel JK, Wiese M (2002) Wiley-VCH, Weinheim, Germany. 2002. xix + 349 pp 17.5 × 24.5 cm. ISBN 3 527 3042 7. $128.00. J Med Chem 46. https://doi.org/10.1021/jm0204688
24. Casey D, Charalambous K, Gee A, Law RV, Ces O (2014) Amphiphilic drug interactions with model cellular membranes are influenced by lipid chain-melting temperature. J R Soc Interface 11:20131062. https://doi.org/10.1098/rsif.2013.1062
25. Seddon AM, Casey D, Law RV, Gee A, Templer RH, Ces O (2009) Drug interactions with lipid membranes. Chem Soc Rev 38:2509. https://doi.org/10.1039/b813853m
26. Orme M (1984) Drug absorption in the gut. Br J Anaesth 56:59–67. https://doi.org/10.1093/bja/56.1.59
27. Flaten GE, Dhanikula AB, Luthman K, Brandl M (2006) Drug permeability across a phospholipid vesicle based barrier: a novel approach for studying passive diffusion. Eur J Pharm Sci 27:80–90. https://doi.org/10.1016/j.ejps.2005.08.007
28. Hauet N, Artzner F, Boucher F, Grabielle-Madelmont C, Cloutier I, Keller G, Lesieur P, Durand D, Paternostre M (2003) Interaction between artificial membranes and enflurane, a general volatile anesthetic: DPPC-enflurane interaction. Biophys J 84:3123–3137. https://doi.org/10.1016/S0006-3495(03)70037-X
29. Seeger HM, Gudmundsson ML, Heimburg T (2007) How anesthetics, neurotransmitters, and antibiotics influence the relaxation processes in lipid membranes. J Phys Chem B 111:13858–13866. https://doi.org/10.1021/jp075346b
30. Cantor RS (1998) The lateral pressure profile in membranes: a physical mechanism of general anesthesia. Toxicol Lett 100:451–458. https://doi.org/10.1016/S0378-4274(98)00220-3

31. Pereira-Leite C, Nunes C, Reis S (2013) Interaction of nonsteroidal anti-inflammatory drugs with membranes: In vitro assessment and relevance for their biological actions. Prog Lipid Res 52:571–584. https://doi.org/10.1016/j.plipres.2013.08.003
32. Boggara MB, Mihailescu M, Krishnamoorti R (2012) Structural association of nonsteroidal anti-inflammatory drugs with lipid membranes. J Am Chem Soc 134:19669–19676. https://doi.org/10.1021/ja3064342
33. Chimote G, Banerjee R (2008) Evaluation of antitubercular drug insertion into preformed dipalmitoylphosphatidylcholine monolayers. Colloids Surfaces B Biointerfaces 62:258–264. https://doi.org/10.1016/j.colsurfb.2007.10.010
34. Kopec W, Khandelia H (2014) Reinforcing the membrane-mediated mechanism of action of the anti-tuberculosis candidate drug thioridazine with molecular simulations. J Comput Aided Mol Des 28:123–134. https://doi.org/10.1007/s10822-014-9737-z
35. Kapoor R, Peyear TA, Koeppe RE, Andersen OS (2019) Antidepressants are modifiers of lipid bilayer properties. J Gen Physiol 151:342–356. https://doi.org/10.1085/jgp.201812263
36. Pignatello R (2013) Drug-biomembrane interaction studies: the application of calorimetric techniques. Woodhead Publishing Books - Elsevier, Oxvord
37. Nademi Y, Amjad Iranagh S, Yousefpour A, Mousavi SZ, Modarress H (2014) Molecular dynamics simulations and free energy profile of Paracetamol in DPPC and DMPC lipid bilayers. J Chem Sci 126:637–647. https://doi.org/10.1007/s12039-013-0556-x
38. Dalhoff AA (2018) Membrane Interactions of Antibacterial Agents. Trends Clin Microbiol 1:4–48
39. Dynarowicz-Łątka P, Seoane R, Miñones J, Velo M, Miñones J (2003) Study of penetration of amphotericin B into cholesterol or ergosterol containing dipalmitoyl phosphatidylcholine Langmuir monolayers. Colloids Surfaces B Biointerfaces 27/3:249–263. https://doi.org/10.1016/S0927-7765(02)00099-1
40. Van Oosten B, Marquardt D, Komljenović I, Bradshaw JP, Sternin E, Harroun TA (2014) Small molecule interaction with lipid bilayers: a molecular dynamics study of chlorhexidine. J Mol Graph Model 48:96–104. https://doi.org/10.1016/j.jmgm.2013.12.007
41. Goldstein DB (1984) The effects of drugs on membrane fluidity. Annu Rev Pharmacol Toxicol 24:43–64. https://doi.org/10.1146/annurev.pa.24.040184.000355
42. Schreier S, Malheiros SVP, De Paula E (2000) Surface active drugs: Self-association and interaction with membranes and surfactants. Physicochemical and biological aspects. Biochim Biophys Acta - Biomembr 1508:210–234. https://doi.org/10.1016/S0304-4157(00)00012-5
43. Pignatello R, Musumeci T, Basile L, Carbone C, Puglisi G (2011) Biomembrane models and drug-biomembrane interaction studies: Involvement in drug design and development. J Pharm Bioallied Sci 3:4–14. https://doi.org/10.4103/0975-7406.76461
44. Wohl CJ, Helms MA, Chung JO, Kuciauskas D (2006) Phospholipid bilayer free volume analysis employing the thermal ring-closing reaction of merocyanine molecular switches. J Phys Chem B 110:22796–22803. https://doi.org/10.1021/jp065406y
45. Flaten GE, Luthman K, Vasskog T, Brandl M (2008) Drug permeability across a phospholipid vesicle-based barrier. 4. The effect of tensides, co-solvents and pH changes on barrier integrity and on drug permeability. Eur J Pharm Sci 34:173–180. https://doi.org/10.1016/j.ejps.2008.04.001
46. Jackson K, Young D, Pant S (2000) Drug – excipient interactions and their affect on absorption is then absorbed by passive diffusion into the. Pharm Sci Technolo Today 3:336–345. https://doi.org/10.1016/s1461-5347(00)00301-1
47. Nitsche JM, Kasting GB (2016) A universal correlation predicts permeability coefficients of fluid- and gel-phase phospholipid and phospholipid-cholesterol bilayers for arbitrary solutes. J Pharm Sci 105:1762–1771. https://doi.org/10.1016/j.xphs.2016.02.012
48. Dahan A, Miller JM, Amidon GL (2009) Prediction of solubility and permeability class membership: provisional BCS classification of the world's top oral drugs. AAPS J 11:740–746
49. Gupta S, Kesarla R, Omri A (2013) Formulation strategies to improve the bioavailability of poorly absorbed drugs with special emphasis on self-emulsifying systems. Int Sch Res Not 2013:848043 (1–16). https://doi.org/10.1155/2013/848043

50. Kasian N, Vashchenko O, Budianska L, Brodskii R, Lisetski L (2019) Cooperative domains in lipid membranes: size determination by calorimetry. J Therm Anal Calorim 136:795–801. https://doi.org/10.1007/s10973-018-7695-8
51. Koukoulitsa C, Durdagi S, Siapi E, Villalonga-Barber C, Alexi X, Steele BR, Micha-Screttas M, Alexis MN, Tsantili-Kakoulidou A, Mavromoustakos T (2011) Comparison of thermal effects of stilbenoid analogs in lipid bilayers using differential scanning calorimetry and molecular dynamics: correlation of thermal effects and topographical position with antioxidant activity. Eur Biophys J 40:865–875. https://doi.org/10.1007/s00249-011-0705-4
52. Mannock DA, Lewis RNAH, McMullen TPW, McElhaney RN (2010) The effect of variations in phospholipid and sterol structure on the nature of lipid–sterol interactions in lipid bilayer model membranes. Chem Phys Lipids 163:403–448. https://doi.org/10.1016/j.chemphyslip.2010.03.011
53. Martini MF, Pickholz M (2012) Molecular dynamics study of uncharged bupivacaine enantiomers in phospholipid bilayers. Int J Quantum Chem 112:3341–3345. https://doi.org/10.1002/qua.24208
54. Caruso B, Sánchez JM, García DA, de Paula E, Perillo MA (2013) Probing the combined effect of flunitrazepam and lidocaine on the stability and organization of bilayer lipid membranes. A differential scanning calorimetry and dynamic light scattering study. Cell Biochem Biophys 66:461–475. https://doi.org/10.1007/s12013-012-9494-3
55. Kyrikou I, Daliani I, Mavromoustakos T, Maswadeh H, Demetzos C, Hatziantoniou S, Giatrellis S, Nounesis G (2004) The modulation of thermal properties of vinblastine by cholesterol in membrane bilayers. Biochim Biophys Acta - Biomembr 1661:222–223. https://doi.org/10.1016/j.bbamem.2003.11.021
56. Ricci M, Oliva R, Del Vecchio P, Paolantoni M, Morresi A, Sassi P (2016) DMSO-induced perturbation of thermotropic properties of cholesterol-containing DPPC liposomes. Biochim Biophys Acta - Biomembr 1858:3024–3031. https://doi.org/10.1016/j.bbamem.2016.09.012
57. Severcan F, Okan Durmus H, Eker F, Akinoglu BG, Haris PI (2000) Vitamin D2 modulates melittin - Membrane interactions. Talanta 53:205–211. https://doi.org/10.1016/S0039-9140(00)00453-7
58. Pashynska V, Stepanian S, Gömöry A, Vekey K, Adamowicz L (2015) Competing intermolecular interactions of artemisinin-type agents and aspirin with membrane phospholipids: Combined model mass spectrometry and quantum-chemical study. Chem Phys 455:81–87. https://doi.org/10.1016/j.chemphys.2015.04.014
59. Pashynska V, Stepanian S, Gömöry Á, Vékey K, Adamowicz L (2017) New cardioprotective agent flokalin and its supramolecular complexes with target amino acids: an integrated mass-spectrometry and quantum-chemical study. J Mol Struct 1146:441–449. https://doi.org/10.1016/j.molstruc.2017.06.007
60. Pashynska V, Stepanian S, Gömöry Á, Adamowicz L (2021) What are molecular effects of co-administering vitamin C with artemisinin-type antimalarials? A model mass spectrometry and quantum chemical study. J Mol Struct 1232:130039. https://doi.org/10.1016/j.molstruc.2021.130039
61. Vashchenko O, Pashynska V, Kosevich M, Panikarska V, Lisetski L (2011) Lyotropic mesophase of hydrated phospholipids as model medium for studies of antimicrobial agents activity. Mol Cryst Liq Cryst 547:155[1845]–163[1853] . https://doi.org/10.1080/15421406.2011.572038
62. Kasian NA, Pashynska VA, Vashchenko OV, Krasnikova AO, Gömöry A, Kosevich MV, Lisetski LN (2014) Probing of the combined effect of bisquaternary ammonium antimicrobial agents and acetylsalicylic acid on model phospholipid membranes: Differential scanning calorimetry and mass spectrometry studies. Mol Biosyst 10:3155–3162. https://doi.org/10.1039/c4mb00420e
63. Vashchenko OV, Kasian NA, Budianska LV (2018) Comparative effects of stearic acid, calcium and magnesium stearates as dopants in model lipid membranes. Funct Mater 25:300–307 . https://doi.org/10.15407/fm25.02.300

64. Vashchenko OV, Shishkina SV, Kasian NA, Budianska LV, Bulavin LA, Soloviov DV, Pashynska VA, Gomory A, Lisetski LN (2019) Formation of antibiotic cycloserine complexes with stearic acid and its calcium and magnesium salts: From quantum mechanical modeling to studies of membranotropic action. Funct Mater 26:673–684 . https://doi.org/10.15407/fm26.04.673
65. Ostroumova OS, Efimova SS, Mikhailova EV, Schagina LV (2014) The interaction of dipole modifiers with amphotericin-ergosterol complexes. Effects of phospholipid and sphingolipid membrane composition. Eur Biophys J 43:207–215. https://doi.org/10.1007/s00249-014-0946-0
66. Grage SL, Afonin S, Kara S, Buth G, Ulrich AS (2016) Membrane thinning and thickening induced by membrane-active amphipathic peptides. Front Cell Dev Biol 4:65 (1–13) . https://doi.org/10.3389/fcell.2016.00065
67. Collins KD (2006) Ion hydration: Implications for cellular function, polyelectrolytes, and protein crystallization. Biophys Chem 119:271–281. https://doi.org/10.1016/j.bpc.2005.08.010
68. Moelbert S, Normand B, De Los RP (2004) Kosmotropes and chaotropes: Modelling preferential exclusion, binding and aggregate stability. Biophys Chem 112:45–57. https://doi.org/10.1016/j.bpc.2004.06.012
69. Russo D (2008) The impact of kosmotropes and chaotropes on bulk and hydration shell water dynamics in a model peptide solution. Chem Phys 345:200–211. https://doi.org/10.1016/j.chemphys.2007.08.001
70. De Xammar Oro JR (2001) Role of co-solute in biomolecular stability: Glucose, urea and the water structure. J Biol Phys 27:73–79. https://doi.org/10.1023/A:1011890506834
71. Leontidis E, Christoforou M, Georgiou C, Delclos T (2014) The ion–lipid battle for hydration water and interfacial sites at soft-matter interfaces. Curr Opin Colloid Interface Sci 19:2–8. https://doi.org/10.1016/j.cocis.2014.02.003
72. Koynova R, Brankov J, Tenchov B (1997) Modulation of lipid phase behavior by kosmotropic and chaotropic solutes. Experiment and thermodynamic theory. Eur Biophys J 25:261–274. https://doi.org/10.1007/s002490050038
73. Vashchenko OV, Kasian NA, Budianska LV, Brodskii RY, Bespalova II, Lisetski LN (2019) Adsorption of ions on model phospholipid membranes. J Mol Liq 275:173–177. https://doi.org/10.1016/j.molliq.2018.11.053
74. Pennington ER, Day C, Parker JM, Barker M, Kennedy A (2016) Thermodynamics of interaction between carbohydrates and unilamellar dipalmitoyl phosphatidylcholine membranes: Evidence of dehydration and interdigitation. J Therm Anal Calorim 123:2611–2617. https://doi.org/10.1007/s10973-016-5288-y
75. Zimmermann R, Küttner D, Renner L, Kaufmann M, Werner C (2012) Fluidity modulation of phospholipid bilayers by electrolyte ions: Insights from fluorescence microscopy and microslit electrokinetic experiments. J Phys Chem A 116:6519–6525. https://doi.org/10.1021/jp212364q
76. Wang C, Ge Y, Mortensen J, Westh P (2011) Interaction free energies of eight sodium salts and a phosphatidylcholine membrane. J Phys Chem B 115:9955–9961. https://doi.org/10.1021/jp112203p
77. Ohtake S, Schebor C, Palecek SP, De Pablo JJ (2005) Phase behavior of freeze-dried phospholipid-cholesterol mixtures stabilized with trehalose. Biochim Biophys Acta - Biomembr 1713:57–64. https://doi.org/10.1016/j.bbamem.2005.05.001
78. Ohtake S, Schebor C, de Pablo JJ (2006) Effects of trehalose on the phase behavior of DPPC-cholesterol unilamellar vesicles. Biochim Biophys Acta - Biomembr 1758:65–73. https://doi.org/10.1016/j.bbamem.2006.01.002
79. Vashchenko OV, Ermak YL, Lisetski LN (2013) Univalent ions in phospholipid model membranes: Thermodynamic and hydration aspects. Biophys (Russian Fed) 58:515–523. https://doi.org/10.1134/S0006350913040180
80. Vashchenko OV, Iermak IL, Krasnikova AO, Lisetski LN (2015) The effects of silver nitrate on the phase state of model multibilayer membranes. Biophys (Russian Fed 60:244–250 . https://doi.org/10.1134/S0006350915020207

81. Vashchenko OV, Kasian NA, Brodskii RY, Budianska LV, Sofronov DS, Lisetski LN (2018) Model lipid bilayers as sensor bionanomaterials for characterization of membranotropic action of water-soluble substances. Funct Mater 25:422–431. https://doi.org/10.15407/fm25.03.422
82. Kasian NA, Vashchenko OV, Budianska LV, Brodskii RY, Lisetski LN (2019) Thermodynamics and kinetics of joint action of antiviral agent tilorone and DMSO on model lipid membranes. Biochim Biophys Acta - Biomembr 1861:123–129. https://doi.org/10.1016/j.bbamem.2018.08.007
83. Gordeliy VI, Kiselev MA, Lesieur P, Pole AV, Teixeira J (1998) Lipid membrane structure and interactions in dimethyl sulfoxide/water mixtures. Biophys J 75:2343–2351. https://doi.org/10.1016/S0006-3495(98)77678-7
84. Westh P (2004) Preferential interaction of dimethyl sulfoxide and phosphatidyl choline membranes. Biochim Biophys Acta - Biomembr 1664:217–223. https://doi.org/10.1016/j.bbamem.2004.06.001
85. Kiselev MA, Gutberlet T, Lesieur P, Hauss T, Ollivon M, Neubert RHH (2005) Properties of ternary phospholipid/dimethyl sulfoxide/water systems at low temperatures. Chem Phys Lipids 133:181–193. https://doi.org/10.1016/j.chemphyslip.2004.10.002
86. Kiselev MA (2007) Combined application of neutron and synchrotron radiation for investigation of the influence of dimethyl sulfoxide on the structure and properties of the dipalmitoylphosphatidylcholine membrane. Crystallogr Rep 52:529–534. https://doi.org/10.1134/S1063774507030352
87. Notman R, Den Otter WK, Noro MG, Briels WJ, Anwar J (2007) The permeability enhancing mechanism of DMSO in ceramide bilayers simulated by molecular dynamics. Biophys J 93:2056–2068. https://doi.org/10.1529/biophysj.107.104703
88. De Menorval MA, Mir LM, Fernández ML, Reigada R (2012) Effects of dimethyl sulfoxide in cholesterol-containing lipid membranes: A comparative study of experiments in silico and with cells. PLoS One 7:e41733. https://doi.org/10.1371/journal.pone.0041733
89. He F, Liu W, Zheng S, Zhou L, Ye B, Qi Z (2012) Ion transport through dimethyl sulfoxide (DMSO) induced transient water pores in cell membranes. Mol Membr Biol 29:107–113. https://doi.org/10.3109/09687688.2012.687460
90. Cheng CY, Song J, Pas J, Meijer LHH, Han S (2015) DMSO induces dehydration near lipid membrane surfaces. Biophys J 109:330–339. https://doi.org/10.1016/j.bpj.2015.06.011
91. Stringfellow DA, Glasgow LA (1972) Tilorone hydrochloride: an oral interferon-inducing agent. Antimicrob Agents Chemother 2:73–78. https://doi.org/10.1128/AAC.2.2.73
92. Krueger RF, Mayer GD (1970) Tilorone hydrochloride: An orally active antiviral agent. Science (80- ) 169:1213–1214 . https://doi.org/10.1126/science.169.3951.1213
93. Chandra P, Wright GJ (1977) Tilorone hydrochloride: the drug profile. Medicinal Chemistry. Springer, Heidelberg, pp 125–148
94. Ekins S, Lingerfelt MA, Comer JE, Freiberg AN, Mirsalis JC, O'Loughlin K, Harutyunyan A, McFarlane C, Green CE, Madrid PB (2018) Efficacy of tilorone dihydrochloride against ebola virus infection. Antimicrob Agents Chemother 62:e01711-e1717. https://doi.org/10.1128/AAC.01711-17
95. Holick MF, Maclaughlin JA, Clark MB, Holick SA, Potts JT, Anderson RR, Blank IH, Parrish JA, Elias P (1980) Photosynthesis of previtamin D3 in human skin and the physiologic consequences. Science (80- ) 210(4466):203–205. https://doi.org/10.1126/science.6251551
96. Verloop A, Koevoet AL, Havinga E (1955) Studies on vitamin D and related compounds III. Short communication on the cis-trans isomerization of calciferol and the properties of "trans"-vitamin D2. Recl des Trav Chim des Pays-Bas 74:1125–1130. https://doi.org/10.1002/recl.19550740911
97. Jacobs HJC, Havinga E (2007) Photochemistry of Vitamin D and its Isomers and of Simple Trienes. In: Jr JNP, Hammond GS, Gollnick K, Grosjean D (eds) Advances in Photochemistry. John Wiley & Sons, Ltd, pp 305–373
98. Fuß W, Lochbrunner S (1997) The wavelength dependence of the photochemistry of previtamin D. J Photochem Photobiol A Chem 105:159–164. https://doi.org/10.1016/S1010-6030(96)04608-4

99. Dmitrenko O, Frederick JH, Reischl W (2001) Previtamin D conformations and the wavelength-dependent photoconversions of previtamin D. J Photochem Photobiol A Chem 139:125–131. https://doi.org/10.1016/S1010-6030(00)00425-1
100. Terenetskaya IP, Bogoslovskii NA, Vysotskii LN, Luknitskii FI (1994) Routes to optimization of previtamin D photosynthesis using irradiation by a sunlamp. Pharm Chem J 28:589–596. https://doi.org/10.1007/BF02219037
101. Maclaughlin JA, Anderson RR, Holick MF (1982) Spectral character of sunlight modulates photosynthesis of previtamin D3 and its photoisomers in human skin. Science (80- ) 216(4549):1001–1003 . https://doi.org/10.1126/science.6281884
102. Tian XQ, Holick MF (1999) A Liposomal Model That Mimics the Cutaneous Production of Vitamin D3: studies of the mechanism of the membrane-enhanced thermal isomerization of previtamin D3 to vitamin D3. J Biol Chem 274:4174–4179. https://doi.org/10.1074/jbc.274.7.4174
103. Moriarty RM, Schwartz RN, Lee C, Curtis V (1980) Formation of vitamin D3 in synthetic lipid multibilayers. A model for epidermal photosynthesis. J Am Chem Soc 102:4257–4259. https://doi.org/10.1021/ja00532a047
104. Borch RF, Yamamoto JK (1985) Photoconversion of 7-dehydrocholesterol to vitamin D3 in synthetic phospholipid bilayers. Biochemistry 24:3338–3344. https://doi.org/10.1021/bi00334a039
105. Holick MF, Tian XQ, Allen M (1995) Evolutionary importance for the membrane enhancement of the production of vitamin D3 in the skin of poikilothermic animals. Proc Natl Acad Sci U S A 92:3124–3126. https://doi.org/10.1073/pnas.92.8.3124
106. Tian XQ, Chen TC, Matsuoka LY, Wortsman J, Holick MF (1993) Kinetic and thermodynamic studies of the conversion of previtamin D3 to vitamin D3 in human skin. J Biol Chem 268:14888–14892. https://doi.org/10.1016/s0021-9258(18)82416-4
107. Kazanci N, Toyran N, Haris PI, Severcan F (2001) Vitamin D2 at high and low concentrations exert opposing effects on molecular order and dynamics of dipalmitoyl phosphatidylcholine membranes. Spectroscopy 15:47–55. https://doi.org/10.1155/2001/890975
108. Tolosa de Talamoni N, Morero R, Canas F (1989) Vitamin D3 administration increases the membrane fluidity of intestinal mitochondria. Biochem Int 19:701–707
109. Castelli F, Gurrieri S, Raudino A, Cambria A (1988) Effect of cholecalcipherol on thermotropic behaviour of phosphatidylethanolamine and its N-methyl derivatives. Chem Phys Lipids 48:69–76. https://doi.org/10.1016/0009-3084(88)90133-8
110. Toyran N, Severcan F (2003) Competitive effect of vitamin D2 and Ca2+ on phospholipid model membranes: An FTIR study. Chem Phys Lipids 123:165–176. https://doi.org/10.1016/S0009-3084(02)00194-9
111. Toyran N, Severcan F (2000) The effect of magnesium ions on vitamin D2-phospholipid model membrane interactions in the presence of different buffer media. Talanta 53:23–27. https://doi.org/10.1016/S0039-9140(00)00378-7
112. Eker F, Durmus HO, Akinoglu BG, Severcan F (1999) Application of turbidity technique on peptide-lipid and drug-lipid interactions. J Mol Struct 482:693–697. https://doi.org/10.1016/S0022-2860(98)00690-5
113. Bondar OP, Rowe ES (1995) Differential scanning calorimetric study of the effect of vitamin D3 on the thermotropic phase behavior of lipids model systems. BBA - Biomembr 1240:125–132. https://doi.org/10.1016/0005-2736(95)00182-4
114. Merz K, Sternberg B (1994) Incorporation of vitamin d3-derivatives in liposomes of different lipid types. J Drug Target 2:411–417. https://doi.org/10.3109/10611869408996817
115. Kasian NA, Vashchenko OV, Gluhova YE, Lisetski LN (2012) Effect of the vitamin D photosynthesis products on thermodynamic parameters of model lipid membranes. Biopolym Cell 28:114–120. https://doi.org/10.7124/bc.000035
116. Lewis RNAH, McElhaney RN (2012) The Mesomorphic phase behavior of lipid bilayers. In: Yeagle PL (ed) The Structure of Biological Membranes: Third Edition. CRC Press, Taylor & Francis Group, Boca Raton, FL, USA, pp 19–90

117. Okamura WH, Midland MM, Hammond MW, Abd. Rahman N, Dormanen MC, Nemere I, Norman AW, (1995) Chemistry and conformation of vitamin D molecules. J Steroid Biochem Mol Biol 53:603–613. https://doi.org/10.1016/0960-0760(95)00107-B
118. Chilaya G, Petriashvili G, Chanishvili A, Terenetskaya I, Kireeva N, Lisetski L (2005) Provitamin D2 and provitamin D3 photo transformations in cholesteric liquid crystal mixtures induced by UV radiation. Mol Cryst Liq Cryst 433:73–85. https://doi.org/10.1080/15421400590956243
119. Gvozdovskyy I, Lisetski L (2007) Rotation of single crystals of chiral dopants at the top of a nematic droplet: Factors affecting the angular velocity. Mol Cryst Liq Cryst 475:113–122. https://doi.org/10.1080/15421400701681331
120. Gvozdovskyy I, Terenetskaya I (2002) Steroid motor: dynamics of cholesteric helix induction in the nematic droplet. Liq Cryst Today 11:8–12. https://doi.org/10.1080/146451802100006824
121. Lisetski LN, Panikarskaya VD, Kasyan NA, Grishchenko LV, Terenetskaya IP (2006) Bioequivalent UV detectors based on cholesteric liquid crystals: Effects of spectral composition and quantitative account for intensity of UV radiation. In: Proceedings SPIE 6023, Tenth International Conference on Nonlinear Optics of Liquid and Photorefractive Crystals, p 60230F
122. Kireyeva N, Panikarska V, Lisetski L, Terenetskaya I (2004) On a possibility to use photoinduced transformations of Provitamin D in cholesteric solvent for bioequivalent UV dosimetry. Biophys Bull 637:118–122
123. Zavora L, Kasian N, Lisetski L, Panikarska V, Terenetskaya I, Orlova T, Torgova S (2008) Photoinduced phase transition smectic A – cholesteric in liquid crystalline matrix contained provitamin D. Liq Cryst Pract Appl 3(25):39–44

# Chapter 5
# Colloidal Particles in Confined and Deformed Nematic Liquid Crystals: Electrostatic Analogy and Its Implications

**O. M. Tovkach, S. B. Chernyshuk, and B. I. Lev**

**Abstract** Liquid crystals doped with inclusions of other solid or liquid phases possess a number of properties, such as intrinsic anisotropy and extreme responsiveness to external stimuli, that make them particularly attractive for biomedical applications, varying from imaging and spectroscopy to biosensing. One of the hallmarks of such systems is an effective host-mediated interaction between colloidal particles. Building upon similarities between electrostatic potential and small deformations of the nematic director, we show how these anisotropic long-range interactions emerge from the interplay of colloids shape and nematic order and derive general expressions for the long-range interaction potentials for particles of arbitrary shape and size in the presence of confining boundaries and electromagnetic fields. To demonstrate the proposed formalism at work, we consider a series of illustrative examples, ranging from ellipsoids and banana-shaped particles in a sandwich-type cell to nanocolloids in a cylindrical capillary. Our results suggest that one can exploit the interplay of shape, symmetry, and liquid-crystalline order to design particles with prescribed long-range elasticity-mediated interactions, which opens promising pathways to novel mesostructured functional materials.

**Keywords** Nematic liquid crystals · Colloids · Elastic interactions · Self-assembly · Multipole expansion · Nanocolloids

O. M. Tovkach
Department of Physics, Georgetown University,
37th and O Streets NW, Washington, DC 20057, USA
e-mail: ot76@georgetown.edu

S. B. Chernyshuk (✉)
Institute of Physics, NAS Ukraine, Prospekt Nauky 46, Kyiv 03028, Ukraine
e-mail: stasubf@gmail.com

B. I. Lev
Bogolyubov Institute for Theoretical Physics, NAS Ukraine,
14-b Metrologichna Street, Kyiv 03680, Ukraine



L. Bulavin and N. Lebovka (eds.), *Soft Matter Systems for Biomedical Applications*,
Springer Proceedings in Physics 266, https://doi.org/10.1007/978-3-030-80924-9_5

## List of Abbreviations and Symbols

| | |
|---|---|
| 2D | Two-dimensional |
| 3D | Three-dimensional |
| LC | Liquid crystal |
| NLC | Nematic liquid crystal |
| NP | Nanoparticle |
| THz | Terahertz ($10^{12}$ Hz) |
| HL | Quasi-hexagonal lattice |
| DL | Dense quasi-hexagonal lattice |

## 5.1 Introduction

Nematic liquid crystals combine spatial anisotropy of the orientational order with the fluidity of conventional liquids. Typical NLCs comprise elongated organic molecules with a strong propensity toward aligning their axes along a specific direction characterized by a unit-length vector **n** — the director. Traditionally widely used in information displays, liquid crystalline materials are now rapidly conquering new realms. One of the most mature non-display applications of LCs is in tunable optic filters and light modulators. These components are crucial, in particular, in hyperspectral medical imaging [1, 2], which has tremendous potential in early detection of pathological conditions and disease, such as different types of cancer [3–5], diabetic foot ulceration [6], hemorrhagic shock [7]. Real-time hyperspectral imaging augments the surgeon's field of view and can be used for tumor residue analysis [4, 8]. Due to their strong birefringence in a wide range of frequencies, liquid-crystalline materials are of keen interest to terahertz optics as well [9–11]. Unlike X-rays, THz waves do not ionize biological tissues and have energies corresponding to low-frequency vibrational and rotational modes of common biomolecules, making them a powerful non-invasive diagnostic tool with unprecedented spatiotemporal sensitivity [12, 13]. Dispersed nanoscale dopants, ranging from quantum dots [14] to metal nanoparticles [15, 16], can further improve the functionality and efficiency of LC-based devices. At the same time, optical properties of NPs strongly depend on their environment. For instance, surface plasmon resonance in golden nanorods is highly sensitive to their spatial arrangement and the dielectric permittivity of the host medium. These parameters are relatively easy to control with the LC matrix [17], which makes LC-NP composites promising materials for plasmonic applications, including biosensing [18–20].

The anisotropy of mechanical properties allows LC-based templates and scaffolds to guide biological cells in specific directions, thereby controlling tissue growth and morphogenesis [21, 22]. Extreme responsiveness to mechanical stimuli endows liquid crystals *per se* with unsurpassed biosensing capabilities. Chemical binding of biomolecules on the surface or in the bulk of an NLC can disrupt orientational order, thereby changing its optical appearance. This mechanism has been proven successful in detecting different proteins [23, 24], lipids [25], glucose [26], DNA [27]. Recently, Kim *et al.* designed a self-reporting and self-regulating biosensor using an LC film doped with microdroplets containing an anti-bacterial agent. Shear stresses induced by swimming bacteria shift an intricate balance of elastic and electrostatic forces at the LC-droplet interface and trigger the controlled release of the agent via a feedback loop [28]. One of the major drawbacks of this type of biosensors is comparatively low sensitivity. A growing number of studies suggests that this problem also can be solved with nanodopants [29–32]. For example, Zhao *et al.* proposed a thrombin sensor utilizing gold nanoparticles functionalized with ~80 binding aptamers each. When loaded with thrombin, the ensuing aggregates are large enough to cause significant distortions of the otherwise uniform director field, thus enabling detection of nanomolar ($10^{-9}$ mol/L) thrombin concentrations [33].

Nematic liquid crystals doped with nano- and microscale inclusions of other materials are representatives of a broad class of NLC colloids. The world of colloidal particles in an NLC host is dramatically different than in isotropic fluids. Particles suspended in a nematic distort its orientational order even on length scales much larger than the particle size. Distortions overlap and interact with each other, giving rise to effective elastic interactions between colloidal particles. Two hallmarks of such deformation-induced interactions are their long-ranged distance dependence and strong spatial anisotropy. These colloidal interactions generally resemble those between electrostatic multipoles, but the ordered host medium often endows them with distinct traits. The elastic interactions are responsible for the formation of different colloidal structures in liquid crystals. A paradigmatic example is a spherical water droplet accompanied by a point defect (hyperbolic hedgehog) with asymptotic dipole symmetry (see Fig. 5.1a). The ensuing director profile gives rise to the elastic dipole-dipole interaction between the droplets, which then assemble into linear chains parallel to the average director field $\mathbf{n}_0$ [34–37]. Droplets of glycerol, on the other hand, induce planar (tangential) anchoring of nematic molecules at the interface, which results in a director field of quadrupole symmetry (Fig. 5.1c) and triggers the aggregation of droplets into inclined chains [36, 38–40]. Similarly, solid microspheres with planar anchoring form chains directed at 30° to $\mathbf{n}_0$ [38].

Colloids floating at the nematic-air interface distort the director in the bulk and deform the LC surface. This elasto-capillary coupling leads to the formation of two-dimensional (2D) hexagonal structures with different lattice constants [41, 42]. Photochemical switching between those structures induced by laser light was directly observed in [43, 44]. Transformation of the interfacial hexagonal lattice into linear chains under the action of magnetic field was studied in [45].

One can create a wide variety of 2D crystals in thin nematic cells. In particular, there have been reported hexagonal lattices of glycerol droplets [46], oblique lattices of silica beads [47], antiferroelectrically ordered crystals of dipolar particles [47, 48], and mixed dipole-quadrupole crystals sandwiched between cell walls [49]. The authors of [50] found colloidal superstructures in mixtures of small and large colloids. In these systems, small particles populate a matrix of topological defects surrounding large colloids. One-dimensional structures bound by delocalized topological defects

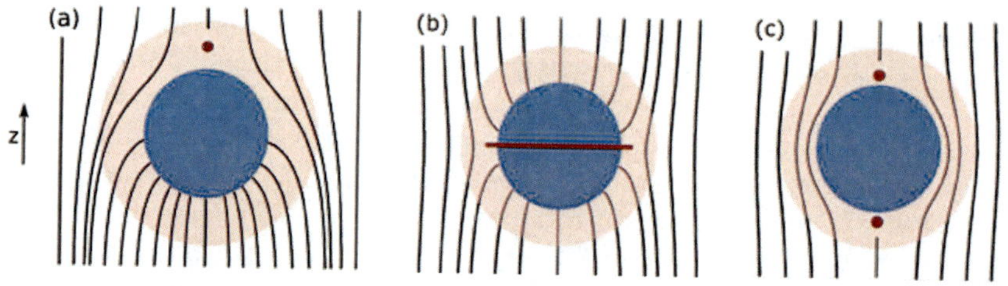


**Fig. 5.1** Possible director configurations around spherical colloidal particles: **a** point defect - hyperbolic hedgehog **b** line defect — Saturn ring disclination; **c** pair of point defects — boojums. Dark pink area (*coat*) contains all topological defects and strong deformations inside. Outside the *coat* deformations are small; they are governed by Laplace equation and can be expanded in multipoles

are also of interest. These so-called colloidal wires, whose existence was confirmed in [51], have great potential as optical waveguides. Two-dimensional colloidal crystals assembled from chiral colloidal dimers in a twisted nematic cell were observed in [52]. If the system is sufficiently large, one can even assemble three-dimensional (3D) structures. For instance, Nych *et al.* developed a step-by-step protocol for the assembly of 3D colloidal crystals of tetragonal symmetry. Surprisingly, these crystals exhibit a highly unusual electro-mechanical response, such as giant electrostriction and collective electro-rotation [53].

Many experimental results were reproduced via numerical minimization of the Landau - de Gennes free energy [47–52, 54–62] as well as by molecular dynamics [63].

In this chapter, we summarise our current understanding of the long-range elastic interactions in NLC colloids. It is worth keeping in mind that short-range interactions are no less important in the context of colloidal assemblies. In NLCs, the short-range interaction is closely intertwined with topological defects emerging in the vicinity of micro-sized colloids, as shown in Fig. 5.1. These singularities in the director field, where it becomes ill-defined, generally have to be studied numerically using Landau-de Gennes formalism. Topological defects and the associated short-range interactions are beyond the scope of this review. For a brief topology oriented discussion of NLC colloids, we refer the reader to [64].

Theoretical understanding of the long-range elastic interactions in NLC colloids is based on their striking similarity to electrostatics. Far from the particle, the director field is governed by the Laplace equation and can be expanded in multipoles. This fact became a starting point for a number of approaches toward the theory of NLC colloids [65–70]. However, only one of them, [65], gives analytical results that *quantitatively* agree with experimental data [71, 72]. The authors of [37] directly measured the elastic interaction between two spherical iron particles by balancing it with a magnetic field and found the predictions of [65] to hold within a few-percent accuracy. Recently, the authors of [60, 61] carried out numerical calculations of the interparticle force between colloids of different sizes. They found their numerical results in good agreement with the experimental data and the predictions of the electrostatic analogy for distances bigger than approximately 3 particle radii.

In practice, liquid crystals are always confined by some surfaces, which can profoundly affect the interparticle interactions. It has been first discovered experimentally that the elastic interactions in a nematic cell are exponentially screened at distances larger than the cell thickness (so-called confinement effect) [54, 73]. Qualitative theoretical investigations of the effects of confinement were performed in [74–78], using the notion of "coat" — introduced in [67] and discussed here in Sect. 5.3 — an area around the particle populated by topological defects and large deformations, as shown schematically in Fig. 5.1. A breakthrough was achieved in Refs. [79–81], where external boundaries and fields were incorporated in the form of Green's functions. Using that approach, the authors derived the interaction potentials for axially symmetric particles in a nematic cell and near one wall with either planar (tangential) or homeotropic (normal) alignment of the LC molecules. The proposed theory [79, 80] is in excellent agreement with the measurements of the confinement effect for two spheres in a homeotropic cell in the range $1 \div 1000\ k_B T$ [54] and a planar cell of varying thickness [73].

While theoretical studies were mostly concerned with spherical or axially symmetric particles, experiments and numerical simulations adopted a more diverse set of shapes and topologies. Using lithographically fabricated polygonal platelets, Lapointe et al. were first to demonstrate the relevance of particles' shape for the interparticle interactions and colloidal self-assembly [82]. In addition to the 2D crystals mentioned above, pentagonal microplatelets in a thin layer of nematic can be assembled into Penrose tilings with *quasicrystalline* order [83, 84]. Other notable examples of non-spherical colloids include rod-like particles [85], handlebodies of high topological genus [86], micron-sized spheres with nanoscale surface roughness [87], key-lock colloids [88], gourd-shaped dimers [89], colloidal boomerangs [90], and shape-shifting elastomeric rods [91]. Interestingly, the presence of chiral inclusions (*e.g.*, spirals) in an intrinsically achiral nematic can induce spontaneous symmetry breaking evident in chiral defect configurations and director profiles [92].

Here, we describe a theoretical framework applicable to colloids of *arbitrary shape and chirality*. Building upon [65, 79–81, 93–95], in Sect. 5.2 we derive general expressions for the pairwise interaction potential in the presence of confining boundaries and electromagnetic fields. Key elements of our model are elastic multipole moments that define the far-field director profile around the particle. In Sect. 5.3, we demonstrate how these coefficients arise from the symmetry of the near-field director, which is dictated by an interplay of the particle shape and the type and strength of nematic anchoring. We thoroughly consider the cases of weak and strong as well as polar and azimuthal anchoring on the particle surface. In the following sections, we apply the proposed formalism to a series of illustrative examples. Namely, Sect. 5.4 discusses in detail the dipole-dipole interaction of banana-shaped particles in a homeotropic and planar nematic cell and contrasts it with that of axially symmetric colloids. In Sect. 5.5, we touch upon elastic monopoles in a confined environment and highlight their distinctive features. The proposed framework is employed to study the behavior of colloids carrying elastic monopole, dipole and/or quadrupole moment in a deformed NLC (nonuniform director field) in Sect. 5.6 and find the distribution of nanoparticles around different types of disclinations in a cylindrical capillary, Sect. 5.7. Finally, Sect. 5.8 elucidates the role of elastic interactions in the coexistence of two quasi-hexagonal structures of glycerol droplets at the nematic-air interface [41, 42].

## 5.2 Elastic Interactions in Nematic Liquid Crystal Colloids

Let us first consider a single colloidal particle suspended in a nematic liquid crystal. We will follow the paper [93]. Anchoring of the LC molecules to the particle's surface distorts the otherwise uniform director field $\mathbf{n} = \mathbf{n}_0$, so that $\mathbf{n}$ varies from point to point. The total free energy of the system can be written as the sum of bulk and surface terms $F = F_{bulk} + F_{surface}$, where the former reads

$$F_{bulk} = \frac{K}{2} \int dV \left[ (\nabla \cdot \mathbf{n})^2 + (\nabla \times \mathbf{n})^2 \right], \tag{5.1}$$

with $K$ denoting the Frank elastic constant. Two comments are in order. First, we adopt the one-constant approximation, $K_{11} = K_{22} = K_{33} = K$; implications of this assumption are discussed in [96]. Second, we also omit the $K_{24}$-term because it does not play a crucial role in further considerations, as was shown in [80]. Surface energy $F_{surface}$ penalizes deviations of the director from the preferred orientation at the particle surface. It will be discussed in detail in Sect. 5.3 [for definition see Eq. (5.20)].

Regardless of the particle's shape, size or anchoring type and strength, deformations of the director field decay with the distance from the particle and eventually become small, $\mathbf{n}(\mathbf{r}) \approx (n_x, n_y, 1)$, where $|n_\mu| \ll 1$, $\mathbf{n}_0 = (0, 0, 1)$ is the NLC ground state, and $\mu = \{x, y\}$ hereafter. Thus, far from the particle the bulk free energy takes a simple harmonic form

$$F_{\text{bulk}} = \frac{K}{2} \int dV \nabla n_\mu \cdot \nabla n_\mu, \tag{5.2}$$

where repeated $\mu$ implies summation over $x$ and $y$, *i.e.* $(\nabla n_\mu)^2 = (\nabla n_x)^2 + (\nabla n_y)^2$. Hereafter, $\int dV$ stands for the integration over the whole volume of the system unless otherwise specified, and we do not take into account the volume of the coat(s). This is justified by several reasons: 1) we are interested mainly in the interparticle interactions which are caused by the *changes* in the director field caused by other particles, and these changes are small inside the coat where the director is governed by strong anchoring to the surface, *i.e.*, the energy of the coat is approximately constant and can be omitted from the theory; 2) the volume inside the coat is much smaller than the volume outside, therefore the replacement of the total energy (5.1) with (5.2) inside the coat does not affect the interparticle interaction considerably (at least when the separation is large enough compared to the particle size); 3) this approximation is in good agreement with experimental data for spherical particles as long as their surfaces are separated by more than ~0.2–0.8 radii depending on the system [97–100]. Note that these studies, in contrast to [60, 61], consider higher-order contributions as explained below.

The Euler-Lagrange equations for $n_\mu$ are of the Laplace type

$$\Delta n_\mu = 0. \tag{5.3}$$

In analogy to classical electrostatics, solutions to these equations can be expanded in multipoles

$$n_\mu(\mathbf{r}) = \frac{q_\mu}{r} + \frac{p^\alpha_\mu r_\alpha}{r^3} + \frac{3Q^{\alpha\beta}_\mu r_\alpha r_\beta}{r^5} - \frac{SpQ_\mu}{r^3} + \ldots, \tag{5.4}$$

where $\alpha$ and $\beta$ take values $x, y, z$ and summation over repeated greek indices is assumed. Quantities $q_\mu$, $p^\alpha_\mu$, $Q^{\alpha\beta}_\mu$ are called elastic charge (monopole), three components of the dipole moment, and five components of the quadrupole moment for each $\mu$, respectively. They are phenomenological far-field characteristics of a given colloidal particle.

Even though we truncated the expansion (5.4) at quadrupolar terms, higher-order contributions may be important as well. It was noticed in [65, 97] that if the leading contribution to $n_\mu$ is dipolar, then anharmonic corrections to $n_\mu$ associated with small variations of $n_z = \sqrt{1 - n_\mu n_\mu}$ scale as $r_\mu/r^7$, suggesting that high-order terms up to $1/r^5$ should be considered. Similarly, if the leading contribution to $n_\mu$ is quadrupolar, anharmonic corrections to $n_\mu$ are of the form $r_\mu/r^{10}$ and high-order terms of up to $1/r^8$ can effectively influence short-range behavior. High-order terms have a profound effect on the spatial anisotropy of the interaction potential at relatively small distances and play an important role in the formation of colloidal crystals [97–100]. Reference [97] also introduced the concept of three zones: in the first zone (at distances $\lesssim$1.1–1.4 particle radii depending on the system) the linear theory is inapplicable; the second zone (between roughly 1.3–1.4 and 2 radii) is where the high-order terms reside; and the third zone ($\gtrsim$2 radii) is dominated by the first non-vanishing term in (5.4). In what follows, we will limit ourselves to the first three terms in the multipole expansion (5.4).

As follows from (5.4), director deviations $n_x$ and $n_y$ have a long-range nature. This means that deformations caused by different particles can overlap even if the particles are located at a large distance from each other. In practice, the overlapping manifests itself in the emergence of effective long-range interactions between colloidal particles mediated by the host medium. As we will see below, these interactions are determined precisely by the multipole moments $q_\mu$, $p^\alpha_\mu$, $Q^{\alpha\beta}_\mu$.

Imagine that somehow we have found all these multipole coefficients. That is, we know two elastic charges $q_\mu$, six components of the dipole moment $p^\alpha_\mu$ and five components of the quadrupole moment $Q^{\alpha\beta}_\mu$ for every $\mu = \{x, y\}$ (10 altogether). Note that the quadrupole moment tensor $\hat{Q}_\mu = \{Q^{\alpha\beta}_\mu\}$ can be always introduced (with a substitution $\tilde{Q}^{\alpha\beta}_\mu = Q^{\alpha\beta}_\mu - \frac{1}{3}\delta_{\alpha\beta} Sp\hat{Q}_\mu$) in such a way that it is symmetric,

$\tilde{Q}_\mu^{\alpha\beta} = \tilde{Q}_\mu^{\beta\alpha}$, and traceless, $\mathrm{Sp}\hat{\tilde{Q}}_\mu = \tilde{Q}_\mu^{\alpha\beta}\delta_{\alpha\beta} = 0$ [101]. Thus, we have 18 multipole parameters. Now we need to build an effective free energy functional, which incorporates these coefficients and gives correct behavior of the director at large distances from the particle.

This aim was first achieved in [65] for the case of axially symmetric particles. The effective free energy functional was written as

$$F_{eff}^{axial-sym} = K\int dV\left[\frac{(\nabla n_\mu)^2}{2} - 4\pi P(\mathbf{x})\partial_\mu n_\mu \; - 4\pi C(\mathbf{x})\partial_z\partial_\mu n_\mu\right], \tag{5.5}$$

where $P(\mathbf{x}) = P\delta(\mathbf{x})$ and $C(\mathbf{x}) = C\delta(\mathbf{x})$ are scalar dipole and quadrupole moment densities and $\partial_\mu n_\mu = \partial_x n_x + \partial_y n_y$.

The generalization of the free energy (5.5) for particles of arbitrary shape and anchoring strength is quite straightforward:

$$F_{eff} = K\int dV\left[\frac{(\nabla n_\mu)^2}{2} - 4\pi q_\mu(\mathbf{x})n_\mu - 4\pi p_\mu^\alpha(\mathbf{x})\partial_\alpha n_\mu \right.$$
$$\left. -4\pi\, Q_\mu^{\alpha\beta}(\mathbf{x})\partial_\alpha\partial_\beta n_\mu\right], \tag{5.6}$$

where $q_\mu(\mathbf{x}) = q_\mu\delta(\mathbf{x})$, $p_\mu^\alpha(\mathbf{x} = p_\mu^\alpha\delta(\mathbf{x})$, $Q_\mu^{\alpha\beta}(\mathbf{x}) = Q_\mu^{\alpha\beta}\delta(\mathbf{x})$ are point-like densities; $\alpha$ and $\beta$ take values $x, y, z$ and summation over repeated greek indices $\mu = x, y$ is assumed. In the case of axially symmetric particles without helical twisting $q_\mu = 0$, $p_\mu^\alpha = 0$ except for $p_x^x = p_y^y = P$, $Q_x^{xz} = Q_x^{zx} = Q_y^{yz} = Q_y^{zy} = C$ and we arrive at Eq. (5.5).

The Euler-Lagrange equations arising from (5.6) are of Poisson type

$$\Delta n_\mu = -4\pi q_\mu(\mathbf{x}) + 4\pi\left[\partial_\alpha p_\mu^\alpha(\mathbf{x}) - \partial_\alpha\partial_\beta Q_\mu^{\alpha\beta}(\mathbf{x})\right]. \tag{5.7}$$

If the liquid crystal is confined by some surface $\Sigma$ such that $n_\mu\Big|_\Sigma = 0$ then solutions to (5.7) are as follows

$$n_\mu = \int_V dV' G_\mu(\mathbf{x}, \mathbf{x}')\left[q_\mu(\mathbf{x}') - \partial'_\alpha p_\mu^\alpha(\mathbf{x}') - \partial'_\alpha\partial'_\beta Q_\mu^{\alpha\beta}(\mathbf{x}')\right], \tag{5.8}$$

where $G_\mu(\mathbf{x}, \mathbf{x}')$ are appropriate Green functions, $\Delta G_\mu(\mathbf{x}, \mathbf{x}') = -4\pi\delta(\mathbf{x} - \mathbf{x}')$ for any $\mathbf{x}, \mathbf{x}' \in V$ and $G_\mu(\mathbf{x}, \mathbf{s}) = 0$ for any $\mathbf{s} \in \Sigma$. In the absence of confinement, $G_\mu(\mathbf{x}, \mathbf{x}') = \frac{1}{|\mathbf{x}-\mathbf{x}'|}$ and (5.8) yields (5.4).

Due to the linearity of the Euler-Lagrange equations (5.7), we can use the superposition principle for the system of $N$ colloidal particles

$$
\begin{aligned}
q_\mu(\mathbf{x}) &= \sum_{i=1}^{N} q_{\mu,i}\delta(\mathbf{x}-\mathbf{x}_i),\\
p_\mu^\alpha(\mathbf{x}) &= \sum_{i=1}^{N} p_{\mu,i}^\alpha\delta(\mathbf{x}-\mathbf{x}_i),\\
Q_\mu^{\alpha\beta}(\mathbf{x}) &= \sum_{i=1}^{N} Q_{\mu,i}^{\alpha\beta}\delta(\mathbf{x}-\mathbf{x}_i).
\end{aligned}
\tag{5.9}
$$

That is, the resulting director deformation is the sum of distortions caused by every single particle. Substituting (5.8) into (5.6) and implying (5.9), we come to the fact that the free energy of the system can be presented as the sum of the self energy part and pair interactions $F_{eff} = U^{self} + U^{interaction}$, where $U^{self} = \sum_i U_i^{self}$ and

$$
U^{interaction} = \sum_i \sum_{j<i} U^{ij}(\mathbf{x}_i, \mathbf{x}_j). \tag{5.10}
$$

The pairwise elastic interaction $U^{ij}$, in turn, is the sum of monopole-monopole, monopole-dipole, monopole-quadrupole, dipole-dipole, dipole-quadrupole and quadrupole-quadrupole interactions, $U^{ij} = U_{\rm qq} + U_{\rm qd} + U_{\rm qQ} + U_{\rm dd} + U_{\rm dQ} + U_{\rm QQ}$, where

$$
U_{\rm qq} = -4\pi K\, q_{\mu,i}\, q_{\mu,j}\, G_\mu(\mathbf{x}_i, \mathbf{x}'_j), \tag{5.11}
$$

$$
U_{\rm qd} = -4\pi K \left\{q_{\mu,i}\, p_{\mu,j}^\alpha\, \partial'_\alpha G_\mu(\mathbf{x}_i, \mathbf{x}'_j) + q_{\mu,j}\, p_{\mu,i}^\alpha\, \partial_\alpha G_\mu(\mathbf{x}_i, \mathbf{x}'_j)\right\}, \tag{5.12}
$$

$$
U_{\rm qQ} = -4\pi K \left\{q_{\mu,i}\, Q_{\mu,j}^{\alpha\beta}\, \partial'_\alpha\partial'_\beta G_\mu(\mathbf{x}_i, \mathbf{x}'_j) + q_{\mu,j}\, Q_{\mu,i}^{\alpha\beta}\, \partial_\alpha\partial_\beta G_\mu(\mathbf{x}_i, \mathbf{x}'_j),\right\} \tag{5.13}
$$

$$
U_{\rm dd} = -4\pi K\, p_{\mu,i}^\alpha\, p_{\mu,j}^\beta\, \partial_\alpha\partial'_\beta G_\mu(\mathbf{x}_i, \mathbf{x}'_j), \tag{5.14}
$$

$$
U_{\rm dQ} = -4\pi K \left\{p_{\mu,i}^\alpha\, Q_{\mu,j}^{\beta\gamma}\, \partial_\alpha\partial'_\beta\partial'_\gamma G_\mu(\mathbf{x}_i, \mathbf{x}'_j) + p_{\mu,j}^\alpha\, Q_{\mu,i}^{\beta\gamma}\, \partial'_\alpha\partial_\beta\partial_\gamma G_\mu(\mathbf{x}_i, \mathbf{x}'_j)\right\}, \tag{5.15}
$$

$$
U_{\rm QQ} = -4\pi K\, Q_{\mu,i}^{\alpha\beta}\, Q_{\mu,j}^{\gamma\delta}\, \partial_\alpha\partial_\beta\partial'_\gamma\partial'_\delta G_\mu(\mathbf{x}_i, \mathbf{x}'_j). \tag{5.16}
$$

Expressions (5.11)–(5.16) give a general form of the pairwise long-range elastic interaction between colloidal particles of arbitrary shape in a confined NLC. They hold in the bulk as well with $G_\mu(\mathbf{x}, \mathbf{x}') = \frac{1}{|\mathbf{x}-\mathbf{x}'|}$.

Self-energy of the colloid (or the energy of the particle-walls interaction) also can be presented as the sum $U_i^{self} = U_{\rm qq}^{self} + U_{\rm qd}^{self} + U_{\rm qQ}^{self} + U_{\rm dd}^{self} + U_{\rm dQ}^{self} + U_{\rm QQ}^{self}$ where all $U_{\rm AB}^{self}$ are given by the formulas (5.11)–(5.16) upon replacing $G_\mu(\mathbf{x}_i, \mathbf{x}'_j)$ with $H_\mu(\mathbf{x}_i, \mathbf{x}'_j)$, where $G_\mu(\mathbf{x}, \mathbf{x}') = \frac{1}{|\mathbf{x}-\mathbf{x}'|} + H_\mu(\mathbf{x}, \mathbf{x}')$ and $\Delta_{\mathbf{x}} H_\mu(\mathbf{x}, \mathbf{x}') = 0$. Note that to regularize the self-energy by excluding the divergent part associated with

$\frac{1}{|\mathbf{x}-\mathbf{x}'|}$, it is necessary to set $\mathbf{x}'_j = \mathbf{x}_i$ after all primed derivatives $\partial'_\xi$ are calculated. Employing this procedure we can find

$$U_{\rm qq}^{self} = -4\pi K\, q_{\mu,i}\, q_{\mu,i}\, H_\mu(\mathbf{x}_i, \mathbf{x}_i), \tag{5.17}$$

$$U_{\rm qd}^{self} = -4\pi K[q_{\mu,i}\, p^\alpha_{\mu,i}\, \partial'_\alpha H_\mu(\mathbf{x}_i, \mathbf{x}'_i) + q_{\mu,i}\, p^\alpha_{\mu,i}\, \partial_\alpha H_\mu(\mathbf{x}_i, \mathbf{x}'_i)]|_{\mathbf{x}_i=\mathbf{x}'_i}, \tag{5.18}$$

$$U_{\rm dd}^{self} = -4\pi K\, p^\alpha_{\mu,i}\, p^\beta_{\mu,i}\, \partial_\alpha \partial'_\beta H_\mu(\mathbf{x}_i, \mathbf{x}'_i)|_{\mathbf{x}_i=\mathbf{x}'_i}, \dots \tag{5.19}$$

and so forth. We leave the rest of this series as an exercise for the reader. It should be noted that the self-energies (5.17)–(5.19) do not include the energy of nonlinear deformations (*i.e.*, internal energy of the coat). While that contribution cannot be computed within the framework of multipole expansion, it can be treated as an unknown constant, as we argued above. In this context, the self-energy refers only to the difference between the energy of the particle in a confined NLC and its energy in the bulk. In other words, it is the energy of interaction between one particle and all confining boundaries. Hence, $H_i^{self} \equiv 0$ in the absence of confinement. In a sandwich-type cell, for instance, $H_i^{self} = H_i^{self}(z)$ with $z$ being normal to the cell walls; minimization of the corresponding self-energy yields an equilibrium position of the particle inside the cell, see [102].

Upon proper substitution of Green's functions $G_\mu(\mathbf{x}_i, \mathbf{x}'_j)$ with $G_\mu^{field}(\mathbf{x}_i, \mathbf{x}'_j)$, Eqs. (5.11)–(5.19) remain valid even in the presence of external electric **E** or magnetic **H** fields, as was shown in [81] for axially symmetric particles. Reference [81] also reported the explicit form of Green's functions $G_\mu^{field}(\mathbf{x}_i, \mathbf{x}'_j)$ for different orientations of the field in a nematic cell with homeotropic and planar boundary conditions.

Equations (5.11)–(5.16) show that the energy of interaction depends on both multipole coefficients *and* Green's functions. The former originate from the interaction between the particle surface and NLC molecules. The latter are determined by the shape of the confining surface $\Sigma$ as well as by external fields **E/H**.

## 5.3 Multipole Coefficients and Particle's Symmetry

In this section we want to establish a connection between the symmetry of the particle and the director deformations produced at large distances.[1]

As mentioned above these distortions are completely described by a set of multipole coefficients $q_\mu$, $p_\mu^\alpha$, $Q_\mu^{\alpha\beta}$. Strictly speaking, we should distinguish two cases here. When the anchoring is weak $Wr_0/K < 1$ ($r_0$ being the average size of the particle ) the deformations are small everywhere outside the particle and the coefficients can be found from the mechanical equilibrium condition (we apply this procedure

[1] This Section lays the foundation for all subsequent examples of elastic multipoles and their interactions. Yet, it is highly technical and can be skipped on first reading. The reader can proceed directly to examples in Subsubsect. 5.3.2.1 or Sect. 5.4.

below). If the anchoring is strong $Wr_0/K > 1$ the deformations in the particle's vicinity are large, even topological defects may appear there. So the coefficients cannot be linked directly to the particle's symmetry. But in this case the notion of the coat suggested in [67] is helpful. The coat is an area, which contains all topological defects and large deformations inside, so that the director field outside is presented in the form of multipole expansion (5.4). The symmetry of the coat matches the symmetry of the director field around the particle and, depending on its size, can be observed experimentally or computed numerically via Landau-de Gennes energy minimization. In fact, one can treat the coat as some imaginary particle with appropriate symmetry and weak anchoring on its surface. Therefore we use only the term "particle" further.

Thus, it is enough to consider one colloidal particle with weak anchoring in a bulk NLC. The free energy of such a system is the sum of two parts: bulk energy (5.2) and the surface anchoring energy. The latter can be presented as

$$F_{\text{surface}} = \oint dS\ W^{\alpha\beta}(\mathbf{s}) n_\alpha(\mathbf{s}) n_\beta(\mathbf{s}) \tag{5.20}$$

where $W^{\alpha\beta}(\mathbf{s})$ is the symmetrical local anchoring tensor at point $\mathbf{s}$ on the particle's surface [103]. The tensor description has the covariant form and describes both polar and azimuthal anchoring simultaneously. But a connection between the particle's symmetry and tensor's properties is not so clear in general case. In order to make our analysis as transparent as possible we should use the following representation of the surface energy

$$F_{surface} = \oint dS\ W^p(\mathbf{s})\big(\boldsymbol{\nu}(\mathbf{s})\cdot\mathbf{n}(\mathbf{s})\big)^2 - \oint dS\ W^a(\mathbf{s})\big(\boldsymbol{\tau}(\mathbf{s})\cdot\mathbf{n}(\mathbf{s})\big)^2. \tag{5.21}$$

This is the generalized Rapini-Popular surface energy with $W^p$ and $W^a$ being the strengths of the polar and azimuthal anchoring energies, respectively. Here $\boldsymbol{\nu}$ is the outer normal to the particle's surface at the point $\mathbf{s}$ and $\boldsymbol{\tau}$ is the unit tangential vector along the local rubbing which also depends on the point $\mathbf{s}$ of the surface. Azimuthal anchoring $W^a > 0$ makes alignment of the director along vector field $\boldsymbol{\tau}(\mathbf{s})$ at the surface. Since the anchoring is weak the total energy can be reduced to

$$F_{harm} = \frac{K}{2}\int dV(\nabla n_\mu)^2 + 2\oint dS\, W^p(\mathbf{s})\nu_z(\mathbf{s})\nu_\mu(\mathbf{s})n_\mu(\mathbf{s}) - 2\oint dS\, W^a(\mathbf{s})\tau_z(\mathbf{s})\tau_\mu(\mathbf{s})n_\mu(\mathbf{s}), \tag{5.22}$$

where we neglected terms like $(\nabla n_z)^2$, $W^p n_\mu n_{\mu'}$ and $W^a n_\mu n_{\mu'}$ because of their smallness. Note that, in fact, $W^p \nu_z \nu_\mu - W^a \tau_z \tau_\mu = W^{z\mu}$ in (5.20). Note that here $\int dV$ denotes integration over the LC volume.

At the same time the director field everywhere outside the particle in the bulk NLC is described by (5.4) so that

$$
\begin{aligned}
(\nabla n_\mu)^2 &= \nabla n_\mu \cdot \nabla n_\mu \\
&= \frac{q_\mu q_\mu}{r^4} + \frac{p_\mu^\alpha p_\mu^\alpha}{r^6} + 3\frac{p_\mu^\alpha r_\alpha p_\mu^\beta r_\beta}{r^8} + 5\frac{Q_\mu^{\alpha\beta} r_\alpha r_\beta Q_\mu^{\gamma\delta} r_\gamma r_\delta}{r^{12}} + 4\frac{Q_\mu^{\alpha\gamma} Q_\mu^{\beta\gamma} r_\alpha r_\beta}{r^{10}} \\
&\quad + 4\frac{q_\mu p_\mu^\alpha r_\alpha}{r^6} + 6\frac{q_\mu Q_\mu^{\alpha\beta} r_\alpha r_\beta}{r^8} + 8\frac{p_\mu^\alpha Q_\mu^{\beta\gamma} r_\alpha r_\beta r_\gamma}{r^{10}} + 4\frac{p_\mu^\alpha Q_\mu^{\alpha\beta} r_\beta}{r^8},
\end{aligned} \tag{5.23}
$$

where the summation over all repeated Greek indices is implied. Then substituting (5.4), (5.23) into (5.22) and performing the integration one can obtain the free energy of the system as a function of the multipole coefficients

$$
F_{harm} = \frac{1}{2}\sum_{uv} a_{uv} m_u m_v + \sum_u c_u m_u, \tag{5.24}
$$

where we introduced vector of the coefficients $\mathbf{m} = (q_\mu, p_\mu^\alpha, Q_\mu^{\alpha\beta}) = (q_x, q_y, p_x^x, p_x^y, p_x^z, p_y^x, p_y^y, p_y^z, Q_x^{xx}, Q_x^{xy} ...)$. Hence $m_u, m_v$ denote unknown multipole coefficients. Quantities $a_{uv}$ arise from the bulk energy, for example,

$$
\begin{aligned}
a_{11(q_x q_x)} &\propto \int_V dV r^{-4} \\
a_{33(p_x^x p_x^x)} &\propto \int_V dV r^{-6} \\
a_{15(q_x p_x^z)} &\propto \int_V dV z r^{-6} \quad \text{etc.}
\end{aligned}
$$

Apparently, all $a_{uu}$ are positive and finite, $a_{uv}$ depend on the particle shape. Each $c_u$ is the sum of two terms $c_u^p$ and $c_u^a$ arising from the polar and azimuthal anchoring, respectively, $c_u = c_u^p + c_u^a$. So, for instance,

$$
\begin{aligned}
c^p_{1(q_x)} &\propto \oint dS\, W^p \nu_z \nu_x, \qquad & c^a_{1(q_x)} &\propto -\oint dS\, W^a \tau_z \tau_x \\
c^p_{3(p_x^x)} &\propto \oint dS\, W^p \nu_z \nu_x s_x, \qquad & c^a_{3(p_x^x)} &\propto -\oint dS\, W^a \tau_z \tau_x s_x \\
&\dots
\end{aligned} \tag{5.25}
$$

$c_u$ depends on both the anchoring and the particle's shape, $\mathbf{s} = (s_x, s_y, s_z)$ is the radius vector from the center of the particle to the point $\mathbf{s}$ at the surface (this center of the particle coincides with the center of the coordinate cystem (CS) from which all $r_\alpha$ are measured). In the remainder of this section, we mathematically prove that if a given $c_u$ vanishes, so does the associated multipole moment. As follows from their definitions (5.25), these quantities, and consequently their corresponding multipoles, vanish if the particle possesses certain symmetries. Table 5.1 summarizes these relations between the particle symmetry and its multipole moments.

Now it is natural to assume that the system under investigation is in equilibrium. Therefore its energy is minimal. Hence one can find the multipole coefficients from the following system of linear equations $\frac{\partial F_{harm}}{\partial m_u} = 0$:

$$a_{uu} m_u + \sum_{v, v \neq u} a_{uv} m_v + c_u = 0, \tag{5.26}$$

or the same in the matrix form:

$$\hat{\mathbf{A}} \mathbf{m} = -\mathbf{c}. \tag{5.27}$$

Here we should make some remarks. First of all this Eq. (5.27) is the exact equation for the multipole coefficients $\mathbf{m}$ for the weak anchoring case. In this case we know exactly weak anchoring coefficient $W_{a,p}(\mathbf{s})$ and vector fields $\boldsymbol{\nu}(\mathbf{s}), \boldsymbol{\tau}(\mathbf{s})$, we can calculate all $c_u$ and $a_{uv}$ and finally we can solve this matrix equation and find all 18 unknown coefficients.

On the other hand, this lengthy procedure is usually unnecessary. Typically, we do not need to calculate coefficients $\mathbf{m}$ - we can infer them from experimental or numerical data. It is thus much more valuable to understand which coefficients vanish and which remain finite. We can then find nonzero coefficients as fitting parameters for the director field measured experimentally or simulated numerically. Therefore, our primary strategy is to understand which coefficients vanish and which remain nonzero *without solving the system* (5.27) *based on symmetry considerations only*! This strategy is realized in the following subsections.

Since $\mathbf{c} = \mathbf{c}^p + \mathbf{c}^a$, a solution of the system (5.27) can be written as the sum

$$\mathbf{m} = \mathbf{m}^p + \mathbf{m}^a \tag{5.28}$$

of two solutions of the following systems

$$\hat{\mathbf{A}} \mathbf{m}^p = -\mathbf{c}^p, \tag{5.29}$$

$$\hat{\mathbf{A}} \mathbf{m}^a = -\mathbf{c}^a. \tag{5.30}$$

Thus the polar and azimuthal anchorings make their contributions to the coefficients independently and we can consider these two cases separately.

**Table 5.1** Multipole coefficients and symmetry of the shape. Here, $\sigma_{ik}$ means that the particle's plane of symmetry coincides with the coordinate $ik$-plane, $\mathcal{I}$ denotes inversion center. If a colloidal particle has at least one of the inhibiting symmetry elements then the appropriate multipole coefficient vanishes. Table is taken from [93].

| Defining integral | Multipole coefficient | Inhibiting symmetry | Multipole coefficient | Inhibiting symmetry |
|---|---|---|---|---|
| $\oint dS\, W^p \nu_z \nu_\mu$ | $q_x$ | $\sigma_{xy},\ \sigma_{yz}$ | $q_y$ | $\sigma_{xy},\ \sigma_{xz}$ |
| $\oint dS\, W^p \nu_z \nu_\mu x$ | $p_x^x$ | $\sigma_{xy},\ \mathcal{I}$ | $p_y^x$ | $\sigma_{xy},\ \sigma_{xz},\ \sigma_{yz},\ \mathcal{I}$ |
| $\oint dS\, W^p \nu_z \nu_\mu y$ | $p_x^y$ | $\sigma_{xy},\ \sigma_{xz},\ \sigma_{yz},\ \mathcal{I}$ | $p_y^y$ | $\sigma_{xy},\ \mathcal{I}$ |
| $\oint dS\, W^p \nu_z \nu_\mu z$ | $p_x^z$ | $\sigma_{yz},\ \mathcal{I}$ | $p_y^z$ | $\sigma_{xz},\ \mathcal{I}$ |
| $\oint dS\, W^p \nu_z \nu_\mu xx$ | $Q_x^{xx}$ | $\sigma_{xy},\ \sigma_{yz}$ | $Q_y^{xx}$ | $\sigma_{xy},\ \sigma_{xz}$ |
| $\oint dS\, W^p \nu_z \nu_\mu yy$ | $Q_x^{yy}$ | $\sigma_{xy},\ \sigma_{yz}$ | $Q_y^{yy}$ | $\sigma_{xy},\ \sigma_{xz}$ |
| $\oint dS\, W^p \nu_z \nu_\mu zz$ | $Q_x^{zz}$ | $\sigma_{xy},\ \sigma_{yz}$ | $Q_y^{zz}$ | $\sigma_{xy},\ \sigma_{xz}$ |
| $\oint dS\, W^p \nu_z \nu_\mu xy$ | $Q_x^{xy}$ | $\sigma_{xz},\ \sigma_{xy}$ | $Q_y^{xy}$ | $\sigma_{xy},\ \sigma_{yz}$ |
| $\oint dS\, W^p \nu_z \nu_\mu xz$ | $Q_x^{xz}$ | – | $Q_y^{xz}$ | $\sigma_{xz},\ \sigma_{yz}$ |
| $\oint dS\, W^p \nu_z \nu_\mu yz$ | $Q_x^{yz}$ | $\sigma_{yz},\ \sigma_{xz}$ | $Q_y^{yz}$ | – |

### 5.3.1 *Polar Anchoring*

Suppose that we have a particle with usual polar anchoring on its surface, $W^a \equiv 0$. Then the multipole coefficients satisfy system (5.29). Here we should say that the phrase "particle symmetry" means that the appropriate symmetry element belongs to the particle shape as well as to the anchoring distribution $W^p(\mathbf{s})$. Thus, in terms of symmetry, particles of symmetrical shape with asymmetric anchoring do not differ from those of asymmetrical shape with symmetric $W^p(\mathbf{s})$.

Assume first that the particle has one plane of symmetry. Say for instance that it coincides with the coordinate $xz$-plane. Then for any point $\mathbf{s} = (x, y, z)$, where $\boldsymbol{\nu} = (\nu_x, \nu_y, \nu_z)$, there exists point $\mathbf{s}' = (x, -y, z)$, where $\boldsymbol{\nu} = (\nu_x, -\nu_y, \nu_z)$, and $W^p(\mathbf{s}) = W^p(\mathbf{s}')$. Then using these symmetry relations one can easily ensure that, for example, $a_{q_x q_x} = K \int dV r^{-4} \neq 0$, $a_{q_x p_x^y} = 4K \int dV y r^{-6} = 0$, $c^p_{q_y} = 2 \int dS\, W^p \nu_z \nu_y s^{-1} = 0$ etc. In the same way $a_{q_x p_x^y} = a_{p_x^x p_x^y} = a_{p_x^y p_x^z} = a_{q_y p_y^y} = a_{p_y^x p_y^y} = a_{p_y^y p_y^z} = 0$ and $c^p_{q_y} = c^p_{p_y^x} = c^p_{p_y^z} = c^p_{p_x^y} = 0$. It is well known that if the leading term in $n_\mu$ decreases as $r^{-n}$ then the leading anharmonic correction will fall off as $r^{-3n}$ [65]. Thus, quadrupolar terms can be neglected here as anharmonic corrections to $n_\mu$. Since in vector $\mathbf{m}$ the multipole coefficients can be arranged in any order, we are able to rewrite the system (5.26) in the following matrix form $\hat{\mathbf{A}}\mathbf{m}^p = -\mathbf{c}^p$:

$$\left(\begin{array}{cccc|cccc} a_{q_x} & b_{q_x p_x^x} & b_{q_x p_x^z} & 0 & 0 & 0 & 0 & 0 \\ b_{q_x p_x^x} & a_{p_x^x} & b_{p_x^x p_x^z} & 0 & 0 & 0 & 0 & 0 \\ b_{q_x p_x^z} & b_{p_x^x p_x^z} & a_{p_x^z} & 0 & 0 & 0 & 0 & 0 \\ 0 & 0 & 0 & a_{p_y^y} & 0 & 0 & 0 & 0 \\ \hline 0 & 0 & 0 & 0 & a_{p_x^y} & 0 & 0 & 0 \\ 0 & 0 & 0 & 0 & 0 & a_{q_y} & b_{q_y p_y^x} & b_{q_y p_y^z} \\ 0 & 0 & 0 & 0 & 0 & b_{q_y p_y^x} & a_{p_y^x} & b_{p_y^x p_y^z} \\ 0 & 0 & 0 & 0 & 0 & b_{q_y p_y^z} & b_{p_y^x p_y^z} & a_{p_y^z} \end{array}\right) \begin{pmatrix} q_x \\ p_x^x \\ p_x^z \\ p_y^y \\ p_x^y \\ q_y \\ p_y^x \\ p_y^z \end{pmatrix} = - \begin{pmatrix} c_{q_x}^p \\ c_{p_x^x}^p \\ c_{p_x^z}^p \\ c_{p_y^y}^p \\ 0 \\ 0 \\ 0 \\ 0 \end{pmatrix}, \tag{5.31}$$

Here $\hat{A}$ is a block diagonal matrix, $\hat{A} = \begin{pmatrix} \hat{A}_{nh} & 0 \\ 0 & \hat{A}_h \end{pmatrix}$. So the system (5.26) splits into two independent subsystems, nonhomogeneous with matrix $\hat{A}_{nh}$ and homogeneous with matrix $\hat{A}_h$. At the same time $\hat{A}$ is a positive-definite matrix. Indeed we can treat components of arbitrary nonzero vector $\mathbf{m}$ as the coefficients of some multipole expansion, then

$$\mathbf{m}^T \hat{A} \mathbf{m} = K \int dV \left(\nabla n_\mu\right)^2 > 0. \tag{5.32}$$

Thus $\det\hat{A} = \det\hat{A}_{nh} \cdot \det\hat{A}_h > 0$. So homogeneous subsystem has only trivial solution. It is easy to ensure that the same scenario occurs for particles of other symmetries. If certain $c_u^p$ is equal to zero then the related multipole coefficient $m_u$ vanishes: $c_u^p = 0 \Rightarrow m_u^p = 0$.

Accordingly, only those multipole coefficients can exist which are allowed by the particle symmetry from the Table 5.1.

Note that the same classification was obtained in the paper [67] on the basis of gradient expansion $\partial n_\mu$ in the center of the particle. But actually the gradient expansion can not be done exactly as $\partial n_\mu \approx 1$ is not a small parameter. Therefore the current approach can be considered as more consistent and correct.

Here we should remark that multipole coefficients $\mathbf{m}$ depend on the chosen coordinate system. In one coordinate system CS1 there will be one set of parameters $\mathbf{m}_1$ and in the CS2 (which can be rotated or shifted by some vector $\mathbf{d}$ with respect to CS1) there will be another set of multipole coefficients $\mathbf{m}_2$, but the total sum (5.4) will be the same in both CSs. In our consideration we have chosen the most appropriate case when symmetry planes coincide with coordinate planes as in this CS the classification is possible and useful. But in any CS the main multipole coefficient $q_\mu$ will be the same as well as in electrostatics - the charge does not depend on the CS [104] while high order moments do depend on the CS.

### 5.3.2 *Azimuthal Anchoring and Chiral Colloids*

As follows from (5.28) and (5.30) the long-ranged director deformations can also arise from the azimuthal anchoring of NLC molecules on the particle surface.

**Table 5.2** Multipole coefficients which are born by azimuthal helical alignment along $z$ axis. Here, $p_{hel} > 0$ and $Q_{hel} > 0$ for right-handed helicity and $p_{hel} < 0$ and $Q_{hel} < 0$ for left-handed helicity (see Fig. 5.2). Table is taken from [93].

| Defining integral | Multipole coefficient | Value | Multipole coefficient | Value |
|---|---|---|---|---|
| $\oint dS\ W^a\tau_z\tau_\mu$ | $q_x$ | 0 | $q_y$ | 0 |
| $\oint dS\ W^a\tau_z\tau_\mu x$ | $p_x^x$ | 0 | $p_y^x$ | $p_{hel}$ |
| $\oint dS\ W^a\tau_z\tau_\mu y$ | $p_x^y$ | $-p_{hel}$ | $p_y^y$ | 0 |
| $\oint dS\ W^a\tau_z\tau_\mu z$ | $p_x^z$ | 0 | $p_y^z$ | 0 |
| $\oint dS\ W^a\tau_z\tau_\mu xx$ | $Q_x^{xx}$ | 0 | $Q_y^{xx}$ | 0 |
| $\oint dS\ W^a\tau_z\tau_\mu yy$ | $Q_x^{yy}$ | 0 | $Q_y^{yy}$ | 0 |
| $\oint dS\ W^a\tau_z\tau_\mu zz$ | $Q_x^{zz}$ | 0 | $Q_y^{zz}$ | 0 |
| $\oint dS\ W^a\tau_z\tau_\mu xy$ | $Q_x^{xy}$ | 0 | $Q_y^{xy}$ | 0 |
| $\oint dS\ W^a\tau_z\tau_\mu xz$ | $Q_x^{xz}$ | 0 | $Q_y^{xz}$ | $Q_{hel}$ |
| $\oint dS\ W^a\tau_z\tau_\mu yz$ | $Q_x^{yz}$ | $-Q_{hel}$ | $Q_y^{yz}$ | 0 |

Simple examples of such particles are uniaxial helicoids - axially symmetric particles like cylinders or cones with the helicoidal alignment along their easy axes $z$ (see Figs. 5.2 and 5.3). For this case we need to take $W^a > 0$ and vector $\tau(\mathbf{s})$ makes screw thread at the surface of the particle. Then using the method suggested in the previous subsection one can find that multipole coefficients $\mathbf{m}^a$ in this case are defined from the Table 5.2. Inasmach as $p_{hel} \propto \oint dSW^a\tau_z\tau_y x$ and $Q_{hel} \propto \oint dSW^a\tau_z\tau_y xz$, so that $p_{hel} > 0$ and $Q_{hel} > 0$ for right-handed helicity and $p_{hel} < 0$ and $Q_{hel} < 0$ for left-handed helicity (see Fig. 5.2).

#### 5.3.2.1 Interaction Between Helicoid Cylinders

Consider cylinders (or other symmetric particles like ellipsoid or sphere) with helicoidal alignment at the surface (see Fig. 5.2). This azimuthal helicoid anchoring gives rise to nonzero dipole moments $p_y^x = p_{hel} = -p_x^y$, though the shape of the cylinder does not produce any dipole moments (see Table 5.1). Under these conditions, (5.11) gives the following dipole-dipole elastic interaction between helicoid cylinders (ellipsoids or spheres):

$$U_{\rm dd} = -4\pi K p_{hel} p'_{hel}(\partial_x\partial'_x G_y(\mathbf{x},\mathbf{x}') + \partial_y\partial'_y G_x(\mathbf{x},\mathbf{x}')). \tag{5.33}$$

The Green's functions $G_x \neq G_y$ are different only when some external field (electric or magnetic) is applied along the axis $x$ or $y$ [81]. When the external fields are absent in any other cases (like in homeotropic or planar nematic cell) $G_x = G_y = G$ and we come to the expression

$$U_{\rm dd} = -4\pi K p_{hel} p'_{hel}\partial_\mu\partial'_\mu G(\mathbf{x},\mathbf{x}'), \tag{5.34}$$

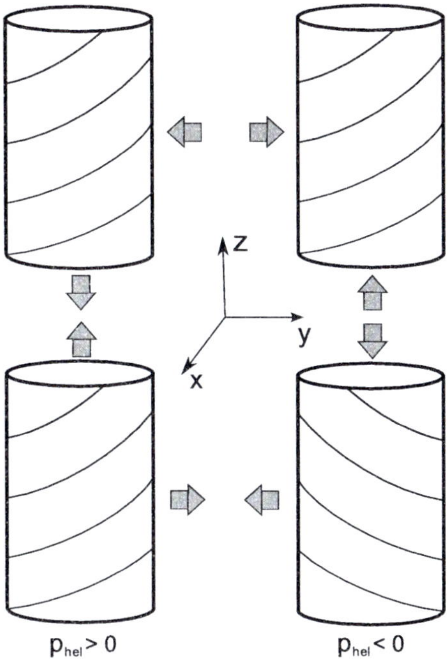


**Fig. 5.2** Helicoid cylinders with the same handedness $p_{hel} p'_{hel} > 0$ attract along $z$ axis and repel in perpendicular direction and vice versa for helicoids with different handedness $p_{hel} p'_{hel} < 0$ (see (5.35)) (reprinted with permission from [93])

which coincides with the dipole-dipole interaction between usual axially symmetric dipole particles ($\partial_\mu \partial'_\mu = \partial_x \partial'_x + \partial_y \partial'_y$). In the bulk nematic liquid crystal, for example, $G(\mathbf{x}, \mathbf{x}') = \frac{1}{|\mathbf{x}-\mathbf{x}'|}$ so that

$$U_{\rm dd}^{bulk} = 4\pi K p_{hel} p'_{hel} \frac{(1 - 3cos^2\theta)}{r^3}, \tag{5.35}$$

where $\theta$ is the angle between $\mathbf{r}$ and $z$ and the director field around the particle has the form:

$$n_x = -p_{hel} \frac{y}{r^3}, \tag{5.36}$$
$$n_y = p_{hel} \frac{x}{r^3}.$$

The formula (5.35) means that helicoids with the same handedness $p_{hel} p'_{hel} > 0$ attract along $z$ axis and repel in perpendicular direction and vice versa for helicoids with different handedness $p_{hel} p'_{hel} < 0$ (see Fig. 5.2). In the nematic cell the interaction, falling off as $r^{-3}$ in the bulk NLC, becomes exponentially screened at distances comparable to the thickness $L$ of the cell. At the same time the borders between the attraction and repulsion zones transform from straight lines into some parabola like curves. These effects are caused only by the confining walls so they do not depend

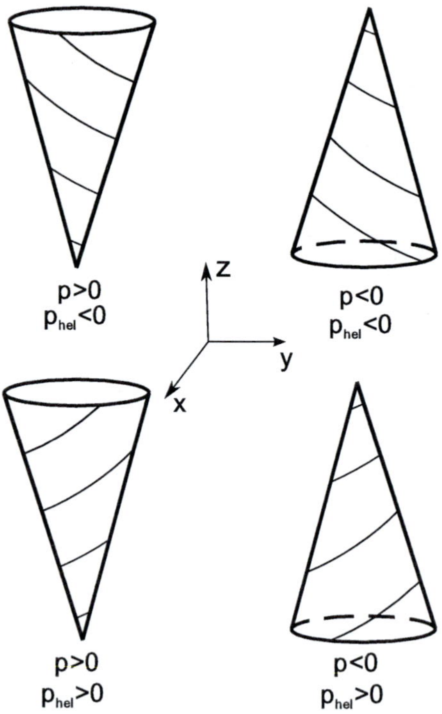


**Fig. 5.3** Helicoid cones with different dipole moments $p$ and $p_{hel}$ produced by the shape and azimuthal helical anchoring respectively (see (5.39)) (reprinted with permission from [93])

on the particles shape and anchoring. More detailed consideration of these issues is presented in [79, 80].

#### 5.3.2.2 Interaction Between Helicoid Cones

Consider cones (or other axially symmetric particles without symmetry plane $\sigma_{xy}$) with helicoidal alignment at the surface (see Fig. 5.3). The shape of the particle produces dipole moments $p^x_x = p^y_y = p$ according to the Table 5.1. Azimuthal helicoid anchoring gives rise to nonzero dipole moments $p^x_y = p_{hel} = -p^y_x$. Then substitution of it to the (5.11) gives the dipole-dipole elastic interaction between helicoid cones:

$$U_{\rm dd} = -4\pi K \left[pp'(\partial_x\partial'_x G_x(\mathbf{x},\mathbf{x}') + \partial_y\partial'_y G_y(\mathbf{x},\mathbf{x}')) \right. \\ \left. + p_{hel} p'_{hel}(\partial_x\partial'_x G_y(\mathbf{x},\mathbf{x}') + \partial_y\partial'_y G_x(\mathbf{x},\mathbf{x}'))\right]. \quad (5.37)$$

In the absence of the external fields $G_x = G_y = G$ and we come to the expression

$$U_{\rm dd} = -4\pi K(pp' + p_{hel}p'_{hel})\partial_\mu\partial'_\mu G(\mathbf{x},\mathbf{x}'). \quad (5.38)$$

In the confined nematic this formula gives the same results as in the [80] but with new coefficient $pp' + p_{hel}p'_{hel}$.

In the bulk nematic liquid crystal $G(\mathbf{x}, \mathbf{x}') = \frac{1}{|\mathbf{x}-\mathbf{x}'|}$ so that

$$U_{\rm dd}^{bulk} = 4\pi K(pp' + p_{hel}p'_{hel})\frac{(1 - 3cos^2\theta)}{r^3}, \tag{5.39}$$

and the director field around the particle has the form:

$$\begin{aligned} n_x &= p\frac{x}{r^3} - p_{hel}\frac{y}{r^3}, \\ n_y &= p\frac{y}{r^3} + p_{hel}\frac{x}{r^3}. \end{aligned} \tag{5.40}$$

Formulas similar to (5.35), (5.36), (5.39), (5.40) were first obtained in [70] but our results predict 3 times stronger interaction. In paper [70] authors made very good classification of different types of dipoles in nematostatics based on the firm fixation of the director field on the surface of the imaginary sphere enclosing the particle and containing all the defects inside. At first glance it seems quite similar to the coat-approach used above. The authors of [70] do not use any anchoring surface energy explicitly and consider the total energy as just the bulk one. But the total energy is the sum of the bulk and surface energies. This, we suppose, is the reason for the discrepancy. The surface terms do play their important role and increase the energy of the system and should be taken into consideration. In the current approach the surface terms are taking into account via terms $-4\pi q_\mu(\mathbf{x})n_\mu - 4\pi p_\mu^\alpha(\mathbf{x})\partial_\alpha n_\mu - 4\pi Q_\mu^{\alpha\beta}(\mathbf{x})\partial_\alpha\partial_\beta n_\mu$ in the effective free energy (5.6). These terms in the effective free energy (5.6) replace surface terms in the real free energy (5.22) so they can be effectively considered as surface born.

## 5.4 Banana-Shaped Particles in a Nematic Cell

As an example of the interaction between non axially-symmetrical colloids we consider now the interaction between banana-shaped particles (see Fig. 5.4) and as well consider the interaction between axially symmetrical particles including particles with helical screw-thread (see Figs. 5.2 and 5.3). Here let us content ourselves with the dipole-dipole interactions in homeotropic and planar nematic cells.

### *5.4.1 Homeotropic Cell*

Coordinate system for this case is depicted in Figs. 5.5a and 5.5b. Green's function then has a form that is well known in electrostatics [104]

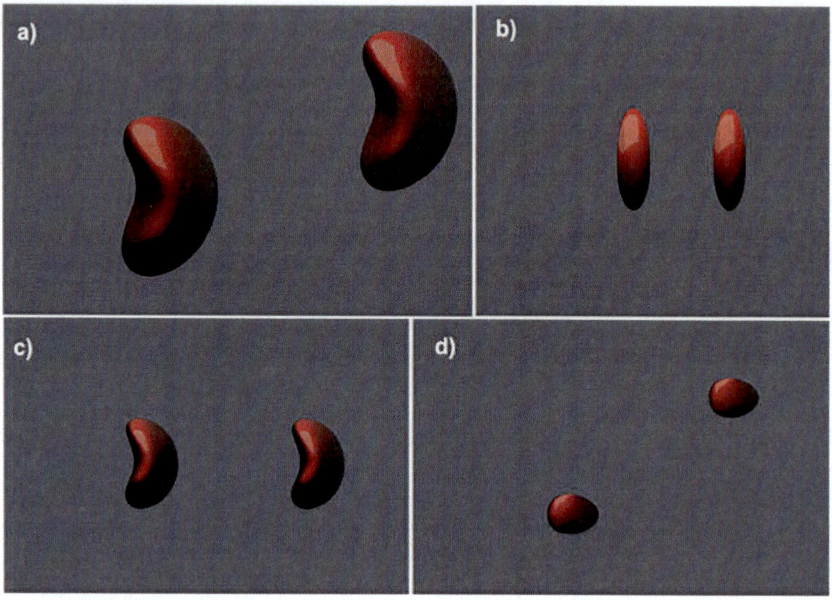


**Fig. 5.4** Banana-shaped colloids: **a** general view, **b** front view (see Figs. 5.6, 5.9), **c** side view (see Fig. 5.5), **d** top view (see Fig. 5.10) (reprinted with permission from [93])

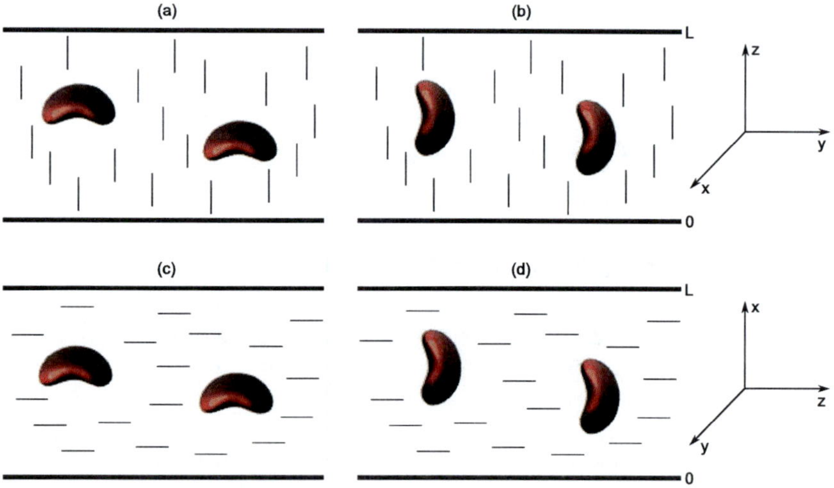


**Fig. 5.5** Banana-shaped particles in the nematic cell, side view (see Fig. 5.4 as well) (reprinted with permission from [93])

$$G_\mu(\mathbf{x},\mathbf{x}') = \tfrac{4}{L}\sum_{n=1}^{\infty}\sum_{m=-\infty}^{\infty} e^{im(\varphi-\varphi')}\sin\tfrac{n\pi z}{L} \times \sin\tfrac{n\pi z'}{L} I_m(\lambda_n\rho_<)K_m(\lambda_n\rho_>), \tag{5.41}$$

where $I_m$, $K_m$ are modified Bessel functions, $\tan\varphi = \frac{y}{x}$, $\tan\varphi' = \frac{y'}{x'}$, $\lambda_n = \frac{n\pi}{L}$, $\rho_<$ is the smaller of $\rho = \sqrt{x^2+y^2}$ and $\rho' = \sqrt{x'^2+y'^2}$.

Every banana-shaped particle has two orthogonal symmetry planes. Suppose first that the particles are oriented in such a way that these planes are parallel to the coordinate $xz$ and $yz$ planes (see Fig. 5.5a), i.e. particles are located primarily perpendicular to the director (we will use symbol $\perp$ for this case). Then using Table 5.1 one can easily find that the allowed dipole coefficients are $p_x^x$ and $p_y^y$. Below we omit the upper indexes and assume $p_y > p_x$. Note that $p_x = p_y$ for axially symmetric particles. It follows from (5.14) that

$$U_{\mathrm{dd},\perp}^{hom} = -4\pi K\left[p_x p_x' \partial_x \partial_x' G + p_y p_y' \partial_y \partial_y' G\right], \tag{5.42}$$

$$U_{\mathrm{dd},\perp}^{hom} = \frac{8\pi K}{L}\left[\left(p_x p_x' + p_y p_y'\right)A_1 + \left(p_x p_x' - p_y p_y'\right)A_2\cos 2\phi\right], \tag{5.43}$$

where $\rho = \sqrt{(y-y')^2+(x-x')^2}$ is the horizontal projection of the distance between particles, $\phi$ is the azimuthal angle between $\rho$ and $x$-axis,

$$A_1 = \sum_{n=1}^{\infty}\lambda_n^2 \sin\frac{n\pi z}{L}\sin\frac{n\pi z'}{L}K_0(\lambda_n\rho), \tag{5.44}$$

$$A_2 = \sum_{n=1}^{\infty}\lambda_n^2 \sin\frac{n\pi z}{L}\sin\frac{n\pi z'}{L}K_2(\lambda_n\rho). \tag{5.45}$$

Before proceeding to a discussion of this interaction in the cell let us consider its features in the bulk liquid crystal. The Green's function for the bulk NLC is $G(\mathbf{x},\mathbf{x}') = \frac{1}{|\mathbf{x}-\mathbf{x}'|}$ so that $U_{\mathrm{dd},\perp}^{\mathrm{bulk}}$ is anisotropic as well,

$$U_{\mathrm{dd},\perp}^{\mathrm{bulk}} = -\frac{4\pi K}{r^3}\left[p_x p_x' + p_y p_y' - 3\sin^2\theta(p_x p_x'\cos^2\phi + p_y p_y'\sin^2\phi)\right], \tag{5.46}$$

where $r$ is the distance between particles, $\theta$ is the polar angle between $\mathbf{r}$ and $z$-axis, $\phi$ is the azimuthal angle between $\rho$ and $x$-axis.

Note that similar formula for bulk NLC was obtained in [70] but our result predicts three times stronger interaction.

Map of the attraction and repulsion zones between two particles with $z = z'$ and $p_y = 3p_x$ in the infinite crystal is depicted by the red dashed lines in Fig. 5.6.

Now assume that the particles are located in the centre of the homeotropic cell $z = z' = \frac{L}{2}$ (solid lines in Fig. 5.6). At small distances $\rho \ll L$ the interaction is the same as in the bulk nematic $U_{\mathrm{dd},\perp}^{hom} \to U_{\mathrm{dd},\perp}^{\mathrm{bulk}}$. But as $\rho$ increases the lateral zones

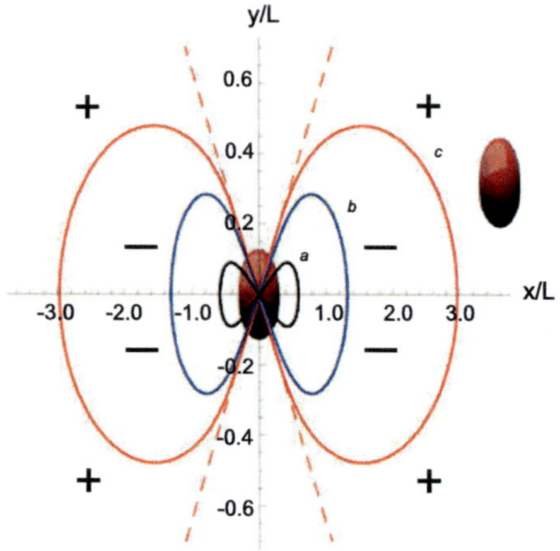


**Fig. 5.6** Map of the attraction and repulsion zones for two identical banana-shaped particles, $p_x = p'_x$ and $p_y = p'_y = \alpha p_x, \alpha > \sqrt{2}$, according to (5.43). Black line **a** corresponds to the case $p_y = 1.5p_x$, blue line **b** corresponds to the $p_y = 2p_x$ and red line **c** corresponds to the $p_y = 3p_x$. The particles are located in the centre of the homeotropic cell $z = z' = \frac{L}{2}$. Their orientation are shown on the Figs. 5.4b and 5.5a. Sign "-" means attraction (inside of the dumbbell-shaped regions), "-" means repulsion. If $p_y < \frac{p_x}{\sqrt{2}}$ then the attraction will appear along the $y$-axis. Dashed lines depict the map of interaction between two particles with $z = z'$ and $p_y = 3p_x$ in the absence of confinement (reprinted with permission from [93])

become closed. So that identical particles attract inside some dumbbell-shaped regions along $x$ axis when $p_y > \sqrt{2}p_x$. These regions shrinks as $|p_y - \sqrt{2}p_x|$ decreases and collapses to the point when $p_y = \sqrt{2}p_x$. The crossover from the attraction to the repulsion when both particles are located along $x$ axes and $p_y > \sqrt{2}p_x$ is shown on the Fig. 5.7.

When $\frac{p_x}{\sqrt{2}} < p_y < \sqrt{2}p_x$ there will be only repulsion for every $\phi$ in the perpendicular plane $\theta = \pi/2$. When $p_y < \frac{p_x}{\sqrt{2}}$ there will be attraction inside some dumbbell-shaped regions along $y$ axis and repulsion everywhere along $x$ axis.

Another important issue is the energy dependence on the distance between particles. It follows from (5.46) that in the bulk nematic host the interaction of dipolar colloidal particles decreases as $\rho^{-3}$ (see dashed line 4 in Fig. 5.8). But in the cell we see completely different picture. The interaction potential falls off as $\rho^{-3}$ only when $\rho < L$. At larger distances $\rho > L$ the potential becomes screened by the cell walls (see solid line 2 in Fig. 5.8). Such screening known as confinement effect was first reported experimentally in [54] and theoretically explained in [74, 80] for spherical particles. This phenomena is related only with the confining surfaces and therefore it occurs despite the particles shape.

But the particles orientation examined above is not the only possible. Their symmetry planes can be parallel to the coordinate $yz$ and $xy$ planes as well (see Fig. 5.5b), i.e. particles lie primarily parallel to the director (we will use || for this case). Then

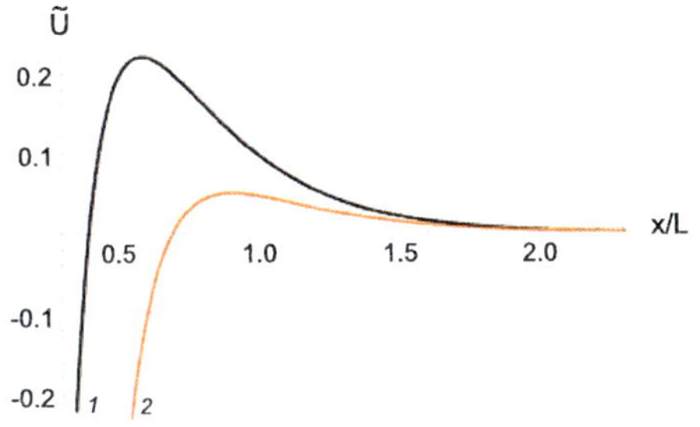


**Fig. 5.7** The crossover from the attraction to the repulsion between two banana-shaped particles located in the middle of the homeotropic cell $z = z' = \frac{L}{2}$ along $x$ axis (see Figs. 5.5a and 5.6), $p_x = p'_x$, $p_y = p'_y = \alpha p_x$ for $\alpha > \sqrt{2}$. $\tilde{U} = U^{hom}_{\text{dd},\perp} L^3/8\pi K(p_x p'_x + p_y p'_y)$, where $U^{hom}_{\text{dd},\perp}$ is given by (5.43) and $\phi = 0$. Solid black line 1 corresponds to the $p_y = p'_y = 1.5 p_x$, orange line 2 corresponds to the $p_y = p'_y = 1.7 p_x$ (reprinted with permission from [93])

we have one dipole coefficient $p^z_y = p \neq 0$ as follows from Table 5.1. Thus

$$U^{hom}_{\text{dd},||} = -4\pi K p p' \partial_z \partial'_z G, \tag{5.47}$$

$$U^{hom}_{\text{dd},||} = -\frac{16\pi K p p'}{L} \sum_{n=1}^{\infty} \lambda_n^2 \cos\frac{n\pi z}{L} \cos\frac{n\pi z'}{L} K_0(\lambda_n \rho). \tag{5.48}$$

As well as in the previous case the interaction given by (5.48) is screened by the cell walls (solid line 1 in Fig. 5.8). But here it exhibits cylindrical symmetry. In particular, parallel dipoles with $z = z'$ attract each other throughout the cell plane. In the unlimited case $G = \frac{1}{|\mathbf{x}-\mathbf{x}'|}$ and (5.48) becomes

$$U^{bulk}_{\text{dd},||} = -\frac{4\pi K p p'}{r^3}(1 - 3\cos^2\theta). \tag{5.49}$$

Similar result for the bulk NLC was obtained in [70] but our result again predicts three times stronger interaction.

For axially symmetrical particles formula (5.38) with the Green function (5.41) gives the result of dipole-dipole interaction between such particles as

$$U^{hom}_{\text{dd},axsym} = \frac{16\pi K(p p' + p_{hel} p'_{hel})}{L} \times \sum_{n=1}^{\infty} \lambda_n^2 \sin\frac{n\pi z}{L} \sin\frac{n\pi z'}{L} K_0(\lambda_n \rho), \tag{5.50}$$

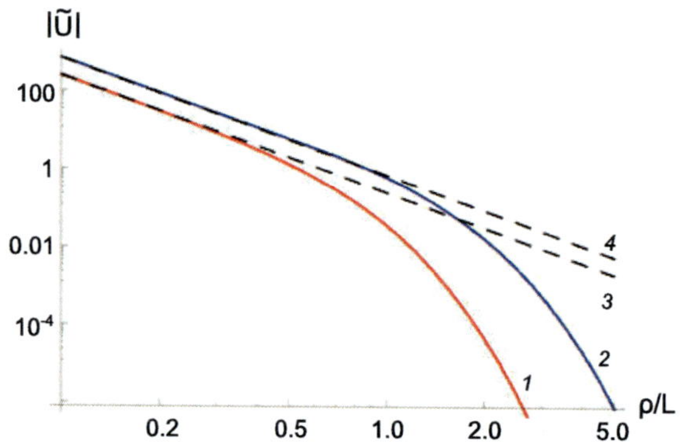


**Fig. 5.8** Log-log plot of the dimensionless energy of the dipole-dipole interaction as a function of the distance between two banana-shaped particles located in the middle of the homeotropic cell $z = z' = \frac{L}{2}$. Solid blue line 2 corresponds to the particles repulsion for the orientation along $y$ axis depicted on the Fig. 5.5a, $p_x = p'_x$, $p_y = p'_y = 2p_x$, $2\phi = \pi$ and $\tilde{U} = U_{\rm dd}L^3/8\pi K(p_x p'_x + p_y p'_y)$, where $U_{\rm dd} > 0$ is given by (5.43). Dashed line 4 is an appropriate power law asymptotic $U_{\rm dd}^{\rm unc}L^3/4\pi K(p_x p'_x + p_y p'_y) \propto \left(\frac{L}{\rho}\right)^3$. Solid red line 1 corresponds to the particles attraction along $y$ axis on the Fig. 5.5b, $p_y^z = p \neq 0$ and $\tilde{U} = U_{\rm dd}L^3/16\pi Kpp'$, where $U_{\rm dd} < 0$ is given by (5.48). Dashed line 3 is the appropriate power law asymptotic $\frac{1}{4}\left(\frac{L}{\rho}\right)^3$ (reprinted with permission from [93])

where $\rho = \sqrt{(y - y')^2 + (x - x')^2}$ is the horizontal projection of the distance between particles. We see that axially symmetric particles either attract or repel each other independent on $\phi$ everywhere inside the homeotropic cell, while the dipole-dipole interaction between banana-shaped particles can be either anisotropic or independent of $\phi$ .

### 5.4.2 *Planar Cell*

Let us choose the coordinate system as shown in Figs. 5.5c and 5.5d. Then the Green's function is as follows:

$$G_\mu(\mathbf{x}, \mathbf{x}') = \tfrac{4}{L} \textstyle\sum_{n=1}^{\infty} \sum_{m=-\infty}^{\infty} e^{im(\varphi-\varphi')} \sin \frac{n\pi x}{L} \times \sin \frac{n\pi x'}{L} I_m(\lambda_n \rho_<) K_m(\lambda_n \rho_>), \quad (5.51)$$

where $L$ is the cell thickness, $I_m$, $K_m$ are modified Bessel functions, $\tan\varphi = \frac{y}{z}$, $\tan\varphi' = \frac{y'}{z'}$, $\lambda_n = \frac{n\pi}{L}$, $\rho_<$ is the smaller of $\rho = \sqrt{z^2 + y^2}$ and $\rho' = \sqrt{z'^2 + y'^2}$. This Green's function was already used by authors of [80] to describe interactions between axially symmetric particles. Their predictions were found in good agreement with the experimental data for a wide range of $L$ [73].

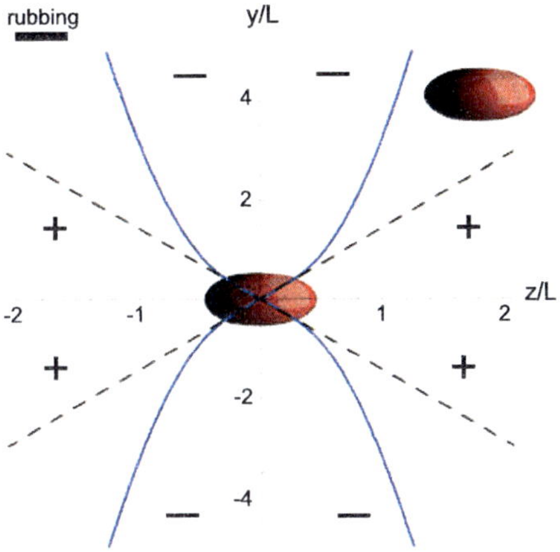


**Fig. 5.9** Map of the attraction and repulsion zones for two identical banana-shaped particles, $p = p'$, according to the (5.53). The particles are located in the centre of the planar cell $x = x' = \frac{L}{2}$. Their orientation are shown on the Figs. 5.4b and 5.5c. Sign "-" means attraction, "+" means repulsion (reprinted with permission from [93])

Imagine first that the particles are oriented as depicted in Fig. 5.5c. Hence the every particle has two symmetry elements affecting on the multipole coefficients existence. They are $\sigma_{xz}$ and $\sigma_{xy}$. Therefore, as follows from Table 5.1, director deviations here can be described by the only dipole coefficient $p_x^z = p \neq 0$. Then

$$U_{\mathrm{dd},||}^{plan} = -4\pi K p p' \partial_z \partial_z' G, \tag{5.52}$$

$$U_{\mathrm{dd},||}^{plan} = \frac{8\pi K p p'}{L} \sum_{n=1}^{\infty} \lambda_n^2 \sin \frac{n\pi x}{L} \times \sin \frac{n\pi x'}{L} \left[ K_0(\lambda_n \rho) + K_2(\lambda_n \rho) \cos 2\theta \right], \tag{5.53}$$

where $\rho = \sqrt{(y-y')^2 + (z-z')^2}$ is the horizontal projection of the distance between particles, $\theta$ is the angle between $\rho$ and $z$-axis. As well as in the homeotropic cell the particles do not "feel" the cell walls at small distances $U_{\mathrm{dd},||}^{plan} \to -\frac{4\pi K p p'}{\rho^3}(1 - 3\cos^2\theta)$. But if $\rho$ increases the interaction falls off exponentially ($K_n(z \to \infty) \propto \frac{e^{-z}}{\sqrt{z}}$) and the borders between zones transform from straight lines into some parabola-like curves (see Fig. 5.9).

Now suppose that the particles symmetry planes are parallel to the coordinate $yz$ and $xz$ planes (see Fig. 5.5d). The allowed dipole coefficients are $p_x^x = p_x$ and $p_y^y = p_y$. Thus

$$U_{\mathrm{dd},\perp}^{plan} = -4\pi K \left[ p_x p_x' \partial_x \partial_x' G + p_y p_y' \partial_y \partial_y' G \right], \tag{5.54}$$

$$U_{\mathrm{dd},\perp}^{plan} = \frac{8\pi K}{L} \left[ -p_x p_x' B_1 + p_y p_y' B_2 - p_y p_y' B_3 \cos 2\theta \right], \tag{5.55}$$

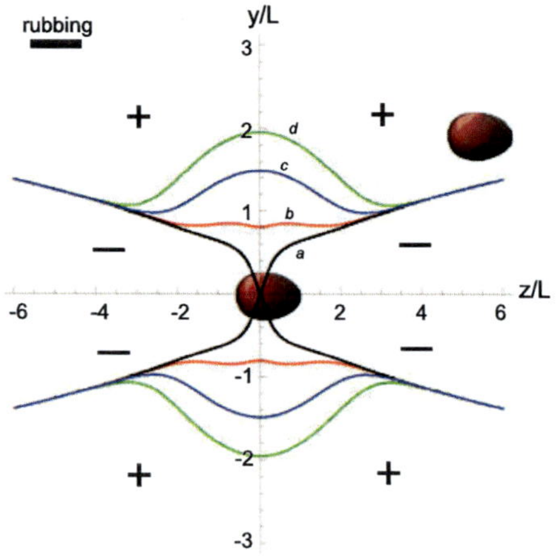


**Fig. 5.10** Map of the attraction and repulsion zones for two identical banana-shaped particles $p_x = p'_x$ and $p_y = p'_y$, according to the (5.55). The particles are located in the centre of the planar cell $x = x' = \frac{L}{2}$. Their orientation are shown on the Figs. 5.4d and 5.5d. Black line **a**: $p_x = p_y$. Red line **b**: $p_x = 2p_y$. Blue line **c**: $p_x = 5p_y$. Green line **d**: $p_x = 10p_y$. Sign "-" means attraction, '"+" means repulsion (reprinted with permission from [93])

where

$$B_1 = 2\sum_{n=1}^{\infty} \lambda_n^2 \cos\frac{n\pi x}{L} \cos\frac{n\pi x'}{L} K_0(\lambda_n \rho), \tag{5.56}$$

$$B_2 = \sum_{n=1}^{\infty} \lambda_n^2 \sin\frac{n\pi x}{L} \sin\frac{n\pi x'}{L} K_0(\lambda_n \rho), \tag{5.57}$$

$$B_3 = \sum_{n=1}^{\infty} \lambda_n^2 \sin\frac{n\pi x}{L} \sin\frac{n\pi x'}{L} K_2(\lambda_n \rho). \tag{5.58}$$

At small distances $B_1 \to \frac{1}{2}\left(\frac{L}{\rho}\right)^3$, $B_2 \to \frac{1}{4}\left(\frac{L}{\rho}\right)^3$ and $B_3 \to \frac{3}{4}\left(\frac{L}{\rho}\right)^3$ and we come to the fact that in this case $U_{\text{dd}} \to U_{\text{dd}}^{\text{bulk}}$ when $\rho \ll L$ as well. Note that here $U_{\text{dd},\perp}^{\text{bulk}}$ is given by (5.46) if we set $x = x'$ ($\phi = \pi/2$), that is, $U_{\text{dd},\perp}^{\text{bulk}} = -\frac{4\pi K}{r^3}\left[p_x p'_x + p_y p'_y - 3p_y p'_y \sin^2\theta\right]$. Say, for instance, $p_x > p_y$. Then it can be easily found that the interaction between such particles in the bulk nematic is completely repulsive or attractive. In the cell we again have both attraction and repulsion (see Fig. 5.10). In turn, since the summation in (5.56) starts from $n = 2$ $B_1$ falls off faster than $B_2$ and $B_3$. Therefore when $\rho \gg L$ the interaction is determined only by the coefficients $p_y$ and $p'_y$. Due to this at large distances these particles will interact as axially symmetrical ones (black lines in Fig. 5.10). On the same grounds, if we set $p_y > p_x$ no attraction will appear along the $y$-axis. Map of the interaction in this case will be quite similar to that one for the axially symmetrical particles.

For axially symmetrical particles formula (5.38) with the Green function (5.51) gives the result of dipole-dipole interaction between such particles as

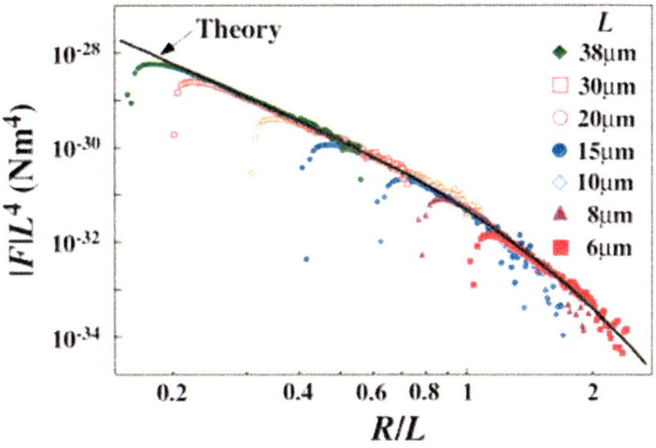


**Fig. 5.11** Dependence of the normalized interparticle force $FL^4$ on the reduced interparticle distance $R/L$ in nematic cell at various thickness $L$. The solid line is the theoretically calculated one from Eqs. (5.59) for $p = p' = 2.04a^2$ and $K = 7.05$ pN (NLC MJ032358) (reprinted with permission from [73])

$$U^{plan}_{\mathrm{dd},axsym} = \frac{16\pi K(pp' + p_{hel}p'_{hel})}{L}(F_1 - F_2\cos^2\theta), \tag{5.59}$$

$$F_1 = \sum_{n=1}^{\infty} \frac{\lambda_n^2}{2} \sin\frac{n\pi x}{L} \sin\frac{n\pi x'}{L}\left[K_0(\frac{n\pi\rho}{L}) + K_2(\frac{n\pi\rho}{L})\right]$$
$$-\lambda_n^2 \cos\frac{n\pi x}{L}\cos\frac{n\pi x'}{L}K_0(\frac{n\pi\rho}{L}), \tag{5.60}$$

$$F_2 = \sum_{n=1}^{\infty} \lambda_n^2 \sin\frac{n\pi x}{L} \sin\frac{n\pi x'}{L}K_2(\frac{n\pi\rho}{L}). \tag{5.61}$$

where $\rho = \sqrt{(y - y')^2 + (x - x')^2}$ is the horizontal projection of the distance between particles and $\theta$ is the angle between $\rho$ and $z$. Formula (5.59) was found to explain very well experimental results [73] where $\theta = 0$ (see Fig. 5.11).

## 5.5 Elastic Monopoles in a Nematic Cell

In this section, we will consider elastic monopoles in a nematic cell and discover that they are insensitive to the type of confinement. Suppose we have two ellipsoidal particles suspended in the cell (Fig. 5.12). For simplicity, let us assume that their orientation is fixed and such that the long axes make angles $\omega$ and $\omega'$ with $\mathbf{n}_0$, $0 < \omega, \omega' < \frac{\pi}{2}$ and lie in the plane of the figure. In practice, configurations of this symmetry have been realized through asymmetric anchoring conditions [105] and via light-induced rotation of photo-responsive colloids [106]. Since ellipsoids have a centre of symmetry, dominant deformations produced by these particles are elastic monopoles: $q_x = q'_x = 0, q_y = q, q'_y = q'$ in the homeotropic and $q_x = q$,

$q'_x = q', q_y = q'_y = 0$ in the planar cell (see Table 5.1). Then, as follows from (5.41) and (5.11), the monopole-monopole interaction in the homeotropic cell is given by

$$U_{\text{qq}}^{\text{hom}} = -\frac{16\pi Kqq'}{L}\sum_{n=1}^{\infty}\sin\frac{n\pi z}{L}\sin\frac{n\pi z'}{L}K_0(\lambda_n\rho), \tag{5.62}$$

where $\rho = \sqrt{(y-y')^2+(x-x')^2}$. In the same way we can find from (5.51) and (5.11) that in the planar cell this interaction is described by

$$U_{\text{qq}}^{\text{plan}} = -\frac{16\pi Kqq'}{L}\sum_{n=1}^{\infty}\sin\frac{n\pi x}{L}\sin\frac{n\pi x'}{L}K_0(\lambda_n\rho), \tag{5.63}$$

where $\rho = \sqrt{(y-y')^2+(z-z')^2}$. Expressions (5.62) and (5.63) demonstrate that the monopole-monopole interaction is the same and does not depend on the type of the nematic cell (see Fig. 5.13) as $z$ on the Fig. 5.12a is the same as $x$ on the Fig. 5.12b. For small distances $\rho \ll L$ both (5.62) and (5.63) converge to the Coulomb-like law

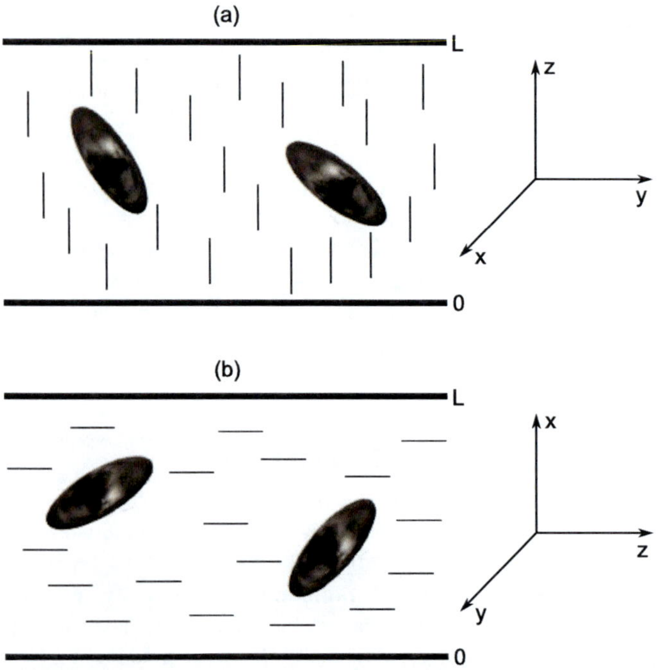


**Fig. 5.12** Ellipsoidal particles in the homeotropic **a** and planar **b** nematic cell (reprinted with permission from [93])

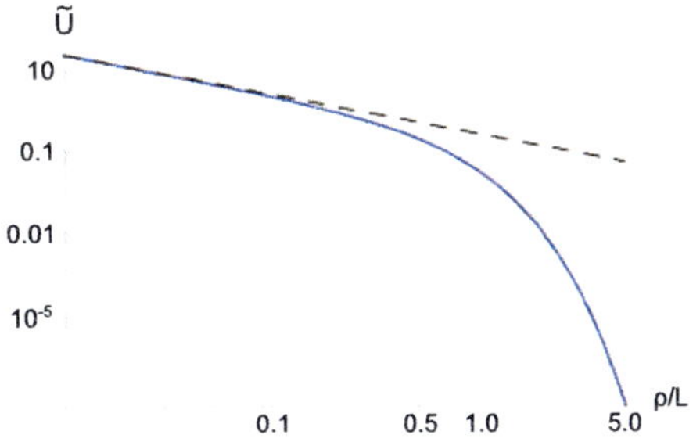


**Fig. 5.13** Monopole-monopole interaction in a nematic cell. Elastic monopoles do not "feel" the type of the cell. Blue line 1 corresponds to $\tilde{U} = -U_{\rm qq}^{\rm plan} L/16\pi Kqq'$. Here $U_{\rm qq}^{\rm plan}$ is given by (5.63). The dashed line 2 is the Coulomb-like $\tilde{U} = \frac{1}{4\rho/L}$ asymptotics for $\rho \ll L$ (reprinted with permission from [93])

$U_{\rm qq} = -4\pi Kqq'\frac{1}{r}$ (see Fig. 5.13). Note that opposite elastic charges repel, and like charges attract.

## 5.6 Colloidal Particles in a Deformed Nematic

In this section, we derive the free energy of a colloid in a deformed (non-uniform) director field. An alternative approach to this question was developed in [107] and obtained results differ a little bit from ours. Consider again a single particle located at point $\mathbf{x}$ in a nematic liquid crystal which is deformed by external boundaries or forces, *i.e.*, the director field is intrinsically non-uniform $\mathbf{n} = \mathbf{n}(\mathbf{r})$. Similarly to the case of two colloids, these intrinsic deformations of the director overlap with those induced by the particle, thereby making its energy position-dependent. If the particle size is much smaller than the characteristic length of the intrinsic director variations, $(\nabla \cdot \mathbf{n})^{-1}$, the particle can be treated as if it was locally in a uniform director field. In the local Cartesian coordinate system where $\mathbf{n}(\mathbf{x}) \approx (n_x, n_y, 1)$, the total energy of the system is as follows:

$$F_{eff} = K \int dV \left[ \frac{(\nabla n_\mu)^2}{2} - 4\pi q_\mu(\mathbf{x}) n_\mu - 4\pi p_\mu^\alpha(\mathbf{x}) \partial_\alpha n_\mu \right.$$
$$\left. -4\pi Q_\mu^{\alpha\beta}(\mathbf{x}) \partial_\alpha \partial_\beta n_\mu \right]. \quad (5.64)$$

Here, the first term is the energy of the LC deformations and the sum of three subsequent terms represents the overlap of deformations. If we subtract the particle's self-energy in the homogeneous director field $\mathbf{n}(\mathbf{x}) = (0, 0, 1)$ and the energy of the bulk deformations caused by external forces and fields, we will find the energy of interaction between the particle and the intrinsic deformations of the host medium, or energy of the particle in the deformed director field $\mathbf{n}(\mathbf{x})$. In the local Cartesian coordinates:

$$F_{\rm self,curv} = -4\pi Kq_\mu n_\mu(\mathbf{x}) - 4\pi Kp_\mu^\alpha \partial_\alpha n_\mu(\mathbf{x}) - 4\pi KQ_\mu^{\alpha\beta} \partial_\alpha \partial_\beta n_\mu(\mathbf{x}). \quad (5.65)$$

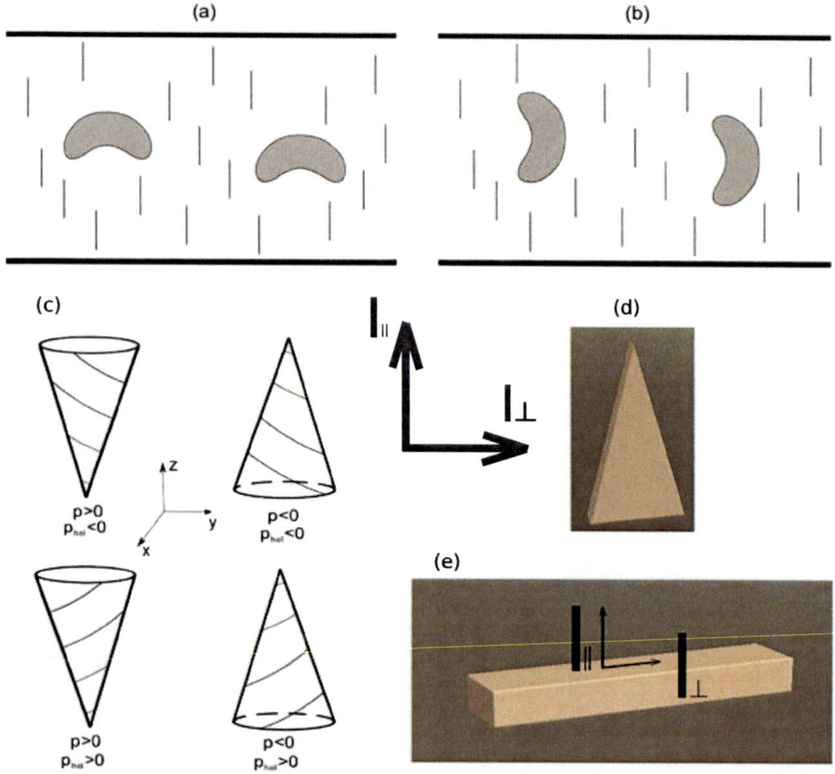


**Fig. 5.14** Examples of dipole and quadrupole particles. Pair of vectors $(\mathbf{l}_{||}, \mathbf{l}_{\perp})$ lies in a vertical symmetry plane and sets local coordinate system XYZ. Vector $\mathbf{l}_{||}$ is parallel to $\mathbf{n}(\mathbf{x})$; **a** banana-shaped dipole particles with two vertical symmetry planes; **b** banana-shaped dipole particles with one vertical and one horizontal symmetry planes; **c** axially symmetric dipole particles with usual elastic moment $p$ and helical elastic moment $p_{hel}$ **d** example of triangular particles with two vertical symmetry planes **e** example of quadrupole particles with two different quadrupole moments

If a colloidal particle is inclined with respect to the local director field $\mathbf{n}(\mathbf{x})$ and has elastic charge $\mathbf{q} = (q_x, q_y, 0)$, the first term has the biggest value in comparison with others.

The second term defines the energy of a dipole. Here we will consider dipole particles which have at least two symmetry planes: either two vertical planes $xz$ and $yz$ (Fig. 5.14(a,c,d)) or one vertical plane $yz$ and one horizontal plane $xy$ (Fig. 5.14(b)) in the local coordinate system. Axially-symmetric particles with screw-thread ((Figs. 5.14(c) and 5.2)) also have a helical elastic dipole moment $p_{hel}$.

The last term represents particles with a quadrupole moment. We will focus on quadrupoles with two symmetry planes or particles with three symmetry planes (Fig. 5.14(a,b,d,e)). Helical quadrupole moment $Q_{hel}$ characterizes, for instance, axially-symmetric cone-like particles with screw-thread (like those shown in

Fig. 5.14(c)). Then, using Tables 5.1 and 5.2, we can explicitly write Eq. (5.65) as follows

$$\begin{aligned} F_{\text{self,curv}} &= -4\pi K q_\mu n_\mu \\ &-4\pi K[p_x^x \partial_x n_x + p_x^y \partial_y n_x] \\ &-4\pi K[p_y^x \partial_x n_y + p_y^y \partial_y n_y + p_y^z \partial_z n_y] \\ &-4\pi K[2Q_x^{xz} \partial_z \partial_x n_x + 2Q_x^{yz} \partial_z \partial_y n_x] \\ &-4\pi K[2Q_y^{yz} \partial_z \partial_y n_y + 2Q_y^{xz} \partial_z \partial_x n_y]. \end{aligned} \tag{5.66}$$

Equation (5.66) can be written in an invariant coordinate independent form:

$$F_{\text{self,curv}} = F_{\text{Coul,curv}} + F_{\text{Dip,curv}} + F_{\text{Quad,curv}}. \tag{5.67}$$

Here,

$$F_{\text{Coul,curv}} = -4\pi K \mathbf{q} \cdot \mathbf{n} \tag{5.68}$$

is valid only in the vicinity of a vector $\mathbf{n}_0$ such that $(\mathbf{n} - \mathbf{n}_0) \cdot \mathbf{n}_0 \approx 0$ and the monopole $\mathbf{q}$ is perpendicular to $\mathbf{n}_0$. In fact, vector $\mathbf{n}_0$ sets the average orientation in this area, the ground state, and defines the elastic charge vector $\mathbf{q}$ in this area. For instance if $\mathbf{n}_0 || z$ then $\mathbf{q} = (q_x, q_y, 0)$.

$$\begin{aligned} F_{\text{Dip,curv}} &= -4\pi K p(\mathbf{l}_{||} \cdot \mathbf{n})\text{div}\mathbf{n} \\ &-4\pi K \Delta p(\mathbf{l}_{||} \cdot \mathbf{n})\mathbf{l}_\perp \cdot (\mathbf{l}_\perp \cdot \nabla)\mathbf{n} \\ &-4\pi K p_{hel}(\mathbf{n} \cdot \text{curl}\mathbf{n}) - 4\pi K p_{ban} \mathbf{l}_\perp \cdot (\mathbf{n} \times \text{curl}\mathbf{n}), \end{aligned} \tag{5.69}$$

where we defined $p = p_x^x$, $p_{hel} = p_y^x = -p_x^y$, $p_{ban} = -p_y^z$, $\Delta p = p_y^y - p_x^x$ with the local axes $z$ and $y$ being along $\mathbf{n} = \mathbf{l}_{||}$ and $\mathbf{l}_\perp$, respectively. Vector $\mathbf{l}_{||} = \mathbf{n}(\mathbf{x})$ is along the particle itself and coincides with the local director field $\mathbf{n}(\mathbf{x})$, and vector $\mathbf{l}_\perp$ is perpendicular to it and lies in the vertical symmetry plane (Fig. 5.14). Strictly speaking, the direction of $\mathbf{l}_\perp$ is not known *a priori* and has to be determined by minimizing the total energy (5.67). Equation (5.69) gives the energy of a dipole colloidal particle with two symmetry planes in a non-uniform director field.

$$\begin{aligned} F_{\text{Quad,curv}} = &-4\pi K Q(\mathbf{n} \cdot \nabla)\text{div}\mathbf{n} \\ &-4\pi K Q_{hel}(\mathbf{l}_{||} \cdot \mathbf{n})\mathbf{n} \cdot (\mathbf{n} \cdot \nabla)\text{curl}\mathbf{n} \\ &-4\pi K \Delta Q \mathbf{l}_\perp \cdot (\mathbf{l}_\perp \cdot \nabla)(\mathbf{n} \times \text{curl}\mathbf{n}), \end{aligned} \tag{5.70}$$

where $Q = 2Q_x^{xz}$, $\Delta Q = 2(Q_x^{xz} - Q_y^{yz})$, $Q_{hel} = 2Q_y^{xz} = -2Q_x^{yz}$. Expression (5.70) defines the energy of a colloidal dipole with two symmetry planes or a quadrupole particle with three symmetry planes in a non-uniform director field (Fig. 5.14).

Let us now apply the general formulas (5.69) and (5.70) to some specific cases of colloids.

For *usual axially-symmetric particles without helical screw-thread (e.g., Fig. 5.1)*, the energy of the particle in the deformed director field takes the form:

$$F_{\text{axial-sym,curv}} = -4\pi K p(\mathbf{l}_{||} \cdot \mathbf{n})\text{div}\mathbf{n} - 4\pi K Q(\mathbf{n} \cdot \nabla)\text{div}\mathbf{n}, \tag{5.71}$$

where both terms should be taken into account for dipole particles (see Fig. 5.1a), and only second term remains for symmetrical quadrupole particles (Fig. 5.1b and c). Note that $Q > 0$ for the Saturn ring configuration (Fig. 5.1 (b)), and $Q < 0$ for boojums (Fig. 5.1 (c)). Equation (5.71) coincides with [65] and shows that a dipole particle moves into regions with splay deformations. As a result, small water droplets congregate at the center of a big nematic drop with normal anchoring at the surface and near the poles/boojums in case of planar boundary conditions [34]. Quadrupole particles behave differently depending on the anchoring type. Since $Q < 0$ for planar conditions and $Q > 0$ for homeotropic, quadrupole particles with planar anchoring will move towards regions of strong splay deformations whereas homeotropic particles will be repelled from them.

For *axially-symmetric particles with helical screw-thread* (Fig. 5.14c), the energy has the form:

$$\begin{aligned} F_{\text{axial-helic,curv}} = &-4\pi K p(\mathbf{l}_{||} \cdot \mathbf{n})\text{div}\mathbf{n} \\ &-4\pi K p_{hel}(\mathbf{n} \cdot \text{curl}\mathbf{n}) \\ &-4\pi K Q(\mathbf{n} \cdot \nabla)\text{div}\mathbf{n} - 4\pi K Q_{hel}(\mathbf{l}_{||} \cdot \mathbf{n})\mathbf{n} \cdot (\mathbf{n} \cdot \nabla)\text{curl}\mathbf{n}, \end{aligned} \tag{5.72}$$

where $p = 0$ and $Q_{hel} = 0$ if the particle has horizontal symmetry plane (Fig. 5.2). For axially *non-symmetric* dipole particles, which have *two vertical symmetry planes* (Fig. 5.14(a,d)) the energy reads

$$\begin{aligned} F_{\text{dip,vv,curv}} = &-4\pi K p(\mathbf{l}_{||} \cdot \mathbf{n})\text{div}\mathbf{n} - 4\pi K \Delta p(\mathbf{l}_{||} \cdot \mathbf{n})\mathbf{l}_{\perp} \cdot (\mathbf{l}_{\perp} \cdot \nabla)\mathbf{n} \\ &-4\pi K Q(\mathbf{n} \cdot \nabla)\text{div}\mathbf{n} - 4\pi K \Delta Q \mathbf{l}_{\perp} \cdot (\mathbf{l}_{\perp} \cdot \nabla)(\mathbf{n} \times \text{curl}\mathbf{n}). \end{aligned} \tag{5.73}$$

For axially *non-symmetric* dipole banana-shaped vertical particles, which have *one vertical and one horizontal symmetry planes* (Fig. 5.14(b)) the energy is as follows

$$\begin{aligned} F_{\text{banana,curv}} = &-4\pi K p_{ban}\mathbf{l}_{\perp} \cdot (\mathbf{n} \times \text{curl}\mathbf{n}) \\ &-4\pi K Q(\mathbf{n} \cdot \nabla)\text{div}\mathbf{n} - 4\pi K \Delta Q \mathbf{l}_{\perp} \cdot (\mathbf{l}_{\perp} \cdot \nabla)(\mathbf{n} \times \text{curl}\mathbf{n}). \end{aligned} \tag{5.74}$$

This framework allows for understanding the behavior of colloids in the vicinity of topological defects in the director field. Following experiment [34], consider a spherical NLC droplet with normal boundary conditions on the surface that induce radial orientation of the director with a defect (radial hedgehog) in the center (Fig. 5.15). In spherical coordinates, the director can be represented as $n_r = 1$, $n_\theta = 0$, $n_\varphi = 0$, so $\text{div}\mathbf{n} = \frac{1}{r}$, where $r$ is the distance from the center of the drop. If the particle has dipole moment parallel to the director, the dominant part of the energy reads $F_1 = -4\pi K p\frac{1}{r} + 4\pi K Q\frac{1}{r^2}$, and the particle therefore moves to the center, as was observed in [34] for $p = 2.05r_0^2 > 0$ and $Q = -0.2r_0^3 < 0$. Likewise, quadrupole particles will assemble in the center of the droplet accordingly to the energy

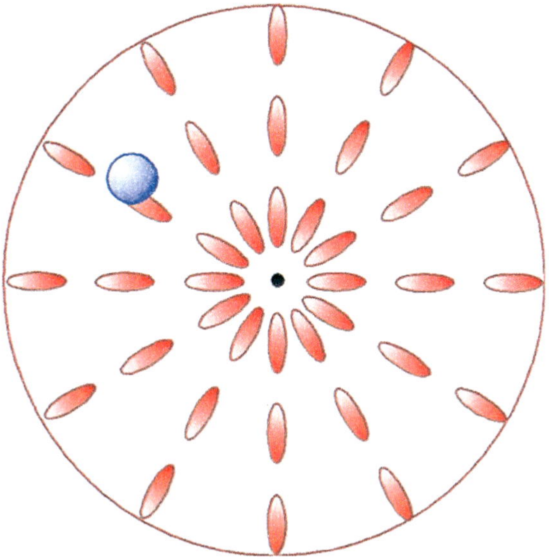

**Fig. 5.15** Director field in a nematic droplet with homeotropic anchoring. Because of the boundary conditions, a topological defect known as a radial hedgehog emerges in the center

$$F_2 = -4\pi K Q\,(\mathbf{n}\cdot\nabla)\,\mathrm{div}\mathbf{n} = 4\pi K Q \frac{1}{r^2}. \tag{5.75}$$

We can also find the energy of the interaction between a particle and a disclination line, another type of defects in nematics. Say, we have a cylinder with homeotropic boundary conditions which give rise to a disclination line along the cylinder axis (Fig. 5.16). Given the symmetry, the director reads $\mathbf{n} = (n_\rho, n_\varphi, n_z) = (1, 0, 0)$ in cylindrical coordinates. Thus, the free energy of a colloid near a radial disclination is as follows

$$\begin{aligned} F_1 &= -4\pi K p \mathrm{div}\mathbf{n} - 4\pi K Q\,(\mathbf{n}\cdot\nabla)\,\mathrm{div}\mathbf{n} \\ &= -4\pi K p \frac{1}{\rho} + 4\pi K Q \frac{1}{\rho^2}, \end{aligned} \tag{5.76}$$

which suggests the attraction of dipoles and negative quadrupoles (for instance, a sphere with boojums) toward the disclination line and repulsion of positive quadrupoles (a sphere with a Saturn ring, for example).

Equation (5.76) can be generalized for disclinations of any topological strength $m$, which defines the angle $\theta = m\phi + C$ of the director rotation in the plane perpendicular to the disclination (say, the angle between $\mathbf{n}$ and axis $x$). The corresponding energies of dipole and quadrupole particles are, respectively,

$$F_1 = -4\pi K p \mathrm{div}\mathbf{n} = -4\pi K p m \frac{1}{\rho} cos\left((m-1)\phi + C\right) \tag{5.77}$$

and

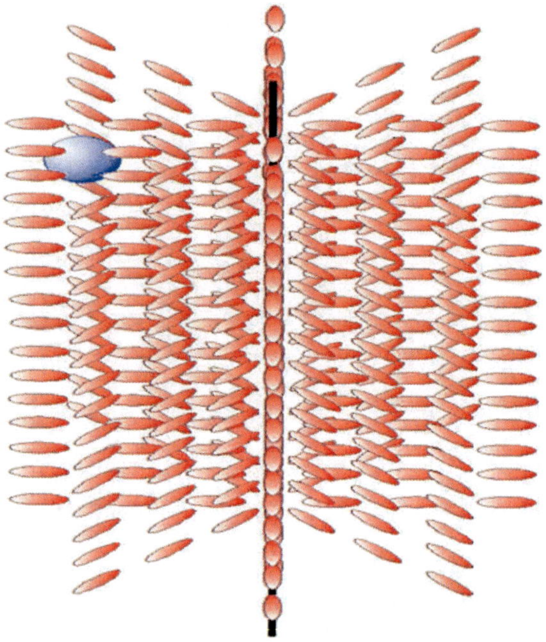

**Fig. 5.16** Small water droplet near a radial disclination line in a nematic liquid crystal

$$F_2 = -4\pi K Q\,(\mathbf{n}\cdot\nabla)\,\mathrm{div}\mathbf{n}$$
$$= \frac{4\pi K Q m}{\rho^2}\left[1 + (m-2)\sin^2((m-1)\phi + C)\right]. \tag{5.78}$$

In most cases, radial configuration considered above is unstable and "escapes in the third dimension" [108], as shown in Fig. 5.17. The ensuing director field can be written in cylindrical coordinates as $n_z = \cos u(\rho), n_\rho = \sin u(\rho), n_\varphi = 0$ with $u(R) = \frac{\pi}{2}$ and $u(0) = 0$, where, in the one-constant approximation, $\tan \frac{u(\rho)}{2} = \frac{\rho}{R}$ with $R$ being the cylinder radius [108]. Then the energy of a dipole particle takes the form

$$F_2 = -4\pi K p \mathrm{div}\mathbf{n}$$
$$= -\frac{16\pi K p R^3}{(R^2+\rho^2)^2} \tag{5.79}$$

which shows that elastic dipoles assemble at the center of the cylinder. For quadrupole particles, we have

$$F_2 = -4\pi K Q\,(\mathbf{n}\cdot\nabla)\,\mathrm{div}\mathbf{n}$$
$$= \frac{64\pi K Q R^4\rho^2}{(R^2+\rho^2)^4}. \tag{5.80}$$

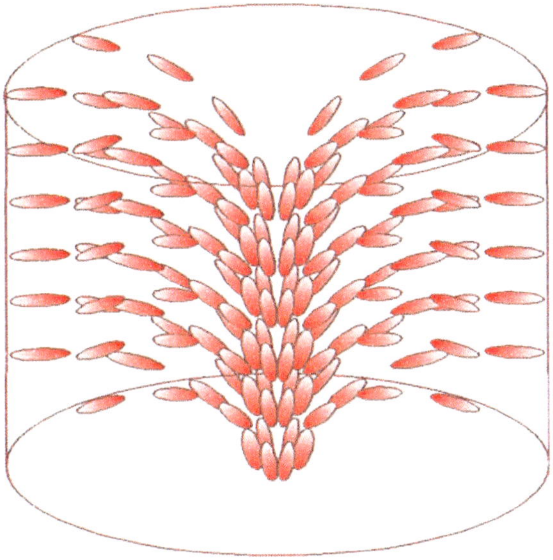

**Fig. 5.17** Radial disclination "escapes in the third dimension" in a cylindrical capillary with homeotropic boundary conditions

Equation (5.80) suggests that positive quadrupoles, $Q > 0$, *e.g.*, spheres with Saturn ring defects (Fig. 5.1 a,b), aggregate at the center of the capillary, which was observed experimentally in [55]. At the same time, particles with negative quadrupole moment $Q < 0$, such as spheres with boojums (Fig. 5.1c) should assemble at a distance $\rho_{min} = \frac{R}{\sqrt{3}}$ from the center. This result has not been confirmed experimentally yet.

## 5.7 Distribution of Nanoparticless in a Deformed Nematic

Thus far, we have focused on micron-sized colloidal particles. In such systems, the energy of the elastic interactions is of the order of hundreds and thousands of $k_B T$. Hence, the temperature does not play a significant role, except for affecting elastic constants and Brownian motion of the particles. Entropy can be safely neglected under these conditions. In this section, we will consider the case of sub-micron ($\sim$0.1 $\mu$m) and nano-sized colloids in a nematic LC. For low concentrations, the interactions between such particles typically do not exceed $\sim$1 $k_B T$, and entropic contributions to the free energy become important.

Let us consider a system of $N$ particles with $f(\mathbf{r})$ being the probability of finding a particle at point $\mathbf{r}$, $f \in (0, 1)$. Equivalently, one can think of $f(\mathbf{r})$ as the local volume fraction of particles,

$$\frac{1}{v_0} \int_V dV f(\mathbf{r}) = N, \tag{5.81}$$

where $V$ and $v_0$ is the system and particle volume, respectively. Then the configurational entropy of the system reads

$$S = -\frac{k_B}{v_0} \int_V dV \left\{ f(\mathbf{r}) \ln f(\mathbf{r}) + [1 - f(\mathbf{r})] \ln[1 - f(\mathbf{r})] \right\}. \tag{5.82}$$

For simplicity, we assume that the concentration is small, so that we can neglect the interparticle interactions and retain only the interaction with global nematic deformations. Thus, the internal energy can be defined as

$$E = \frac{1}{v_0} \int_V dV \epsilon(\mathbf{r}) f(\mathbf{r}), \tag{5.83}$$

where $\epsilon(\mathbf{r})$ is generally given by Eq. (5.67). For axially symmetric colloids with dipole elastic moment $p$ and quadrupole elastic moment $Q$, it takes the form (5.71),

$$\epsilon(\mathbf{r}) = -4\pi K p (\mathbf{l}_{||}\mathbf{n}) \mathrm{div}\mathbf{n} - 4\pi K Q (\mathbf{n}\nabla) \mathrm{div}\mathbf{n}. \tag{5.84}$$

Note that nanoparticles, because of their size, only weakly distort the NLC. Therefore, the corresponding multipole moments are determined by the particle's actual shape; for spheres, in particular, $p = 0$. Given (5.82) and (5.83), the grand potential of nanocolloids in a nematic host can be written as follows

$$\begin{aligned} \Omega_p = &\frac{1}{v_0} \int_V dV \left\{ (\epsilon(\mathbf{r}) - \mu) f(\mathbf{r}) \right\} \\ &+ \frac{k_B T}{v_0} \int_V dV \left\{ f(\mathbf{r}) \ln f(\mathbf{r}) + [1 - f(\mathbf{r})] \ln[1 - f(\mathbf{r})] \right\}, \end{aligned} \tag{5.85}$$

where chemical potential $\mu$ enforces the constraint $\int dV f(\mathbf{r}) = v_0 N$. Minimizing (5.85), we arrive at the distribution function for nanoparticles in a deformed nematic liquid crystal,

$$f(\mathbf{r}) = \frac{1}{1 + e^{(\epsilon(\mathbf{r}) - \mu)\beta}} \approx f_0 e^{-\frac{\epsilon(\mathbf{r})}{k_B T}}, \tag{5.86}$$

where $\beta = \frac{1}{k_B T}$, $f_0 = e^{\mu\beta} \ll 1$ for small concentrations and can be found from the condition $\int dV f(\mathbf{r}) = v_0 N$.

Then, for instance, the distribution of small spherical nano-colloids near a disclination line of strength $m$ is given by

$$f(\mathbf{r})^{discl}_{quad} = f_0 \exp\left[ -\frac{4\pi K Q m \left[ 1 + (m-2) \sin^2((m-1)\phi + C) \right]}{\rho^2 k_B T} \right], \tag{5.87}$$

There $\theta = m\phi + C$ is the director rotation angle in the plane perpendicular to the disclination (say, the angle between $\mathbf{n}$ and axis $x$) and $\phi$ is the azimuthal angle.

Near the cores of disclination lines at nanoscale distances, nematic ordering melts, and the order parameter, as well as the elastic constant $K$, tend to zero. Our theory is inapplicable in such regions.

Similarly, near an escaped radial disclination but far enough from the core we have

$$f(\mathbf{r})_{quad}^{capillar} = f_0 \exp\left[-\frac{64\pi K Q R^4 \rho^2}{(R^2+\rho^2)^4 k_B T}\right] \qquad (5.88)$$

with $\rho$ being the distance from the particle to the disclination line.

## 5.8 Coexistence of Two Colloidal Structures at the NLC-Air Interface

References [41, 42] experimentally observed the coexistence of two structures formed by glycerol droplets at the nematic-air interface. Such droplets form usual quasi-hexagonal (HL) and dense quasi-hexagonal (DL) structures (Fig. 5.18). The experimental setup comprises glycerol droplets of radius $R = 2.5\ \mu$m floating on the NLC-air interface in a cone-like Petri dish (Fig. 5.18a). Glycerol slowly flows out, ad a patch with the HL structure is shrinking. Initially, the average inter-particle distance in the HL packing is $r_3 = 10\ \mu$m. When it becomes approximately $r = r_c = \frac{\sqrt{3}r_3}{2} \approx 8.66\ \mu$m, the DL structure with the average lattice constant $r_1 = 6\ \mu$m emerges.

Due to planar anchoring of the LC molecules to the droplets surface, they are accompanied by surface boojums residing at their south poles. One can notice, however, that in the dense structure the boojums are shifted from the vertical axis. Indeed, Fig. 5.18(c) demonstrates that in the HL packing boojums show in the center of the droplet under crossed polarizers, whereas in Figs. 5.18(b) and (d) they are rotated equally for all particles in parallel planes. In Ref. [67], it was shown that such a rotation by an angle $\theta$ gives rise to elastic monopoles and Coulomb-like attraction between the particles, which can be expressed as $U_{qq} \propto -b^2 sin^2(2\theta)/r < 0$ for small angles. At the same time, the self-energy of the particle scales as $U_{self} \propto a \sin^2(\theta) > 0$. Since Coulomb attraction is proportional to the number of neighbors and inversely proportional to the spacing of the particles, one can conjecture that at some distance both terms will be balanced. Further compression will increase the attraction and cause collapse to the dense structure, which will be eventually stabilized by the elastic dipole-dipole repulsion (5.39), $U_{elb,thick}(r) = K\frac{\alpha^2 R^4}{r^3}$ with $R$ denoting the droplet radius. To quantify this transition, recall the total free energy of the system of colloids

$$U_{total} = \sum_i \left( U_{i,self} + \frac{1}{2}\sum_{j\neq i} U_{ij} \right) = \sum_i F_i, \qquad (5.89)$$

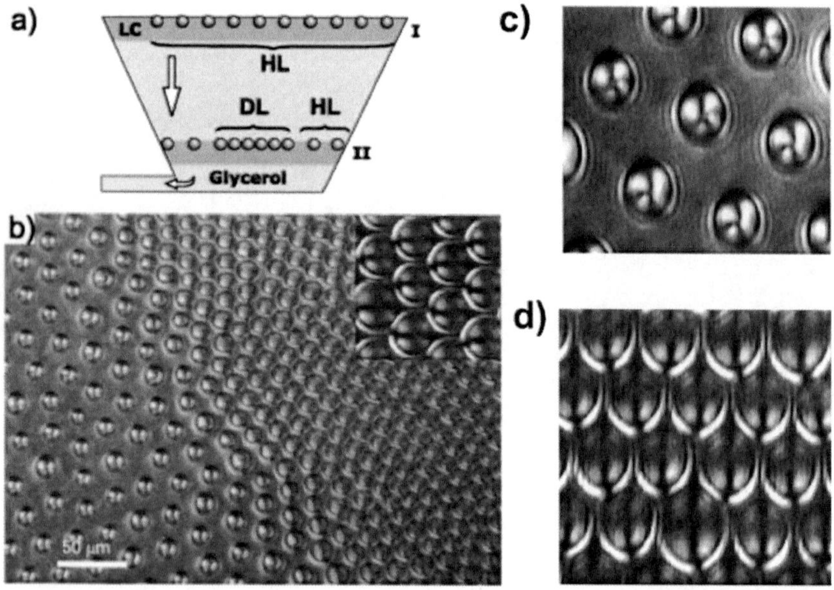


**Fig. 5.18** Two equilibrium quasi-hexagonal structures at the nematic-air interface which we refer to as the usual hexagonal (HL) and dense hexagonal (DL), respectively. The DL lattice emerges when the average spacing between the particles becomes smaller than some critical value $r_c$ (reprinted with permission from [42])

where $F_i$ is the average energy per particle, $F_i = U_{i,self} + \frac{1}{2}\sum_{neighbors} U_{ij}$. Taking for simplicity only the nearest 6 neighbors, we can write $F_i$ as a function of the average particle spacing $r$ and the rotation angle $\theta$ in the following form

$$F(r,\theta) = aKRsin^2(\theta) - 3Kb^2R^2sin^2(2\theta)G(r) + \frac{3K\alpha^2R^4}{r^3} + \frac{3f_{el}^2}{2\pi\sigma_{LCA}}\ln(\frac{r}{\lambda}), \tag{5.90}$$

where $G(r) = \frac{1}{r} - \frac{1}{\sqrt{r^2+4h^2}}$ is the Green function for a located at a distance $h$ from the interface. $G(r)$ is, in fact, a superposition of the Coulomb attraction between the particles, ~1/r, and repulsion between the particle and its mirror reflection, $\sim 1/\sqrt{r^2+4h^2}$, which accounts for the presence of the interface. In practice, one could approximate $G(r)$ by $G_0(r) = 1/r$, but it is instructive to keep both contributions. We consider that all particles have the same elastic charge (say $q_x$ only) as all boojums are rotated equally for all particles in parallel planes. Parameters $a$, $b$ and $\alpha$ are unknown dimensionless constants which we need to find from the comparison with experimental data. Self energy of the glycerol droplet with the boojum is proportional to the $a$, elastic charge of the glycerol droplet with skewed boojum is proportional to the $bR$ and elastic dipole moment of the glycerol droplet is proportional to the $\alpha R^2$. We know that $\alpha$ was found to be $\alpha = 0.2$ from the experiment [41] so that only $a$ and $b$ are unknown. The last term in (5.90) describes capillary attraction caused by deformations of the NLC-air interface, the force $f_{el} \approx 10K$ replaces standard gravitational buoyancy, $\sigma_{LCA}$ is the surface tension and $\lambda \approx 2$ mm is the capillary length [41].

Setting $h \approx 2R$, we see that mechanical equilibrium, $\frac{\partial F}{\partial \theta} = 0$, implies that

$$\cos(2\theta) = \begin{cases} 1, & r > r_b \\ \frac{a}{12b^2RG(r)} = \frac{G(r_b)}{G(r)}, & r \le r_b \end{cases} \tag{5.91}$$

where $r_b$ is the largest critical distance at which boojums start to move from the poles, $r_b$ can be found from the condition $\frac{a}{12b^2RG(r_b)} = 1$. Substituting (5.91) into (5.90), we arrive at the average per particle energy as a function of the particle spacing $r$ only

$$F(r) = F_1(r) + \frac{3K\alpha^2R^4}{r^3} + \frac{3f_{el}^2}{2\pi\sigma_{LCA}}\ln(\frac{r}{\lambda}), \tag{5.92}$$

where

$$F_1(r) = \begin{cases} 0, & r > r_b \\ 6Kb^2R^2G(r_b) - 3Kb^2R^2\left[G(r) + \frac{G^2(r_b)}{G(r)}\right], & r \le r_b. \end{cases} \tag{5.93}$$

Now we need to to find out under what conditions this function has two local minima and one maximum. In dimensionless units $x = r/R$, the condition $\frac{\partial F}{\partial r} = 0$ reduces to the equation

$$f(x) = f_1(x) + 3\alpha^2\left(\frac{1}{x \cdot x_3^3} - \frac{1}{x^4}\right) = 0, \tag{5.94}$$

where

$$f_1(x) = \begin{cases} 0, & x > x_b \\ b^2G'(x)\cdot\left[\frac{G^2(x_b)}{G^2(x)} - 1\right], & x \le x_b \end{cases} \tag{5.95}$$

Experimental data [41, 42] suggests that $x_1 = \frac{r_1}{R} = 2.4$, $x_2 = x_c = \frac{r_c}{R} = 3.46$, $x_3 = \frac{r_3}{R} = 4$ for droplets of radius $R = 2.5\,\mu$m, $\alpha = 0.2$. Thus, we need to find such $b$ and $x_b$ that $f(x_1) = 0$ and $f(x_2) = 0$. This is equivalent to finding necessary $a$ and $b$. Omitting technicalities, we report that such solutions exist. Namely, for the unscreened Coulomb potential $G_0(x) = 1/x$, we have $b_{Coul}^2 = 0.028$, $x_{b,Coul} = 3.7$ (that is, $a = a_{Coul} = 0.091$ and the boojum rotation angle $\theta_{Coulomb} = 24.7°$ in the DL), and for the full Green function $G(x) = \frac{1}{x} - \frac{1}{\sqrt{x^2+16}}$, which accounts for the interface, $b^2 = 0.023$, $x_{b,Green} = 3.66$ ($a = a_{Greenl} = 0.025$ in this case and the boojum rotation angle $\theta_{Green} = 31.7°$ in the DL).

The average per particle energy $F(r)$, calcualted from (5.92) for the experimental data [42], as a function of the average interparticle distance is plotted in Fig. 5.19.

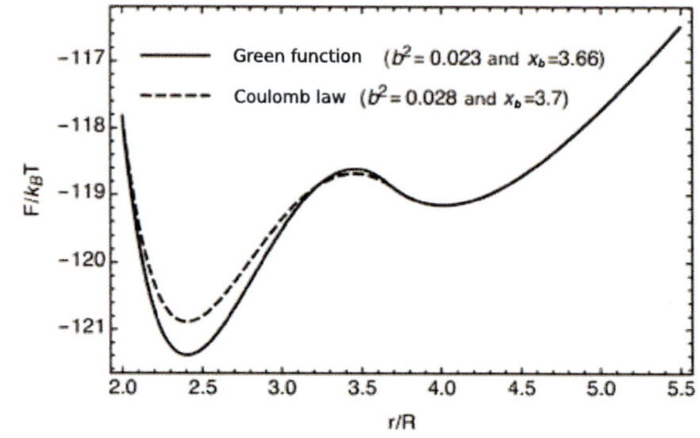


**Fig. 5.19** Calculated per particle energy $F(r)$ given by (5.92) as a function of the average distance between the droplets. Dotted line - Coulomb attraction between skewed boojums, solid line - attraction with Green function from (5.90). Here parameters $R = 2.5\ \mu$m, $K = 7$ pN, $\sigma_{LCA} = 3.8 \cdot 10^{-2}$ J/m$^2$, $\alpha = 0.2$, $f_{el} = 88.4$ pN, $\lambda = 2$ mm are taken from [42]

It features two distinct minima separated by a $\sim k_B T$ barrier, consistently with the coexistence of HL and DL lattices.

In the paper [42] another dipolar approach explanation of DL structure has been proposed. We think that the current approach describes the experiment more correctly. In Ref. [42] there is no explanation of the existence of the global minimum with respect to the boojum rotation angle and the particle separation distance both at the same time. Here we have found such global minimum. Therefore, we consider that this approach describes the situation in a more correct way.

Dense packing of glycerol droplets at the nematic-air interface is caused by spontaneous rotational symmetry breaking with the emergence of elastic charges and Coulomb-like elastic attraction between rotated topological defects - boojums at the bottom of glycerol droplets.

## 5.9 Conclusions

Colloidal particles suspended in a nematic liquid crystal cause deviations of the director from its ground state. Far from the particle, these distortions can be written in the form of the multipole expansion. The notion of multipoles, well familiar from classical electrostatics, has become one of the cornerstones of our current understanding of the host-mediated interactions in NLC colloids.

Building upon the electrostatic analogy, we developed a general framework of colloidal nematostatics applicable to particles of arbitrary shape, size, chirality, and anchoring strength and type. Unlike many alternative approaches, the formalism we propose in this paper easily incorporates the effects of confinement and the presence of external electric or magnetic fields. Our theory is centered around elastic multipoles — coefficients in the multipole expansion of the far-field director. Considering two limiting cases of anchoring strength, weak ($Wr_0 \ll K$) and strong ($Wr_0 \gg K$), we demonstrated how these coefficients emerge as asymptotics of the near-field solution. In the former case, $Wr_0 \ll K$, the proposed approach allows us to find exact analytical expressions for all multipole moments. While that is generally impossible when $Wr_0 \gg K$, one can still infer the overall structure of the expansion (that is, which coefficients vanish and which do not) just from the symmetry of the director in the immediate vicinity of the particle. Since this information is typically available in experiments, the formalism we developed here can be readily applied for particles of arbitrary shapes, sizes, and anchoring types.

The proposed framework relies on two assumptions: (i) differences in the NLC elastic constants can be neglected; (ii) deviations of the director field from its ground state are small, $\mathbf{n} = \mathbf{n}_0 + \delta\mathbf{n}$ with $|\mathbf{n}_0| \approx 1$ and $|\delta\mathbf{n}| \ll 1$. This assumption is well justified for submicron particles but is limited to large interparticle distances in the case of micron and supramicron colloids. The vicinity of a large particle features strong director deformations and topological defects, which render the electrostatic analogy invalid. As such, our theory is inapplicable to entangled and knotted colloidal structures [51, 109–111], where disclination loops extend over many particle sizes. These structures are born out of short-range interactions. It is worth noting that even in the case of localized defects, the nature and extent of short-range interactions (in other words, the size and shape of the coat) are not fully understood. Indirectly, this question was addressed in [99]. There was shown that lattice constants and angles of dipolar 2D and 3D structures could be consistently reproduced with multipole interactions incorporating higher-order terms and spherically symmetric hard-core repulsion. Using this model, the authors estimated that the radius of such a sphere-shaped coat around a hedgehog dipole (Fig. 5.1a) is about $1.2r_0$, where $r_0$ is the particle size, and roughly matches the distance to the defect. It would be interesting to explore the transition between near-field and far-field director in three dimensions to develop a more detailed idea of the coat's size and shape and to quantify the limits of electrostatic analogy.

To demonstrate our approach at work, we considered the dipole-dipole interaction between banana-shaped particles in a nematic cell with both planar and homeotropic alignment of the LC molecules at the boundaries. Given that a banana-like colloid possesses two orthogonal planes of symmetry, one can intuitively expect that in the mechanical equilibrium either both of these planes are normal to the cell walls (orientation A) or just one (orientation B). We found that the A-oriented particles placed in the middle of a homeotropic cell interact anisotropically. Each "banana" in this case has two "butterfly wings" on its longer sides within which the interaction is attractive. The less asymmetric the particles are, the smaller the "wings", so when the axial symmetry is fully restored, the interaction is isotropic. The "wings" have finite size, as $r \to \infty$ the interaction becomes completely repulsive. This behavior contrasts with the interaction of banana-like colloids in the bulk LC, where the angular and distance dependencies are completely decoupled. In a planar cell, both axially symmetric colloids and B-oriented "bananas" interact anisotropically. However, the symmetric particles repel along the direction perpendicular to the rubbing, whereas the asymmetric ones attract at small distances.

We have also found that elastic monopoles are insensitive to the type of confinement: nematostatic analog of the Coulomb law has the same form regardless of anchoring at the cell walls and converges to the $1/r$ scaling in the limit of infinite thickness $L \to \infty$.

General expressions for the self-energy of colloidal particles carrying elastic monopole, dipole and/or quadrupole moment in a nonuniform director field were derived within the proposed framework and employed to find the distribution of nanoparticles in a cylindrical capillary. We showed that particles with homeotropic boundary conditions assemble at the center of the escaped radial disclination, whereas those with planar anchoring aggregate at a distance $\rho_{min} = \frac{R}{\sqrt{3}}$ from the capillary axis.

To elucidate the role of elastic interactions in many-particle systems, we considered the coexistence of two colloidal structures — usual hexagonal and dense hexagonal — formed by glycerol droplets at the nematic-air interface [41, 42]. Our calculations demonstrate how the dense configuration emerges from spontaneous rotational symmetry breaking, which triggers the onset of Coulomb-like attraction between skewed boojums — topological defects at the bottom of glycerol droplets.

These examples strikingly demonstrate how one could exploit the interplay of shape, symmetry, and order to design colloids with prescribed long-range interactions, thereby possibly revealing new types of colloidal superstructures. An interesting and yet unexplored example is that of spheres with conically degenerate surface anchoring — elastic hexadecapoles [100]. These particles appear to have very localized, isotropic short-range interactions and highly anisotropic long-range potential, which underlies the formation of quasi diamond-like assemblies. Such low-symmetric structures have great potential for photonic applications. Another emerging direction, which likely will gain significant attention in the future, focuses on charged colloids. Recently, Everts et al. demonstrated, both theoretically and experimentally, how the interplay of elastic and electrostatic interactions between nearly-spherical particles gives rise to qualitatively different interaction regimes [112].

**Acknowledgements** One of the authors (B.I.L) is sincerely grateful for the support of the Department of Physics and Astronomy of the National Academy of Sciences of Ukraine — project (N 120U100857) "Mathematical models of non-equilibrium processes in open systems" and the National Research Foundation of Ukraine — project (N 0120U105146) "Models of non-equilibrium process in colloidal systems".

## References

1. Woltman SJ, Jay GD, Crawford GP (2007) Liquid-crystal materials find a new order in biomedical applications. Nat Mater 6:929–938. https://doi.org/10.1038/nmat2010
2. Lu G, Fei B (2014) Medical hyperspectral imaging: a review. J Biomed Opt 19:1–24. https://doi.org/10.1117/1.JBO.19.1.010901
3. Akbari H, Halig L, Schuster DM, Fei B, Osunkoya A, Master V, Nieh P, Chen G (2012) Hyperspectral imaging and quantitative analysis for prostate cancer detection. J Biomed Opt 17:076005. https://doi.org/10.1117/1.JBO.17.7.076005

4. Panasyuk SV, Yang S, Faller DV, Ngo D, Lew RA, Freeman JE, Rogers AE (2007) Medical hyperspectral imaging to facilitate residual tumor identification during surgery. Cancer Biol Ther 6:439–446. https://doi.org/10.4161/cbt.6.3.4018
5. Sorg BS, Moeller BJ, Donovan O, Cao Y, Dewhirst MW (2005) Hyperspectral imaging of hemoglobin saturation in tumor microvasculature and tumor hypoxia development. J Biomed Opt 10:044004. https://doi.org/10.1117/1.2003369
6. Greenman RL, Panasyuk S, Wang X, Lyons TE, Dinh T, Longoria L, Giurini JM, Freeman J, Khaodhiar L, Veves A (2005) Early changes in the skin microcirculation and muscle metabolism of the diabetic foot. The Lancet 366:1711–1717. https://doi.org/10.1016/S0140-6736(05)67696-9
7. Cancio LC, Batchinsky AI, Mansfield JR, Panasyuk S, Hetz K, Martini D, Jordan BS, Tracey B, Freeman JE (2006) Hyperspectral imaging: a new approach to the diagnosis of hemorrhagic shock. J Trauma Acute Care Surg 60:1087–1095. https://doi.org/10.1097/01.ta.0000217357.10617.3d
8. Zuzak KJ, Naik SC, Alexandrakis G, Hawkins D, Behbehani K, Livingston EH (2007) Characterization of a near-infrared laparoscopic hyperspectral imaging system for minimally invasive surgery. Anal Chem 79:4709–4715. https://doi.org/10.1021/ac070367n
9. Abdulhalim I (2011) Non-display bio-optic applications of liquid crystals. Liq Crys Today 20:44–60. https://doi.org/10.1080/1358314X.2011.563975
10. Wang J, Tian H, Wang Y, Li X, Cao Y, Li L, Liu J, Zhou Z (2018) Liquid crystal terahertz modulator with plasmon-induced transparency metamaterial. Opt Express 26:5769–5776. https://doi.org/10.1364/OE.26.005769
11. Shen Z, Zhou S, Ge S, Duan W, Ma L, Lu Y, Hu W (2019) Liquid crystal tunable terahertz lens with spin-selected focusing property. Opt Express 27:8800–8807. https://doi.org/10.1364/OE.27.008800
12. Yang X, Zhao X, Yang K, Liu Y, Liu Y, Fu W, Luo Y (2016) Biomedical applications of terahertz spectroscopy and imaging. Trends Biotechnol 34:810–824. https://doi.org/10.1016/j.tibtech.2016.04.008
13. D'Arco A, Di Fabrizio M, Dolci V, Petrarca M, Lupi S (2020) Thz pulsed imaging in biomedical applications. Condens Matter 5:25. https://doi.org/10.3390/condmat5020025
14. Mirzaei J, Reznikov M, Hegmann T (2012) Quantum dots as liquid crystal dopants. J Mater Chem 22:22350–22365. https://doi.org/10.1039/C2JM33274D
15. Garbovskiy YA, Glushchenko AV (2010) Liquid crystalline colloids of nanoparticles: preparation, properties, and applications. Solid State Phys 62:1–74. https://doi.org/10.1016/B978-0-12-374293-3.00001-8
16. Shen Y, Dierking I (2019) Perspectives in liquid-crystal-aided nanotechnology and nanoscience. Appl Sci 9:2512. https://doi.org/10.3390/app9122512
17. Zhang Y, Liu Q, Mundoor H, Yuan Y, Smalyukh II (2015) Metal nanoparticle dispersion, alignment, and assembly in nematic liquid crystals for applications in switchable plasmonic color filters and e-polarizers. Acs Nano 9:3097–3108. https://doi.org/10.1021/nn5074644
18. Brolo AG (2012) Plasmonics for future biosensors. Nat Photonics 6:709–713. https://doi.org/10.1038/nphoton.2012.266
19. Špačková B, Wrobel P, Bocková M, Homola J (2016) Optical biosensors based on plasmonic nanostructures: a review. Proc IEEE 104:2380–2408. https://doi.org/10.1109/JPROC.2016.2624340
20. Si G, Zhao Y, Leong ESP, Liu YJ (2014) Liquid-crystal-enabled active plasmonics: a review. Materials 7:1296–1317. https://doi.org/10.3390/ma7021296
21. Turiv T, Krieger J, Babakhanova G, Yu H, Shiyanovskii SV, Wei QH, Kim MH, Lavrentovich OD (2020) Topology control of human fibroblast cells monolayer by liquid crystal elastomer. Sci Adv 6:eaaz6485. https://doi.org/10.1126/sciadv.aaz6485
22. Sung B, Kim MH (2018) Liquid-crystalline nanoarchitectures for tissue engineering. Beilstein J Nanotechnol 9:205–215. https://doi.org/10.3762/bjnano.9.22
23. Gupta VK, Skaife JJ, Dubrovsky TB, Abbott NL (1998) Optical amplification of ligand-receptor binding using liquid crystals. Science 279:2077–2080. https://doi.org/10.1126/science.279.5359.2077

24. Aliño VJ, Sim PH, Choy WT, Fraser A, Yang KL (2012) Detecting proteins in microfluidic channels decorated with liquid crystal sensing dots. Langmuir 28:17571–17577. https://doi.org/10.1021/la303213h
25. Hartono D, Qin WJ, Yang KL, Yung LYL (2009) Imaging the disruption of phospholipid monolayer by protein-coated nanoparticles using ordering transitions of liquid crystals. Biomaterials 30:843–849. https://doi.org/10.1016/j.biomaterials.2008.10.037
26. Khan M, Park SY (2014) Liquid crystal-based proton sensitive glucose biosensor. Anal Chem 86:1493–1501. https://doi.org/10.1021/ac402916v
27. Lai SL, Yang KL (2011) Detecting DNA targets through the formation of DNA/CTAB complex and its interactions with liquid crystals. Analyst 136:3329–3334. https://doi.org/10.1039/C1AN15173H
28. Kim YK, Wang X, Mondkar P, Bukusoglu E, Abbott NL (2018) Self-reporting and self-regulating liquid crystals. Nature 557:539–544. https://doi.org/10.1038/s41586-018-0098-y
29. Yang S, Liu Y, Tan H, Wu C, Wu Z, Shen G, Yu R (2012) Gold nanoparticle based signal enhancement liquid crystal biosensors for DNA hybridization assays. ChemComm 48:2861–2863. https://doi.org/10.1039/C2CC17861C
30. Tan H, Yang S, Shen G, Yu R, Wu Z (2010) Signal-enhanced liquid-crystal DNA biosensors based on enzymatic metal deposition. Angew Chem Int 49:8608–8611. https://doi.org/10.1002/anie.201004272
31. Nandi R, Loitongbam L, De J, Jain V, Pal SK (2019) Gold nanoparticle-mediated signal amplification of liquid crystal biosensors for dopamine. Analyst 144:1110–1114. https://doi.org/10.1039/C8AN02171F
32. Li X, Li G, Yang M, Chen LC, Xiong XL (2015) Gold nanoparticle based signal enhancement liquid crystal biosensors for tyrosine assays. Sens Actuators B Chem 215:152–158. https://doi.org/10.1016/j.snb.2015.03.054
33. Zhao D, Peng Y, Xu L, Zhou W, Wang Q, Guo L (2015) Liquid-crystal biosensor based on nickel-nanosphere-induced homeotropic alignment for the amplified detection of thrombin. ACS Appl Mater Interfaces 7:23418–23422. https://doi.org/10.1021/acsami.5b08924
34. Poulin P, Stark H, Lubensky TC, Weitz DA (1997) Novel colloidal interactions in anisotropic fluids. Science 275:1770–1773. https://doi.org/10.1126/science.275.5307.1770
35. Poulin P, Cabuil V, Weitz DA (1997) Direct measurement of colloidal forces in an anisotropic solvent. Phys Rev Lett 79:4862–4865. https://doi.org/10.1103/PhysRevLett.79.4862
36. Poulin P, Weitz DA (1998) Inverted and multiple nematic emulsions. Phys Rev E 57:626–637. https://doi.org/10.1103/PhysRevE.57.626
37. Noël CM, Bossis G, Chaze AM, Giulieri F, Lacis S (2006) Measurement of elastic forces between iron colloidal particles in a nematic liquid crystal. Phys Rev Lett 96:217801. https://doi.org/10.1103/PhysRevLett.96.217801
38. Smalyukh II, Lavrentovich OD, Kuzmin AN, Kachynski AV, Prasad PN (2005) Elasticity-mediated self-organization and colloidal interactions of solid spheres with tangential anchoring in a nematic liquid crystal. Phys Rev Lett 95:157801. https://doi.org/10.1103/PhysRevLett.95.157801
39. Smalyukh II, Kuzmin AN, Kachynski AV, Prasad PN, Lavrentovich OD (2005) Optical trapping of colloidal particles and measurement of the defect line tension and colloidal forces in a thermotropic nematic liquid crystal. Appl Phys Lett 86:021913. https://doi.org/10.1063/1.1849839
40. Kotar J, Vilfan M, Osterman N, Babič D, Čopič M, Poberaj I (2006) Interparticle potential and drag coefficient in nematic colloids. Phys Rev Lett 96:207801. https://doi.org/10.1103/PhysRevLett.96.207801
41. Smalyukh II, Chernyshuk S, Lev BI, Nych AB, Ognysta U, Nazarenko VG, Lavrentovich OD (2004) Ordered droplet structures at the liquid crystal surface and elastic-capillary colloidal interactions. Phys Rev Lett 93:117801. https://doi.org/10.1103/PhysRevLett.93.117801
42. Nych AB, Ognysta UM, Pergamenshchik VM, Lev BI, Nazarenko VG, Muševič I, Škarabot M, Lavrentovich OD (2007) Coexistence of two colloidal crystals at the nematic-liquid-crystal-air interface. Phys Rev Lett 98:057801. https://doi.org/10.1103/PhysRevLett.98.057801

43. Yamamoto T, Yamamoto J, Lev BI, Yokoyama H (2002) Light-induced assembly of tailored droplet arrays in nematic emulsions. Appl Phys Lett 81:2187–2189. https://doi.org/10.1063/1.1508816
44. Lev B, Chernyshuk SB, Yamamoto T, Yamamoto J, Yokoyama H (2008) Photochemical switching between colloidal photonic crystals at the nematic-air interface. Phys Rev E 78:020701. https://doi.org/10.1103/PhysRevE.78.020701
45. Lev B, Nych A, Ognysta U, Reznikov D, Chernyshuk S, Nazarenko V (2002) Nematic emulsion in a magnetic field. JETP Lett 75:322–325. https://doi.org/10.1134/1.1485260
46. Nazarenko VG, Nych AB, Lev BI (2001) Crystal structure in nematic emulsion. Phys Rev Lett 87:075504. https://doi.org/10.1103/PhysRevLett.87.075504
47. Muševič I, Škarabot M, Tkalec U, Ravnik M, Žumer S (2006) Two-dimensional nematic colloidal crystals self-assembled by topological defects. Science 313:954–958. https://doi.org/10.1126/science.1129660
48. Škarabot M, Ravnik M, Žumer S, Tkalec U, Poberaj I, Babič D, Osterman N, Muševič I (2007) Two-dimensional dipolar nematic colloidal crystals. Phys Rev E 76:051406. https://doi.org/10.1103/PhysRevE.76.051406
49. Ognysta U, Nych A, Nazarenko V, Muševič I, Škarabot M, Ravnik M, Žumer S, Poberaj I, Babič D (2008) 2d interactions and binary crystals of dipolar and quadrupolar nematic colloids. Phys Rev Lett 100:217803. https://doi.org/10.1103/PhysRevLett.100.217803
50. Škarabot M, Ravnik M, Žumer S, Tkalec U, Poberaj I, Babič D, Muševič I (2008) Hierarchical self-assembly of nematic colloidal superstructures. Phys Rev E 77:061706. https://doi.org/10.1103/PhysRevE.77.061706
51. Ravnik M, Škarabot M, Žumer S, Tkalec U, Poberaj I, Babič D, Osterman N, Muševič I (2007) Entangled nematic colloidal dimers and wires. Phys Rev Lett 99:247801. https://doi.org/10.1103/PhysRevLett.99.247801
52. Tkalec U, Ravnik M, Žumer S, Muševič I (2009) Vortexlike topological defects in nematic colloids: Chiral colloidal dimers and 2d crystals. Phys Rev Lett 103:127801. https://doi.org/10.1103/PhysRevLett.103.127801
53. Nych A, Ognysta U, Škarabot M, Ravnik M, Žumer S, Muševič I (2013) Assembly and control of 3d nematic dipolar colloidal crystals. Nat Commun 4:1489. https://doi.org/10.1038/ncomms2486
54. Vilfan M, Osterman N, Čopič M, Ravnik M, Žumer S, Kotar J, Babič D, Poberaj I (2008) Confinement effect on interparticle potential in nematic colloids. Phys Rev Lett 101:237801. https://doi.org/10.1103/PhysRevLett.101.237801
55. Kossyrev P, Ravnik M, Žumer S (2006) Branching of colloidal chains in capillary-confined nematics. Phys Rev Lett 96:048301. https://doi.org/10.1103/PhysRevLett.96.048301
56. Ji Fukuda, Stark H, Yoneya M, Yokoyama H (2004) Interaction between two spherical particles in a nematic liquid crystal. Phys Rev E 69:041706. https://doi.org/10.1103/PhysRevE.69.041706
57. Aoki KM, Lev BI, Yokoyama H (2001) Cluster formation of colloids in nematics. Mol Cryst Liq Cryst 367:537–544. https://doi.org/10.1080/10587250108028674
58. Andrienko D, Tasinkevych M, Dietrich S (2005) Effective pair interactions between colloidal particles at a nematic-isotropic interface. Europhys Lett 70:95. https://doi.org/10.1209/epl/i2004-10455-3
59. Tasinkevych M, Andrienko D (2010) Colloidal particles in liquid crystal films and at interfaces. Condens Matter Phys 13:33603. https://doi.org/10.5488/CMP.13.33603
60. Kishita T, Takahashi K, Ichikawa M, Ji Fukuda, Kimura Y (2010) Arrangement dependence of interparticle force in nematic colloids. Phys Rev E 81:010701. https://doi.org/10.1103/PhysRevE.81.010701
61. Kishita T, Kondo N, Takahashi K, Ichikawa M, Ji Fukuda, Kimura Y (2011) Interparticle force in nematic colloids: Comparison between experiment and theory. Phys Rev E 84:021704. https://doi.org/10.1103/PhysRevE.84.021704
62. Stark H (1999) Director field configurations around a spherical particle in a nematic liquid crystal. Eur Phys J B 10:311–321. https://doi.org/10.1007/s100510050860

63. Andrienko D, Tasinkevych M, Patricio P, Allen MP, da Gama MMT (2003) Forces between elongated particles in a nematic colloid. Phys Rev E 68:051702. https://doi.org/10.1103/PhysRevE.68.051702
64. Muševič I (2018) Nematic liquid-crystal colloids. Materials 11:24. https://doi.org/10.3390/ma11010024
65. Lubensky TC, Pettey D, Currier N, Stark H (1998) Topological defects and interactions in nematic emulsions. Phys Rev E 57:610–625. https://doi.org/10.1103/PhysRevE.57.610
66. Lev BI, Tomchuk PM (1999) Interaction of foreign macrodroplets in a nematic liquid crystal and induced supermolecular structures. Phys Rev E 59:591–602. https://doi.org/10.1103/PhysRevE.59.591
67. Lev BI, Chernyshuk SB, Tomchuk PM, Yokoyama H (2002) Symmetry breaking and interaction of colloidal particles in nematic liquid crystals. Phys Rev E 65:021709. https://doi.org/10.1103/PhysRevE.65.021709
68. Pergamenshchik VM, Uzunova VO (2007) Coulomb-like interaction in nematic emulsions induced by external torques exerted on the colloids. Phys Rev E 76:011707. https://doi.org/10.1103/PhysRevE.76.011707
69. Pergamenshchik VM, Uzunova VA (2010) Colloidal nematostatics. Condens Matter Phys 13:33602. https://doi.org/10.5488/CMP.13.33602
70. Pergamenshchik VM, Uzunova VA (2011) Dipolar colloids in nematostatics: tensorial structure, symmetry, different types, and their interaction. Phys Rev E 83:021701. https://doi.org/10.1103/PhysRevE.83.021701
71. Yada M, Yamamoto J, Yokoyama H (2004) Direct observation of anisotropic interparticle forces in nematic colloids with optical tweezers. Phys Rev Lett 92:185501. https://doi.org/10.1103/PhysRevLett.92.185501
72. Takahashi K, Ichikawa M, Kimura Y (2008) Force between colloidal particles in a nematic liquid crystal studied by optical tweezers. Phys Rev E 77:020703. https://doi.org/10.1103/PhysRevE.77.020703
73. Kondo N, Iwashita Y, Kimura Y (2010) Dependence of interparticle force on temperature and cell thickness in nematic colloids. Phys Rev E 82:020701. https://doi.org/10.1103/PhysRevE.82.020701
74. Ji Fukuda, Žumer S (2009) Confinement effect on the interaction between colloidal particles in a nematic liquid crystal: An analytical study. Phys Rev E 79:041703. https://doi.org/10.1103/PhysRevE.79.041703
75. Fukuda J, Lev BI, Yokoyama H (2003) Effect of confining walls on the interaction between particles in a nematic liquid crystal. J Phys Condens Matter 15:3841. https://doi.org/10.1088/0953-8984/15/23/301
76. Ji Fukuda, Lev BI, KM, Yokoyama H, (2002) Interaction of particles in a deformed nematic liquid crystal. Phys Rev E 66:051711. https://doi.org/10.1103/PhysRevE.66.051711
77. Oettel M, Dominguez A, Tasinkevych M, Dietrich S (2009) Effective interactions of colloids on nematic films. Eur Phys J E 28:99–111. https://doi.org/10.1140/epje/i2008-10360-1
78. Pergamenshchik VM, Uzunova VA (2009) Colloid-wall interaction in a nematic liquid crystal: the mirror-image method of colloidal nematostatics. Phys Rev E 79:021704. https://doi.org/10.1103/PhysRevE.79.021704
79. Chernyshuk SB, Lev BI (2010) Elastic interaction between colloidal particles in confined nematic liquid crystals. Phys Rev E 81:041701. https://doi.org/10.1103/PhysRevE.81.041701
80. Chernyshuk SB, Lev BI (2011) Theory of elastic interaction of colloidal particles in nematic liquid crystals near one wall and in the nematic cell. Phys Rev E 84:011707. https://doi.org/10.1103/PhysRevE.84.011707
81. Chernyshuk SB, Tovkach OM, Lev BI (2012) Theory of elastic interaction between colloidal particles in a nematic cell in the presence of an external electric or magnetic field. Phys Rev E 85:011706. https://doi.org/10.1103/PhysRevE.85.011706
82. Lapointe CP, Mason TG, Smalyukh II (2009) Shape-controlled colloidal interactions in nematic liquid crystals. Science 326:1083–1086. https://doi.org/10.1126/science.1176587

83. Dontabhaktuni J, Ravnik M, Žumer S (2014) Quasicrystalline tilings with nematic colloidal platelets. Proc Natl Acad Sci USA 111:2464–2469. https://doi.org/10.1073/pnas.1312670111
84. Senyuk B, Liu Q, Bililign E, Nystrom PD, Smalyukh II (2015) Geometry-guided colloidal interactions and self-tiling of elastic dipoles formed by truncated pyramid particles in liquid crystals. Phys Rev E 91:040501. https://doi.org/10.1103/PhysRevE.91.040501
85. Tkalec U, Škarabot M, Muševič I (2008) Interactions of micro-rods in a thin layer of a nematic liquid crystal. Soft Matter 4:2402–2409. https://doi.org/10.1039/B807979J
86. Senyuk B, Liu Q, He S, Kamien RD, Kusner RB, Lubensky TC, Smalyukh II (2013) Topological colloids. Nature 493:200–205. https://doi.org/10.1038/nature11710
87. Varney MC, Zhang Q, Smalyukh II (2015) Stick-slip motion of surface point defects prompted by magnetically controlled colloidal-particle dynamics in nematic liquid crystals. Phys Rev E 91:052503. https://doi.org/10.1103/PhysRevE.91.052503
88. Silvestre NM, Tasinkevych M (2017) Key-lock colloids in a nematic liquid crystal. Phys Rev E 95:012606. https://doi.org/10.1103/PhysRevE.95.012606
89. Senyuk B, Varney MC, Lopez JA, Wang S, Wu N, Smalyukh II (2014) Magnetically responsive gourd-shaped colloidal particles in cholesteric liquid crystals. Soft Matter 10:6014–6023. https://doi.org/10.1039/C4SM00878B
90. Peng C, Turiv T, Zhang R, Guo Y, Shiyanovskii SV, Wei QH, De Pablo J, Lavrentovich OD (2016) Controlling placement of nonspherical (boomerang) colloids in nematic cells with photopatterned director. J Phys Condens Matter 29:014005. https://doi.org/10.1088/0953-8984/29/1/014005
91. Evans JS, Sun Y, Senyuk B, Keller P, Pergamenshchik VM, Lee T, Smalyukh II (2013) Active shape-morphing elastomeric colloids in short-pitch cholesteric liquid crystals. Phys Rev Lett 110:187802. https://doi.org/10.1103/PhysRevLett.110.187802
92. Senyuk B, Pandey MB, Liu Q, Tasinkevych M, Smalyukh II (2015) Colloidal spirals in nematic liquid crystals. Soft Matter 11:8758–8767. https://doi.org/10.1039/C5SM01539A
93. Tovkach OM, Chernyshuk SB, Lev BI (2012) Theory of elastic interaction between arbitrary colloidal particles in confined nematic liquid crystals. Phys Rev E 86:061703. https://doi.org/10.1103/PhysRevE.86.061703
94. Terentjev EM (1995) Disclination loops, standing alone and around solid particles, in nematic liquid crystals. Phys Rev E 51:1330–1337. https://doi.org/10.1103/PhysRevE.51.1330
95. Ruhwandl RW, Terentjev EM (1997) Long-range forces and aggregation of colloid particles in a nematic liquid crystal. Phys Rev E 55:2958–2961. https://doi.org/10.1103/PhysRevE.55.2958
96. Tovkach O, Chernyshuk S, Lev B (2015) Colloidal interactions in a homeotropic nematic cell with different elastic constants. Phys Rev E 92:042505. https://doi.org/10.1103/PhysRevE.92.042505
97. Chernyshuk S (2014) High-order elastic terms, boojums and general paradigm of the elastic interaction between colloidal particles in the nematic liquid crystals. Eur Phys J E 37:6. https://doi.org/10.1140/epje/i2014-14006-5
98. Chernyshuk S, Tovkach O (2016) Colloidal particles as elastic triads in nematic liquid crystals. Liq Cryst 43:2410–2421. https://doi.org/10.1080/02678292.2016.1216619
99. Chernyshuk SB, Tovkach OM, Lev BI (2014) Elastic octopoles and colloidal structures in nematic liquid crystals. Phys Rev E 89:032505. https://doi.org/10.1103/PhysRevE.89.032505
100. Senyuk B, Puls O, Tovkach OM, Chernyshuk SB, Smalyukh II (2016) Hexadecapolar colloids. Nat Commun 7:1–7. https://doi.org/10.1038/ncomms10659
101. Landau LD, Lifshitz EM (1994) The Classical Theory of Fields. Butterworth-Heinemann, Oxford
102. Xiao K, Chen X, Wu CX (2019) Fréedericksz-like positional transition. Phys Rev Res 1:033041. https://doi.org/10.1103/PhysRevResearch.1.033041
103. Shiyanovskii SV, Glushchenko A, Reznikov Y, Lavrentovich OD, West JL (2000) Tensor and complex anchoring in liquid crystals. Phys Rev E 62:R1477–R1480. https://doi.org/10.1103/PhysRevE.62.R1477
104. Jackson LD (1999) Classical Electrodynamics. Hoboken, Wiley

105. Lee BK, Kim SJ, Kim JH, Lev B (2017) Coulomb-like elastic interaction induced by symmetry breaking in nematic liquid crystal colloids. Sci Rep 7:1–8. https://doi.org/10.1038/s41598-017-16200-z
106. Yuan Y, Liu Q, Senyuk B, Smalyukh II (2019) Elastic colloidal monopoles and reconfigurable self-assembly in liquid crystals. Nature 570:214–218. https://doi.org/10.1038/s41586-019-1247-7
107. Pergamenshchik VM (2014) Elastic multipoles in the field of the nematic director distortions. Eur Phys J E 37:1–15. https://doi.org/10.1140/epje/i2014-14121-3
108. de Gennes PG, Prost J (1993) The Physics of Liquid Crystals. Clarendon Press, Oxford
109. Tkalec U, Ravnik M, Čopar S, Žumer S, Muševič I (2011) Reconfigurable knots and links in chiral nematic colloids. Science 333:62–65. https://doi.org/10.1126/science.1205705
110. Jampani VSR, Škarabot M, Ravnik M, Čopar S, Žumer S, Muševič I (2011) Colloidal entanglement in highly twisted chiral nematic colloids: Twisted loops, hopf links, and trefoil knots. Phys Rev E 84:031703. https://doi.org/10.1103/PhysRevE.84.031703
111. Čopar S, Tkalec U, Muševič I, Žumer S (2015) Knot theory realizations in nematic colloids. Proc Natl Acad Sci USA 112:1675–1680. https://doi.org/10.1073/pnas.1417178112
112. Everts JC, Senyuk B, Mundoor H, Ravnik M, Smalyukh II (2021) Anisotropic electrostatic screening of charged colloids in nematic solvents. Sci Adv 7(eabd0662). https://doi.org/10.1126/sciadv.abd0662

# Part II Mechanisms and Molecular Interactions

# Chapter 6
# DNA-Polyamine Interactions: Insight from Molecular Dynamics Simulations on the Sequence-Specific Binding of Spermidine$^{3+}$

**Francesca Mocci, Aatto Laaksonen, Leon Engelbrecht, Tudor Vasiliu, and Sergiy Perepelytsya**

**Abstract** DNA is a polyanion stabilized in vivo by positively charged counterions, including metal ions and small organic molecules, e.g., putrescine$^{2+}$, spermidine$^{3+}$, and spermine$^{4+}$. In this chapter, after a brief review of previous studies on such DNA-counterion interactions, we focus the attention on the interactions between spermidine$^{3+}$ and DNA. In this context, we present our original molecular dynamics simulation study to establish the specificity of the binding of polyamines to different nucleotide sequence motifs. Spermidine$^{3+}$ molecules tend to be localized in the minor groove of the DNA double helix around the regions with AATT and ATAT nucleotide sequences. In the major groove, the polyamine does not bind to the AT-rich sequences, but instead localizes on the CG-region. The positioning of polyamines on the DNA surface also determines the DNA-DNA contacts due to the formation of polyamine cross-links. The cross-linking spermidine$^{3+}$ molecules are localized parallel to the sugar-phosphate backbone of the double helix, neutralizing the negatively charged

F. Mocci (✉) · L. Engelbrecht
Department of Chemical and Geological Sciences, University of Cagliari, 09042 Monserrato, Italy
e-mail: fmocci@unica.it

A. Laaksonen
Division of Physical Chemistry, Department of Materials and Environmental Chemistry, Arrhenius Laboratory, Stockholm University, 10691 Stockholm, Sweden

Laboratory of Materials-Oriented and Chemical Engineering, Centre of Advanced Research in Bionanoconjugates, Nanjing Tech University, Nanjing 210009, P. R. China

A. Laaksonen
e-mail: aatto.laaksonen@mmk.su.se

A. Laaksonen · T. Vasiliu
Centre of Advanced Research in Bionanoconjugates and Biopolymers "Petru Poni" Institute of Macromolecular Chemistry, 700487 Iasi, Romania
e-mail: vasiliu.tudor@icmpp.ro

S. Perepelytsya
Bogolyubov Institute for Theoretical Physics, National Academy of Sciences of Ukraine, Kyiv 03143, Ukraine
e-mail: perepelytsya@bitp.kiev.ua



L. Bulavin and N. Lebovka (eds.), *Soft Matter Systems for Biomedical Applications*, Springer Proceedings in Physics 266, https://doi.org/10.1007/978-3-030-80924-9_6

phosphate groups. The presented results are important for understanding the role of spermidine$^{3+}$ in the biological functions of DNA and also have implications for possible technological applications.

## List of Abbreviations

| | |
|---|---|
| A | Adenine |
| A-T | Adenine-Thymine base pair |
| AMBER | Assisted Model Building with Energy Refinement |
| ApT | Adenine-phosphate-Thymine |
| ASAXS | Anomalous Small Angle X-ray Scattering |
| C | Cytosyne |
| CHARMM | Chemistry at Harvard Macromolecular Mechanics |
| CG | Coarse Grained |
| C-G | Cytosyne-Guanine base pair |
| DDD | Drew Dickerson Dodecamer |
| DNA | Deoxyribonucleic Acid |
| FF | Force Field |
| G | Guanine |
| GAFF | Generalized Amber Force Field |
| GROMACS | GROningen MAchine for Chemical Simulations |
| MD | Molecular Dynamics |
| NAB | Nucleic Acid Builder |
| RDF | Radial Distribution Function |
| RNA | Ribonucleic Acid |
| RMS | Root Mean Square |
| SAXS | Small angle X-ray scattering |
| SNIC | Swedish National Infrastructure for Computing |
| T | Thymine |
| TIP3P | Transferable Intermolecular Potential with 3 Points |
| TpA | Thymine-phosphate-Adenine |
| VMD | Visual Molecular Dynamics |

## 6.1 Introduction

Polyamines occur ubiquitously in living organisms, where they have important cellular functions [1]. Natural polyamines, putrescine$^{2+}$, spermidine$^{3+}$, and spermine$^{4+}$, are normally present in the intracellular environment in micromolar concentrations, and the increase or decrease of their concentrations may be related to different diseases, such as Alzheimer's, Huntington, cancer and others [2–4].

In particular, spermidine$^{3+}$ molecules are known to have important properties for autophagy and longevity [5]. Natural polyamines participate in various biological processes of cell metabolism, associated with cell proliferation and apoptosis [6, 7]. In the cell nuclei, polyamines are involved in the mechanisms of DNA compaction in chromatin and in gene regulation [8]. They are also known to play a key role in the molecular mechanisms of DNA and RNA compaction in viral capsids, and in the regulation of viral infection [9, 10]. To elucidate the role of polyamines in the molecular mechanisms behind the biological activity of nucleic acids, it is necessary to characterize the structure and dynamics of their complexes with DNA.

In the cell nucleus, DNA is mostly organized in the familiar double helix structure. The nucleotide bases are located in the inner part of the macromolecule forming complementary hydrogen (H-) bonded pairs: adenine–thymine (A-T), and guanine-cytosine (G-C). The two spiraling sugar-phosphate backbones act as frames at the periphery of the macromolecule [11–13]. The outer radius of the DNA double helix is about 10 Å and the extended length of natural DNA contained in human cells may reach up to two meters [11]. In the *B*-DNA conformation, which is considered dominant in vivo, the wrapping of the two DNA chains in a double helix, generates two grooves of different width, indicated as minor (the narrower) and major (the larger) grooves, having average widths of about 12 Å and 18 Å, respectively [11]. The width of the grooves is modulated by the sequence of nucleotides and may fluctuate by up to 2 Å. The conformational mechanics of the DNA double helix is of great interest for the understanding of the mechanisms of biological functioning and is widely studied [14].

Starting from the very first works by James Watson and Francis Crick [15], and by Rosalind Franklin [16], the structure of the double helix was known to be stabilized by water molecules and counterions (metal ions and polyamines). The counterions neutralize the negatively charged phosphate groups of the DNA backbone, reducing the repulsion of the opposite strands of the double helix. Due to the hydrophobic effect, the nucleotide bases "hide" from water by positioning in the inner part of the double helix. The ion-hydration shell around the double helix may be considered an integral part of the double helix structure [17]. The counterions may be localized inside the grooves or may just interact with the macromolecule backbone [18–20]. Taking into consideration that the natural polyamine molecules have an elongated, flexible structure, localizing them around DNA is a challenging task. In this regard, Molecular Dynamics (MD) simulations are one of the few methods that can give important insights into the detailed polyamine distribution around DNA.

In this chapter, we give a brief overview on the fundamental findings concerning the interactions of natural polyamine molecules with the DNA double helix. Then we present our MD simulations intended to study the interaction of spermidine$^{3+}$ molecules with the DNA double helixes having different nucleotide sequences. The analysis of DNA-DNA interactions induced by spermidine$^{3+}$ is also described.

The structure of the chapter is as follows. In the Sect. 6.2, the most relevant experimental and theoretical findings concerning the interactions of the counterions with DNA macromolecule are reviewed. In Sect. 6.3, the procedure of using MD simulations to study DNA with polyamines is described. Based on MD simulation results, the distribution of spermidine$^{3+}$ molecules around the DNA double helix is described in Sect. 6.4, and the sequence-specificity of polyamine-DNA interactions is described in Sect. 6.4 based on the simulation data. The interaction modes between DNA oligomers are described in Sect. 6.6. The obtained results are summarized in Sect. 6.7.

## 6.2 Features of DNA-Counterion Interactions

DNA is a strong polyelectrolyte due to the negatively charged phosphate groups along the macromolecule backbone [21]. Inside the cell nucleus, the DNA negative charge is partially neutralized by the positive charges of the histone proteins, around which DNA is wound [12, 13, 22]. The rest of the negative charge of DNA is neutralized by small positively charged ions (counterions) that are present in the cell media (Fig. 6.1a). Under physiological conditions, these are usually potassium, sodium, magnesium and organic positively charged molecules, like polyamines [12, 13]. The natural polyamines are putrescine$^{2+}$, spermidine$^{3+}$ and spermine$^{4+}$ (Fig. 6.1b). The structure of the nucleosome core particle is determined by the electrostatic interactions between DNA, the protein core and the surrounding counterions. Therefore, understanding the distribution of counterions and their specific binding to the structural motifs of the double helix are of paramount importance.

In aqueous solution the counterions condense around DNA, forming a diffuse atmosphere surrounding the macromolecule [21, 23, 24]. The effect of counterion condensation has been predicted within the framework of simple theoretical models

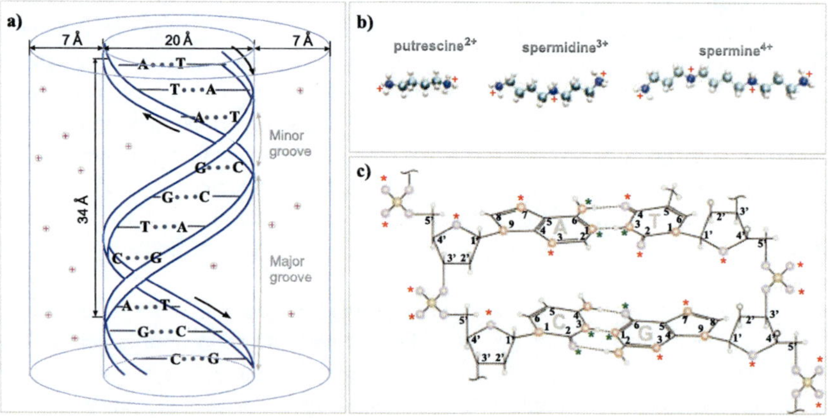


**Fig. 6.1** **a** Counterion cloud around the DNA double helix. **b** Natural polyamines: putrescine$^{2+}$, spermidine$^{3+}$ and spermine$^{4+}$. **c** Binding sites of DNA for the metal counterions (on the basis of [18]). Red asterisks denote the characteristic binding sites for all kind of metal ions, green asterisks mark those characteristics for bivalent ions.

that represent DNA as a chain of negatively charged beads surrounded by a positively charged continuum [21]. The results show that the condensed counterions reduce the charge of each phosphate group of *B*-DNA by a factor of about 0.76. This result has been confirmed by different experiments [25–29], and by more sophisticated models considering different features of DNA structure [30]. Starting from the results using the very first polyelectrolyte models of DNA, the polyamine molecules are known to interact strongly with the DNA double helix [21]. However, studies of the interaction of polyamines with DNA based on such simplified models have limitations, related to the specific structural features of DNA, the conformational flexibility of both DNA and polyamine molecules, and the neglect of electrostatic correlations between the counterions. Such limitations do not allow establishing the binding sites and sequence specific effects of polyamine binding.

The anomalous small angle X-ray scattering (SAXS) data of DNA water solutions with $Rb^+$, $Sr^{2+}$ ions show that the counterions are localized within a cloud around the DNA macromolecule with thickness up to about 7 Å, depending on the type of counterion [25–29]. Increasing the charge of counterions, the counterion cloud shrinks to a smaller size [25]. Inside the cloud, the counterions can move around the macromolecule and penetrate the internal compartments of the double helix: the minor groove, the major groove and the region around the phosphate groups of the double helix backbone (Fig. 6.1a).

The interaction of counterions with specific binding sites of DNA can be characterized by X-ray crystallography [31–34], vibrational spectroscopy [35–37], calorimetric measurement, small angle X-ray scattering (SAXS) [25, 38] nuclear magnetic resonance [39–41] and other experiments [18]. The results of experimental studies established that the counterions tend to be localized in the DNA minor groove near the atoms $N_3$ of purine and $O_2$ of pyrimidine, in the major groove near the atoms $N_7$, $N_4$, $O_4$ of purine and $O_4$ of pyrimidine, and near the oxygen atoms of the phosphate groups of the double helix backbone (Fig. 6.1c). The specificity of the ions to the binding sites defined above is determined by the counterion type and by the nucleotide's sequence [42]. The monovalent metal ions ($Na^+$, $K^+$) can bind to all mentioned atoms of the DNA double helix, while bivalent metal ions, for example $Cu^{2+}$, can also bind to the internal atoms in nucleotide pairs [18, 35]. In some cases, the counterions can even bind to the atoms that are usually involved in H-bonding in the complementary nucleotide pairs (the atoms marked by green asterisks on Fig. 6c). In spite of broad range of experimental results describing DNA-counterion interaction, the molecular mechanisms of counterion binding are still far from being completely understood. In particular, the problem of interactions of molecular counterions with DNA requires special consideration.

In this regard, MD simulations have become an important tool for the study of interaction of counterions with DNA. MD simulation studies of DNA-counterion systems started in the early '80s with the simulation of small DNA fragments in an implicit water solvent with counterions [43–48]. From the very beginning, the method faced the problem of the proper description of electrostatic interactions that is critically important for such kinds of systems. Toward the end of the '80s this problem was effectively solved within the framework of the Ewald summation method [49, 50].

The historical perspective of the method development and the problems of modeling of DNA-counterion systems are described in the reviews [20] and [51]. Simulations of nucleic acids are becoming more and more reliable with the continuous refinement of the force fields optimized to properly reproduce the details of the nucleic acid structure and dynamics. The most used force field for simulating nucleic acids are certainly AMBER - whose latest versions [52, 53] can be used to simulate a wide variety of nucleic acids conformation such as nucleic acid quadruple helix [54] - and CHARMM [55] and the related polarizable force field, developed by MacKerell and coworkers (see [56] and references therein), which can also be applied to a variety of DNA conformations [57]. For a comparison of the accuracy of these FF for *B*-DNA we refer the interested reader to the relatively recent work of Dans et al. [58]. Also, the development of reliable set of parameters for the ions in various solvent models, has been fundamental to have a proper description of ions-DNA interaction [59, 60] which is generally more challenging for divalent monoatomic ions [61–63].

Toward the end of the last century, MD simulations provided important information about the main binding sites on DNA for different metal counterions ($Na^+$, $K^+$, $Cs^+$, $Mg^{2+}$ and others) [64–66]. The counterions stay in the minor groove for a long time (up to tens of ns), while in the major groove and near the phosphate groups outside the macromolecule, the residence time is much shorter than that in the minor groove [67–73].

Also, recent studies on counterion distribution, performed using microsecond MD simulation trajectories [70–72], have confirmed that the counterions that are localized in the grooves (especially in the minor groove of the double helix) exhibit nucleotide sequence specificity. In this regard, the effect of ion hydration is found to be important for different counterions [65, 72, 74, 75]. Extensive studies of DNA hydration have shown that water molecules have different organization in different regions of DNA [76–78]. In particular, in the minor groove of DNA, the hydration shell is more structured than in other regions and is characterized by the longest residence times (up to 100 ns) [76]. In the A-tract sequence (uninterrupted sequence of at least 4 A-T base pairs without any TpA step) the formation of the hydration spine in the minor groove was experimentally observed in crystals [79] and confirmed by many experiments [80, 81] and simulations in solutions [64, 69]. In the major groove, the residence times of water are much shorter than in the minor groove, and near the phosphate groups the structure of hydration shell is the most friable [76]. The interplay between water molecules in the hydration shell, counterions and DNA should be understood to elucidate the interaction of different ions in different binding DNA sites [72].

Due to their high positive charge, the polyamine molecules interact strongly with DNA through electrostatic attractions. Calculations of the binding free energy of interaction of putrescine$^{2+}$, spermidine$^{3+}$, and spermine$^{4+}$ molecules with the DNA double helix show that the stability of the complexes increases with the polyamine charge, and that this has important consequences for the formation of dangerous adducts [82]. The preferred polyamines binding sites are the DNA grooves, where they are usually stacked [83, 84]. The interactions with polyamines induce conformational transformations in the flexible structure of the double helix backbone. In particular, localization of polyamines in the minor groove induces its narrowing [85]. The double helix conformation transitions, such as the *B-A* and *B-Z*, and changes at higher organization levels, may also be induced by naturally occurring polyamines [86]. The effect of DNA condensation upon complexation with added spermidine$^{3+}$ and spermine$^{4+}$ is well known [87–89]. This effect occurs with any type of counterion having charge of 3 + or more (for poly(A) sequence also with divalent ions [90]), becomes more noticeable with increasing counterion concentration, and was shown to be related with the formation of counterion bridges between neighboring DNA macromolecules [91, 92]. The effect of condensation induced by spermidine$^{3+}$ has also been observed for a chromatin macromolecular chain [93]. Considering the presence of polyamines in the cell nucleus, it is clear that the influence of polyamine molecules on the structure of DNA has a direct link to the mechanisms of biological processes of DNA. In this regard, it is important to determine how the nucleotide sequences affect the binding of polyamines to DNA.

Sequence specificity in the interaction of polyamines with DNA has long been disputed. Early experimental studies and MD simulations indicated that the homogeneous negatively charged backbone of the double helix is the most important binding site for the polyamines, and the specificity to the nucleotide sequence of DNA was not considered essential in recognition [83, 84, 94, 95]. However, in recent studies, the specificity of DNA-polyamine interaction to the nucleotide sequence has been found to be important [92, 96–103]. Indeed, calorimetric experiments [97] show that the stability of the complexes of putrescine$^{2+}$, cadaverine$^{2+}$, spermidine$^{3+}$ and spermine$^{4+}$ molecules with the double helix depends on the organisms from which the DNA is extracted. This dependence is related to the different AT/GC content in the DNA of different organisms, with the polyamine binding being greater for DNA with higher AT content [97]. Furthermore, both spectroscopic and simulation studies have shown that the attraction between two DNA oligomers induced by spermine$^{4+}$ is larger in the presence of AT-rich sequences compared to GC-rich ones [92].

Recent MD studies [103] have shown that putrescine$^{2+}$, spermidine$^{3+}$, and spermine$^{4+}$ prefer to be localized in the DNA minor groove of the AT rich regions. The localization of polyamines in this region of the double helix is related to natural narrowing of the minor groove. Since the widths of the DNA grooves are modulated by the sequence of nucleotide bases, the positioning of polyamines in the minor groove is sequence-dependent [103]. Differently, the more complex polyamine trilysine, also relevant for the DNA compaction on histones, did not show any clear preference for the minor or major groove [104]. Presently, the DNA sequence-specific localization of polyamine molecules is still far from being fully understood due to the lack of experimental and simulation data to characterize the polyamine distribution around specific nucleotide sequences. To close this gap, computational studies of DNA-polyamine systems in the case of different nucleotide sequences are still required.

Taking into consideration the described experimental and theoretical results on the interactions of counterions, and polyamines in particular (also comprising those constituted by the positively charged side chain of the amino acids), with the DNA double helix, there are still many aspects to be clarified. These include, but are not limited to, the understanding of how the nucleotide sequence affects their binding to DNA, and in turn how the polyamine binding affects DNA structure and DNA-DNA, DNA-proteins or DNA-drugs interactions. All these aspects are of great relevance for the understanding of the complex organization and activity of the molecule that is at the foundation of what we call life and have important applications in the nanotechnological field. While the details of these interactions need to be studied at the atomistic level, or even at electronic level with quantum mechanical methods, the effect of these interactions propagates to time and space scales much larger than those accessible to atomistic and electronic methods. To access larger scales, such as those involving the nucleosome core particles aggregation, or even beyond, it is necessary to use coarse-grained (CG) models, where one particle is constituted by several atoms. The development of CG models for DNA-counterions interactions started more than two decades ago [105] and is nowadays a very active area of research [20, 106–109], active for both *B*-DNA and non-conventional DNA structures such as quadruplexes [110, 111].

In the following, as an example of the research in some of the directions outlined above, we discuss the results of our MD simulations concerning the interaction of spermidine$^{3+}$ with various DNA oligomers having the same nucleotides contents but different sequences, and the DNA-DNA interactions induced by the polyamines.

## 6.3 Molecular Dynamics Simulations of DNA with Polyamines

Several systems have been simulated, all of which contained two fragments of *B*-DNA duplexes of 22 base pairs each with different nucleotides sequence: A) CGCGAATTCGCGCGAATTCGCG, B) CGCGATATCGCGCGAATTCGCG, C) CGCGATATCGCGCGATATCGCG. The starting structure of the polynucleotide A was created by repeating the Drew Dickerson dodecamer (DDD) [79] sequence, which contains an A-tract, in such a way that the two A-tracts are in phase with the helix turn. To check the relevance of the A-tract in the interactions with counterions, the original A-tract sequence (but not the base pair content) has been substituted in B with the alternating ATAT sequence. A-tracts and alternating AT sequences are known to have different structural features, which also affect their interactions with ions. In the sequence C, only one of the two A-tracts present in A has been substituted with the alternating AT sequence. The presence of two AT rich tracts in each sequence, and the presence of two DNA duplexes in each simulation box, allows for reasonable confidence in the results concerning sequence specificity.

Three systems containing two identical DNA molecules in an aqueous solution of spermidine$^{3+}$ (RandA, RandB, and RandC) or of KCl (FreeA, FreeB, and FreeC) have been simulated. The DNA duplexes were placed in the center of the cell, separated by a distance of 40 Å (Fig. 6.2a). In the case of the systems with spermidine$^{3+}$, the polyamines were inserted in the simulation cell at random positions with random

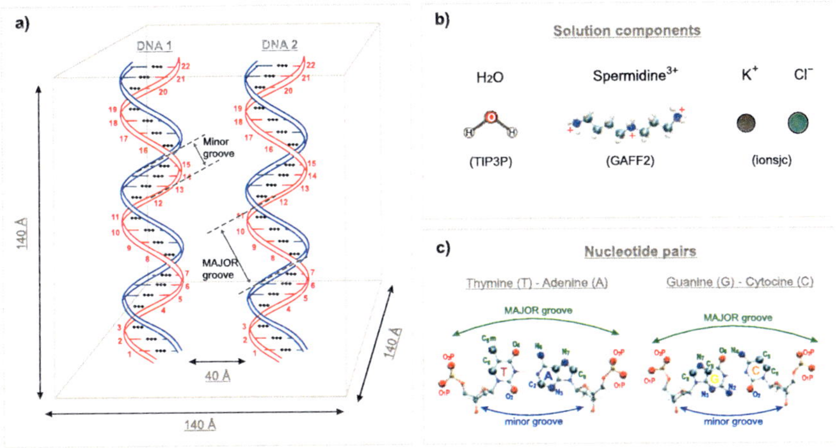


**Fig. 6.2** Representation of the main molecular components of the simulated systems. **a** Schematic illustration of the relative orientation of the two DNA fragment in the starting configuration. The numbering of the nucleotide pairs and the simulation box size are indicated. **b** Molecular structure of counterions and solvent, together with water model and force field specification. **c** The nucleotide pairs of DNA. The atoms that have been used as the reference atoms in the calculation of the radial distribution functions are indicated

orientations before adding water molecules and ions. In the starting configuration, the minimum distance between polyamines and DNA was no less than 10 Å. Detailed information about the systems is shown in Table 6.1.

The concentration of spermidine$^{3+}$ in the system was chosen large enough to induce the condensation of DNA macromolecules present in the simulation box. According to experimental data, DNA condensation can be observed with counterions having a charge greater than 3+ and with a counterion-DNA charge ratio of charge $\geq 1$ [86, 89]. In order to observe, in a reasonable simulation time, the DNA condensation in our MD simulations, the ratio of polyamines-DNA charge was set equal to 1.5.

**Table 6.1** Composition of the simulation systems

| Name | Number of DNA molecules | DNA sequence | Water | Ions ($K^+/Cl^-$) | $Spr^{3+}$ |
|---|---|---|---|---|---|
| RandA | 2 | CGCGAATTCGCGCGAATTCGCG | 69885 | 198/240 | 42 |
| RandB | 2 | CGCGATATCGCGCGAATTCGCG | 69873 | 198/240 | 42 |
| RandC | 2 | CGCGATATCGCGCGATATCGCG | 69948 | 198/240 | 42 |
| FreeA | 2 | CGCGAATTCGCGCGAATTCGCG | 70331 | 282/198 | 0 |
| FreeB | 2 | CGCGATATCGCGCGAATTCGCG | 70333 | 282/198 | 0 |
| FreeC | 2 | CGCGATATCGCGCGATATCGCG | 70346 | 282/198 | 0 |

DNA and spermidine$^{3+}$ model parametrization was performed using Amber Tools 18 [112]. The DNA structures were built in the Arnott *B*-DNA canonical structure using the nucleic acid builder (NAB) tool included in the Amber 18 package [112], while the spermidine$^{3+}$ structure was prepared using Avogadro [113, 114]. The bonded and non-bonded parameters were obtained from the AMBER force field BSC1 DNA Force field [52] for DNA, and for spermidine$^{3+}$the GAFF2 version of the Generalized Amber Force Field [115], distributed with Amber 18. The TIP3P model was used for water molecules [116], while $K^+$ and $Cl^-$ ions were modeled using the ionsjc parameter set [59].

All simulations were performed using the GROMACS 2018 simulation package [117]. The Nosé-Hoover thermostat [118, 119] was used to control the temperature at 293 K, while the Parrinello-Rahman barostat [120] was used to control the pressure, which was set at 1 bar. The lengths of all bonds with hydrogen atoms were constrained using the LINCS algorithm [121]. The long-range electrostatic interactions were treated using the smooth particle mesh Ewald method [122]. The switching and cut-off distances for the long-range interactions were both set to 10 Å, respectively, and the Fourier spacing was 1.2 Å. For each system, a simulation trajectory with a length of 500 ns was obtained.

The analysis and visualization of the simulation trajectories were carried out using the VMD software package [123]. The distribution of the polyamine molecules around the DNA double helix and the orientation of the macromolecules with respect to each other have been studied using radial distribution functions (RDFs). The RDFs were calculated using the plug-in [124] implemented to VMD [123] according to the following definition:

$$g(r) = \lim_{\Delta r \to 0} \frac{p(r)}{4\pi r^2 \Delta r N_p / V} \tag{6.1}$$

where $p(r)$ is the average number of particles that is found at the distance $r$ within the slice $\Delta r$, $N_p$ is the number of pairs of selected atoms for which the radial distribution functions are calculated, $V$ is the system volume. The thickness of the spherical slice $\Delta r$ in our calculations has been taken equal to 0.1 Å. The average number of particles in the volume of slice constrained by the radii $r_1$ and $r_2$ (coordination number) can be determined by the direct integration of the radial distribution function:

$$n_{12} = \int_{r_1}^{r_2} \frac{N_p}{V} r^2 g(r) \mathrm{d}r \tag{6.2}$$

The radial distribution functions describing the localization of polyamines with respect to the atomic groups of the double helix have been calculated by taking into consideration the heavy atoms of the spermidine$^{3+}$ molecule. For the calculation of the RDFs, the atoms $N_3$, $O_2$ in the minor groove, $N_7$, $O_6$ in the major groove, and the oxygen atoms $O_1P$ and $O_2P$ of the phosphate groups were taken as reference atoms (Fig. 6.2c). Thus, for each DNA molecule in the simulation box, three sets of the radial distribution function for spermidine$^{3+}$ have been obtained: the RDFs with respect to the minor groove ($RDF_{MIN}$), the RDFs with respect to the major groove ($RDF_{MAJ}$), and the RFDs with respect to the phosphate groups ($RDF_{PH}$). The RDFs have been calculated for each DNA molecule in the simulation box.

To analyze the orientation of DNA oligomers with respect to each other, the radial distribution functions of the terminal nucleotide pairs of one DNA oligomer with respect to each nucleotide pair of another DNA oligomer have been calculated. Thus, two sets of RDFs have been obtained for each system: the RDFs of the nucleotide pairs 1 and 22 of DNA 1 (Fig. 6.2a) with respect to each of 22 nucleotides of DNA 2 ($RDF_{DNA21}$), and vice versa, the RDFs for nucleotide pairs 1 and 22 of DNA 2 with respect to each of 22 nucleotides of DNA 1 ($RDF_{DNA12}$).

To characterize the structure of the DNA double helix, the width of the minor and major grooves were calculated as the shortest distance between the phosphate groups of the opposite strands of the double helix (Fig. 6.2a) using the 3DNA algorithms [125] implemented to do_x3dna plug-in [126] of VMD software [123].

## 6.4 Distribution of Spermidine$^{3+}$ Around DNA

To characterize the distribution of polyamine molecules around the DNA double helix, the radial distribution functions (RDFs) of polyamines with respect to the atoms in the grooves of the double helix and the phosphate groups of each nucleotide pair have been calculated. The RDFs and their integrals averaged over all nucleotides of two DNA molecules for the case of RandA, RandB, and RandC systems with spermidine$^{3+}$ molecules are shown on Fig. 6.3.

The RDFs are characterized by several maxima within the distance up to about 6 Å, related to the positions of atoms in the backbone of spermidine$^{3+}$ molecule. The peaks of the RDFs describing the interaction between polyamines and the phosphate groups are sharp, indicating that the complexes of spermidine$^{3+}$ with DNA are rather rigid and persistent. The root mean square (RMS) deviations, indicated in the plot as error bars, reveal the essential variation of the RDFs with the nucleotide sequence. In the case of the $RDF_{MIN}$, characterizing the distribution of polyamines with respect to the atoms of the DNA minor groove, the shapes of the average RDFs are rather different for the different nucleotide sequences (Fig. 6.3). In the case of RandA system (two A-tracts in the sequence), a distinct peak at the distance of about 2.8 Å is observed. This peak characterizes direct contact of polyamine molecule with the atoms of nucleotide bases. At the distance of about 3.5 Å, a shoulder is observed. In the case of RandB and RandC systems (with only one or no A-tract in the sequence), the first maximum decreases, and the shoulder transforms into a rather prominent maximum. This RDF behavior indicates that, for the polyamines, the direct contact in the minor groove is most probable in A-tracts containing sequence.

The shapes of RDFs characterizing the localization of polyamines with respect to the atoms of the DNA major groove ($RDF_{MAJ}$) are rather similar for the three

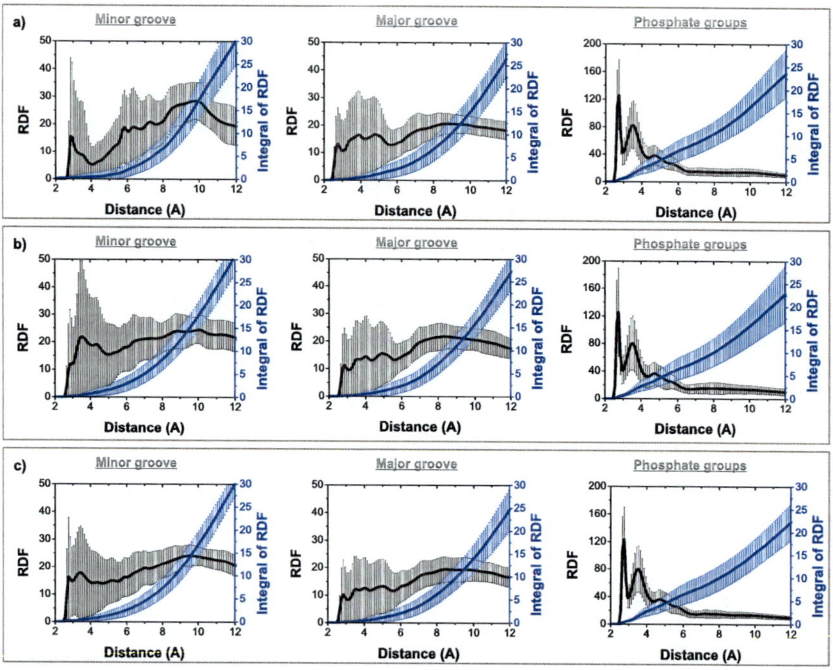


**Fig. 6.3** Radial distribution functions (black), and corresponding integrals (blue), for polyamines with respect to the reference atoms in the minor groove ($RDF_{MIN}$), major groove ($RDF_{MAJ}$), and phosphate groups ($RDF_{PH}$) of DNA simulation systems for: **a** RandA with the DNA nucleotide sequence CGCGAATTCGCGCGAATTCGCG, **b** RandB with the DNA nucleotide sequence CGCGATATCGCGCGAATTCGCG, **c** RandC with the DNA nucleotide sequence CGCGATATCGCGCGATATCGCG. The radial distribution functions have been averaged over all nucleotides in each simulation system. The error bars show the root mean square deviations

studied sequences (Fig. 6.3). There is one distinct peak at ~ 2.8 Å, followed by several less well-defined peaks separated by less than 1 Å. These peaks correspond to the atoms in the polyamine backbone, whose separation is due to the length of a chemical bond. Such RDF structures may appear due to the orthogonal orientation of polyamine molecules with respect to the double helix.

The RDFs for polyamines relatively to oxygen atoms of the phosphate groups ($RDF_{PH}$) have the most prominent structures among those studied (Fig. 6.3). The first $RDF_{PH}$ peak is much higher than in the case of the $RDF_{MIN}$ and $RDF_{MAJ}$, and it is located at the distance of 2.7 Å, slightly closer than in the case of the grooves of the double helix. This is a result of electrostatic attraction of the positively charged polyamine molecules and the negatively charged phosphate groups of the DNA double helix. The second peak at the distance of 3.5 Å is significantly lower than the first (nearly half), but it is still very high. Then, several very weak maxima are observed with increasing distance. This structure suggests that spermidine$^{3+}$ molecules bind very tightly to the DNA phosphate groups. The presence of a high

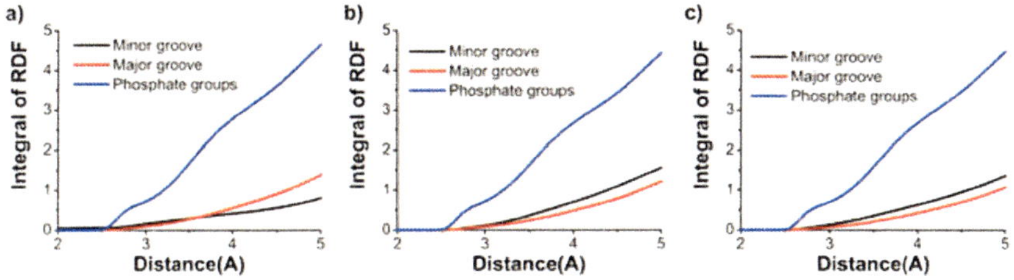


**Fig. 6.4** The integrals of radial distribution functions for polyamines with respect to the atoms in the minor groove, major groove, and phosphate groups of the DNA double helix. **a** RandA system with the DNA nucleotide sequence CGCGAATTCGCGCGAATTCGCG. **b** RandB system with the DNA nucleotide sequence CGCGATATCGCGCGAATTCGCG. **c** RandC system with the DNA nucleotide sequence CGCGATATCGCGCGATATCGCG

first peak indicates that the polyamine molecules, binding to the phosphate groups, are oriented parallel to the backbone of the double helix. At the same time, maxima in the region from 3 to 5 Å are compatible with polyamine molecule orientations orthogonal to the helical axis.

The integrals of the RDFs (blue lines in Fig. 6.3) in the case of the $RDF_{MIN}$ and $RDF_{MAJ}$ within the range up to 5 Å increase slowly with increasing distance, and then the curve becomes sharper. On the contrary, the $RDF_{PH}$ increase almost linearly from the very beginning. At distances < 5 Å the average number of particles near the phosphate groups is about 5, much higher than in the grooves (Fig. 6.4). In the case of RandA system, the numbers of particles in the minor and major groove are almost equal (Fig. 6.4a), while in the case of RandB and RandC it is always larger in the minor groove than in the major groove (Fig. 6.4b,c).

Taking into consideration the results obtained from the analysis of RDFs and from a visual examination of the simulation trajectories, the following modes of polyamine localization with respect to the DNA double helix may be proposed (Fig. 6.5). The polyamine molecules can be localized in the minor groove, major groove, and near the phosphate groups outside the double helix. The interaction of polyamines with the phosphate groups of DNA is the most probable and, in this case, the polyamine molecules are mostly oriented parallel to the backbone of the double helix (Fig. 6.5a). In the case of RandA system, polyamine localization in the minor groove parallel to the double helix axis (Fig. 6.5b) is the most frequent in the A-tract regions. The orientation of spermidine$^{3+}$, orthogonal to the helical axis, is also commonly observed in the case of both the minor and major groove in all considered simulation systems. The modes of polyamine localization essentially depend on the particular nucleotide sequence and on the relative orientation of DNA oligomers.

In Fig. 6.6, four sets of snapshots describing the modes of polyamine localization around the DNA double helix are represented, using as example the RandA system. Each set represents the position, relative to a DNA molecule, of a given polyamine molecule along the entire trajectory. Different coloring of the polyamine molecule is used to represent different points in time, as indicated in the legend. It should be noted

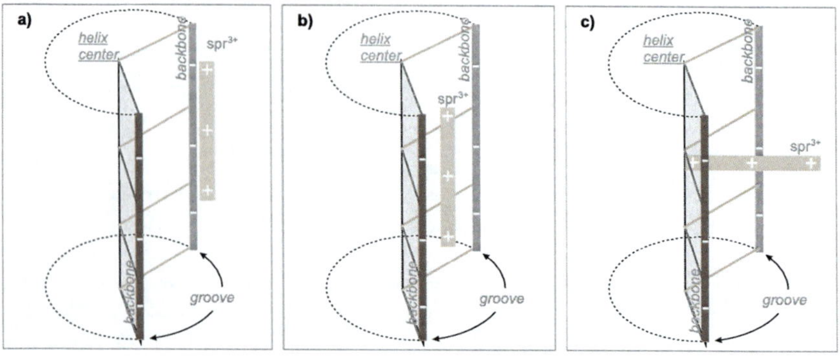


**Fig. 6.5** Modes of polyamine localization around DNA double helix. **a** Parallel orientation near the phosphate groups outside the double helix. **b** Parallel orientation of polyamine in the minor or major groove of the double helix. **c** Orthogonal orientation of polyamine molecule partially penetrated in the minor groove or major groove of the double helix

that the DNA backbone moves during the trajectory, and for clarity of representation we only show the initial canonical *B*-DNA conformation. In the first set of snapshots, top row, the position of a polyamine molecule varies with time and all the three modes of polyamine orientation described in Fig. 6.5 can be seen. At the beginning (the first 100 ns), spermidine$^{3+}$ is oriented orthogonal to the macromolecule, partially penetrated into the minor groove of the double helix that corresponds to the interaction mode schematized on Fig. 6.5c. Then, it goes inside the minor groove (the mode described on Fig. 6.5b), where it remains about 50 ns. In the final part of the simulation trajectory, the polyamine molecule emerges from the minor groove and sticks to the backbone of the macromolecule outside the double helix.

Visual inspection shows that the spermidine$^{3+}$ molecule often stays for almost the entire simulation in some defined binding site, adopting only one of the interaction modes described on Fig. 6.5. The second row of snapshots in Fig. 6.6 describes the case of a polyamine molecule localized inside the major groove of the double helix during the entire simulation trajectory. This case of DNA-polyamine interaction is related to the interaction mode schematized on Fig. 6.5b. The third row of snapshots (Fig. 6.6) illustrates the interaction mode with the polyamine molecule outside the double helix, as schematize in Fig. 6.5 a). The bottom row (Fig. 6.6) shows the case of the orthogonal orientation of polyamines, corresponding to the mode in Fig. 6.5.c).

## 6.5 Sequence Specificity of DNA-Polyamine Binding

To study the sequence specificity of polyamine binding to DNA, the integral values of the average RDFs reported in Fig. 6.3 have been analyzed in more detail, considering separately the RDF with respect to each DNA base pair. Considering that the first RDF minimum is at a distance $\geq 3$ Å, depending on the region of the double helix,

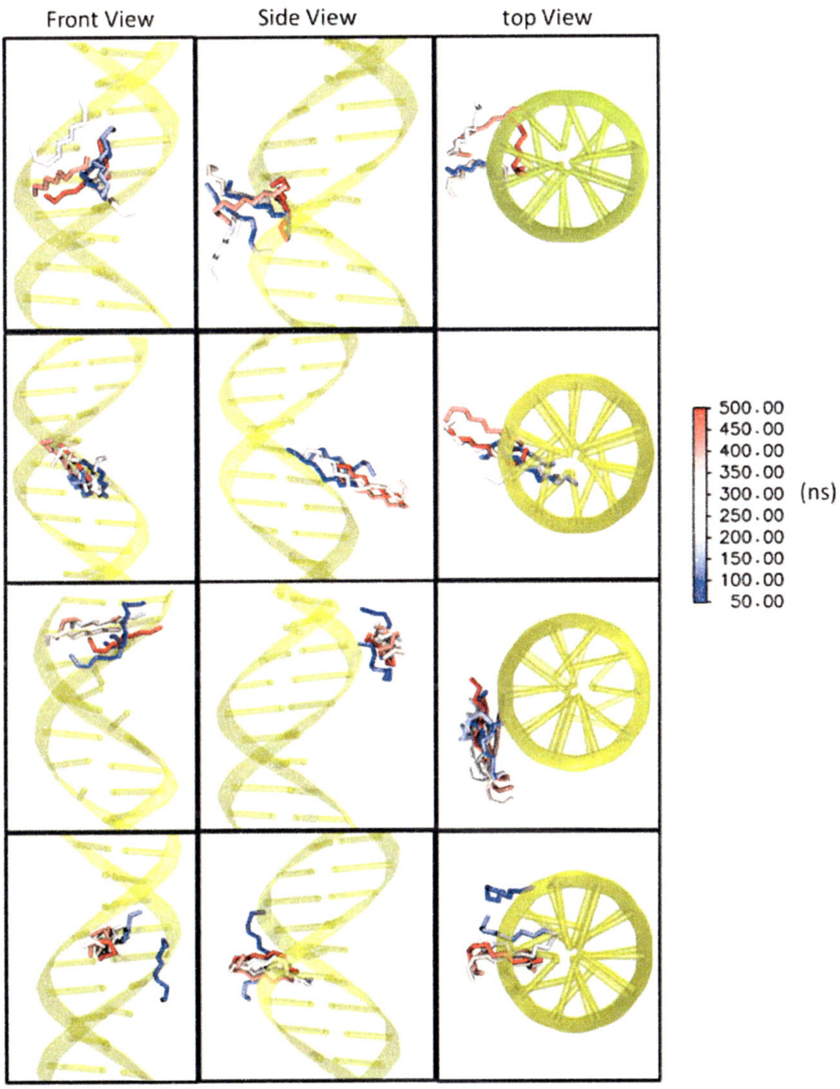


**Fig. 6.6** Snapshots representing the orientation with respect to DNA for selected spermidine$^{3+}$ molecule using front, side and top views. Each line represents the trajectory of a single polyamine molecule, and the coloring map indicating the time is shown on the right. The top line shows the case, where different modes of polyamine localization are adopted at different times. The second line illustrates the parallel orientation mode for a polyamine in the major groove. The third line shows the parallel orientation mode of spermidine$^{3+}$ near the phosphate groups outside the double helix. At the bottom, the orthogonal orientation mode with the polyamine molecule partially penetrated in the minor groove is shown

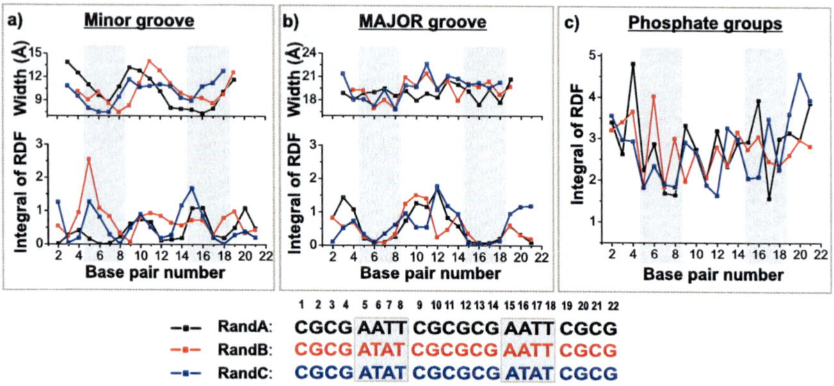


**Fig. 6.7** The dependence of DNA-polyamine RDF integrals calculated up to the distance of 4 Å from DNA reference atoms, for three nucleotide sequences. **a** The integrals of RDFs with respect to the reference DNA atoms located in the minor groove. The width of the minor groove is shown at the top. **b** The integrals of RDFs with respect to the reference DNA atoms located in the major groove. The width of the major groove is shown at the top. **c** The integrals of RDFs relatively to the reference atoms of the DNA phosphate groups. The three self-complementary sequences are shown at the bottom. For visual guidance, the AATT and ATAT region are marked by gray color

the integrals have been calculated up to 4 Å, taken as a representative distance for the comparison of coordination numbers obtained from different RDFs.

The dependence of coordination numbers on the nucleotide sequence is shown in Fig. 6.7 together with the groove width to help with correlating the two properties, since the specificity of polyamine binding to some nucleotide sequences of DNA is related to this structural parameter [103].

In the minor groove (Fig. 6.7a), the values of the coordination numbers show a clear preference for the A-T rich regions. In the case of the systems with the DNA molecules that have one or two ATAT nucleotide sequences (RandB and RandC), the coordination numbers of polyamines have maxima in these regions. In the case of RandA, having two AATT sequences, the highest coordination is observed in one of the A-tract regions, while it remains very low in the other A-tract region comprising the base pairs 5 to 8. Such asymmetry in the ion-DNA interaction comparing the two A-tracts of a palindromic sequence is due to the asymmetric interactions between the two duplexes following the two DNA molecules condensation. In the A-tract of RandB, the polyamine coordination number does not display high peaks, ranging between 0.5 and 1. Note that spermidine$^{3+}$ molecules are observed in a minor groove in the central part of DNA fragments in the case of all simulated systems with the polyamines. Comparing the minor groove width with the ion occupancy, we can conclude that in the regions with the larger minor groove (those with the alternating C-G base pairs) the ion occupation does not reach the high peaks observed in the regions with A-T base pairs. Furthermore, in the alternating A-T nucleotide sequence, the coordination number reaches higher values than in the A-tracts.

In the case of the major groove (Fig. 6.7b), the values of RDF integrals decrease to almost zero in the regions with AATT and ATAT nucleotide sequences in all simulated systems. At the same time, in the CGCG region in the middle of the DNA fragments, an increase of the coordination numbers is observed, reaching values similar to those observed for the maxima in the AT regions of the minor groove. An increase of coordination numbers is also observed at the ends of the macromolecules. Comparing the major groove width, averaged over two DNA molecules (at the top of Fig. 6.7b), with the spermidine$^{3+}$ coordination number, no correlation is found. The lack of polyamines localized in the major groove near A-T nucleotide pairs may be related to the number of atomic groups acting as H-bond acceptors. Due to the $CH_3$ group of thymine, the number of acceptor groups is lower than in the case of C-G nucleotide pair. As a result, the polyamine molecules in the major groove form less H-bonds with the acceptor atoms of DNA in the regions near A-T nucleotide pairs. The same effect was observed in the simulations of DNA with spermine$^{4+}$ [92].

In the region near the phosphate groups, the coordination numbers change randomly with the sequence of nucleotides. However, some tendency to increase at the ends of DNA fragments is observed. Taking into consideration that, in the center of the DNA fragment, the electrostatic field is stronger than at the ends, an increase in the values of RDFs integrals is expected for the central part of DNA. Such redistribution of the polyamines may be related to the DNA-DNA interaction that will be discussed in the next subsection.

To evaluate the effect of the polyamine on the widths of the minor and major grooves, these parameters were calculated both with and without polyamine and are shown on Fig. 6.8 for each of the simulated DNA molecules. It can be seen that, in presence of polyamines (RandA, RandB, and RandC simulation systems), the width of the minor groove significantly decreases in the regions with AATT and ATAT nucleotide sequences (Fig. 6.8a). The width of the minor groove in these regions is within 6 – 8 Å. In the center and at the ends of the DNA fragment, where the C-G nucleotide pairs are present, the minor groove is rather wide in the case of all systems with polyamines (about 13 – 14 Å), with the exception of one of the RandC molecules. Despite the symmetrical localization of AATT and ATAT nucleotide regions, the narrowing of the grooves is not symmetrical, and it is not uniform in general.

The effect of the spermidine$^{3+}$ on the narrowing of the minor groove can be inferred by comparing the same parameter calculated on the same sequence in systems without the polyamine (FreeA, FreeB, and FreeC simulation systems). The results (Fig. 6.8b) show that, even in the absence of polyamines, the minor groove widths of the double helixes are narrower in the regions with AATT and ATAT nucleotide sequences (about 9 Å). Notably, in the case of AATT region, the width of the minor groove is about 0.5 Å narrower than in the ATAT nucleotide sequence. This property of the double helix structure is well known for such specific nucleotide sequences [11–13].

The major groove width of the double helix also depends on sequence of nucleotides, but it is not as prominent as in the case of the minor groove. The mean value of the groove width is about 19 Å. In the case of the systems without polyamines (FreeA, FreeB, and FreeC), the narrowing to about 16 Å of the major groove occurs in

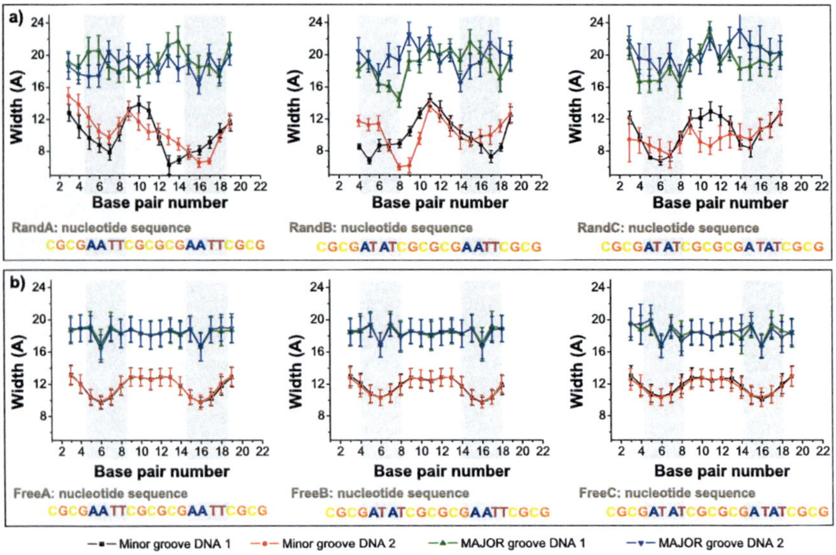


**Fig. 6.8** The dependencies of the width of the minor and major grooves on the nucleotide sequence in the case. **a** The systems of DNA with spermidine$^{3+}$ molecules: RandA, RandB, and RandC. **b** The systems of DNA without spermidine$^{3+}$ molecules: FreeA, FreeB, and FreeC. The groove widths have been averaged over the simulation trajectories and the error bars show the root mean square deviations. The parameters have been calculated for each DNA oligomers in the simulation box. The sequences of nucleotide pairs of DNA oligomer in the case of each system are shown at the bottom. The AATT and ATAT region are marked by gray color

the center of AATT and ATAT regions (Fig. 6.8b). This narrowing is within the RMS deviation of the groove width, but it is repeating systematically. In the case of the systems with polyamines (RandA, RandB, and RandC), the major groove becomes narrower in general and the dependence on the sequence of nucleotides is not regular (Fig. 6.8b).

## 6.6 DNA-DNA Interactions

In the present MD study, the simulated systems contained two DNA oligomers in the simulation box. The DNA duplexes are not fixed and interact with each other. Moreover, in the case of the system of DNA with polyamines (RandA, RandB, and RandC), the effect of DNA condensation that is observed experimentally [89] should also be observed in our simulations.

Visual examination of the simulation trajectories has shown that the DNA-DNA contacts occur mostly through the formation of the complexes at the ends of the oligomers. Therefore, to study the interactions between DNA oligomers, the RDFs of the end nucleotide pairs of one of the two DNA oligomers with respect to each nucleotide pair of the other oligomer have been calculated. The obtained RDFs in the case of the systems of DNA with the polyamines are shown in the Fig. 6.9. The maxima of obtained RDFs characterize the contacts of the nucleotide pairs of one

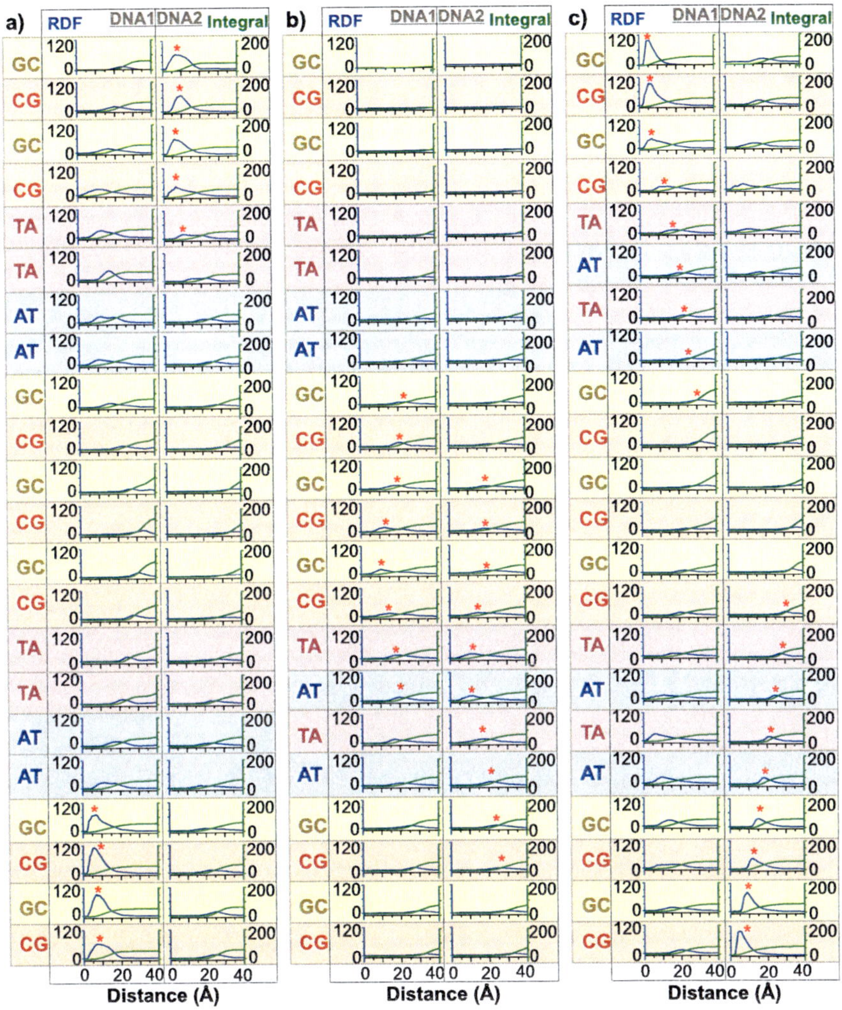


**Fig. 6.9** The radial distribution functions and their running integrals for the end nucleotide pairs of one DNA oligomer with respect to each nucleotide pair of another DNA oligomer. The simulation systems with polyamines: **a** RandA with the DNA nucleotide sequence CGCGAATTCGCGC-GAATTCGCG; **b** RandB with the DNA nucleotide sequence CGCGATATCGCGCGAATTCGCG; **c** RandC with the DNA nucleotide sequence CGCGATATCGCGCGATATCGCG. The RDFs and their integrals are shown by blue and green lines, respectively, in the double plot for each nucleotide pair. The sequence of nucleotide pairs is shown on the left by different colors. The maxima of the RDFs that correspond to the stable DNA-DNA contacts are marked by red asterisks

DNA molecule with the ends of another DNA molecule. The position and intensity of the maxima are the result of the structure of DNA-DNA complexes. The shortest distance between the DNA molecules, observed by the first maximum of the RDFs, is about 10 Å corresponding to the radius of the double helix. The DNA oligomers that are localized at such distance have direct contact. For DNA in RandA and RandC systems, high maxima of the RDFs are observed at the ends of the macromolecules (the maxima are marked by a red asterisk in Fig. 6.9). The intensity of these maxima gradually decreases as the pair of nucleotides is located away from the end, towards the center of the macromolecule, and the position of these maxima move away to larger distances. Such fading away of the RDFs maxima indicates that different DNA molecules interact through terminal nucleotides to form the complexes. In the case of DNA in RandB system, the DNA-DNA contacts are observed between nucleotides that are closer to the central part of the macromolecule (Fig. 6.9), therefore, the RDFs used for the analysis of the contacts with the ends of DNA molecules do not show the prominent maxima. Visual examination of the trajectories shows that DNA-DNA-sperimidine$^{3+}$ complexes are formed by 3 – 5 nucleotides of each DNA molecule, mediated by spermidine$^{3+}$ (Fig. 6.11). In the complex, the polyamine molecules are localized parallel to the DNA backbone and form the cross-links between negatively charged phosphate groups. One positively charged polyamine molecule may be coordinated by two or three chains of negatively charged phosphate groups of the DNA backbone of the same or different macromolecules.

In the case of the systems of DNA without polyamine molecules (FreeA, FreeB, and FreeC), the RDFs of the end nucleotides of one DNA molecule with respect to the nucleotides of another nucleotide molecule show the presence of prominent maxima only in the case of FreeA system (Fig. 6.10). The positions of these maxima are even closer than 10 Å and diminish with increasing distance as in the case of the RandA and RandC systems. The formation of such maxima is not related to the electrostatic screening of the phosphate groups as in the case of the systems with polyamines, but it is due to stacking interactions of the end nucleotide bases of different macromolecules (Fig. 6.11). The formation of DNA-DNA contacts in the case of the other systems without polyamine molecules (FreeB and FreeC) has not been observed.

Thus, the obtained results show that the interactions between the DNA oligomers have been observed mostly in the systems with spermidine$^{3+}$, but in some cases the DNA-DNA contacts were also present in the systems without polyamines. However, it should be noted that the structure of the complexes, and the mechanisms of their formation, are different in the case of the systems with polyamines and those without them. In the case of DNA with polyamines, the two DNA oligomers bind to each other due to the formation of polyamine cross-links between different macromolecules (Fig. 6.12a). The central part of the cross-links is the spermidine$^{3+}$ molecule that is coordinated by negatively charged phosphate groups belonging to the backbone of different macromolecules. In the case of the systems without polyamines, the mechanism of formation of the DNA-DNA complex is related to the formation of stable stacking contacts between the end nucleotide pairs of different macromolecules (Fig. 6.12b). Such types of DNA contacts are possible due to the fact that, in the

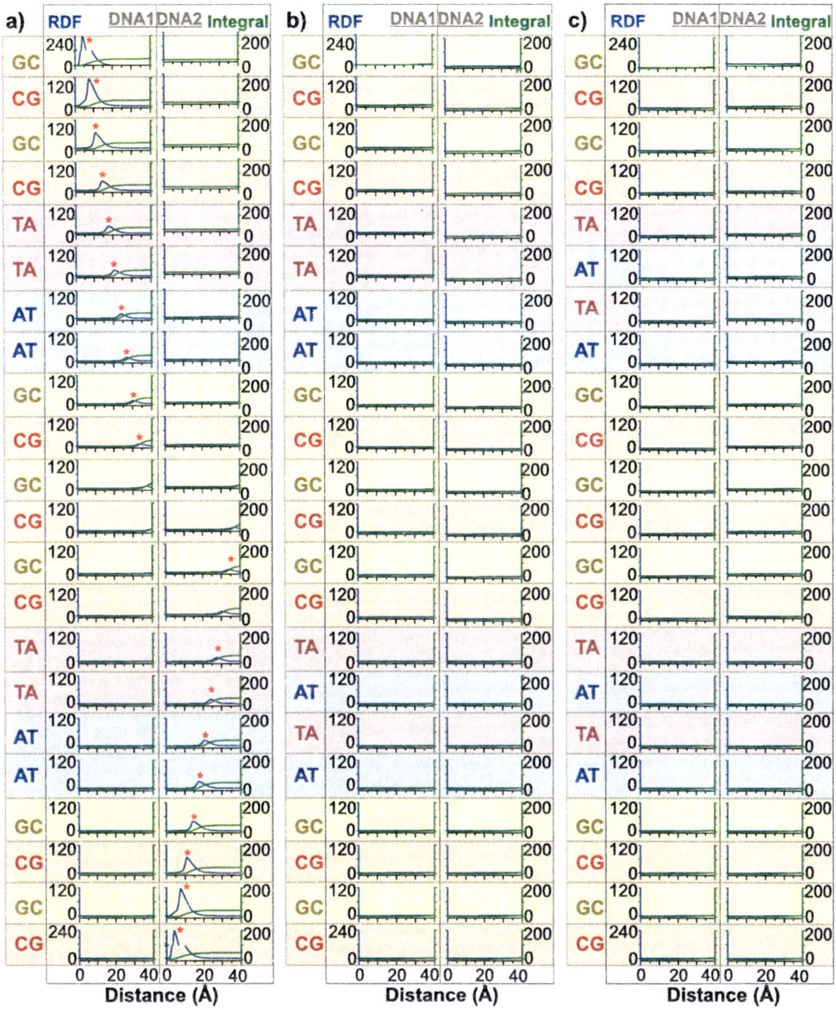


**Fig. 6.10** The radial distribution functions and their integrals for the end nucleotide pairs of one DNA oligomer with respect to each nucleotide pair of another DNA oligomer. The simulation systems without polyamines: **a** FreeA with the DNA nucleotide sequence CGCGAATTCGCGCGAATTCGCG; **b** FreeB with the DNA nucleotide sequence CGCGATATCGCGCGAATTCGCG; **c** FreeC with the DNA nucleotide sequence CGCGATATCGCGCGATATCGCG. The RDFs and their integrals are shown by blue and green lines in the double plot for each nucleotide pair. The sequence of nucleotide pairs is shown on the left by different colors. The maxima of the RDFs that correspond to the stable DNA-DNA contacts are marked by red asterisks

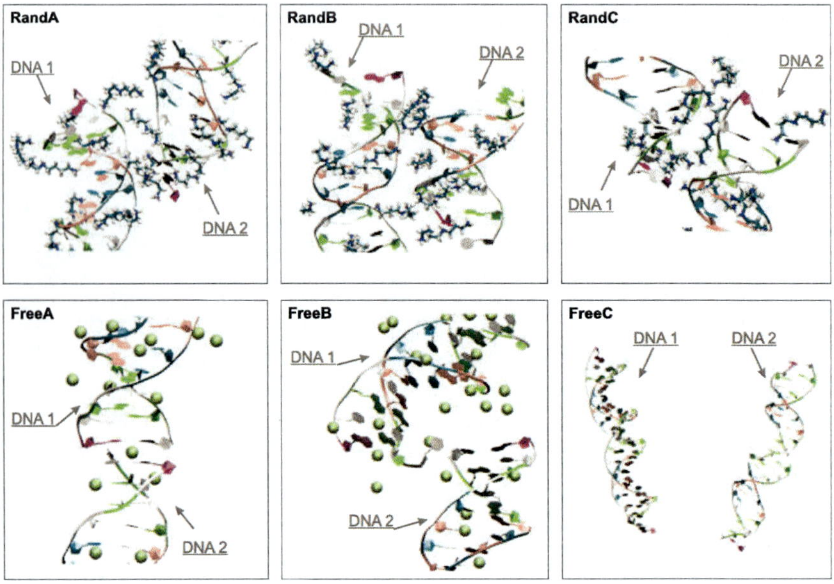


**Fig. 6.11** The snapshots of the complexes of DNA oligomers in the case of different simulation systems

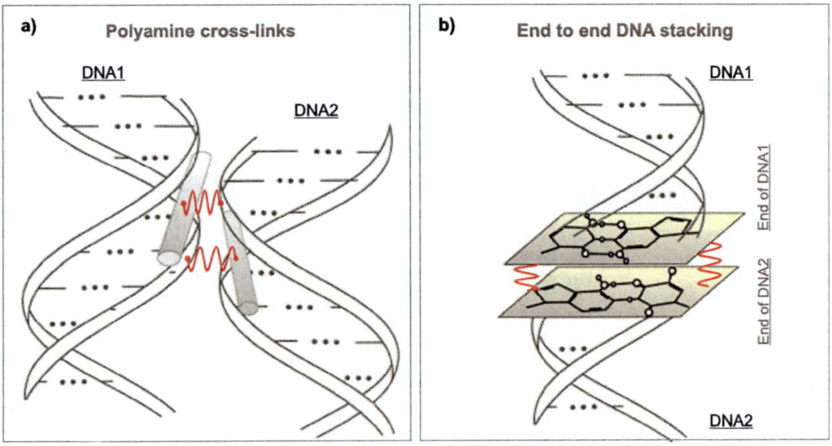


**Fig. 6.12** Schemes of the modes of DNA-DNA complexes. **a** Polyamine mediated complex of two DNA molecules. The polyamines are coordinated by the phosphate groups of different DNA strands, which is schematically shown by red springs. **b** End-to-end complex of two DNA molecules. The complex is formed due to stacking interaction of end nucleotides of different DNA oligomers, which is schematically shown by red spring

present simulation, short DNA fragments have been used and are likely not relevant in biological systems due to the essential length of natural DNA. However, this effect may be interesting for technological applications where short DNA fragments may be used. In particular, a liquid-crystalline phase of short DNA fragments having end-to-end contacts has been observed at high concentrations of polynucleotides [127]. The systems we have considered here are indeed simulated in conditions in which the formation of such a liquid crystalline phase can occur.

## 6.7 Discussion and Conclusions

MD simulations have been carried out for six different systems, each containing two DNA oligomers, with and without spermidine$^{3+}$ molecules. The different nucleotide sequences of the simulated DNA duplexes allowed studying the sequence specificity of the binding of spermidine$^{3+}$ molecules. The results show that spermidine$^{3+}$ molecules condense on DNA and strongly interact with the double helix. The polyamines penetrate to the minor and major grooves of the double helix and bind to the phosphate groups on the outside of the macromolecule. At the considered spermidine$^{3+}$ concentration, the polyamine molecules in the minor or major groove sides tend to arrange orthogonal to the helical axis. The binding of spermidine$^{3+}$ molecules to the negatively charged phosphate groups of the DNA double helix has been found as the most probable interaction mode.

The analysis of the radial distribution functions shows a clear sequence specificity in the polyamine binding to DNA. In particular, in the minor groove of the DNA double helix, the occupancy of spermidine$^{3+}$ molecules are higher around AATT and ATAT nucleotide sequences, while, for the same A-T-rich tracts in the major groove, the mean number of polyamines was found to be very low. As observed in previous studies [64, 66, 69], in absence of polyamine the minor groove is narrower in A-T rich region, being slightly narrower for the A-tracts compared to the alternating sequences. These points are important for the polyamine binding. The obtained results agree with those of previous simulations [103], where such preferential binding in AATT nucleotide region has been explained by the enhanced electrostatic attraction due to the narrowing of the minor groove. The minor groove width in the A-T rich region decreases in the system with spermidine$^{3+}$. The low occupancy of polyamines in the major groove of A-T rich sequence may be explained by the presence of the methyl group of thymine that reduces the number of H-bond acceptor atoms for the formation of the H-bonds between polyamines and DNA. Furthermore, the methyl groups of thymine act as steric blocks for the polyamines [92].

The high concentration of polyamines in the simulated systems (much higher than that found under the physiological conditions) induces DNA condensation, in agreement with extensive experimental observations [86, 87, 89, 99]. In the regions where the DNA-DNA contacts occur, the polyamine molecules interacting with the double helix form cross-links and tend to be oriented parallel to the phosphate groups of the double helix backbone that may belong to different macromolecules. The polyamines are often localized in the minor groove of one DNA oligomer, attached by the phosphate groups of the double helix backbone, and at the same time contacting the phosphate groups of another DNA molecule. Taking into consideration that the localization of the spermidine$^{3+}$ molecules in the minor groove is more probable in the AATT and ATAT regions, the simulations suggest that the formation of the polyamine cross-links may be regulated by the sequence of nucleotides. Using such sequence-controlled formation of the cross-links in the DNA-DNA complexes, macromolecular aggregates with desired properties may be engineered.

In conclusion, the present MD simulation study shows that spermidine$^{3+}$ molecules preferentially bind to AATT and ATAT nucleotide sequences in the DNA minor groove, but not in the major groove. The positioning of polyamines on DNA determines the formation of the DNA-DNA contacts due to the cross-links, where the polyamines are localized parallel to the backbones of different macromolecules. The obtained results are important for understanding the role of spermidine$^{3+}$ in the biological mechanisms and function of DNA as well as for possible technological applications.

**Acknowledgements** Grants of computing time from the Swedish national infrastructure for computing (SNIC) are gratefully acknowledged. S. Perepelytsya acknowledges the support from the National Academy of Sciences of Ukraine (project 0117U000240). A. Laaksonen acknowledges Swedish Research Council for financial support, and partial support from a grant from Ministry of Research and Innovation of Romania (CNCS - UEFISCDI, project number PN-III-P4-ID-PCCF-2016-0050, within PNCDI III). F. Mocci acknowledges financial support from Progetto Fondazione di Sardegna (Grant CUP: F72F20000230007) and Regione Autonoma della Sardegna (RASSR81788-2017).

## References

1. Igarashi K, Kashiwagi K (2010) Modulation of cellular function by polyamines. Int J Biochem Cell Biol 42:39–51. https://doi.org/10.1016/j.biocel.2009.07.009
2. Gerner EW, Meyskens FL (2004) Polyamines and cancer: Old molecules, new understanding. Nat Rev Cancer 4:781–792. https://doi.org/10.1038/nrc1454
3. Petri G, Antonello M, Fernando C (2016) Modulation of learning and memory by natural polyamines. Pharmacol Res 112:99–118. https://doi.org/10.1016/j.phrs.2016.03.023
4. Madeo F, Eisenberg T, Pietrocola F, Kroemer G (2018) Spermidine in health and disease. Science 359:eean2788
5. Petri G, Antonello M, Fernando C (2010) Modulation of learning and memory by natural polyamines Gustavo. Autophagy 8627:1–4. https://doi.org/10.4161/auto.6.1.10600
6. Pegg AE (2013) Toxicity of polyamines and their metabolic products. Chem Res Toxicol 26:1782–1800. https://doi.org/10.1021/tx400316s
7. Moschou PN, Roubelakis-Angelakis KA (2014) Polyamines and programmed cell death. J Exp Bot 65:1285–1296. https://doi.org/10.1093/jxb/ert373
8. Visvanathan A, Ahmed K, Even-Faitelson L, Lleres D, Bazett-Jones DP, Lamond AI (2013) Modulation of higher order chromatin conformation in mammalian cell nuclei can be mediated by polyamines and divalent cations. PLoS One 8:e67689. https://doi.org/10.1371/journal.pone.0067689
9. Mounce BC, Olsen ME, Vignuzzi M, Connor JH (2017) Polyamines and their role in virus infection. Microbiol Mol Biol Rev 81:1–12. https://doi.org/10.1128/mmbr.00029-17

10. Firpo MR, Mounce BC (2020) Diverse functions of polyamines in virus infection. Biomolecules 10:1–13. https://doi.org/10.3390/biom10040628
11. Saenger W (1984) Principles of nucleic acid structure. Springer-Verlag, New York
12. Schlick T (2002) Molecular modeling and simulation: an interdisciplinary guide. Springer, US, New York
13. Sivolob AV (2011) Physics of DNA: A study guide. Publishing and Printing Center "Kyiv University," Kyiv
14. Blagoi YP, Veselkov ON, Volkov S, Hovorun D, Yevstigneev M, Zhurakivsky RO, Korneliuk OI, Maleev VY, Semenov MO, Sorokin VO, Kharkianen VM, L.M. C, Shestopalova AV, (2011) Physical principles of molecular organization and structural dynamics of biopolymers, vol N. Karazin Kharkiv National University, Kharkov
15. Watson JD, Crick FHC (1953) Molecular structure of nucleic acids: A structure for deoxyribose nucleic acid. Nature 171:737–738
16. Franklin RE, Gosling RG (1953) The structure of sodium thymonucleate fibres. II. The cylindrically symmetrical Patterson function. Acta Crystallogr 6:678–685. https://doi.org/10.1107/s0365110x53001940
17. Maleev VI, Semenov MA, Gasan AI, Kashpur VA (1993) Physical properties of the DNA-water system. Biofizika 38:768–790
18. Blagoi YP, Galkin VL, Gladchenko VL, Kornilova SV, Sorokin VA, Shkorbatov AG (1991) The complexes of nucleic acids and metals in the solutions. Naukova Dumka, Kyiv
19. Atzori A, Liggi S, Laaksonen A, Porcu M, Lyubartsev AP, Saba G, Mocci F (2016) Base sequence specificity of counterion binding to DNA: What can MD simulations tell us? Can J Chem 94:1181–1188. https://doi.org/10.1139/cjc-2016-0296
20. Mocci F, Laaksonen A (2012) Insight into nucleic acid counterion interactions from inside molecular dynamics simulations is "worth its salt." Soft Matter 8:9268–9284. https://doi.org/10.1039/c2sm25690h
21. Manning GS (1978) The molecular theory of polyelectrolyte solutions with applications to the electrostatic properties of polynucleotides. Q Rev Biophys. https://doi.org/10.1017/S0033583500002031
22. Korolev N, Allahverdi A, Lyubartsev AP, Nordenskiöld L (2012) The polyelectrolyte properties of chromatin. Soft Matter 8:9322–9333. https://doi.org/10.1039/c2sm25662b
23. Oosawa F (1971) Polyelectrolytes. Marcel Dekker. Inc., New York
24. Frank-kamenetskii MD, Anshelevich VV, Lukashin AV (1987) Polyelectrolyte model of DNA. Sov Phys Usp 30:317–330. https://doi.org/10.1070/PU1987v030n04ABEH002833
25. Das R, Mills TT, Kwok LW, Maskel GS, Millett IS, Doniach S, Finkelstein KD, Herschlag D, Pollack L (2003) Counterion distribution around DNA probed by solution X-ray scattering. Phys Rev Lett 90:4. https://doi.org/10.1103/PhysRevLett.90.188103
26. Andresen K, Das R, Park HY, Smith H, Kwok LW, Lamb JS, Kirkland EJ, Herschlag D, Finkelstein KD, Pollack L (2004) Spatial distribution of competing ions around DNA in solution. Phys Rev Lett 93:1–4. https://doi.org/10.1103/PhysRevLett.93.248103
27. Das P, Schuster GB (2005) Effect of condensate formation on long-distance radical cation migration in DNA. Proc Natl Acad Sci USA 102:14227–14231. https://doi.org/10.1073/pnas.0506778102
28. Dias RS, Magno LM, Valente AJM, Das D, Das PK, Maiti S, Miguel MG, Lindman B (2008) Interaction between DNA and cationic surfactants: Effect of DNA conformation and surfactant headgroup. J Phys Chem B 112:14446–14452. https://doi.org/10.1021/jp8027935
29. Lipfert J, Doniach S, Das R, Herschlag D (2014) Understanding nucleic acid-ion interactions. Annu Rev Biochem 83:813–841. https://doi.org/10.1146/annurev-biochem-060409-092720
30. Kornyshev AA, Lee DJ, Leikin S, Wynveen A (2007) Structure and interactions of biological helices. Rev Mod Phys 79:943–996. https://doi.org/10.1103/RevModPhys.79.943
31. Tereshko V, Minasov G, Egli M (1999) A "hydrat-ion" spine in a B-DNA minor groove. J Am Chem Soc 121:3590–3595. https://doi.org/10.1021/ja984346+
32. Minasov G, Tereshko V, Egli M (1999) Atomic-resolution crystal structures of B-DNA reveal specific influences of divalent metal ions an conformation and packing. J Mol Biol 291:83–99. https://doi.org/10.1006/jmbi.1999.2934

33. Tereshko V, Wilds CJ, Minasov G, Prakash TP, Maier MA, Howard A, Wawrzak Z, Manoharan M, Egli M (2001) Detection of alkali metal ions in DNA crystals using state-of-the-art X-ray diffraction experiments. Nucleic Acids Res 29:1208–1215. https://doi.org/10.1093/nar/29.5.1208
34. Egli M, Tereshko V (2004) Lattice- and sequence-dependent binding of Mg2+ in the crystal structure of a B-DNA dodecamer. In: Nucleic Acids: Curvature and Deformation. vol 884, pp 87–109. ACS Symposium Series. https://doi.org/10.1021/bk-2004-0886.ch004
35. Duguid J, Bloomfield VA, Benevides J, Thomas GJ (1993) Raman spectroscopy of DNA-metal complexes. I. Interactions and conformational effects of the divalent cations: Mg, Ca, Sr, Ba, Mn Co, Ni, Cu, Pd, and Cd. Biophys J 65:1916–1928. https://doi.org/10.1016/S0006-3495(93)81263-3
36. Duguid JG, Bloomfield VA, Benevides JM, Thomas GJ (1995) Raman spectroscopy of DNA-metal complexes. II. The thermal denaturation of DNA in the presence of Sr2+, Ba2+, Mg2+, Ca2+, Mn2+, Co2+, Ni2+, and Cd2+. Biophys J 69:2623–2641. https://doi.org/10.1016/S0006-3495(95)80133-5
37. Deng H, Bloomfield VA (1999) Structural effects of cobalt-amine compounds on DNA condensation. Biophys J 77:1556–1561. https://doi.org/10.1016/S0006-3495(99)77003-7
38. Pollack L (2011) SAXS studies of ion-nucleic acid interactions. Annu Rev Biophys 40:225–242. https://doi.org/10.1146/annurev-biophys-042910-155349
39. Hud NV, Feigon J (1997) Localization of divalent metal ions in the minor groove of DNA A-tracts. J Am Chem Soc 119:5756–5757. https://doi.org/10.1021/ja9704085
40. Marincola FC, Denisov VP, Halle B (2004) Competitive Na+ and Rb+ binding in the minor groove of DNA. J Am Chem Soc 126:6739–6750. https://doi.org/10.1021/ja049930z
41. Marincola FC, Virno A, Randazzo A, Mocci F, Sabaa G, Laia A (2009) Competitive binding exchange between alkali metal ions (K+, Rb+, and Cs+) and Na+ ions bound to the dimeric quadruplex [d(G4T4G4)]2: A 23Na and 1HNMR study. Magn Reson Chem 47:1036–1042. https://doi.org/10.1002/mrc.2509
42. Hud N V, Engelhart AE (2008) Sequence-specific DNA – metal ion interactions. In: Hud N V. (ed) Nucleic Acid–Metal Ion Interactions. Royal Society of Chemistry, Cambridge, pp 71–113
43. Corongiu G, Clementi E (1981) Simulations of the solvent structure for macromolecules. I. Solvation of B-DNA double helix at T = 300 K. Biopolymers. https://doi.org/10.1002/bip.1981.360200309
44. Seibel GL, Singh UC, Kollman PA (1985) A molecular dynamics simulation of double-helical B-DNA including counterions and water. Proc Natl Acad Sci U S A. https://doi.org/10.1073/pnas.82.19.6537
45. Clementi E, Corongiu G, Detrich JH, Khanmohammadbaigi H, Chin S, Domingo L, Laaksonen A, Nguyen NL (1985) Parallelism in computational chemistry: Applications in quantum and statistical mechanics. Phys B+C. https://doi.org/10.1016/0378-4363(85)90142-1
46. Van Gunsteren WF, Berendsen HJC, Geurtsen RG, Zwinderman HRJ (1986) A molecular dynamics computer simulation of an eight-base-pair DNA fragment in aqueous solution: Comparison with experimental two-Ddimensional NMR data. Ann N Y Acad Sci. https://doi.org/10.1111/j.1749-6632.1986.tb20962.x
47. Nilges M, Clore GM, Gronenborn AM, Brunger AT, Karplus M, Nilsson L (1987) Refinement of the solution structure of the DNA hexamer 5'd(GCATGC)2 combined use of nuclear magnetic resonance and restrained molecular dynamics. Biochemistry. https://doi.org/10.1021/bi00386a068
48. Singh UC, Wiener SJ, Kollman P (1985) Molecular dynamics simulations of d(C-G-C-G-A)·d(T-C-G-C-G) with and without "hydrated" counterions. Proc Natl Acad Sci U S A 82:755–759. https://doi.org/10.1073/pnas.82.3.755
49. Laaksonen A, Nilsson LG, Jönsson B, Teleman O (1989) Molecular dynamics simulation of double helix Z-DNA in solution. Chem Phys. https://doi.org/10.1016/0301-0104(89)80002-3
50. Cheatham TE, Miller JL, Fox T, Darden TA, Kollman PA (1995) Molecular dynamics simulations on solvated biomolecular systems: The particle mesh Ewald method leads to stable trajectories of DNA, RNA, and proteins. J. Am. Chem. Soc.

51. Maffeo C, Yoo J, Comer J, Wells DB, Luan B, Aksimentiev A (2014) Close encounters with DNA. J Phys Condens Matter 26:413101. Doi:https://doi.org/10.1088/0953-8984/26/41/413101
52. Ivani I, Dans PD, Noy A, Pérez A, Faustino I, Hospital A, Walther J, Andrio P, Goñi R, Balaceanu A, Portella G, Battistini F, Gelpí JL, González C, Vendruscolo M, Laughton CA, Harris SA, Case DA, Orozco M (2016) Parmbsc1: A refined force field for DNA simulations. Nat Methods 13:55–58. https://doi.org/10.1038/nmeth.3658
53. Zgarbova M, Šponer J, Otyepka M, Cheatham TE, Galindo-Murillo R, Jurečka P (2015) Refinement of the sugar – phosphate backbone torsion beta for AMBER force field improves the description of Z- and B-DNA. J Chem Theory Comput 11:5723–5736. https://doi.org/10.1021/acs.jctc.5b00716
54. Rebič M, Laaksonen A, Šponer J, Uličný J, Mocci F (2016) Molecular dynamics simulation study of parallel telomeric DNA quadruplexes at different ionic strengths: evaluation of water and ion models. J Phys Chem B 120:7380–7391. https://doi.org/10.1021/acs.jpcb.6b06485
55. Hart K, Foloppe N, Baker CM, Denning EJ, Nilsson L, Mackerell AD (2012) Optimization of the CHARMM additive force field for DNA: Improved treatment of the BI/BII conformational equilibrium. J Chem Theory Comput 8:348–362. https://doi.org/10.1021/ct200723y
56. Lemkul JA, MacKerell AD (2017) Polarizable force field for DNA based on the classical Drude oscillator: I. Refinement using quantum mechanical base stacking and conformational energetics. J Chem Theory Comput 13:2053–2071. https://doi.org/10.1021/acs.jctc.7b00067
57. Lemkul JA (2020) Same fold, different properties: polarizable molecular dynamics simulations of telomeric and TERRA G-quadruplexes. Nucleic Acids Res 48:561–575. https://doi.org/10.1093/nar/gkz1154
58. Dans PD, Ivani I, Hospital A, Portella G, González C, Orozco M (2017) How accurate are accurate force-fields for B-DNA? Nucleic Acids Res 45:4217–4230. https://doi.org/10.1093/nar/gkw1355
59. Joung IS, Cheatham TE (2008) Determination of alkali and halide monovalent ion parameters for use in explicitly solvated biomolecular simulations. J Phys Chem B 112:9020–9041. https://doi.org/10.1021/jp8001614
60. Joung IS, Cheatham TE (2009) Molecular dynamics simulations of the dynamic and energetic properties of alkali and halide ions using water-model-specific ion parameters. J Phys Chem B 113:13279–13290
61. Allnér O, Nilsson L, Villa A (2012) Magnesium ion-water coordination and exchange in biomolecular simulations. J Chem Theory Comput 8:1493–1502. https://doi.org/10.1021/ct3000734
62. Panteva MT, Giambaşu GM, York DM (2015) Force Field for Mg2+, Mn2+, Zn2+, and Cd2+ ions that have balanced interactions with nucleic a cids. J Phys Chem B 119:15460–15470. https://doi.org/10.1021/acs.jpcb.5b10423
63. Leonarski F, D'Ascenzo L, Auffinger P (2016) Binding of metals to purine N7 nitrogen atoms and implications for nucleic acids: A CSD survey. Inorganica Chim Acta 452:82–89. https://doi.org/10.1016/j.ica.2016.04.005
64. Young MA, Jayaram B, Beveridge DL (1997) Intrusion of counterions into the spine of hydration in the minor groove of B-DNA: Fractional occupancy of electronegative pockets. J Am Chem Soc. https://doi.org/10.1021/ja960459m
65. Lyubartsev AP, Laaksonen A (1998) Molecular dynamics simulations of DNA in solution with different counter-ions. J Biomol Struct Dyn 16:579–592. https://doi.org/10.1080/07391102.1998.10508271
66. McConnell KJ, Beveridge DL (2000) DNA structure: What's in charge? J Mol Biol. https://doi.org/10.1006/jmbi.2000.4167
67. Várnai P, Zakrzewska K (2004) DNA and its counterions: A molecular dynamics study. Nucleic Acids Res 32:4269–4280. https://doi.org/10.1093/nar/gkh765
68. Ponomarev SY, Thayer KM, Beveridge DL (2004) Ion motions in molecular dynamics simulations on DNA. Proc Natl Acad Sci U S A 101:14771–14775. https://doi.org/10.1073/pnas.0406435101

69. Mocci F, Saba G (2003) Molecular dynamics simulations of A·T-rich oligomers: Sequence-specific binding of Na+ in the minor groove of B-DNA. Biopolymers. https://doi.org/10.1002/bip.10334
70. Lavery R, Maddocks JH, Pasi M, Zakrzewska K (2014) Analyzing ion distributions around DNA. Nucleic Acids Res 42:8138–8149. https://doi.org/10.1093/nar/gku504
71. Dans PD, Danilane L, Ivani I, Drsata T, Lankas F, Hospital A, Walther J, Pujagut RI, Battistini F, Gelpi JL, Lavery R, Orozco M (2016) Long-timescale dynamics of the Drew-Dickerson dodecamer. Nucleic Acids Res 44:4052–4066. https://doi.org/10.1093/nar/gkw264
72. Pasi M, Maddocks JH, Lavery R (2015) Analyzing ion distributions around DNA: Sequence-dependence of potassium ion distributions from microsecond molecular dynamics. Nucleic Acids Res 43:2412–2423. https://doi.org/10.1093/nar/gkv080
73. Mocci F, Laaksonen A, Lyubartsev A, Saba G, Chimiche S, Cagliari U, Universitaria C (2004) Molecular dynamics Investigation of 23Na NMR relaxation in oligomeric DNA aqueous solution. J Phys Chem B 108:16295–16302
74. Perepelytsya S (2018) Hydration of counterions interacting with DNA double helix: a molecular dynamics study. J Mol Model 24:171. https://doi.org/10.1007/s00894-018-3704-x
75. Perepelytsya SM (2020) Positively and negatively hydrated counterions in molecular dynamics simulations of DNA double helix. Ukr J Phys 65:510–520. https://doi.org/10.15407/ujpe65.6.510
76. Saha D, Supekar S, Mukherjee A (2015) Distribution of residence time of water around DNA base pairs: governing factors and the origin of heterogeneity. J Phys Chem B 119:11371–11381. https://doi.org/10.1021/acs.jpcb.5b03553
77. Duboué-Dijon E, Fogarty AC, Hynes JT, Laage D (2016) Dynamical disorder in the DNA hydration shell. J Am Chem Soc 138:7610–7620. https://doi.org/10.1021/jacs.6b02715
78. Laage D, Elsaesser T, Hynes JT (2017) Water dynamics in the hydration shells of biomolecules. Chem Rev 117:10694–10725. https://doi.org/10.1021/acs.chemrev.6b00765
79. Drew HR, Wing RM, Takano T, Broka C, Tanaka S, Itakura K, Dickerson RE (1981) Structure of a B-DNA dodecamer: conformation and dynamics. Proc Natl Acad Sci U S A 78:2179–2183. https://doi.org/10.1073/pnas.78.6.2179
80. Liepinsh E, Leupin W, Otting G (1994) Hydration of DNA in aqueous solution: NMR evidence for a kinetic destabilization of the minor groove hydration of d-(TTAA)2 versus d-(AATT)2 segments. Nucleic Acids Res 22:2249–2254. https://doi.org/10.1093/nar/22.12.2249
81. Johannesson H, Halle B (1998) Minor Groove Hydration of DNA in Solution: Dependence on Base Composition and Sequence. J Am Chem Soc 120:6859–6870
82. Bignon E, Chan CH, Morell C, Monari A, Ravanat JL, Dumont E (2017) Molecular dynamics insights into polyamine–DNA binding modes: Implications for cross-link selectivity. Chem - A Eur J 23:12845–12852. https://doi.org/10.1002/chem.201702065
83. Korolev N, Lyubartsev AP, Nordenskiöld L, Laaksonen A (2001) Spermine: An "invisible" component in the crystals of B-DNA. A grand canonical Monte Carlo and molecular dynamics simulation study. J Mol Biol 308:907–917. https://doi.org/10.1006/jmbi.2001.4642
84. Korolev N, Lyubartsev AP, Laaksonen A, Nordenskiöld L (2004) A molecular dynamics simulation study of polyamine- and sodium-DNA. Interplay between polyamine binding and DNA structure. Eur Biophys J 33:671–682. https://doi.org/10.1007/s00249-004-0410-7
85. Ouameur AA, Tajmir-Riahi HA (2004) Structural analysis of DNA interactions with biogenic polyamines and cobalt(III)hexamine studied by fourier transform infrared and capillary electrophoresis. J Biol Chem 279:42041–42054. https://doi.org/10.1074/jbc.M406053200
86. Estévez-Torres A, Baigl D (2011) DNA compaction: Fundamentals and applications. Soft Matter 7:6746–6756. https://doi.org/10.1039/c1sm05373f
87. Gosule LC, Schellman JA (1978) DNA condensation with polyamines. I. Spectroscopic studies. J Mol Biol 121:311–326. https://doi.org/10.1016/0022-2836(78)90366-2
88. Chattoraj DK, Gosule LC, Schellman JA (1978) DNA condensation with polyamines. II. Electron microscopic studies. J Mol Biol 121:327–337. https://doi.org/10.1016/0022-2836(78)90367-4

89. Bloomfield VA (1996) DNA condensation. Curr Opin Struct Biol 6:334–341. https://doi.org/10.1016/S0959-440X(96)80052-2
90. Srivastava A, Timsina R, Heo S, Dewage SW, Kirmizialtin S, Qiu X (2020) Structure-guided DNA–DNA attraction mediated by divalent cations. Nucleic Acids Res 48:7018–7026. https://doi.org/10.1093/nar/gkaa499
91. Dai L, Mu Y, Nordenskiöld L, Van Der Maarel JRC (2008) Molecular dynamics simulation of multivalent-ion mediated attraction between DNA molecules. Phys Rev Lett 100:1–4. https://doi.org/10.1103/PhysRevLett.100.118301
92. Yoo J, Kim H, Aksimentiev A, Ha T (2016) Direct evidence for sequence-dependent attraction between double-stranded DNA controlled by methylation. Nat Commun 7:1–7. https://doi.org/10.1038/ncomms11045
93. Zinchenko A, Berezhnoy NV, Wang S, Rosencrans WM, Korolev N, Van Der Maarel JRC, Nordenskiöld L (2018) Single-molecule compaction of megabase-long chromatin molecules by multivalent cations. Nucleic Acids Res 46:635–649. https://doi.org/10.1093/nar/gkx1135
94. van Dam L, Korolev N, Nordenskiöld L (2002) Polyamine-nucleic acid interactions and the effects on structure in oriented DNA fibers. Nucleic Acids Res 30:419–428. https://doi.org/10.1093/nar/30.2.419
95. Korolev N, Lyubartsev AP, Laaksonen A, Nordenskiöld L (2002) On the competition between water, sodium ions, and spermine in binding to DNA: A molecular dynamics computer simulation study. Biophys J 82:2860–2875. https://doi.org/10.1016/S0006-3495(02)75628-2
96. Patel MM, Anchordoquy TJ (2006) Ability of spermine to differentiate between DNA sequences-Preferential stabilization of A-tracts. Biophys Chem 122:5–15. https://doi.org/10.1016/j.bpc.2006.02.001
97. Kabir A, Suresh Kumar G (2013) Binding of the biogenic polyamines to deoxyribonucleic acids of varying base composition: base specificity and the associated energetics of the interaction. PLoS One 8:e70510. https://doi.org/10.1371/journal.pone.0070510
98. Kabir A, Kumar GS (2014) Probing the interaction of spermine and 1-naphthyl acetyl spermine with DNA polynucleotides: A comparative biophysical and thermodynamic investigation. Mol Biosyst 10:1172–1183. https://doi.org/10.1039/c3mb70616h
99. Iacomino G, Picariello G, D'Agostino L (2012) DNA and nuclear aggregates of polyamines. Biochim Biophys Acta - Mol Cell Res 1823:1745–1755. https://doi.org/10.1016/j.bbamcr.2012.05.033
100. Lindemose S, Nielsen PE, Møllegaard NE (2005) Polyamines preferentially interact with bent adenine tracts in double-stranded DNA. Nucleic Acids Res. https://doi.org/10.1093/nar/gki319
101. Yoo J, Aksimentiev A (2016) The structure and intermolecular forces of DNA condensates. Nucleic Acids Res 44:2036–2046. https://doi.org/10.1093/nar/gkw081
102. Luan B, Aksimentiev A (2010) Electric and electrophoretic inversion of the DNA charge in multivalent electrolytes. Soft Matter 6:243–246. https://doi.org/10.1039/b917973a
103. Perepelytsya S, Uličný J, Laaksonen A, Mocci F (2019) Pattern preferences of DNA nucleotide motifs by polyamines putrescine2+, spermidine3+ and spermine6. Nucleic Acids Res 47:6084–6097. https://doi.org/10.1093/nar/gkz434
104. Chan CH, Monari A, Ravanat JL, Dumont E (2019) Probing interaction of a trilysine peptide with DNA underlying formation of guanine-lysine cross-links: Insights from molecular dynamics. Phys Chem Chem Phys 21:23418–23424. https://doi.org/10.1039/c9cp04708e
105. Lyubartsev AP, Laaksonen A (1999) Effective potentials for ion-DNA interactions. J Chem Phys 111:11207–11215. https://doi.org/10.1063/1.480476
106. Dans PD, Walther J, Hansel G, Orozco M (2016) Multiscale simulation of DNA. Curr Opin Struct Biol 37:29–45. https://doi.org/10.1016/j.sbi.2015.11.011
107. Minhas V, Sun T, Mirzoev A, Korolev N, Lyubartsev AP, Nordenskiöld L (2020) Modeling DNA flexibility: comparison of force fields from atomistic to multiscale levels. J Phys Chem B 124:38–49. https://doi.org/10.1021/acs.jpcb.9b09106
108. Naômé A, Laaksonen A, Vercauteren DP (2015) A coarse-grained simulation study of the structures, energetics, and dynamics of linear and circular DNA with its ions. J Chem Theory Comput 11:2813–2826. https://doi.org/10.1021/acs.jctc.5b00113

109. Potoyan DA, Savelyev A, Papoian GA (2013) Recent successes in coarse-grained modeling of DNA. Wiley Interdiscip Rev Comput Mol Sci 3:69–83. https://doi.org/10.1002/wcms.1114
110. Rebič M, Mocci F, Laaksonen A, Uličný J (2015) Multiscale simulations of human telomeric G-quadruplex DNA. J Phys Chem B 119:105–113. https://doi.org/10.1021/jp5103274
111. Rebič M, Mocci F, Uličný J, Lyubartsev AP, Laaksonen A (2017) Coarse-Grained simulation of rodlike higher-order quadruplex structures at different salt concentrations. ACS Omega 2:386–396. https://doi.org/10.1021/acsomega.6b00419
112. Case DA, Ben-Shalom IY, Brozell SR, Cerutti DS, Cheatham TEI, Cruzeiro VWD, Darden TA, Duke RE, Ghoreishi D, Gilson MK, Gohlke H, Goetz AW, Greene D, Harris R, Homeyer N, Huang Y, Izadi S, Kovalenko A, Kurtzman T, Lee TS, LeGrand S, Li P, Lin C, Liu J, Luchko T, Luo R, Mermelstein DJ, Merz KM, Miao Y, Monard G, Nguyen C, Nguyen H, Omelyan I, Onufriev A, Pan F, Qi R, Roe DR, Roitberg A, Sagui C, Schott-Verdugo S, Shen J, Simmerling CL, Smith J, Ferrer RS, Swails J, Walker RC, Wang J, Wei H, Wolf RM, Wu X, Xiao L, York DM, Kollman PA (2018) AMBER 2018. University of California, San Francisco
113. Hanwell MD, Curtis DE, Lonie DC, Vandermeersch T, Zurek E, Hutchison GR (2012) Avogadro: an advanced semantic chemical editor, visualization, and analysis platform. J Cheminform 4:17. https://doi.org/10.1186/1758-2946-4-17
114. Avogadro: an open-source molecular builder and visualization tool. Version 1.2.
115. Wang J, Wolf RM, Caldwell JW, Kollman PA, Case DA (2004) Development and testing of a general Amber force field. J Comput Chem 25:1157–1174. https://doi.org/10.1002/jcc.20035
116. Jorgensen WL, Chandrasekhar J, Madura JD, Impey RW, Klein ML (1983) Comparison of simple potential functions for simulating liquid water. J Chem Phys 79:926–935. https://doi.org/10.1063/1.445869
117. Abraham MJ, Murtola T, Schulz R, Páll S, Smith JC, Hess B, Lindahl E (2015) GROMACS: High performance molecular simulations through multi-level parallelism from laptops to supercomputers. SoftwareX 1–2:19–25 . https://doi.org/10.1016/j.softx.2015.06.001
118. Nosé S (1984) A molecular dynamics method for simulations in the canonical ensemble. Mol Phys 52:255–268. https://doi.org/10.1080/00268978400101201
119. Hoover WG (1985) Canonical dynamics: equilibrium phase-space distributions. Phys Rev A 31:1695–1697. https://doi.org/10.1103/PhysRevA.31.1695
120. Parrinello M, Rahman A (1981) Polymorphic transitions in single crystals: A new molecular dynamics method. J Appl Phys 52:7182–7190. https://doi.org/10.1063/1.328693
121. Hess B, Bekker H, Berendsen HJC, Fraaije JGEM (1997) LINCS: A linear constraint solver for molecular simulations. J Comput Chem 18:1463–1472. https://doi.org/10.1002/(SICI)1096-987X(199709)18:12%3c1463::AID-JCC4%3e3.0.CO;2-H
122. Essmann U, Perera L, Berkowitz ML, Darden T, Lee H, Essmann U, Perera L, Berkowitz ML, Darden T, Lee H, Pedersen LG (1995) A smooth particle mesh Ewald method. J Chem Phys 103:8577. https://doi.org/10.1063/1.470117
123. Humphrey W, Dalke A, Schulten K (1996) VMD: Visual molecular dynamics. J Mol Graph 14:33–38. https://doi.org/10.1016/0263-7855(96)00018-5
124. Levine BG, Stone JE, Kohlmeyer A (2011) Fast analysis of molecular dynamics trajectories with graphics processing units-Radial distribution function histogramming. J Comput Phys 230:3556–3569. https://doi.org/10.1016/j.jcp.2011.01.048
125. Lu XJ, Olson WK (2003) 3DNA: a software package for the analysis, rebuilding and visualization of three-dimensional nucleic acid structures. Nucleic Acids Res 31:5108–5121. https://doi.org/10.1093/nar/gkg680
126. Kumar R, Grubmüller H (2015) Do-x3dna: A tool to analyze structural fluctuations of dsDNA or dsRNA from molecular dynamics simulations. Bioinformatics 31:2583–2585. https://doi.org/10.1093/bioinformatics/btv190
127. Nakata M, Zanchetta G, Chapman BD, Jones CD, Cross JO, Pindak R, Bellini T, Clark NA (2007) End-to-end stacking and liquid crystal condensation of 6-to 20-base pair DNA duplexes. Science (80- ) 318:1276–1279 . https://doi.org/10.1126/science.1143826

# Chapter 7
# The Influence of Sequence Dependence and External Solvents on DNA Conformation

**Hui-Ting Xu, Nan Zhang, Ming-Ru Li, and Feng-Shou Zhang**

**Abstract** DNA sequence has important biological significance, its effect on groove size affects the expression of protein specific and non-specific binding groove. In liquid environments, different solvents interact with DNA differently. In this chapter, we will focus on the microscopic mechanism of internal bases and external solvents on DNA conformation. In the ordinary water solvent, with the increase of G base content, the DNA structure is transformed by B form—(A-B)—A form. In glycol solvents, DNA maintains B-form configuration, while in ethanol solvent and low activity water solvent, DNA exhibits a shorter and denser A-form.

## List of Abbreviations and Symbols

| | |
|---|---|
| EG | glycol |
| EA | ethanol |
| RDF | radial distribution function |
| RMSD | Mean square root coordinate deviation |

H.-T. Xu · N. Zhang · M.-R. Li · F.-S. Zhang (✉)
The Key Laboratory of Beam Technology and Material Modification of Ministry of Education, College of Nuclear Science and Technology, Beijing Normal University, Beijing 100875, China
e-mail: fszhang@bnu.edu.cn

H.-T. Xu
e-mail: xuhuiting@mail.bnu.edu.cn

N. Zhang
e-mail: zhangnan@mail.bnu.edu.cn

M.-R. Li
e-mail: mruli@mail.bnu.edu.cn

Beijing Radiation Center, Beijing 100875, China

F.-S. Zhang
Center of Theoretical Nuclear Physics, National Laboratory of Heavy Ion Accelerator of Lanzhou, Lanzhou 730000, China



L. Bulavin and N. Lebovka (eds.), *Soft Matter Systems for Biomedical Applications*, Springer Proceedings in Physics 266, https://doi.org/10.1007/978-3-030-80924-9_7

## 7.1 Introduction

DNA has a variety of forms. Under different external conditions, DNA can present different configurations, such as type A, type B, type Z, type C, etc., but the vast majority of DNA in organisms presents type B structure. The flexibility and polymorphism of DNA structure are closely related to their gene expression and biological function. It is speculated that DNA template presents B-type structure in the process of DNA replication, while DNA undergoes reversible B to A transformation in the process of RNA transcription [1]. In eukaryotic and prokaryotic cells, DNA can undergo B to A transformation through the binding of acid soluble spore protein [2]. The binding of DNA with some drugs can lead to freezing of B to A transformation, which may lead to DNA biological dysfunction [3]. Therefore, it is very important to study the change of DNA conformation.

The conformational changes of DNA are environment dependent and sequence dependent. Environmental factors include temperature, humidity, salt concentration, pH value, solvent polarity, etc. In addition to the influence of environmental factors, DNA conformation changes also depend on the base sequence [4, 5]. DNA sequence has important biological and physical significance. Its effect on groove size affects the expression of protein specific and non-specific binding groove. By analyzing the DNA crystal structure data set of free and binding proteins and the typical sequence of free DNA in the solution of NMR [6]. It was found that the size of the DNA groove was significantly related to the BI/BII configuration state of the DNA skeleton [7–9] , and the BI/BII configuration was highly dependent on the DNA sequence [8, 10]. In B-DNA, the phosphoric acid base uses two conformations, BI and BII. In addition, BI phosphate base-related base pairs are positioned at the center of the double helix structure, The BII conformation is accompanied by the offset of the base pair to the relative center, toward the major groove. The two DNA structures in Fig. 7.1 intuitively show the relationship between BI or BII-rich region and DNA groove. This base-pair displacement is directly related to groove depth, a high density of BII steps leading to shallow major grooves and deep minor grooves [11]. Each of the 16 B-DNA dinucleotides is characterized by a NMR $^{31}$P chemical shifts in solution, and therefore by a specific BII percentage [12] in Table 7.1. Other helix parameters describe changes in the B-DNA structure, such as rising or tilting [13], which are too weak to significantly be related to the main chain state (or other structural parameters). Therefore, the BI/BII ratio reflects the most closely related DNA helical descriptor. The properties of the DNA were described using an experimental scale of XTR in terms of Twist, Roll, and X-disp (base pair displacement), which quantifies the inherent flexibility of the ten dinucleotide steps [12]. On this scale, each complementary dinucleotide step is characterized by a no-parameter fraction (its experimental average BII population for phosphates), reflecting the flexibility of the associated phosphates. The XTR at the dinucleotide level is a representation of the sequence dependency of DNA intrinsic plasticity. Due to the tight coupling between the spiral parameter and BII phosphate populations, this score also reflect the intrinsic flexibility of the corresponding twist, roll and X-disp. According to the

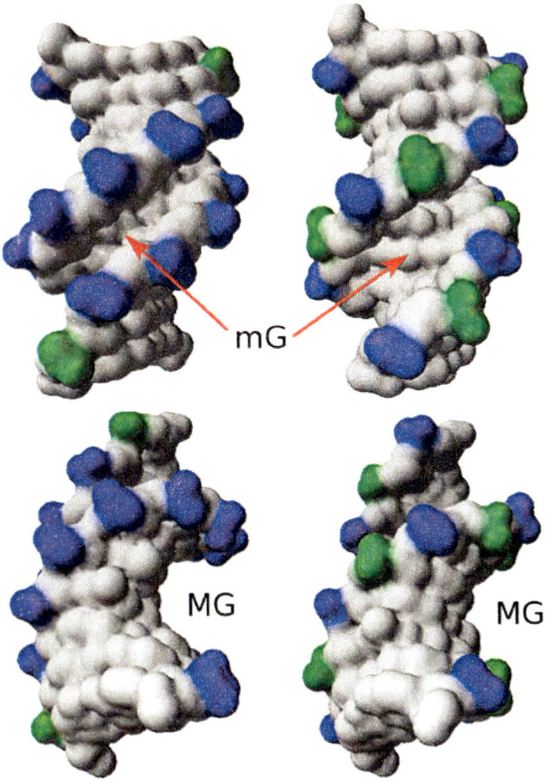


**Fig. 7.1** The BI-rich (PDB code 1EHV, left) and BII-rich (PDB code 3GGI, right) decamer X-ray structures are in the same orientations. BI and BII phosphate groups are in blue and green respectively. The minor groove was labeled as mG and the major groove was labeled as MG (reprinted with permission from [11])

scale, ApA · TpT, ApT · ApT, TpA · TpA and ApC · GpT exist restricted flexibility. In contrast, GpG · CpC, GpC · GpC, CpG · CpG and CpA · TpG exist increased flexibility. The steps which are mixed in terms of base composition have flexibilities either restricted, GpA · TpC being intermediate. The effect of the base sequence of DNA on its state density also alters the conductive behavior of DNA molecules. Due to the cumulative interactions and differences between adjacent base pairs, the DNA molecules dominated by the cytosine base pair will tend to be conductors, while the DNA molecules dominated by the adenine base pair will tend to be semiconductors [14].

As a common solvent, water provides a solution environment for biological macromolecules. Because the centers of positive and negative charges are not coincident, water molecules are polar molecules. The polarity of water molecules often affects the solubility of ions in water and the structure of biomacromolecules in solution. The phosphate group of DNA double helix skeleton is negatively charged, and the repulsive force is formed between the negative charges, which is not conducive to the stability of DNA structure. Because the phosphate skeleton of DNA molecules has a large number of negative charges, polar water molecules will form a hydrated

**Table 7.1** The DNA sequence is expressed in terms of the ten dinucleotides base steps of frequency N, summarized from a previous study. The intrinsic flexibility of each dinucleotide is quantified by its experimental scale called TRX, The higher the TRX score illustrate the greater the intrinsic flexibility of the step (reprinted with permission from [12])

| | N | %BII | TRX score |
|---|---|---|---|
| CpG · CpG | 25 | **43 · 43** | **43** |
| CpA · TpG | 28 | **52 · 31** | **42** |
| GpG · CpC | 11 | **47 · 37** | **42** |
| GpC · GpC | 22 | **25 · 25** | **25** |
| GpA · TpC | 25 | **31 · 11** | **22** |
| TpT · TpT | 12 | **14 · 14** | **14** |
| ApG · CpT | 19 | **18 · 0** | **9** |
| ApA · TpT | 17 | **11 · 0** | **5** |
| ApC · GpT | 23 | **8 · 0** | **4** |
| ApT · ApT | 22 | **0 · 0** | **0** |

shell around the DNA skeleton. And it will form a water ridge in the spiral groove, bridging the two chains and playing a role in stabilizing the configuration. Both the hydrated shell structure and the water ridge depend on the properties of water. In recent years, many studies on alcohol solutions have found that alcohols can also change the properties of water [15]. Alcohols and water can be mutually soluble in any proportion. How do different concentrations of alcohols affect the properties of aqueous solutions? In the mixed solution of alcohol and water, due to the change of surrounding solution structure, what kind of change will happen to the configuration of DNA?

In this chapter, The factors that affect the conformational changes of DNA were discussed, and DNA configuration transformation caused by base sequence and solvents were studied. We analyzed the of DNA structural parameters and the distribution of ions around it with the increase of base G content in the sequence. For different sequences, the binding of ions is different [16]. Therefore, studying the distribution of ions around DNA is helpful to understand the mechanism of interaction between ions and nucleic acids. In addition, the study of DNA in different solvents is meaningful for practical storage. We discussed the properties of the mixed solution of different alcohols (ethanol/glycol) and water. On this basis, the flexibility and polymorphism of DNA in alcohol solvent were analyzed. The format of this chapter is as follows: the first section is introduction and the second section introduces the simulation method. In the third section, the change of DNA configuration and the distribution of ions caused by bases sequence are discussed. In the fourth section, the effects of alcohol solvent and low activity water on DNA conformation are researched. The conclusion of the fifth section is at the end of this chapter. Through molecular dynamics simulation, the influence of internal base sequence and external environment on DNA in liquid environment and the interaction mechanism were studied.

## 7.2 Methods

The software used in this article is GOMACS [17], it can do fast parallel computing. DNA structure building software is NAB [18]. The GROMACS software was developed by a research team at the University of Grogan in the Netherlands, primarily for the simulation of biological systems, molecular dynamics software, can also be used for simulation of chemical material systems, such as carbon nanotubes. The advantages of GROMACS software over other molecular dynamics simulation software are: (1) The good algorithm; (2) the high calculation efficiency, (3) the high reliability of the force field, and (4) the good maintenance service, etc.

About the DNA molecules used in the simulation, this paper uses NAB [18] software to construct a double helix DNA of a given sequence composition and EcoRI dodecamer (named 171d) from Protein Data Bank (PDB) [26]. Using software to build standard Type B DNA and standard Type A DNA in a sequence of four nucleotide ABCD repeatedly arranged 5'G-G-D-ABCD-ABCD-ABCD-G-G-3' as a framework, five standard A-type and five standard B-DNA are constructed according to the amount of base G content contained in the sequence (here ABCD is AAAA, GAAA, GGAA, GGGA, GGGG), respectively. The five standard Type B DNA are the initial conformations of the simulation, followed by AAAA, GAAA, GGAA, GGGA, GGGG, GGGG representing five initial conformations. The five standard Type A DNAs are designed to contrast with simulated evolutionary structures.

The force field is selected AMBERS0 force field [19], The water model is TIP3P water [20]. A combination of the two can better describe the interaction between solutes, solvents and ions in the solution. The interaction between molecules or between atoms within molecules contains two parts, the electrostatic potential energy and the Lennard-Jones potential energy. For long-range electrostatic interactions, this paper uses the Particle Mesh Ewald plus method [21]. The cutoff radius of the Lennard-Jones and Coulombic interactions is 14 Å. LINCS [22] algorithm is used to constrain the bond lengths at their equilibrium values, which per-mitted the use of 2 fs timestep for the integration of Newton equations of motion. The initial type B DNA and solvent water in each simulation are placed in a cubic box that meets the periodic boundary conditions. The distance from the DNA molecule to the edge of the box is at least 1.2 nm. Since DNA is polyanion, in order for the system to remain electrically neutral, counterions need to be added to each simulation box to neutralize the negative charge on the DNA. The counterions used in this article are either $K^+$ or $Na^+$. When counterions are added, the counterions replace the water molecules in the solution and are randomly and evenly distributed around the DNA. Finally, there are about 6,780 water molecules in each cubic cell. Considering the DNA molecules in the solution, the equilibrium ions and solvent molecules, the density of the system is approximately 1017 kg/m$^3$. The temperature of the system is set to 300 K and the pressure is set to 1 bar.

Before the final molecular dynamics simulation, the system goes through three pre-equilibrium stages to achieve pre-equilibrium to reduce the time of subsequent molecular dynamics evolution. First, the simulated DNA is placed in a water box at room temperature, energy minimization is treated to obtain a stable configuration, and the algorithm used is the steepest drop method. This is because there may be a relatively close atomic distance in the initial structure, the force between atoms is too large, is not conducive to the stability of the configuration, so the energy minimization treatment is generally carried out to make the initial structure to a low energy state. Finally, using the structure of energy minimization as the initial structure, the position-restrictive NVT simulation of the solute DNA molecules is simulated, which is a warming process, the temperature of the system should eventually be close to the set temperature of 300 K.

The temperature regulation used Nose-Hoover thermostat [23]. The coupling time constant is 0.5 ps. At the beginning, 500 ps NVT ensemble and 1 ns NPT ensemble simulation with 2 fs integration timestep are performed to allow solvent molecules and ions to form the outer shells around DNA obtaining a initial balance. After that, the product simulations of 100 ns (every 0.5 ps forming one snapshot) are carried out to data analyses or structural analysis,Curves 5.3 [24] software is used to calculate various structural parameters of DNA. The software is based on a series of definitions of DNA structural parameters proposed by Lavery et al., gives a quantitative description of the DNA structural parameters. We take into account the edge effect and the wear and consumption at both ends of the simulation process sequence, and the analysis ignores the two base pairs at both ends of the sequence, comparing only the average of the structural parameters of the DNA internal base pairs. In addition, the view software used as a secondary to this article is Canion program, it can be used to calculate the distribution of conterion around DNA oligomer in the 2D (dimension) plane, based on a curvilinear helicoidal coordinate (CHC) system [25]. The position of the ions is given in terms of its radial (R), longitudinal (D), and angle (A) coordinates.

## 7.3 Influencing Factors of DNA Conformational Transition

Under physiological conditions, DNA is mainly stable in the B-shaped structure, but when the external environment changes, the configuration of DNA changes, for example $B \sim A$ [33], $B \sim Z$ [34], $B \sim S$ [35], $B \sim P$ etc. There are many factors affecting DNA conformation, such as temperature [36, 37], solvent activity [38], cations [39, 40], solution PH value, external force, pressure, organic solvent [41] and so on, these factors can not lead to the degree of DNA conformation change, or even structural destruction. Typically, water plays an important role in the structural stability of DNA, with a B-type structure in a solution with a high relative humidity and a type A in a solution with a lower relative humidity. [42] Many physical and chemical methods are used to study the interaction of water with DNA [43, 44]. Water molecules can shield the repulsion between negative electrophoresis groups on the backbone by hydrated shell layers [45], It is also possible to form a ridge of two strands of DNA in a minor groove, which is conducive to stabilizing the double helix structure of DNA. Any factor that disrupts the water environment around DNA can also affect changes in the structure of DNA.

The right-handed double-helix Watson-Crick model is the most common Type B structure [27]. In addition to this classical structure, some other structural types of DNA have been observed. For example, A, B, C, D, Z and other structural types [28, 29]. Among them, A-type, B-type and Z-type are the most common configurations; C, D, E are a transition state for A- and B-type structures. A-DNA is obtained from X-diffraction experiments on the sodium fibrous structure of DNA when the relative humidity is less than 75% [30]. Type A DNA is a right-hand double helix structure, two chains in reverse parallel, a complete screw contains 11 base pairs, the pitch is 2.8 nm, the diameter of the spiral is 2.55 nm, the vertical distance between the base pairs is 0.28 nm. The base pair plane of type A DNA is tilted by a 20° skew relative to the spiral axis and moves toward the outside of the spiral, which causes the spiral axis not to pass through the base pair plane, but to be in a major groove, so looking down from the top of the spiral axis, the center is a hole. From the overall structure of A-type DNA, the spiral is wide and short, the spiral surface is deep and narrow, and the groove is shallow and wide (almost not easily distinguished). B-DNA is a two-chain parallel right-handed double helix structure, a complete helix has 10 base pairs, the pitch is 3.4 nm, the helix diameter is 2 nm, the vertical distance between the base pairs is 0.34 nm. The base pair plane of Type B DNA is perpendicular to the spiral axis and layered on the inner side of the spiral, and the two base pair planes are parallel to each other, with a rotation angle of 36° relative to the spiral axis [31, 32]. From the overall structure of Type B DNA, the spiral is thin and long, the spiral surface is wide, the groove is narrow, and the depth of the two grooves is almost the same. Factors that affect the change in its configuration can be classified as the following:

(1) solvent polarity. The polarity of solvent water can be changed by proportionally reducing or amplifying the charge ratio of oxygen and hydrogen atoms in the water, at which point the strength of the hydrogen bond in the water also changes, and the role of DNA and water changes. Changes in the polarity of solvents affect the role of ions and DNA in the solution [46], making the structure of the DNA changes to varying degrees [38].

(2) Temperature. At room temperature, DNA is generally present as a B-type structure, and as the temperature increases, the structure of DNA changes to varying degrees, resulting in a (A-B) mixed structure [47]. When the temperature exceeds a certain value, the DNA molecule denatures, and when the temperature recovers, the DNA molecule reverts to a double helix [48].

(3) Cations. Cations play a role in stabilizing the double helix structure of DNA, but they also have an effect on the change of DNA conformation. The functional function of metal ions in the shape of nucleic acid structure can be referenced [49]. Metal ions can interact directly with nucleic acids or indirectly interact with nucleic acids through hydrated shell structures. Metal ions are primarily stable DNA structures, but can also affect DNA conformation when changes in the external environment (e.g. water polarity, temperature mentioned above).

(4) Organic solvents. Some alcohols can also affect changes in the structure of DNA, such as ethanol, where different concentrations of ethanol can vary depend on the structure of DNA [50].

(5) Base sequence. The conformational changes of DNA have environmental dependence and sequence dependence. In addition to the effects of external environment on DNA, the sequence size and composition of the DNA itself also affects the choice of DNA conformation in the solution, where the tendency of DNA A- or B-type has sequence dependence.

This chapter mainly discusses the influence of internal factors of base sequence and external factors of solvent on DNA: (1) The sequence dependent of DNA conformation; (2) Structure and dynamics properties of liquid alcohol solvent; (3) Dynamics and flexibility of DNA structure in different solvents.

### *7.3.1 Sequence Dependent-Induced DNA Conformational Transition*

After the synthesis of the first double helix single crystal structure of B-type DNA, a series of questions about the effect of sequence on DNA structure, solvation, conformational stability and axial bending have been raised [51]. In the past decades, many follow-up studies have focused on the effect of sequence on DNA structure. The smallest structural unit of DNA carrying three-dimensional structural information is dinucleotide base pair, which is expressed in the form of 5'- dXpY-3', where x and Y may be any one of A, T, G or C. Four possible choices of X, Y result in 16 XpY permutations, 10 of which are different. People first studied the structural characteristics of local single dimer [52], and it was found that ten different dimers have different A-and B-form affinities. For example, dimers AA and CG have B-type affinity, while GC and GG have A-type affinity. Considering the influence of adjacent bases, the structure of any single XpY may be affected by the nearest base pair. As a result, a series of studies on DNA trimers [53] and tetramers [54] have been carried out, such as conformational propensity, influence of adjacent bases, sequence related hydration effect and ion distribution.

Various experimental and theoretical studies have shown that DNA has flexible and polymorphic structures [55]. It is generally accepted that base sequences have an important effect on the nature of DNA and play an important role in protein/drug-DNA interactions that are critical in the cellular process [56, 57]. Therefore, the study of sequence-dependent DNA conformation flexibility at the molecular level is expected to raise awareness of this issue. In this section, the effect of sequence on DNA conformation change and the distribution of ions and water molecules around it was studied at the molecular level. With the increase of base G content in the sequence, the degree of DNA structure change and the sequence dependence of ions and water molecules around it were analyzed. The specific method is: according to the increasing trend of base G content in the sequence, AAAA, GAAA, GGAA, GGGA,

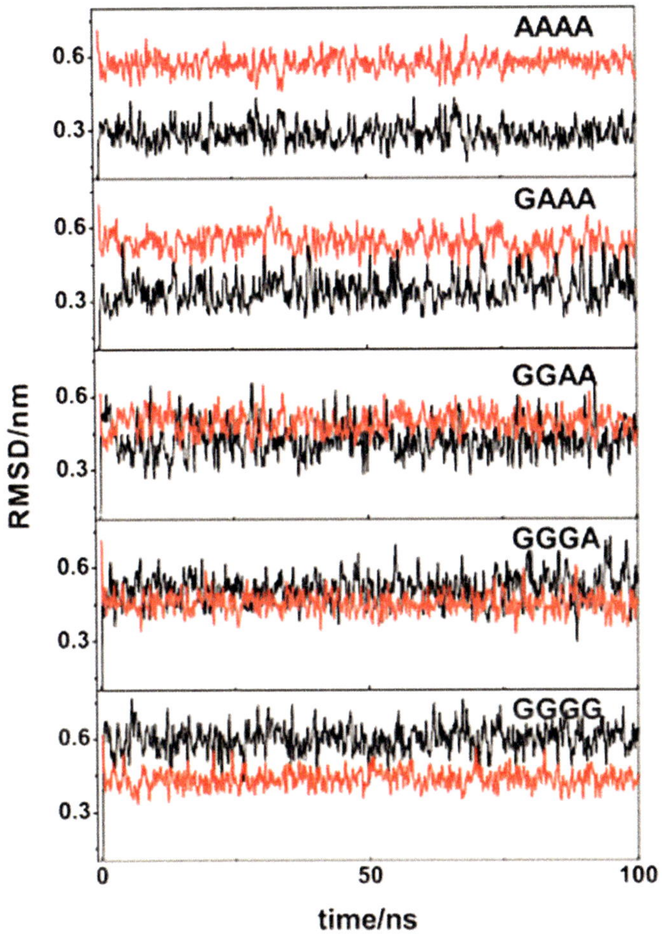


**Fig. 7.2** With the increase of G content, the root mean square deviation (RMSD) of DNA, red and black respectively represent the RMSD of relative standard A-DNA and B-DNA (reprinted with permission from [58])

GGGG five B-forms DNA were constructed and put into five identical environments (potassium balance ion or sodium balance ion, and aqueous solution) respectively for a period of dynamic evolution to observe the conformational changes of different DNA. [58] Fig. 7.2 show the mean square coordinate deviation (RMSD) of the five DNA polymers, which varies with the increase of G content in the NaCl solution environment.

For AAAA polymers, AAAA oligomers strongly prefer B-DNA. In the solution, it was noted that the poly (A) sequence does not change into a type A structure, even under extreme conditions [59]. When GAAA exists, the relative deviation from standard A-DNA is reduced. However, in this case, GAAA still keeps the nearly B-formation. For GGGA and GGAA oligomers, the red and black fluctuating curve is almost at the same level, indicating the presence of (A-B min) state in DNA. In the case of GGGG, the configuration of DNA is converted into stable A-DNA. This shows that with the increase of G content in the oligomer, the DNA conformation shows sequence dependence, the average structure parameters are evolving to the standard A form, and the change of B—to A—form occurs.

The average structure parameters of DNA with various G-base content of 10 ns are calculated and listed in Fig. 7.3, and compared with the parameters of standard A-type and B-type DNA [60]. It can be seen that the average structure parameters of DNA oligomer show the evolution to the standard a type when the G content in DNA oligomer increases in ionic atmosphere. With the increase of G content, the A affinity of DNA increased gradually. For AAAA and GAAA oligomers, most of the structural parameters are shown in the B-DNA family. In contrast, GGGA and GGGG oligomers are shown in the A-DNA family. For AAAA and GGGG oligomers, they showed the strongest affinity for B and a, respectively. This phenomenon is consistent with that AA and CG have B-form affinity and GC and GG have A-form affinity [61]. In addition, most of the structural parameters of GGAA and GGGA oligomers are between B-DNA and A-DNA, and their structures are similar to the intermediate structures of B-form or A-form. The properties of the ions are independent of their contribution to this conformational change. This indicates that the composition of DNA base is the main reason for DNA conformation change in physiological environment, followed by salt ions.

Early studies have shown that B-DNA conformation is stable by Base-stacking interactions rather than base-pairing hydrogen bond interactions [62]. The morphology changes in this chapter are mainly caused by the base composition, which is mainly caused by the base stacking interaction. Tolstorukov and his colleagues [63] demonstrated that the strong B affinity of the AA/TT step was due to the attractive interactions between clusters of thymus incesium-cyanosine methyl (CH3) clusters, sugars (CH2) clusters, and thymus C6 atomic hydrophobic clusters. However, the GG/CC step reduces static interaction and therefore makes it easy to perform B to A transition in water [64].

## *7.3.2 Distribution of Ions Around DNA*

The phosphate group of DNA double helix skeleton is negatively charged, and the repulsive force is formed between the negative charges, which is not conducive to the stability of DNA structure. The hydration of water molecules in solution and the electrostatic interaction of ions can shield this repulsive force and play an important role in the stability of DNA structure. However, there is a competition between these two roles, which makes DNA have different conformations and further affects the expression of DNA function. In addition, for different sequences, the distribution of ions and water molecules around them are also different [65]. Therefore,

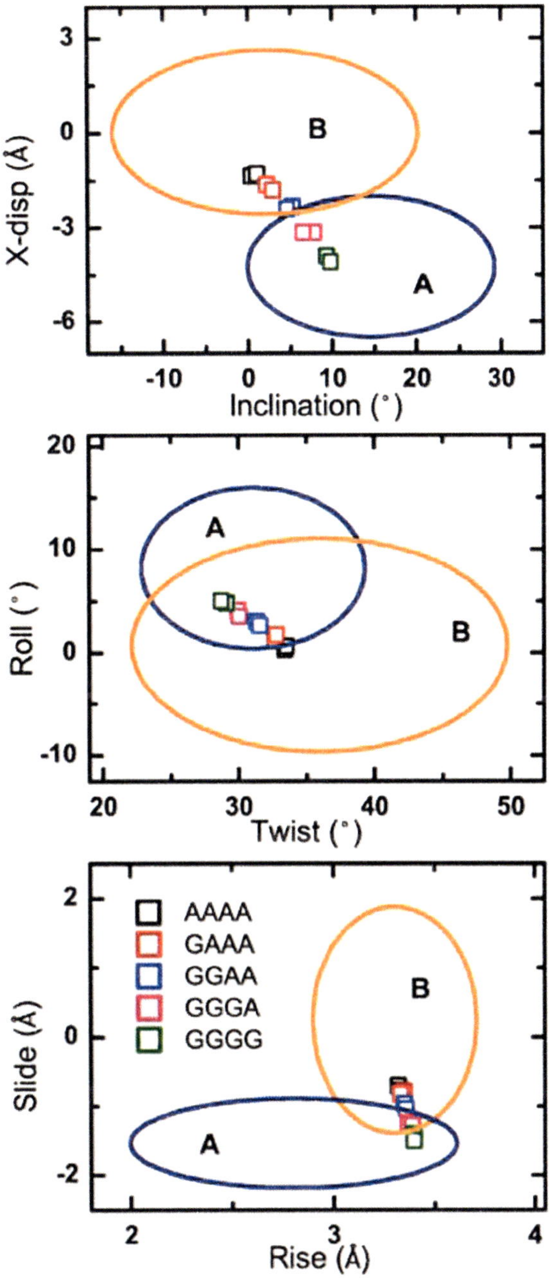


**Fig. 7.3** Averaged structural parameters for various G base content DNAs under $Na^+$ and $K^+$ ion solutions. Regions of the points are marked as follows: AAAA, black squares; GAAA, red squares; GGAA, blue squares; GGGA, pink squares; GGGG, green squares. The orange and navy ovals show regions typical for B- and A-DNA families respectively (reprinted with permission from [58])

studying the effect of sequence on the distribution of ions around DNA is helpful to further understand the mechanism of interaction between ions and DNA with different sequences.

A unique feature of nucleic acids is that they are negatively charged. DNA is an anionic polyelectrolyte, each nucleotide carries a unit of negative charge, resulting in a high charge density resulting in a strong electrostatic repulsion. In order to avoid DNA fragmentation, it is necessary to reduce the electrostatic repulsion force to maintain the stability of DNA structure. The hydration of water molecules in solution and the electrostatic interaction of equilibrium ions can shield the electrostatic repulsion between phosphate groups on DNA skeleton. These environmental effects may regulate the binding of proteins and small molecules to dsDNA, and then affect the structure and function of DNA. Therefore, it is of great significance to study the distribution of ions around DNA and the mechanism of their interaction with DNA. It has been found that the binding of metal ions to DNA is related to the structure of nucleic acid and the sequence of bases [66]. What is the effect of base sequence on the distribution of ions around DNA? Will the distribution of ions in solution be regular with the increase of base G content in DNA sequence? In this section, we analyze the molecular dynamics trajectory file to study the sequence dependence of the distribution of ions around DNA. Radial distribution function (RDF) is used to describe the distribution of the $Na^+/K^+$ bind with DNA in solution, the electrostatic interaction between counterions and DNA also showed sequence-dependent binding. In this chapter, we analyze the binding of ions and DNA electronegative sites using radial distribution function (RDFs) in Figs. 7.4 and 7.5.

The electrostatic interaction between O1P and counterions is a sequence-dependent combination from Fig. 7.4, the first peak is gradually reduced with the increase of G content, and the affinity of the electronegative sites on the phosphate group is higher than that of $K^+$ mean that the electrostatic attraction is higher than that of $K^+$. The interaction of O2P also presents a peak, but their combination does not show a change in sequence dependence like O1P. The RDFs of counterions relative

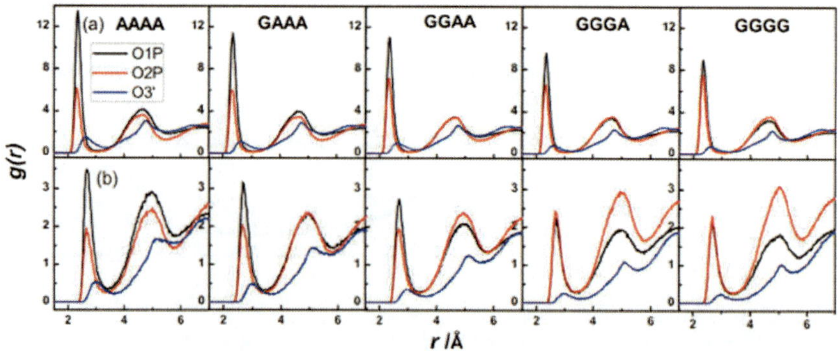


**Fig. 7.4** RDFs of cations ($Na^+$ or $K^+$) relative to the electronegative atoms of the phosphate groups of DNA. (a) $Na^+$-DNA, (b) $K^+$-DNA (reprinted with permission from [58])

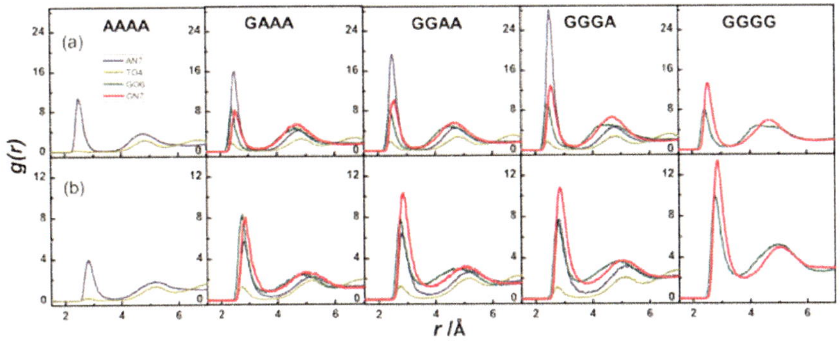


**Fig. 7.5** RDFs of $Na^+$ (a) and $K^+$ (b) relative to the electronegative atoms of the major groove of DNA. The negatively charged sites are N7 of the adenine, O4 of the thymine, O6 of the guanine and O7 of the guanine, respectively (reprinted with permission from [58])

to electronegative atom in DNA major groove are shown in Fig. 7.5, the negative sites show the sequence dependence of binding to the ions, and the first peak gradually strengthens with the increase of G base. This shows that with the increase of the G base, the concentration of ions in the major grooves of the DNA oligomer increases.

For ion-DNA systems, the case of ion binding in minor grooves is a bit complicated that with the increase of G content, the combination of cations and minor grooves decreases, except for GGAA oligomer in the atmosphere of Na+ ions. Ions are more concentrated in the major groove of DNA olihomers than in minor grooves, where a small number of ions interact with electronegative atoms in the minor grooves of the oligomers. The main reason is: (1) the minor groove of the oligomers has a strong hydration density, occupies a larger volume, making it difficult for ions to enter the small groove. (ii) The population of the major groove is generally significantly larger than the minor groove [67].

For AAAA olimogers, a large number of hydrophobic clusters cause strong attraction interactions between bases, resulting in the burial of hydrophilic groups. Therefore, in the DNA major groove, the static interaction between the counterions and the negative atoms is weak, resulting in more ions staying near the phosphoric group. With the increase of G base in DNA oligomers, the attraction interactions of bases hydrophobic groups and the bace stacking interactions are weakened, resulting in anisotropy ratio of the bases increasing and the static strength of the major groove increasing. As a result, the negative charge atoms on the base are exposed, and ions near the phosphoric group and minor groove move toward the main groove, enhancing the interaction between the ions and the electronegative atoms on the major groove and weakening the role of the minor grooves and phosphoric group. This is the main reason for ions sequence dependence. Consistent with recent studies that found that the maximum molar concentration corresponds to the major groove, not the phosphoric backbone or minor groove [25].

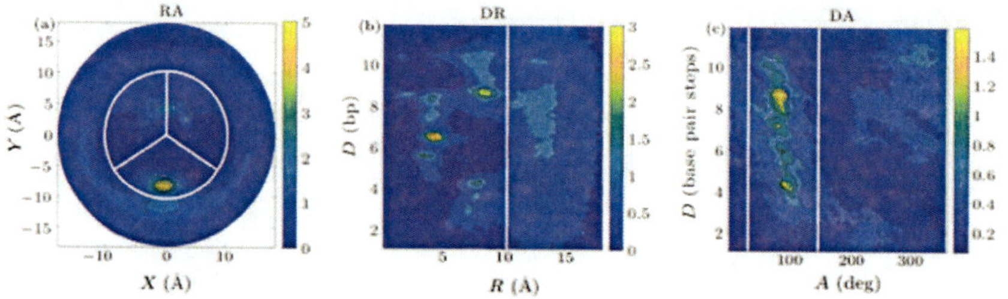


**Fig. 7.6** Two-dimensional $Na^+$ molarity (aM) distribution around DNA in aqueous solution, as a function of RA, DR and DA. The molarity increases going from blue to red, in discrete, non-uniform steps chosen to highlight the structure of the distribution. The white circle in RA and vertical bars in DR indicates the phosphorous radius (R = 10.25 Å), while the RA and DA plots show the minor groove limits 33–147° (reprinted with permission from [70])

Savelyev and colleagues [68] studied an counterions-anions cluster in a 16-base DNA oligomer and concluded that about half of the $K^+$ were associated with $Cl^-$, while almost all of the $Na^+$ were clusterless. In the case of different G content sequences, sodium ions performed a higher affinity for DNA than that of $K^+$. We inferred that the concentration of $Cl^-$ ions in the $K^+$ environment around DNA was low, due to their easy formation of K-Cl clusters, which weakened the K-DNA interaction. In contrast, a small number of clusters of Na-Cl in the system cause more $Na^+$ to condense around the DNA. This may explain the difference in affinity for DNA from $Na^+$ and $K^+$.

Whether as the carrier of genetic information or as the basic unit of emerging molecular devices, the base is the basis of DNA functional expression. This chapter shows that with the increase of G base content, the DNA structure changes B type—(A-B) mix—A type. The change of DNA conformation at physiological salt concentration is caused by the base composition, and the changes in counterions distribution is mainly caused by DNA elasticity.

In addition, we also discussed the distribution of ions around DNA in different solvents. The distribution of sodium ions around DNA fragments was discussed in water, ethanol and glycol, and the ion distribution around nucleic acids was analyzed by canion program [65]. Now we can learn more about it in 2D representation. Figure 7.6 shows the RA, DR and DA diagrams of $Na^+$ ions around the oligomer. We found a strong ion binding region in the major groove of d $(AATT)_2$ and the minor groove of C·G base pair. The outer region of the d $(AATT)_2$ fragment and the phosphate backbone were shown as secondary ion binding regions. In the figure of RA, it can be seen that $Na^+$ is closer to the spiral axis on the side of the major groove. In addition, there is only one strongest binding site in the center of the minor groove, while we see two molarity peaks in the major groove. The more dispersed molar distribution is located in the outer region of the minor groove and surrounds the negatively charged phosphate group to shield the electrostatic repulsion. DR figure showed that the strongest ion nucleotide interaction occurred in the major groove of the central d $(AATT)_2$ segment. A strong local region with high $Na^+$ molar

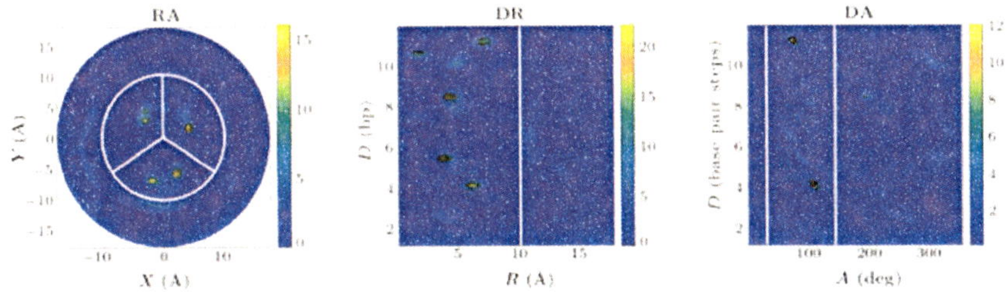

**Fig. 7.7** Two-dimensional $Na^+$ molarity (aM) distribution around DNA in ethylene glycol solution, as a function of RA, DR and DA plane. The molarity increases going from blue to red, in discrete, non-uniform steps chosen to highlight the structure of the distribution. The white circle in RA and vertical bars in DR indicates the phosphorous radius (R = 10.25 Å), while the RA and DA plots show the minor groove limits 33–147° (reprinted with permission from [70])

concentration was found in the minor groove, in which $Na^+$ mainly coordinated with C·G, and more diffusion binding regions were located in the outer region of the secondary groove section d $(AATT)_2$.

The ion analysis of EG solution is shown in Fig. 7.7. The d (CGCGAATTCGCG) sites in EG solution is a strong ion binding region, and the outer region d $(ATT)_2$ fragment in the minor groove is a secondary ion binding region. Looking at the RA diagram, we can also see that the highest molar concentration in the minor groove is located in a continuous region across the center of the groove, while the density of the major groove shows four different peaks. DR diagram showed that the strongest ion nucleotide interaction occurred in the inner region of the groove of d $(CGCGAATTCGCG)_2$ site and the diffuse bonding region appeared in the outer region of the minor groove of d $(ATT)_2$ segment in the center.

$Na^+$ ions are mainly bound to oxygen phosphate in EA solvent. One of the important reasons for the change of DNA to A-form configuration in alcohol is the competition between hydration and direct cation coupling with free oxygen phosphate in DNA skeleton. In EA solution, $Na^+$ is strongly coordinated with free oxygen phosphate to form chemical bond Na-O (P) and limit electrostatic repulsion on the backbone. Therefore, it is possible for DNA to transform into a more compact A-form DNA in ethanol. DNA is a very flexible polymer, the analysis of ion distribution is very important to understand the conformation of DNA in different solvents under physiological conditions.

## 7.4 Configuration Transformation of DNA in Different Solvents

### 7.4.1 The Structural and Dynamic Properties of Alcohol Solvents

Information about the molecular interactions between water and alcohols under a series of alcohol mass scores is important for us to understand the complex interactions of DNA in alcohol solvents. This chapter studies the changes in intermolecular structure caused by the increase in alcohol mass fractions. Molecular dynamics simulation were performed of the mixture of mass fractions Ci of 0, 20, 40, 60, 80, and 100% of glycol (EG) and water and the structural and dynamic characteristics were analyzed. Extensively studied ethanol (EA) and water mixed solutions were also calculated [69].

The RDF of the water molecules in the EG and water mixtures are shown in Fig. 7.8(a) and (b). There is HB interaction between water molecules whether the mixture is water-rich or not. With the increase of The EG mass fraction, the number of first shell water molecules and the thickness of the first shell increase around the target water molecules and indicate that local water molecules appear to be more

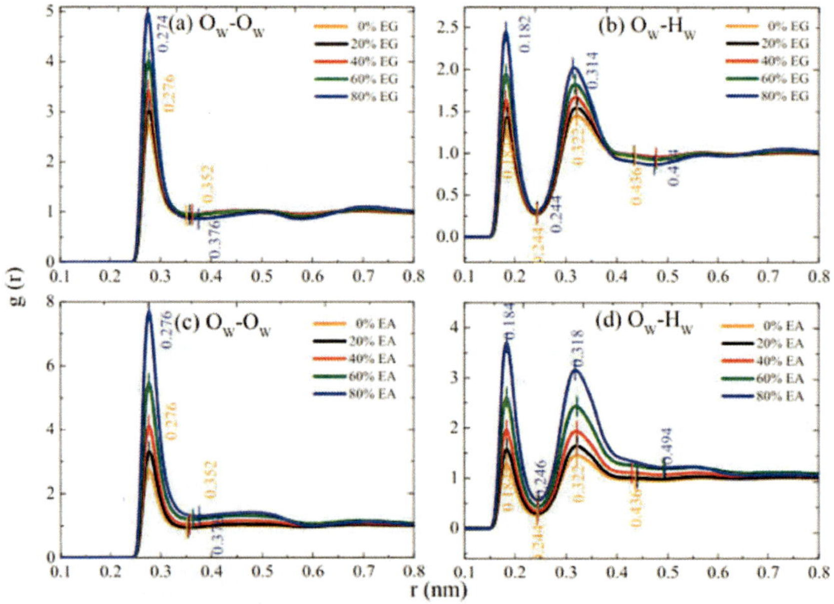


**Fig. 7.8** The radial distribution functions (a) to (b) of water-water in a series of concentrations of EG or EA solutions. (a) Water oxygen (OW)-water oxygen (OW), (b) water oxygen (OW)-water hydrogen (HW) (reprinted with permission from [69])

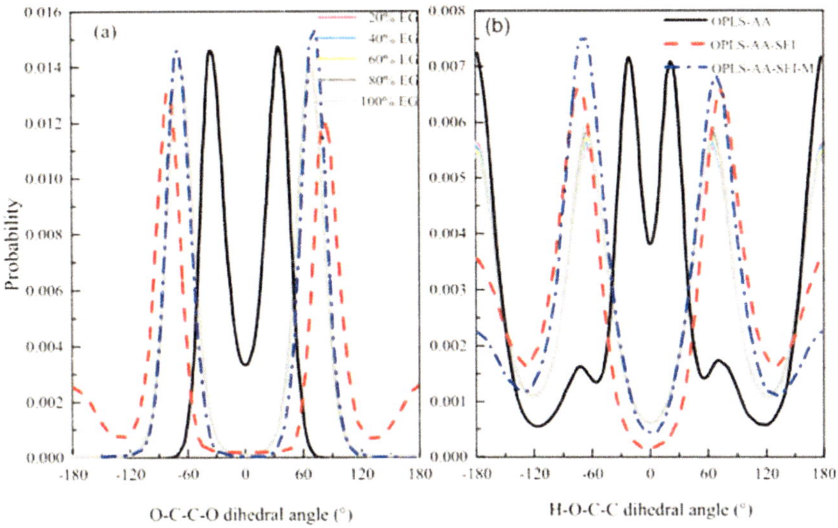


**Fig. 7.9** Probability distribution of OCCO and HOCC angles in different field force. Mass fraction 20, 40, 60, 80, 100% are shown as magenta, cyan, yellow, gray, and light gray, respectively. OPLS-AA (solid), OPLS-AA-SEI (dash), OPLS-AA-SEI-M (dash dot) (reprinted with permission from [69])

concentrated with the Increase of EG mass fractions. The RDF of the water molecule in the EA and water mixture indicates that the integral area of the first shell of the EA solution is greater than the integral area of the EG solution at the same mass fraction. It is shown that water molecules are more concentrated in the EA solution than EG solutions. The reason for this difference is that the proportion of weak polar groups in EA is larger, which makes the water molecules in the EA solution bind more to themselves. The molecular structure characteristics of EG molecules are determined by dihedral angles, an OCCO and two HOCCs. Each dihedral angle has three conformations: gauche conformations (g+/g−), and trans conformation (t).

The probability distribution of one OCCO and two HOCCs under different EG mass scores are given in Fig. 7.9, and compares the results of other force fields. Studies show that when Ci increases from 0 to 100%, the intermolecular structure of EG-EG is not significantly affected by the EG mass score until Ci increases to 80%. The probability distribution of OCCO is g conformation, not change with the EG mass fraction, the difference only appears on the dihedral angle of HOCC. The probability distribution of OCCO depends primarily on the force field. As the EG mass fraction increases, the absolute value of the maximum probability distribution of its g conformation increases, the probability increases in Fig. 7.9(b), on the contrary, the t peak position does not change, but the probability decreases.

In summary, the EG mass score does not affect the probability distribution of OCCO, and has less effect on the t/g probability distribution of HOCC. The dihedral angular probability distribution of OCCO and HOCC represents the molecular

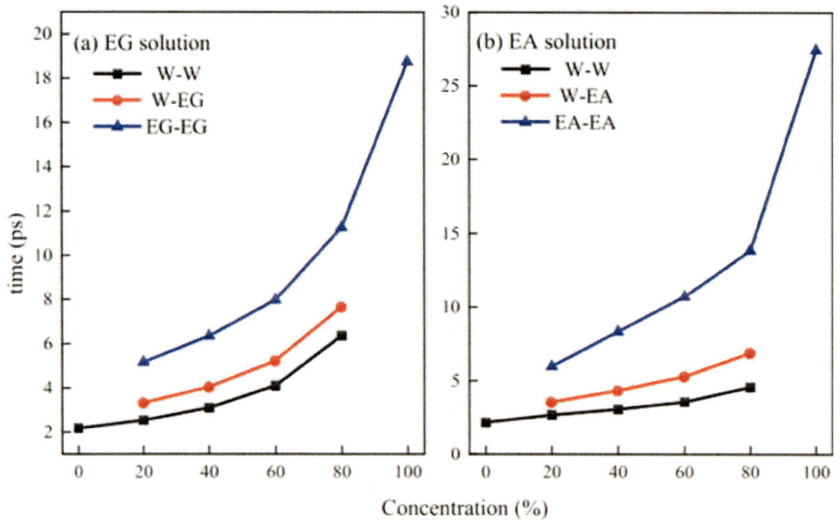


**Fig. 7.10** (a) The HB lifetime for t = 2 ps in EG solution. (b) The HB lifetime in EA solution (reprinted with permission from [69])

structure of the EG, which is determined primarily by the force field, not by the mass fraction of the EG.

In the study of hydrogen bonds, it is found that the fracture rate of HB decreases with the increase of mass score due to the enhancement of the effective strength of HB in Fig. 7.10. The intermolecular HB binding has the following priority: water-water greater than EG-water greater than EG-EG and water-water greater than water-EA greater than EA-EA. Compared to EG solution, longer hydrogen bond vitality is found in the EA solution. This is because the EG (EA) molecule is heavier than water, and according to the energy average theorem, at the same temperature, GA(EA) rotates more slowly and the HB does not break quickly. Longer vitality in EA solution is related to the corresponding HB mesh pattern. The HB linear chain of EA is more stable than the three-dimensional HB mesh of EG and water.

The study of alcohol structure helps to understand the mechanism of DNA interaction in its solution. In the study of the proportional mixing of EG/EA and water, when Ci increased from 0 to 100%, the intermolecular structure of the EG-EG was not significantly affected by the EG mass fraction until Ci increased to 80%. The dihedral angular probability distribution of OCCO and HOCC represents the molecular structure of the EG, which is determined primarily by the force field, not by the mass fraction of the EG. Except EG-EG, the intermolecular structure of solvent-solvent shows different degrees of clustering and the concentration of water molecules is in order: EA solution greater than EG solution greater than pure water solution. The study of hydrogen bonds found that the intermolecular HB binding has the following priority: water-water greater than EG-water greater than EG-EG and water-water greater than water-EA greater than EA-EA. With the increase of the number of EG molecules, the rate of HB break decreases with the increase of mass score.

### 7.4.2 *Alcohols Solvent and Low Activity Water Induced DNA Conformational Transitions*

Solvents are found in many types, such as acid or alkali solvents, organic or inorganic solvents, and hydrophilic or hydrophobic solvents. It is clearly undesirable that strong acids and alkali solutions are corrosive or that DNA is insoluble in most organic solvents. The organic solvent glycol is an exception. DNA has been observed to maintain its biological activity and double-stranded structure in glycol salt solutions for a long time. [71] Mixing other molecules (alcohols) with pure water proportionally adjusts the hydrogen bond strength of the solution, and when charged particles are added to the aqueous solution, the electrostatic interaction between charged particles and water molecules destroys the direction of water molecules and reduces the chance of hydrogen bonds between water molecules. In addition, compared with aqueous solution, ethylene glycol solution has the following advantages. (i) DNA is more resistant to radiation damage in glycol than in water and does not produce an equivalent amount of hydroxyradical free radicalination in a radiation-affected glycol solution. (ii) Double-stranded DNA may active in a glycol solution at an appropriate temperature range of 10 to 25 °C, where microorganisms usually do not grow. As a result, it may be cheaper and more efficient to store DNA in biological reservoirs than water in bio-banks. Here, in order to study the effects of alcohol solvents on DNA conformation, the polymorphism and flexibility of DNA fragments (PDB ID:171d) in water, glycol (EG) and ethanol (EA) solutions were simulated using molecular dynamics [70]. Figure 7.11 shows that in water and EG solvents, the groove and base pair structural parameters are close to the value of B-DNA, and the value of the axis-base pair is located in the middle region between A and B DNA, but closer to B, belonging to the B-DNA state. In terms of EA, slide maintains the value of B-DNA, and other parameters are close to A-DNA or in the middle of the state between A and B. In the EA solution, the DNA fragment transition state is specified as the A-DNA state.

In order to investigate the potential causes of solvent induced DNA structure transfer, we refer to the ion distribution around the DNA fragment that we discussed earlier. In the EG solution, the d $(CGAATTCG)_2$ sites is a strong ion binding zone, and the area outside the d $(ATT)_2$ of minor grooves is shown as an ion secondary binding zone. In the EA solution, The $Na^+$ ions are bound around the phosphate oxygen atoms, and an important reason for the configuration of DNA from B to A is the competition between direct cation coupling and hydration on the phosphate free oxygen atoms on the DNA skeleton. In aqueous solution or EG, the free oxygen phosphate atom is mainly protected by the solvent molecule by binding to the hydrogen bond with a positively charged proton, and the $Na^+$ ion is more dispersed, showing

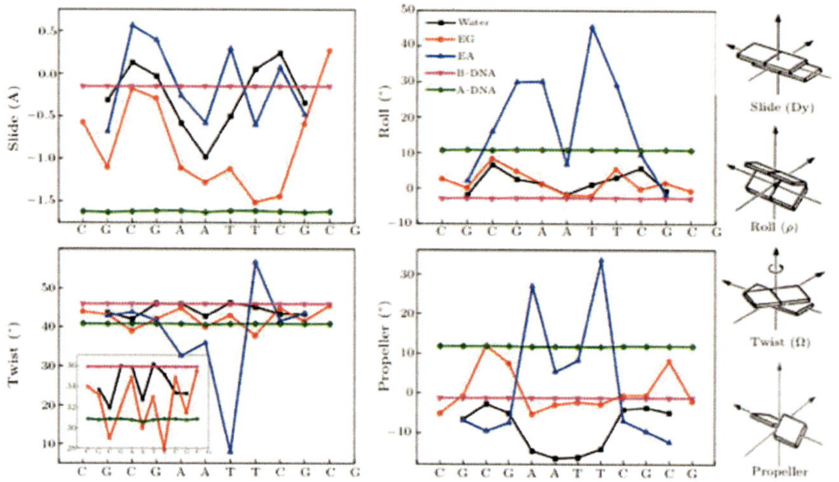


**Fig. 7.11** Averaged DNA structure parameters during the last 100 ns simulations. They are the slide, roll and twist angle of inter base pair parameters, intra-base pairs parameter propeller. The horizontal lines indicate the reference values of typical A (green) and B (pink) forms (reprinted with permission from [70])

B-DNA. In EG solution, ethylene glycol molecules give priority to the formation of a ring structure around the phosphate group, through the interaction of hydrogen to affect the configuration of DNA, glycol molecules similar to water molecules to screen the negative charge of the phosphoric acid between the repulsion between the action, to prevent this rejection effect of the spiral expansion and extension, so that the configuration of DNA to maintain The B-type. In the EA solution, no water molecule maintains the hydration structure of phosphate oxygen, and the $Na^+$ ion is strongly matched with a free oxygen phosphate atom to form a chemical bond of Na-O (P) that inhibits electrostatic repulsion on the skeleton. As a result, DNA exhibits shorter, denser A-type. From the distribution diagram of EG solvent, the law of interaction between cations and A · T is different from that of C · G.

In addition to alcohol addition, we also used water models with different bond lengths to study the effects of solvent activity on DNA structure. The effects of solvent activity were studied using molecular dynamics (MD) by combining the DNA fragments of $d(GCGAATTCGCG)_2$ into SPC water solvents in different models with varying bond lengths from 1.1 to 0.7 Å. DNA underwent the transformation of B from—(A-B) mix—A form [72].

In ethanol solvent and low activity water solvent, the DNA changed to the A-type configuration. There are similar reasons behind it, both of which are due to the weak hydrogen bond in this kind of solvent, the hydration layer structure is easy to be destroyed, and the ions in the solution are easy to enter the hydration shell to replace the position of water molecules, and directly contact with DNA to neutralize the negative charge and change the configuration of DNA. The results show that the equilibrium ions and solvents play an important role in maintaining the stable structure of DNA.

## 7.5 Conclusions

DNA has a variety of configurations, DNA configuration changes depend on the environment and sequence. DNA presents different conformations in various solvents. We constructs five initial B-type DNA according to the increase of base G content in the sequence, and then the dynamic evolution in the ion balance solution, through its trajectory file analysis, the following conclusions are obtained: (1) DNA configuration, with the increase of the base G content in the sequence, the pro-A tendency gradually strengthens, the pro-B gradually weakened, that is, the ability to convert from B-type to A-type improved and there is A-type DNA in GGGG. (2) Ion distribution, ion distribution also has sequence dependence with the increase of G base in DNA oligomers, the mutual attraction of hydrophobic groups between bases and bases stacking interactions weakened, resulting in anisotropy ratio of the base increased, the static strength in major groove increased. As a result, the negative charge atoms on the base are exposed, and ions near the phosphate backbone and minor groove move toward the major groove, enhancing the interaction between the ions and the electronegative atoms in the major groove.

The study of alcohol structure helps to understand the mechanism of DNA interaction in its solution. In the study of the proportional mixing of EG/EA and water, when Ci increased from 0 to 100%, the intermolecular structure of the EG-EG was not significantly affected by the EG mass fraction until Ci increased to 80%. The study of hydrogen bonds found that the intermolecular HB binding has the following priority: water-water greater than EG-water greater than EG-EG and water-water greater than water-EA greater than EA-EA. Comparing the simulated evolution of DNA in EG and EA solvents, it is found that in EG and EA solvents found that in EG solutions, ethylene glycol molecules are similar to water molecules in screening for rejection between negatively charged phosphates, preventing this rejection from spreading and extending the spiral so that the DNA's configuration maintains type B. In the EA solution, no water molecule maintains the hydration structure of phosphate oxygen, and the $Na^+$ is strongly matched with a free oxygen phosphate atom to form a chemical bond of Na-O (P) that inhibits electrostatic repulsion on the skeleton. As a result, DNA exhibits shorter, denser A-type. A decrease solvent-water activity has the similar reason that could lead to configurational changes in A-DNA. There is the same reason for the change of DNA to A configuration from over-active to low-active aqueous solvents.

Studying the effects of sequence and environment on changes in DNA conformation and ion distribution can help to understand the structure and function of DNA and the origin of DNA conformation polymorphism at the molecular level. At the same time, we can select or change the sequence of DNA to achieve the desired DNA conformation, to meet the implementation of various biological functions of DNA.

The structural characteristics and ion distributions directly related to the sequence obtained in this chapter are important in identifying DNA binding to proteins or drugs, developing DNA structures, and understanding the nature of reversible A-B conversion. Discussion of different solvents offers the possibility that glycol stores DNA in biological banks more cheaply and more efficiently than water solvents in bio-stores.

**Acknowledgements** Thanks to Prof. L. Bulavin, N. Atamas, and K. Cherevko for the fruitful discussions. This work was supported by the National Natural Science Foundation of China under Grants No. 11635003, No.11025524, No. 11161130520, the National Basic Research Program of China under Grant No. 2010CB832903, and the European Commission's 7th Framework Programme (Fp7-PEOPLE-2010-IRSES) under Grant Agreement Project No. 269131.

## References

1. Hamilton LD (1968) DNA: models and reality. Nature 218:633–637. https://doi.org/10.1038/218633a0
2. Mohr SC, Sokolov NV, He CM, Setlow P (1991) Binding of small acid-soluble spore proteins from Bacillus subtilis changes the conformation of DNA from B to A, pp 88:77–81. https://doi.org/10.1073/pnas.88.1.77
3. Forester TR, McDonald IR (1991) Molecular dynamics studies of the behaviour of water molecules and small ions in concentrated solutions of polymeric B-DNA. Mol Phys 72:643–660. https://doi.org/10.1080/00268979100100481
4. Okonogi TM, Alley SC, Reese AW, Hopkins PB, Robinson BH (2000) Sequence-dependent dynamics in duplex DNA. Biophys J 78:2560–2571. https://doi.org/10.1016/S0006-3495(00)76800-7
5. Arauzo-Bravo MJ, Sarai A (2008) Indirect readout in drug-DNA recognition: role of sequence-dependent DNA conformation. Nucleic Acids Res 36:376–386. https://doi.org/10.1093/nar/gkm892
6. Gorenstein DG (1994) Conformation and dynamics of DNA and protein-DNA complexes by 31P NMR. Chem Rev 94:1315–1338
7. Moravek Z, Neidle S, Schneider B (2002) Protein and drug interactions in the minor groove of DNA. Nucleic Acids Res 30:1182–1191. https://doi.org/10.1093/nar/30.5.1182
8. Heddi B, Abi-Ghanem J, Lavigne M, Hartmann B (2010) Sequence-dependent DNA flexibility mediates DNase I cleavage. J Mol Biol 395:123–133. https://doi.org/10.1016/j.jmb.2009.10.023
9. Svozil D, Kalina J, Omelka M, Schneider B (2008) DNA conformations and their sequence preferences. Nucleic Acids Res 36:3690–3706. https://doi.org/10.1093/nar/gkn260
10. Heddi B, Foloppe N, Bouchemal N, Hantz E, Hartmann B (2006) Quantification of DNA BI/BII backbone states in solution. Implications for DNA overall structure and recognition. J Am Chem Soc 128:9170–9177. https://doi.org/10.1021/ja061686j
11. Oguey C, Foloppe N, Hartmann B (2010) Understanding the sequence-dependence of DNA groove dimensions: implications for DNA interactions. PLoS One 5:e15931. https://doi.org/10.1371/journal.pone.0015931
12. Heddi B, Oguey C, Lavelle C, Foloppe N, Hartmann B (2009) Intrinsic flexibility of B-DNA: the experimental TRX scale. Nucleic Acids Res 38:1034–1047. https://doi.org/10.1093/nar/gkp962
13. Olson WK, Gorin AA, Lu X-J, Hock LM, Zhurkin VB (1998) DNA sequence-dependent deformability deduced from protein-DNA crystal complexes. Proc Natl Acad Sci USA 95:11163–11168. https://doi.org/10.1073/pnas.95.19.11163

14. Hui X, Ai-Min G, Song-Shan M (2007) The influence of base pair sequence on electronic structure of DNA molecules. Acta Physica Sinica 56:1208–1213. https://doi.org/10.7498/aps.56.1208
15. Zhang N, Li MR, Zhang FS (2018) Ethylene glycol solution-induced DNA conformational transitions. Chin Phys B 27:113102–113102. https://doi.org/10.1088/1674-1056/27/11/113102
16. Lavery R, Maddocks JH, Pasi M, Zakrzewska K (2014) Analyzing ion distributions around DNA. Nucleic Acids Res 42:8138–8149. https://doi.org/10.1093/nar/gku504
17. Hess B, Kutzner C, van der Spoel D, Lindahl E (2008) GROMACS 4: algorithms for highly efficient, load-balanced, and scalable molecular simulation. J Chem Theory Comput 4:435–447. https://doi.org/10.1021/ct700301q
18. Macke TJ, Case DA (1997) Molecular modeling of nucleic acids. Am Chem Soc, 379–393. https://doi.org/10.1021/bk-1998-0682.ch024
19. Perez A, Marcháin I, Svozil D, Sponer J, Cheatham TE, Laughton CA, et al (2007) Refinement of the AMBER force field for nucleic acids: improving the description of / conformers. Biophys J 92:3817–3829. https://doi.org/10.1529/biophysj.106.097782
20. Jorgensen WL (1981) Transferable intermolecular potential functions for water, alcohols, and ethers. Application to liquid water. J Am Chem Soc 103:335–340. https://doi.org/10.1021/ja00392a016
21. York DM, Darden TA, Pedersen LG (1993) The effect of long-range electrostatic interactions in simulations of macromolecular crystals: A comparison of the Ewald and truncated list methods. J Chem Phys 99:8345–8348. https://doi.org/10.1063/1.465608
22. Hess B, Bekker H, Berendsen HJC, Fraaije JGEM (1997) LINCS: A linear constraint solver for molecular simulations. J Chem Phys 18:1463–1472. https://doi.org/10.1002/(SICI)1096-987X(199709)18:12<1463::AID-JCC4>3.0.CO;2-H
23. Nose S, Klein ML (1983) Constant pressure molecular dynamics for molecular systems. Mol Phys 50:1055–1076. https://doi.org/10.1080/00268978300102851
24. Lavery R, Moakher M, Maddocks JH, Petkeviciute D, Zakrzewska K (2009) Conformational analysis of nucleic acids revisited: Curves+. Nucleic Acids Res 37:5917–5929. https://doi.org/10.1093/nar/gkp608
25. Pasi M, Maddocks JH, Lavery R (2015) Analyzing ion distributions around DNA: sequence-dependence of potassium ion distributions from microsecond molecular dynamics. Nucleic Acids Res 43:2412–2423. https://doi.org/10.1093/nar/gkv080
26. Schweitzer BI, Mikita T, Kellogg GW, Gardner KH, Beardsley GP (1994) Solution structure of a DNA dodecamer containing the anti-neoplastic agent arabinosylcytosine: combined use of NMR, restrained molecular dynamics, and full relaxation matrix refinement. Biochemistry 33:11460–11475. https://doi.org/10.1021/bi00204a008
27. Watson JD, Crick FHC (1953) Molecular structure of nucleic acids: a structure for deoxyribose nucleic acid. Nature 171:737–738. https://doi.org/10.1038/171737a0
28. Ghosh A, Bansal M (2003) A glossary of DNA structures from A to Z. Acta Crystallogr D 59:620–626. https://doi.org/10.1107/S0907444903003251
29. Dickerson RE, Ng H-L (2001) DNA structure from A to B, pp 98:6986–6988. https://doi.org/10.1073/pnas.141238898
30. Franklin RE, Gosling RG (2010) The structure of sodium thymonucleate fibres. II. The cylindrically symmetrical Patterson function. Acta Crystallogr 6:678–685. https://doi.org/10.1107/S0365110X53001940
31. Fuller W, Wilkins WHF, Wilson HR, Hamilton LD (1965) The molecular configuration of deoxyribonucleic acid: IV. X-ray diffraction study of the A form. J Mol Biol 12:60–76. https://doi.org/10.1016/S0022-2836(65)80282-0
32. Langridge R, Wilson HR, Hooper CW, Wilkins MHF, Hamilton LD (1960) The molecular configuration of deoxyribonucleic acid: I. X-ray diffraction study of a crystalline form of the lithium salt. J Mol Biol 2:19-IN11. https://doi.org/10.1016/S0022-2836(60)80004-6
33. Waters JT, Lu X-J, Galindo-Murillo R, Gumbart JC, Kim HD, Cheatham TE et al (2016) Transitions of double-stranded DNA between the A- and B-forms. J Phys Chem B 120:8449–8456. https://doi.org/10.1021/acs.jpcb.6b02155

34. Rai A, Bush ME, Gracic A, Kim J, Lerner MG, Seewald AK et al (2016) Molecular dynamics investigations of Z[WC] DNA and its potential role in the B to Z-DNA transition. Biophys J 110:404a. https://doi.org/10.1016/j.bpj.2015.11.2183
35. Zhang X, Le S, Chen H, Doyle P, Yan J (2013) Sequence dependence of the "B-to-S" DNA overstretching transition. Biophys J 104:262a. https://doi.org/10.1016/j.bpj.2012.11.1471
36. Hackl EV, Blagoi YP (2005) The effect of temperature on DNA structural transitions under the action of Cu2+ and Ca2+ ions in aqueous solutions. Biopolymers 77:315–324. https://doi.org/10.1002/bip.20225
37. Kundu S, Mukherjee S, Bhattacharyya D (2012) Effect of temperature on DNA double helix: An insight from molecular dynamics simulation. J Biosci 37:445–455. https://doi.org/10.1007/s12038-012-9215-5
38. Gu B, Zhang FS, Wang ZP, Zhou HY (2008) Solvent-induced DNA conformational transition. Phys Rev Lett 100: https://doi.org/10.1103/PhysRevLett.100.088104
39. Marc G, Olivier B, Oliver M, Catherine E, Brigitte H, Franca F (2012) Mg2+ in the major groove modulates B-DNA structure and dynamics. PLoS One 7. https://doi.org/10.1016/S0959-440X(00)00205-0
40. Cheng Y (2006) Similarities and differences in interaction of K+ and Na+ with condensed ordered DNA. A molecular dynamics computer simulation study. Nucleic Acids Res 34:686–696. https://doi.org/10.1093/nar/gkj434
41. Noy A, Perez A, Laughton CA, Orozco M (2007) Theoretical study of large conformational transitions in DNA: the B?A conformational change in water and ethanol/water. Nucleic Acids Res 35:3330–3338. https://doi.org/10.1093/nar/gkl1135
42. Wilkins MHF (1963) Molecular Configuration of Nucleic Acids: From extensive diffraction data and molecular model building a more detailed picture is emerging. Science 140:941–950. https://doi.org/10.1126/science.140.3570.941
43. Texter J (1978) Nucleic acid-water interactions. Prog Biophys Mol Biol 33:83–97. https://doi.org/10.1016/0079-6107(79)90026-9
44. MaB T, Hearst JE (1968) On the hydration of DNA. I. Preferential hydration and stability of DNA in concentrated trifluoroacetate solution. Biopolymers 6:1325–1344. https://doi.org/10.1002/bip.1968.360060908
45. Saenger W, Hunter WN, Kennard O (1986) DNA conformation is determined by economics in the hydration of phosphate groups. Nature 324:385–388. https://doi.org/10.1038/324385a0
46. Santoso Y, Joyce CM, Potapova O, Reste LL, Hohlbein J, Torella JP et al (2010) Conformational transitions in DNA polymerase I revealed by single-molecule FRET. Proc Natl Acad Sci U S A 107:715–720. https://doi.org/10.1073/pnas.0910909107
47. Shen X, Gu B, Che SA, Zhang FS (2011) Solvent effects on the conformation of DNA dodecamer segment: A simulation study. J Chem Phys 135: https://doi.org/10.1063/1.3610549
48. Mrevlishvili GM, Metreveli NO, Razmadze GZ, Mdzinarashvili TD, Khvedelidze MM (1998) Partial heat capacity change - fundamental characteristic of the process of thermal denaturation of biological macromolecules (proteins and nucleic acids). Thermochim Acta 308:41–48. https://doi.org/10.1016/S0040-6031(97)00328-6
49. Muller J (2010) Functional metal ions in nucleic acids. Metallomics 2:318–327. https://doi.org/10.1039/c000429d
50. Wen J, Shen H, Zhai YR, Zhang FS (2016) Simulation of DNA in water/ethanol mixture. Physica A 450:515–522. https://doi.org/10.1016/j.physa.2015.12.091
51. Dickerson RE, Drew HR (1981) Structure of a B-DNA dodecamer: II. Influence of base sequence on helix structure. J Mol Biol 149:761–786. https://doi.org/10.1016/0022-2836(81)90357-0
52. Yonetani Y, Kono H (2009) Sequence dependencies of DNA deformability and hydration in the minor groove. Biophys J 97:1138–1147. https://doi.org/10.1016/j.bpj.2009.05.049
53. Hays FA, Teegarden A, Jones ZJR, Harms M, Raup D, Watson J et al (2005) How sequence defines structure: a crystallographic map of DNA structure and conformation 102:7157–7162. 10.1073/pnas.0409455102

54. Dixit SB, Beveridge DL, Case DA, Cheatham TE, Giudice E, Lankas F et al (2005) Molecular Dynamics Simulations of the 136 Unique Tetranucleotide Sequences of DNA Oligonucleotides. II: Sequence Context Effects on the Dynamical Structures of the 10 Unique Dinucleotide Steps. Biophys J 89:3721–3740. https://doi.org/10.1529/biophysj.105.067397
55. Hays FA, Teegarden A, Jones ZJR, Harms M, Raup D, Watson J et al (2005) How sequence defines structure: a crystallographic map of DNA structure and conformation. Proc Natl Acad Sci U S A 102:7157–7162. https://doi.org/10.1073/pnas.0409455102
56. Otokiti EO, Sheardy RD (1997) Sequence effects on the relative thermodynamic stabilities of B-Z junction-forming DNA oligomeric duplexes. Biophys J 73:3135–3141. https://doi.org/10.1016/S0006-3495(97)78339-5
57. Gueron M, M, Demaret JP, Filoche M (2000) A unified theory of the B-Z transition of DNA in high and low concentrations of multivalent ions. Biophys J 78:1070–1083. https://doi.org/10.1016/S0006-3495(00)76665-3
58. Li MR, Zhang N, Xu HT, Zhang FS (2020) Sequence dependence of the conformational transitions of DNA. Chem Phys Lett 748: https://doi.org/10.1016/j.cplett.2020.137344
59. Saenger P (2013) Principles of nucleic acid structure. Springer-Verlag, New York
60. Lu X-J, Shakked Z, Olson WK (2000) A-form conformational motifs in ligand-bound DNA structures. J Mol Biol 300:819–840. https://doi.org/10.1006/jmbi.2000.3690
61. Dickerson RE, Chiu TK (1997) Helix bending as a factor in protein/DNA recognition. 44:361–403. https://doi.org/10.1002/(SICI)1097-0282(1997)44:4<361::AID-BIP4>3.0.CO;2-X
62. Yakovchuk P, Protozanova E, Frankkamenetskii MD (2006) Base-stacking and base-pairing contributions into thermal stability of the DNA double helix. Nucleic Acids Res 34:564–574. https://doi.org/10.1093/nar/gkj454
63. Tolstorukov MY, Ivanov VI, Malenkov GG, Jernigan RL, Zhurkin VB (2001) Sequence-dependent B-A transition in DNA evaluated with dimeric and trimeric scales. Biophys J 81:3409–3421. https://doi.org/10.1016/S0006-3495(01)75973-5
64. Il'Icheva IA, Vlasov PK, Esipova NG, Tumanyan VG (2010) The intramolecular impact to the sequence specificity of BA transition: low energy conformational variations in AA/TT and GG/CC steps. J Biomol Struct Dyn 27:677–693. https://doi.org/10.1080/07391102.2010.10508581
65. Pasi M, Maddocks JH, Lavery R (2015) Analyzing ion distributions around DNA: sequence-dependence of potassium ion distributions from microsecond molecular dynamics. Nucleic Acids Res 43:2412–2423. https://doi.org/10.1093/nar/gkv080
66. Egli M (2002) DNA-cation interactions: quo vadis? Chem Biol 9:277–286. https://doi.org/10.1016/S1074-5521(02)00116-3
67. Lu XJ, Shakked Z, Olson WK (2000) A-form conformational motifs in ligand-bound DNA structures. J Mol Biol 300:819–840. https://doi.org/10.1006/jmbi.2000.3690
68. Savelyev A, Papoian G (2006) Electrostatic, steric, and hydration interactions favor Na(+) condensation around DNA compared with K(+). J Am Chem Soc 128:14506–14518. https://doi.org/10.1021/ja0629460
69. Zhang N, Li M-R, Zhang F-S (2019) Structure and dynamics properties of liquid ethylene glycol from molecular dynamics simulations. Chem Phys Lett 718:12–21. https://doi.org/10.1016/j.cplett.2019.01.025
70. Zhang N, Li MR, Xu HT, Zhang FS (2020) Polymorphism and flexibility of DNA in alcohols. Chin Phys Lett 37: https://doi.org/10.1088/0256-307x/37/8/088701
71. Anagnostopoulos CA, Spizizen J (1961) Requirements for transformation in Bacillus subtilis. J Bacteriol 81:741–746. https://doi.org/10.1016/B978-012373944-5.00036-5
72. Li MR, Zhang N, Zhang FS (2018) Computational investigation of the conformation transitions of DNA in modified water models. J Mol Liq 271:175–181. https://doi.org/10.1016/j.molliq.2018.08.129

# Chapter 8
# Mechanisms of Heteroassociation of Ceftriaxone and Doxorubicin Drugs with Bovine Serum Albumin

**Oksana Dmytrenko, Mykola Kulish, Olena Pavlenko, Andrii Lesiuk, Andriy Momot, Tetiana Busko, Mykola Kaniuk, Tymofii Nikolaienko, and Leonid Bulavin**

**Abstract** Quantum chemical modeling of the electronic and oscillatory structures of doxorubicin (DOX) and ceftriaxone (Cef) molecules was performed. Parameters of the resonance energy transfer between a bovine serum albumin (BSA) biomolecule and molecules of DOX and Cef were calculated. The obtained results are consistent with the molecular docking study. The binding parameters of the BSA-DOX conjugate in an aqueous solution were determined. Changes in thermodynamic parameters of the solutions indicate the presence of van der Waals interaction and hydrogen bonds in the conjugate with a significant contribution of electrostatic forces. The conditions of excitation energy transfer in BSA-Cef complexes were established. The values of the binding constants indicate a static mechanism of fluorescence quenching. The interaction in BSA-Cef complexes has a spontaneous hydrophobic character. The analysis of the components of the amide band I of IR transmission spectrum indicates insignificant conformational transformations of BSA molecules in the complexes with Cef.

O. Dmytrenko · M. Kulish · O. Pavlenko · A. Lesiuk (✉) · A. Momot · T. Busko · T. Nikolaienko · L. Bulavin
Faculty of Physics, Taras Shevchenko National University of Kyiv , 64/13 Volodymyrska Street, Kyiv 01601, Ukraine

O. Dmytrenko
e-mail: o_dmytrenko@ukr.net

M. Kulish
e-mail: mpkulish@ukr.net

T. Busko
e-mail: tbusko@ukr.net

T. Nikolaienko
e-mail: tim_mail@ukr.net

M. Kaniuk
Palladin Institute of Biochemistry of the National Academy of Sciences of Ukraine (NASU) , 9 Leontovicha Street , Kyiv 01030, Ukraine
e-mail: kanyuk@nas.gov.ua



L. Bulavin and N. Lebovka (eds.), *Soft Matter Systems for Biomedical Applications*, Springer Proceedings in Physics 266, https://doi.org/10.1007/978-3-030-80924-9_8

## List of Abbreviations and Symbols

### *Abbreviations*

| | |
|---|---|
| BSA | bovine serum albumin |
| HSA | human serum albumin |
| Cef | ceftriaxone |
| DOX | doxorubicin |
| Trp | tryptophan |
| AuNP | gold nanoparticle |
| FL | fluorescence |
| DFT | density functional theory |
| TD SCF | time-dependent self consistent field |
| B3LYP | Becke, 3-parameter, Lee–Yang–Parr |
| HOMO | the highest occupied molecular orbital |
| LUMO | the lowest unoccupied molecular orbital |
| IR | infrared |
| SERS | surface enhanced Raman spectroscopy |
| SV | Stern-Volmer |

### *Symbols*

| | |
|---|---|
| $C$ | concentration |
| $F$ | intensity |
| $\xi$ | absorbance |
| $\lambda$ | wavelength |
| $K$ | characteristic constant |
| $n$ | number of binding sites |

## 8.1 Introduction

Doxorubicin (adriamycin) (DOX) belongs to the family of anthracycline antibiotics, which are widely used in chemotherapy of malignant neoplasms and other types of cancer. It is assumed that the most important mechanism of doxorubicin impact on malignant cells is the intercalation of DOX molecules into the DNA biomolecule, which causes inhibition of the synthesis of DNA macromolecules, RNA and proteins. The formation of the free radicals capable of neutralizing DNA molecules, lipids peroxidation, the crosslinks formation in DNA and other processes of inhibition in the presence of DOX have also been considered [1–3].

Ceftriaxone (Cef) belongs to the third generation of antibiotics characterized by the high antimicrobial, antifungal efficacy and pronounced bactericidal action. The latter implies the inhibition of the cell walls synthesis of gram-positive and gram-negative bacteria. It is realized through suppressing the synthesis of transpeptides, one of the important enzymes of bacteria, which provides crosslinking between neighboring tetrapeptides [4–6]. The enhancement of the antibacterial properties of ceftriaxone is due to its electronic structure and conformational state, which can be changed by adding a variety of ligands, primarily metals [7–9] and nanoparticles [10–13], to the drug. The formation of conjugates with ceftriaxone results in a significant change of the drug activity with respect to the proteins, lipids and hydrocarbon components of the membranes of bacterial cell [14].

Taking into account the high toxicity of drugs and consequently high probability for the normal cells to be damaged, the important issue is the targeted delivery of drugs to malignant cells. This problem is solved in various ways, including creating less toxic prodrugs on the basis of anthracycline antibiotics [15] or by the localization of drug molecules on nano- and magnetic particles [16–18]. At present, the natural way of drugs transportation using plasma albumin as the main protein which contacts with all cells of the body remains perspective. At the same time, albumin is characterized by antioxidant properties and the ability to form complexes with various ligands due to disulfide interaction. The antitoxic effect of biomolecules is also manifested in albumin complexes with drugs.

Currently, a large number of studies on the properties of ceftriaxone, doxorubicin and its analogues, their interaction with bovine serum albumin (BSA) and human serum albumin (HSA) [19–27], as well as the interaction of protein biomolecules with other drugs [28–35], have been carried out using electrophotometric methods. These methods are based on changes in the optical density of drug solutions due to the presence of the albumin [20, 21], or fluorescence quenching of albumin solutions with increasing concentration of ligands [28–35]. They allow to obtain important parameters of complexation, binding constant and a number of binding sites of ligand molecules to prodrugs [28–35].

Despite a large number of studies on the properties of doxorubicin molecules, especially the electronic structure of various protonated forms, and ceftriaxone, the mechanisms of interaction in complexes of protein biomolecules with drugs, also in the presence of metal ions or metal nanoparticles, have not been sufficiently studied yet.

The purpose of this work is to perform quantum chemical modeling of the electronic structure, optical transitions in the oscillatory structure of doxorubicin and ceftriaxone molecules and the establishment of mechanisms of interaction in complexes of drug molecules with BSA.

## 8.2 Preparation of Samples and Experimental Methods

The crystallized powder of 99% of the bovine serum albumin manufactured by Sigma Aldrich/Merck and dosage forms of doxorubicin and ceftriaxone (18% of active substance (see instructions)) were used in the experiment. The active substances were dissolved in distilled water. Solutions with a constant concentration of BSA $C_{BSA} = 0.01$ mM and a variable concentration of ceftriaxone $C_{Cef} = 0$ M; $10^{-6}$ M; $2 \times 10^{-6}$ M; $4 \times 10^{-6}$ M; $6 \times 10^{-5}$ M; $8 \times 10^{-5}$ M; $1 \times 10^{-5}$ M; $1.2 \times 10^{-5}$ M were obtained by the method of sequential dilution.

For the solutions with doxorubicin, concentrations $C_{BSA} = 10^{-4}$ M = const of BSA and $C_{DOX} = 0.1 \times 10^{-5}$M; $0.5 \times 10^{-5}$M; $1 \times 10^{-5}$M; $2 \times 10^{-5}$M; $3 \times 10^{-5}$M; $5 \times 10^{-5}$M; $6 \times 10^{-5}$M; $8 \times 10^{-5}$M; $10 \times 10^{-5}$M of the drug.

Quantum chemical calculations were performed on the basis of the Gaussian 03 software package. Forms of molecular orbitals of the doxorubicin molecule were obtained by DFT (B3LYP, 6-31G) method, the energies of electron transitions were obtained by TD SCF (B3LYP, 6-31G).

The study of molecular fluctuations in the system of drug—bovine serum albumin (BSA) was performed using spectroscopy of absorption of infrared (IR) light and Raman light scattering (Raman). As in the previous articles [18, 32], the samples were prepared in the form of films by means of drying solutions of doxorubicin with BSA of different concentration at room temperature on glass substrates covered with a thin layer of Au nanostructures made by the Turkevich method. The ratio of BSA to doxorubicin concentrations varied from 0 to 1, thus, the obtained concentrations of doxorubicin and BSA in the solutions $C_{DOX} = 17.7 \times 10^{-5}$ M, $C_{BSA} = 1.77 \times 10^{-5}$ M, $3.54 \times 10^{-5}$ M, $5.31 \times 10^{-5}$ M, $7.08 \times 10^{-5}$ M, $8.85 \times 10^{-5}$ M, $12.39 \times 10^{-5}$ M, $17.7 \times 10^{-5}$ M.

Micro-Raman spectra were measured at room temperature in the backscattering geometry using a Horiba Jobin Yvon T64000 (Horiba, France) spectrometer equipped with an UV-Visible-NIR Olympus BX41 confocal microscope (Olympus, Japan) and a thermoelectrically cooled CCD detector. An Ar-Kr laser with a wavelength $\lambda = 488$ nm was used to excite the Raman spectra. The excitation radiation was focused on the surface of the test sample using an Olympus lens with a 10-increase in the area of $\sim 5\,\mu m^2$. The power density of the excitation radiation on the sample varied in the range of $0.5 \div 100\,kW/cm^2$. The accumulation of the signal was 30 s. In addition, Raman spectra were recorded using a micro-novel spectrometer Renishaw InVia Raman Microscope (Renishaw plc, United Kingdom).

A standard Bruker Vertex 70v Fourier infrared spectrometer (Bruker, Germany) was used to measure the reflection spectra in the spectral range of 100–3500 $cm^{-1}$ with a resolution of 4 $cm^{-1}$ in the infrared region.

The measurements of light absorption spectra of the drug solution in the visible optical region were performed on a two-channel spectrophotometer UV-visible-near-IR region "UNICO SpectroQuest 4802" (United Products & Instruments, Inc., USA) in the spectral range $350 \div 700$ nm with a step of 1 nm on fluorine COULTER® EPICS® XLTM Flow Cytometer (Beckman Coulter, Inc., USA), as well as on a portable laser fluorimeter “Fluorotest Nano” (IFN NASU, Ukraine) with excitation at a wavelength of 532 nm.

Molecular docking was performed in AutoDock Vina 1.1.2 software package [36], where the iterative local search method is used as a global optimization algorithm. The algorithm uses random starting conformations of the protein molecule and ligand [36].

## 8.3 Results and Discussion for Doxorubicin

### *8.3.1 Quantum Chemical Modeling of the Electronic Structure of Doxorubicin Molecules*

Figure 8.1 shows experimentally obtained spectra of optical density and fluorescence for aqueous solutions of doxorubicin.

It can be seen that the main absorption peaks are at 475 nm (more intensive one) and at 500 nm. To establish the nature of the absorption, quantum chemical calculations of the electronic transitions of the doxorubicin molecule were performed by the TD SCF method, B3LYP (6-31G).

The forms of molecular orbitals (obtained by DFT method) involved in transitions are shown in Fig. 8.2.

Thus, according to calculations, the main contribution to the intensity is provided by the electronic transition from HOMO-1 to LUMO +2, which are delocalized along the phenolic rings of the molecule. The wavelength of this transition is 434 nm, and it differs from the experimental peak by 40 nm, which can happen due to the presence of a solvent. The second electronic transition, according to the calculations, is the result of the transition between HOMO-1 and LUMO orbitals. The wavelength,

**Fig. 8.1** Optical density (1) and fluorescence (FL) (2) spectra of aqueous solutions of doxorubicin, $C_{DOX} = 10^{-4}$ M, $\lambda_{ex} = 488$ nm

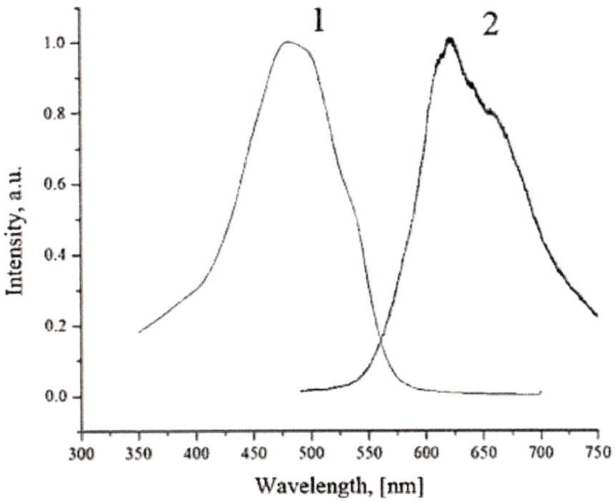

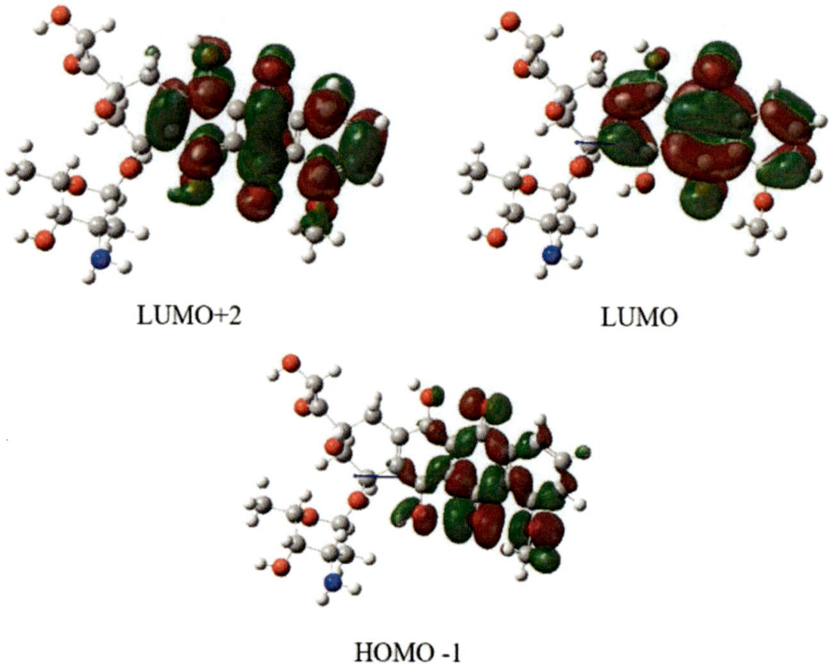


**Fig. 8.2** Molecular orbitals of doxorubicin molecule involved in electronic transitions

corresponding to this transition is 526 nm and is shifted by 26 nm with respect to the experimental value.

This absorption band indicates that doxorubicin is in a monocationic protonated state (the amine group $NH_2$ turns into the cationic form $NH_3^+$ [37]. The maximum of the FL band is near $\lambda_{max} = 650$ nm.

Figure 8.3 shows the experimental spectra of Raman scattering of doxorubicin films on substrates with gold nanoparticles and IR transmission of the aqueous solution of DOX. Gold nanoparticles of 13–17 nm in diameter, prepared in V.E. Lashkaryov Institute of Semiconductor Physics of NASU, was used in SERS (Surface-enhanced Raman Spectroscopy) technique of the experiment in order to enhance typically weak Raman scattering of biological materials.

The most intense bands in the Raman spectrum of DOX molecules are around 1220, 1255 $cm^{-1}$, corresponding to $\upsilon$ (C-O) + $\upsilon$ (C-O + H, $\upsilon$ (C = H)), near 1464 $cm^{-1}$ ($\upsilon$ ( C = C) + $\upsilon$ (C-C) + $\upsilon$ (C = O… H)), 1642 $cm^{-1}$ ($\upsilon$ (C = O)… H) [38]. In the IR transmission spectrum, there are intense bands of oscillating modes with frequencies of 806 $cm^{-1}$. The study of conjugates of doxorubicin with other molecules shows that the oscillating spectra of DOX molecules are rearranged, which makes it possible to detect the binding position in the corresponding complexes [39–43].

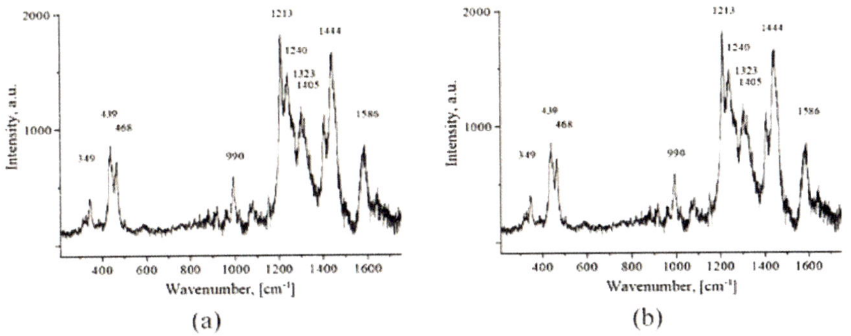


**Fig. 8.3** Raman spectra **a** of doxorubicin films on substrates of gold nanoparticles and **b** IR transmission of an aqueous solution of DOX

### 8.3.2 *Interaction in DOX-BSA Conjugates*

The study of optical absorption of the DOX-BSA conjugates at a constant concentration $C_{DOX}$ of the drug and variable content of BSA $C_{BSA}$ demonstrates the decrease of the optical density of doxorubicin in aqueous solutions [43, 44]. The optical density drops due to the formation of DOX-BSA complexes of one molecule of doxorubicin and one molecule of albumin, as evidenced by calculations of the number of binding sites $n = 1$. The optical density spectra of the DOX-BSA aqueous solution becomes more complex after the addition of spherical gold nanoparticles with the diameter of 13 nm, which is a consequence of the effective interaction of citrated nanoparticles with protein macromolecules, resulting in a change of longitudinal and transverse plasmon resonance spectra.

The addition of gold nanoparticles to the aqueous solution of DOX also altered the spectra of Raman scattering (Fig. 8.4).

There are significant changes of intensity and bands near 349, 468, 990, 1219, 1240, 1323, 1405, 1444, and 1568 cm$^{-1}$, which are typical for DOX molecules. This rearrangement of the SERS spectrum indicates an interaction between the components in the DOX-AuNPs conjugates.

The interaction between DOX and BSA is also manifested in fluorescence quenching after adding DOX to the aqueous solution of protein biomolecules (Fig. 8.5).

It can be seen from the FL spectrum of the DOX aqueous solution that the addition of BSA results in quenching of the fluorescence, which grows with the increase of the albumin concentration. Obviously, a similar FL quenching can also be observed in DOX-BSA conjugates with variable concentration of DOX.

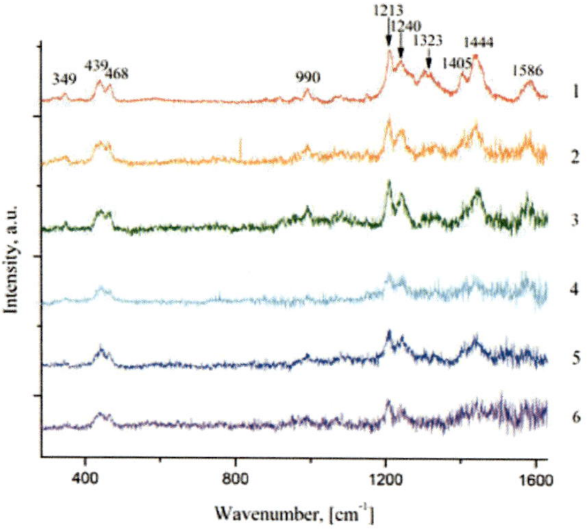


**Fig. 8.4** SERS spectra of the films precipitated from an aqueous solution of DOX with substrates of gold nanoparticles at constant concentration $C_{DOX} = 1.77 \times 10^{-4}$ M; and different concentration $C_{BSA} = 0(1)$; $0.177 \times 10^{-4}$ M(2); $0.53 \times 10^{-4}$ M(3); $0.707 \times 10^{-4}$ M(4); $1.24 \times 10^{-4}$ M(5); $1.77 \times 10^{-4}$ M (6) of BSA ($\lambda_{ex} = 488$ nm)

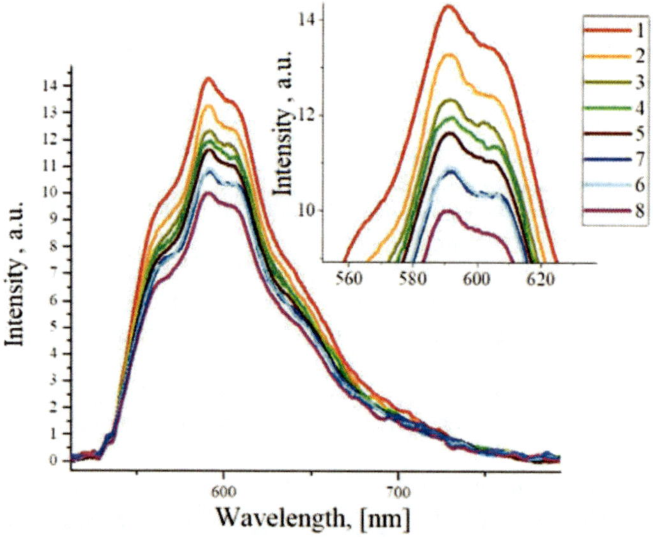


**Fig. 8.5** Fluorescence spectra of an aqueous solution of DOX-BSA conjugates at $C_{DOX} = 0.885 \times 10^{-4}$ M(1) and $C_{BSA} = 0.885 \times 10^{-5}$ M(2), $1.77 \times 10^{-5}$ M (3), $2.655 \times 10^{-5}$ M(4), $3.54 \times 10^{-5}$ M (5), $4.43 \times 10^{-5}$ M (6), $6.2 \times 10^{-5}$ M (7), $8.85 \times 10^{-5}$ M(8) (reprinted with permission from [18])

### 8.3.3 Excitation Energy Transfer in the BSA-DOX System

The fluorescence quenching in the DOX-BSA solution takes place due to the transfer of energy from the donor (BSA) to the acceptor (DOX) after excitation of the protein molecules. The process is possible under certain conditions. The degree of energy transfer, as follows from the theory of resonance energy transfer [45, 46], depends on the distance $r$ between the donor and the acceptor molecules and the distance $R_0$ at which the efficiency of the energy transfer is 50%:

$$E = 1 - \frac{F}{F_0} = \frac{R_0^6}{R_0^6 + r^6}, \tag{8.1}$$

where $F_0$, $F$ are the fluorescence intensity at the maximum of the emission band $\lambda_{max}$ of the donor (BSA) in the absence and presence of quencher (DOX), respectively.

The distance $r$ at which the energy can be transferred is in the range $0.5R_0 < r < 1.5R_0$. The value of $R_0$ depends on the value of the overlap integral $J$ between the FL spectra of the donor (BSA) and the absorption of the acceptor (DOX) in accordance with the equation:

$$R_0^6 = 8,8 \times 10^{-25} k^2 \Phi J N^{-4}, \tag{8.2}$$

where $k^2$ is the factor of the spatial orientation of the molecules, which is equal to 2/3, $N$ is the refractive index of the medium, which for the DOX-BSA system is equal to 1.336, $\Phi$ is a fluorescence quantum yield of the donor molecule and equals to 0.118.

The overlap integral of the spectra is given by the following relationship:

$$J = \frac{\sum F(\lambda)\xi(\lambda)\lambda^4 \Delta\lambda}{\sum F(\lambda)\Delta\lambda}, \tag{8.3}$$

where $F(\lambda)$ and $\xi(\lambda)$ are the FL and absorption spectra (molar extinction coefficient).

The spectra of $F(\lambda)$ and $\xi(\lambda)$ for aqueous solutions of pure BSA and DOX are shown in Fig. 8.6. The overlap integral of the fluorescence spectra $F(\lambda)$ of BSA and the molecular extinction coefficient $\xi(\lambda)$ of DOX from Eq. (8.3) is equal to $J = 1.5 \times 10^{-16}\,\text{cm}^3\text{M}^{-1}$.

For the specified values the calculated distance $R_0$ is equal to 2.2 nm. To assess the energy transfer we analysed the FL bands of the aqueous solution of pure BSA and the one with presence of DOX. The obtained by the Eq. (8.1) values of $F/F_0$ yield the transfer efficiency $E = 0.28$. For the above-mentioned values of $E$ and $R_0$, we obtained $r = 2.81$ nm. Since the emission of BSA macromolecules occurs due to the FL of the tryptophan amino acid residues Trp212 and Trp134, the distance $r$ represents the average value of the distance of DOX molecules to the tryptophan residues.

**Fig. 8.6** Overlap of the FL $F(\lambda)$ and absorption $\xi(\lambda)$ spectra of the aqueous solutions of BSA (red curve) and DOX (black curve), respectively, with concentrations $C_{BSA} = 1 \times 10^{-5}$ M, $C_{DOX} = 1 \times 10^{-5}$ M

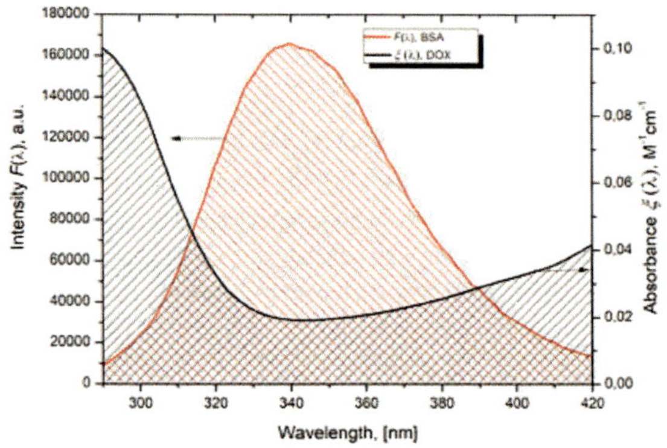


## 8.3.4 *Molecular Docking Study of the BSA-DOX Conjugates*

Figure 8.7 presents the result of modeling of the BSA-DOX system.

The tryptophan amino acid residues, which are responsible for the photoluminescent properties of BSA, are placed in different parts of the protein biomolecule. If the first amino acid residue is localized in the first domain near the surface of the BSA macromolecule, the second amino acid residue can be found in the hydrophobic pocket of the second domain of BSA. The DOX molecule is attached to the protein in the position at distances of 14.6 A and 20.4 A to the amino acid residues Trp212 and Trp134, respectively. It is localized among the polar charged amino acid residues of the biomolecule.

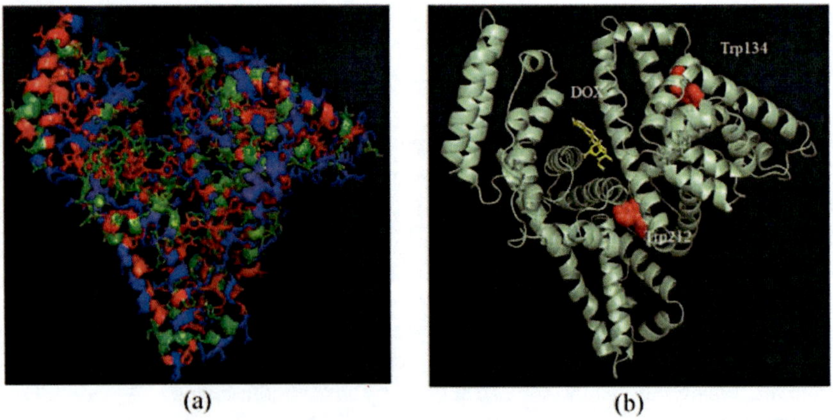


**Fig. 8.7** The hydrophobic (red), polar uncharged (green) and charged (blue) amino acid residues of the BSA biomolecule [47] **(a)** and localization of doxorubicin molecules (yellow) and tryptophan amino acid residues Trp212 and Trp134 (red) **(b)**

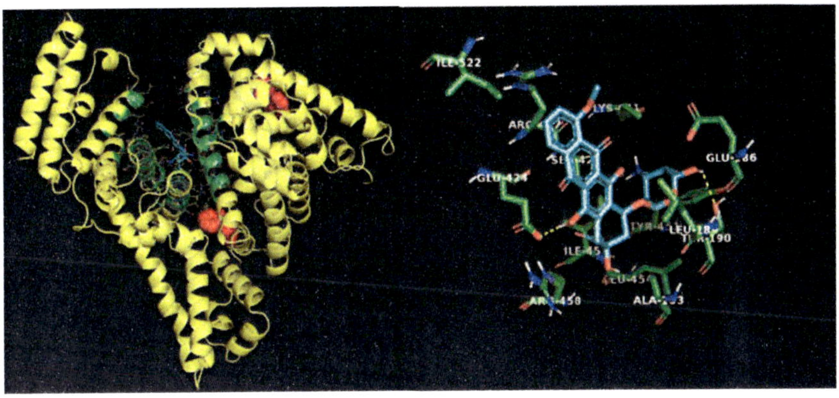


**Fig. 8.8** The 4 A—surrounding of the doxorubicin molecule in the BSA molecule [46]

The BSA amino acid neighborhood of the DOX molecule is shown in Fig. 8.8. The doxorubicin molecule is surrounded by the residues Glu-186, Leu-189, Thr-190, Ala-193, Glu-424, Lys-431, Ser-428, Ile-522, Ile-455, Arg-458, Tyr-451 and Leu-454 [10]. As followed from the docking modeling, the binding energy of the DOX molecule with the surrounding amino acid residues was equal to $-3.5 \times 10^4$ J/mol. This interaction energy can be also found from the values of the equilibrium coupling constant at different temperatures, using the study of FL quenching of the BSA-DOX aqueous solutions.

### *8.3.5 Parameters of the Interaction in BSA-DOX Conjugates*

In the case of complexation of BSA with DOX in the aqueous solutions, the maximum FL intensity droped from $F_0$ to $F$ with increasing concentration of the quencher (DOX).

According to Stern-Volmer theory [22, 23], the rate of FL reduction is determined by the quenching rate constant $K_q$ or the Stern-Volmer constant $K_{SV}$ and can be obtained from the following equation:

$$\frac{F_0}{F} = 1 + K_q \tau_0 [Q] = 1 + K_{SV}[Q], \tag{8.4}$$

where $\tau_0$ is the lifetime of the BSA molecule in the excited state, which is equal to $1 \times 10^{-8}$ s. The constants $K_q$ and $K_{SV}$ can be found from the FL spectra of an aqueous solution of pure BSA and BSA-DOX conjugates at different temperatures and concentrations of doxorubicin $C_{DOX}$ = [Q].

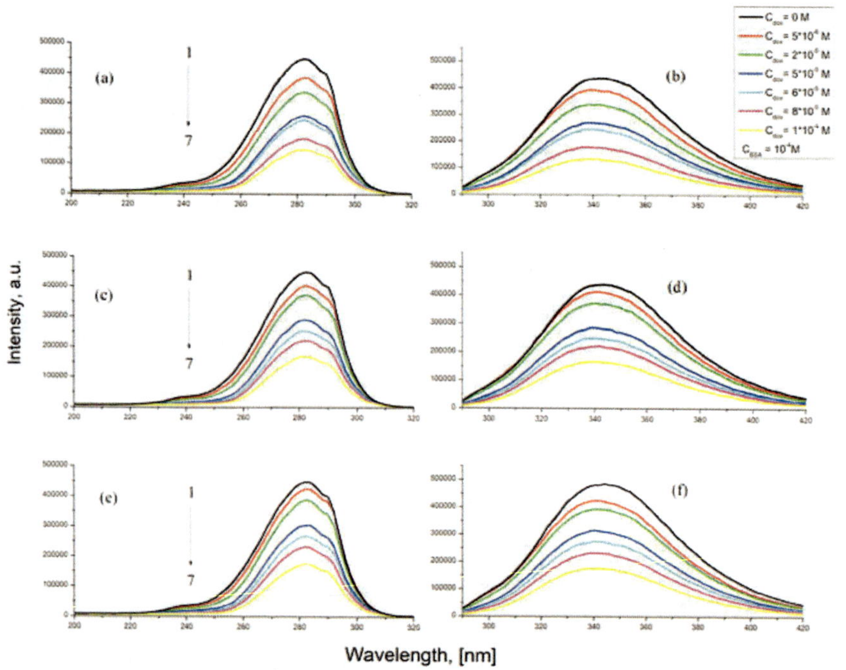


**Fig. 8.9** Excitation fluorescence spectra (a, c, e) and fluorescence emission (b, d, f) for aqueous solutions of BSA-DOX conjugates at $C_{BSA} = 10^{-4}\,\text{M} = \text{const}$ and $C_{DOX} = 0\,\text{M}(1)$; $0.5 \times 10^{-5}\,\text{M}(2)$, $3 \times 10^{-5}\,\text{M}(3)$, $5 \times 10^{-5}\,\text{M}(4)$, $6 \times 10^{-5}\,\text{M}(5)$, $8 \times 10^{-5}\,\text{M}(6)$, $10 \times 10^{-5}\,\text{M}(7)$ at temperatures $T = 293\,\text{K}$(a, b), $303\,\text{K}$(c, d), $313\,\text{K}$(d, e), $\lambda_{em} = 345\,\text{nm}$, $\lambda_{ex} = 280\,\text{nm}$

The spectra of FL emission and excitation at different temperatures are presented in Fig. 8.9. For the solutions of pure BSA and with addition of DOX, the FL bands in the emission spectra are wide enough and the positions of their maxima at all temperatures are maintained at $\lambda = 345\,\text{nm}$. In the excitation fluorescence spectra, the main absorption maximum is at about 280 nm. When DOX molecules are added to the BSA solution, the peak of intensity decreases and the position of its maximum shifts to the red region.

The experimental points of $F_0/F$ for several concentrations $C_{DOX} = [Q]$ of DOX and their linear approximations according to the Stern-Volmer Eq. (8.4) at different temperaures are depicted in Fig. 8.10. From the slope of the approximating lines we obtained the values of $K_q$ and $K_{SV} = K_q \tau_0$. The values are given in Table 8.1.

Comparing the values of $K_q$ with $K_{q,max} = 2 \times 10^{10}\text{M}^{-1}\text{s}^{-1}$, which is the maximum value of the quenching rate constant related to the intermolecular diffusion of the components, we concluded that in the BSA-DOX aqueous solution there is not a dynamic but a static mechanism of FL quenching.

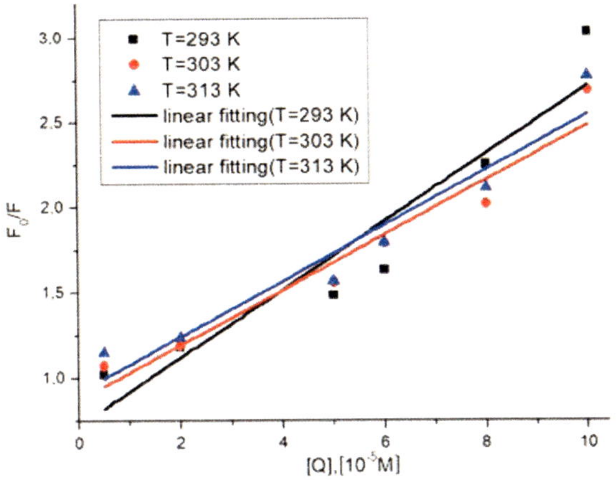


**Fig. 8.10** Stern-Volmer dependences at temperatures $T = 293$ K (■), 303 K (●), 313 K (▲)—experimental data at temperatures $T = 293$, 303, 313 K, respectively

**Table 8.1** Quenching rate constant and Stern-Volmer constant at different temperatures in an aqueous solutions of the DOX-BSA system

| $T$(K) | Extinguishing parameters | | |
|---|---|---|---|
| | Quenching rate constant $K_q\,(M^{-1}s^{-1})$ | Stern-Volmer constants $K_{SV}\,(M^{-1})$ | $R$ |
| 293 | $(1.99 \pm 0.003) \times 10^{12}$ | $(1.99 \pm 0.003) \times 10^4$ | 0.9493 |
| 303 | $(1.62 \pm 0.002) \times 10^{12}$ | $(1.62 \pm 0.002) \times 10^4$ | 0.9735 |
| 313 | $(1.63 \pm 0.0022) \times 10^{12}$ | $(1.63 \pm 0.0022) \times 10^4$ | 0.9656 |

The static mechanism implies formation of complexes of BSA and DOX molecules due to intermolecular interaction. Such an interaction can be characterized by two parameters—binding constant $K_A$ and the number $n$ of binding sites of the ligand (DOX) to the albumin macromolecule (BSA). These values can be calculated from the Hill equation [22, 23]:

$$\lg\frac{F_0}{F} = \lg K_A + n\lg[Q]. \tag{8.5}$$

Points in Fig. 8.11 represent experimental values of $\lg(F_0/F - 1)$ for different values of $\lg[Q]$ in BSA-DOX aqueous solutions at different temperatures. These data were approximated by linear Hill dependences.

The values of the $K_A$ and $n$ obtained from the linear approximation of the experimental data according to Eq. (8.5) are given in Table 8.2.

It can be seen that the interaction in BSA-DOX conjugates varies with temperature. The number of binding sites at all temperatures is approximately equal to one ($n = 1$).

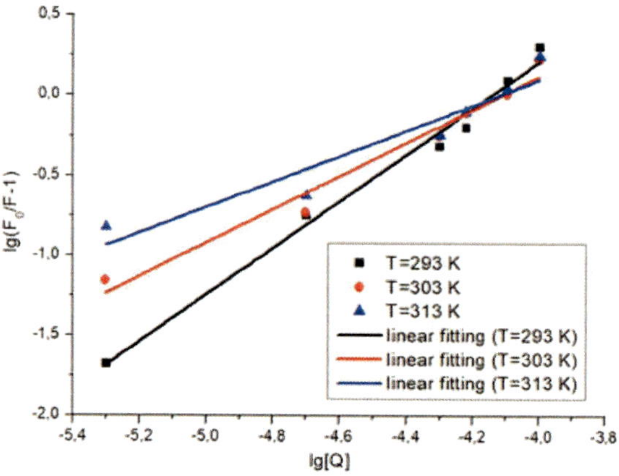


**Fig. 8.11** Hill dependences at 293 K, 303 K, 313 K (experimental data at $T =$ 293 K (■), 303 K(●), 313 K (▲), respectively, the lines correspond the Hill equation

**Table 8.2** Parameters of intermolecular interaction in BSA-DOX conjugates at different temperatures

| | Binding parameters | | |
|---|---|---|---|
| $T$(K) | $\lg K_A$ | $n$ | $R$ |
| 293 | $6.04 \pm 0.14$ | $1.45 \pm 0.076$ | 0.9945 |
| 303 | $4.28 \pm 0.19$ | $1.04 \pm 0.088$ | 0.9895 |
| 313 | $3.27 \pm 0.26$ | $0.79 \pm 0.12$ | 0.958 |

### *8.3.6 Mechanisms of Intermolecular Interaction in BSA-DOX Conjugates*

In accordance with van Hoff's theory [48], the values of the binding constants between the molecules in the complexes at each temperature allow us to calculate the changes of the thermodynamic characteristics of the solutions: enthalpy $\Delta H$, free energy $\Delta G$ and entropy $\Delta S$. The relationship between the binding constants $K_A$ and the mentioned thermodynamic characteristics is described by van Hoff's equations. Thus, for a given temperature $T$ of the solution, the change of the Gibbs potential

$$\Delta G = -RT \ln K_A, \tag{8.6}$$

where the gas constant $R = 8.314\ \mathrm{J \cdot mol^{-1} \cdot K^{-1}}$.

The change of the enthalpy at close values of inverse temperatures is assumed to be constant and can be found from the following equation:

$$\ln K_A = -\frac{\Delta H}{RT} + \frac{\Delta S}{R}. \tag{8.7}$$

The set of values of $\Delta G$ and $\Delta H$ obtained from Eq. (8.6) and Eq. (8.6) allows us to calculate the change in entropy using the wellknown relationship

**Table 8.3** Thermodynamic characteristics for the BSA-DOX aqueous solutions at different temperatures

| | Thermodynamic parameters | | |
|---|---|---|---|
| $T$(K) | $\Delta G(\times 10^3 \text{J/mol})$ | $\Delta S(\text{J} \cdot \text{mol}^{-1} \cdot \text{K}^{-1})$ | $\Delta H(\times 10^3 \text{J/mol})$ |
| 293 | $-36.0 \pm 1.9$ | $-708.8 \pm 43.0$ | |
| 303 | $-25.0 \pm 2.3$ | $-721.7 \pm 44.6$ | $-243.7 \pm 12.6$ |
| 313 | $-18.3 \pm 3.4$ | $-720.0 \pm 51.8$ | |

$$\Delta G = \Delta H - T \Delta S. \tag{8.8}$$

For the found values of the $K_A$ from Table 8.2, changes of the specified thermodynamic characteristics were calculated and are given in Table 8.3.

For the BSA-DOX conjugate, negative values of $\Delta S$ and $\Delta H$ indicate a high contribution to the interaction energy of the van der Waals forces and hydrogen bonds. At the same time, a negative value of $\Delta G$ indicates the spontaneous nature of the interaction and the presence of electrostatic forces in the intermolecular interaction [48]. The value of $\Delta G$ is close to the results of molecular docking simulation.

### 8.3.7 *Conformational Transformations of BSA Macromolecules in Their Interaction with DOX Molecules*

Figure 8.12 shows the spectra of IR transmission of aqueous solutions of DOX and BSA-DOX at different concentrations of protein component. The most interesting in the spectra is the region with bands near 1656 cm$^{-1}$ and 1543 cm$^{-1}$, which correspond to the oscillatory modes υ (C = 0) in the amides of polypeptide bonds of

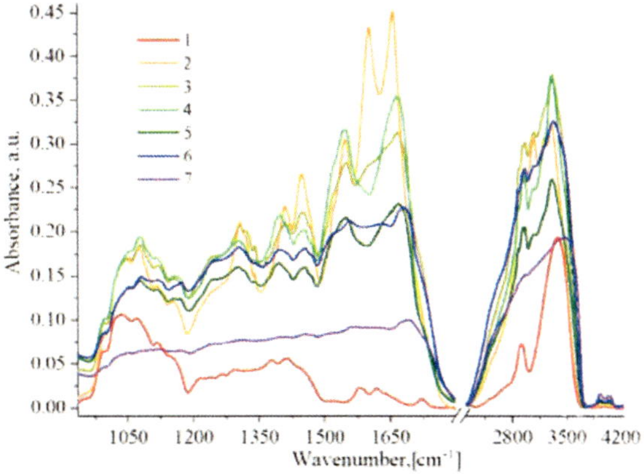


**Fig. 8.12** Fourier spectra of IR transmission of aqueous solutions of DOX molecules ($C_{DOX} = 17.7 \times 10^{-5}$ M, $C_{BSA} = 0$ M(1) and conjugates BSA-DOX $C_{BSA} = 1.77 \times 10^{-5}$ M(2), $3.54 \times 10^{-5}$ M(3), $5.31 \times 10^{-5}$ M(4), $7.08 \times 10^{-5}$ M(5), $12.4 \times 10^{-5}$ M(6), $17.7 \times 10^{-5}$ M(7)(pH = 7, $T$ = 293 K)

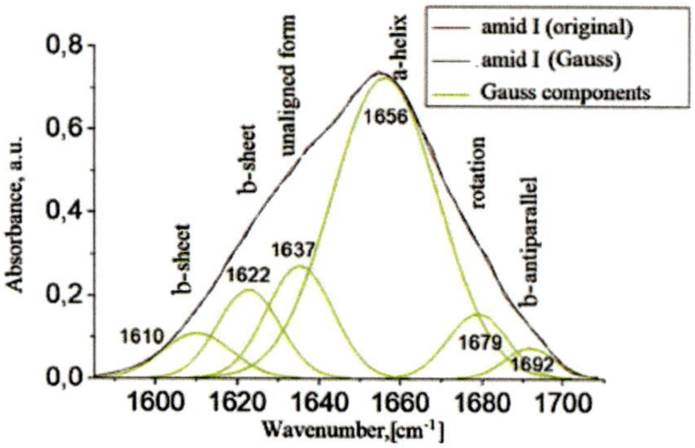


**Fig. 8.13** Fourier spectra of IR transmission for BSA-DOX aqueous solution

biomolecules. These bands are sensitive to changes in the secondary and tertiary structures of proteins. The first band (amide I) is particularly sensitive to conformational transformations of BSA macromolecules, which also occur after formation of conjugates with drug molecules due to the intermolecular interaction. By decomposing this band into components which correspond to different configurations of BSA, it becomes possible to establish a conformational rearrangement of the secondary structure of the protein biomolecules after adding DOX [23].

Fourier-IR transmission spectra in the region of the amide I band at about 1653 $cm^{-1}$ for pure BSA and BSA-DOX conjugates and its decomposition into separate components are shown in Fig. 8.13.

Since the BSA conformation, related to the a-helix, is characterized by a component of about 1656 $cm^{-1}$, its content in pure BSA is 63%; it decreases to 44% in conjugates.

The conformation of the b-sheet corresponds to a component of about 1626 $cm^{-1}$. Its contribution to pure BSA is 16%, and for the conjugate BSA-DOX it falls to 10%. The contribution of the rotational conformation, which corresponds to the component of about 1638 $cm^{-1}$, increased from 12 to 23%. The unaligned structure with a corresponding component around 1678 $cm^{-1}$ increased its contribution from 6 to 22%. The contribution of b-antiparallel at 1691 $cm^{-1}$ decreased from 3 to 1%. The shift of components maxima after adding DOX to the BSA solution is insignificant. Thus, there is a significant rearrangement of the secondary structure of protein biomolecules as a result of the intermolecular interaction in the BSA-DOX conjugate.

## 8.4 Results and Discussion for Ceftriaxone

### *8.4.1 Absorption Spectrum of Ceftriaxone*

Figure 8.14 shows experimental spectra of optical density for the aqueous solution of ceftriaxone. Let us consider the localization of the molecular orbitals responsible for the electronic transitions of the molecule in the optical absorption spectra.

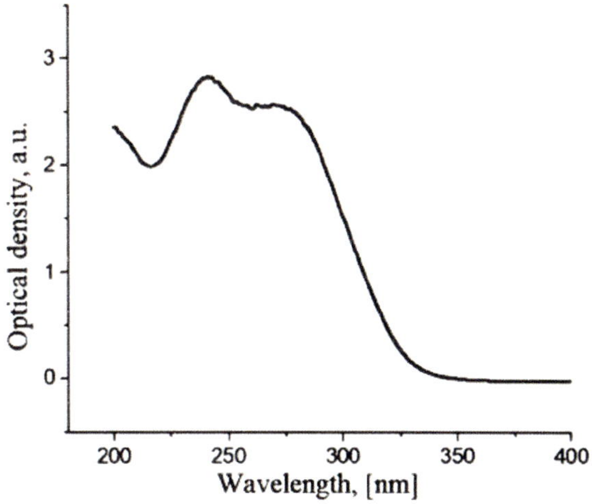


**Fig. 8.14** Optical absorption spectrum spectra of an aqueous solution ($C_{Cef} = 10\,\text{M}$) of ceftriaxone

It can be seen that the main absorption peaks are at 240 nm (more intense one) and at 277 nm. Quantum chemical calculations of electronic transitions of the ceftriaxone molecule (obtained by TD SCF, B3LYP (6-31G)) indicate that the main contribution to the intensity is provided by the electronic transition from HOMO-2 to LUMO, which are localized in the region from the phenolic ring and cover sulfur atoms (see Fig. 8.15).

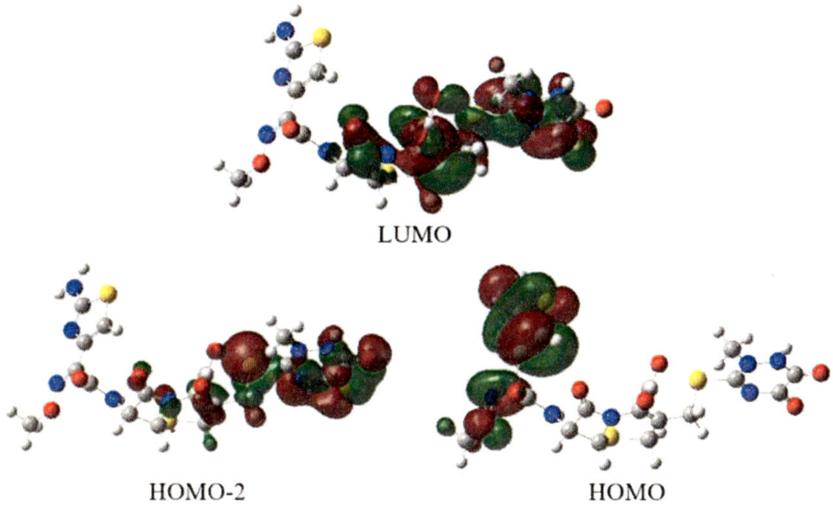


**Fig. 8.15** Molecular orbitals of ceftriaxone molecule involved in electronic transitions (TD SCF, B3LYP (6-31G))

The wavelength of this transition is 379 nm and differs significantly from the experimental data, which can happen due to the presence of solvent. The second electronic transition with a wavelength of 405 nm is the result of a transition between HOMO and LUMO orbitals.

### 8.4.2 *Molecular Docking Study of the BSA-Cef Conjugates*

Figure 8.16 shows the results of docking modeling of the molecular system BSA-Cef. It can be seen that the ceftriaxone molecule is surrounded by residues of different amino acids near the two tryptophan residues, which are responsible for the fluorescent properties of a BSA protein molecule. The Cef is at distances $r_1 = 1.18$ nm and $r_2 = 1.78$ nm to the residues Trp-212 and Trp-134, respectively. The molecule of ceftriaxone is localized among the polar charged amino acid residues. The specified environment within a distance of 0.4 nm consists of amino acid residues Pro-146, His-145, Arg-144, Asp-108, Ser-192, Len-189, Glu-186, Lys-187, Thr-190, Ala-193, Arg-196, Arg-458, Glu-424, Len-454, Ser-428, Lys-431, Val-432, Thr-434, Tyr-451, Arg-435 (see Fig. 8.17). Calculated binding energy of ceftriaxone molecule with the listed amino acid residues is $E = -2.1 \times 10^4$ J/mol.

The considered interaction makes possible the resonant excitation energy transfer from tryptophan amino acid residues as a donor to the antibiotic molecule as an acceptor.

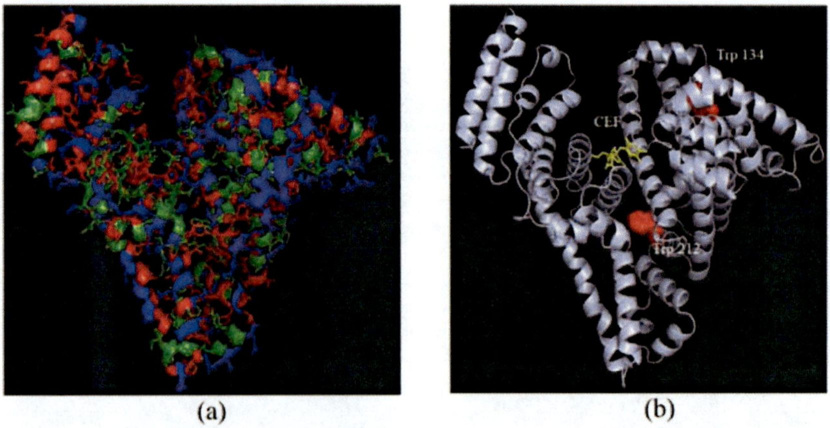


**Fig. 8.16** Hydrophobic (red), polar uncharged (green) and charged (blue) amino acid residues of the BSA biomolecule [47] (**a**) and localization of ceftriaxone (yellow) and tryptophan amino acid residues Trp-134, Trp-212 in the protein (**b**)

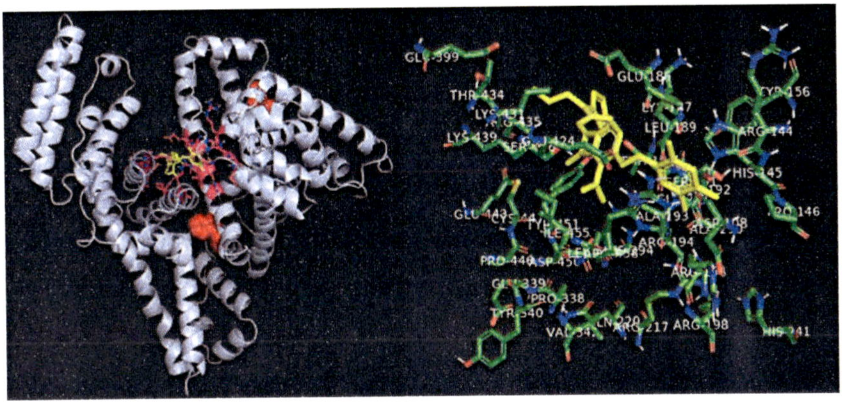

Fig. 8.17 The closest environment of the amino acid residues of BSA biomolecule around Cef [47]

### 8.4.3 *Nonradiative Transfer of Excitation Energy in the BSA-Cef Molecular System*

The conditions for energy transfering from the donor to the acceptor molecule are determined similarly to the previous case using Forster's theory [45, 46]. Figure 8.18 represents the overlap of the FL spectra $F(\lambda)$ of aqueous solutions of BSA and the molar extinction coefficient $\xi(\lambda)$ of Cef. The value of the overlap integral $J$ in accordance with Eq. (8.3) is equal to $0.88 \times 10^{-16}\,\text{cm}^3\text{M}^{-1}$. The distance $R_0 = 2.2\,\text{nm}$ was obtained from Eq. (8.2). The values of FL intensities $F_0$ and $F$ were

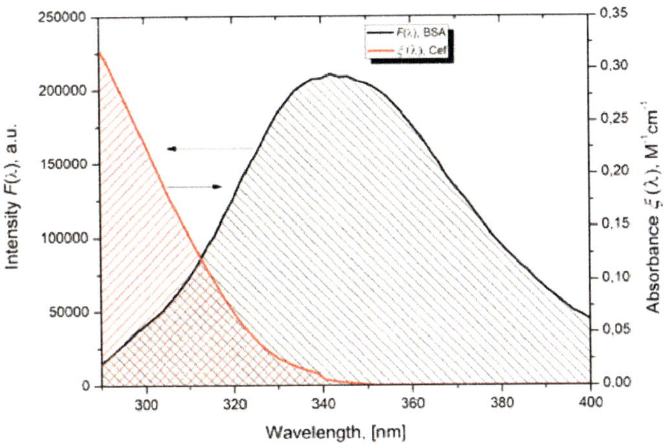


Fig. 8.18 Fluorescence spectra $F(\lambda)$ and the molecular absorption coefficient $\xi(\lambda)$ in the region of their overlap for aqueous solutions of BSA ($C_{BSA} = 5 \times 10^{-7}$ M) and ceftriaxone ($C_{Cef} = 1.2 \times 10^{-5}$ M), respectively

measured for aqueous solutions of pure BSA ($C_{BSA} = 5 \times 10^{-7}$ M = const) and its mixtures with ceftriaxone ($C_{Cef} = 1.2 \times 10^{-5}$ M) respectively, thus, using Eq. (8.1), the rate of energy transfer between the donor and the acceptor $E = 0.114$.

This degree of energy transfer is provided when the average distance $r$ from the amino acid residues of tryptophan Trp-212 and Trp-134 to Cef molecule is 3.5 nm.

### *8.4.4 Parameters of Intermolecular Interaction in the BSA-Cef System*

Under Forster's conditions, the FL quenching of BSA can be expected after adding ceftriaxone to the protein aqueous solution. As a result of the excitation energy transfer from BSA to Cef, the quenching of the FL increased as the concentration of antibiotic in the aqueous solution grows. At the same time, the quenching rate constant for increasing content of ceftriaxone molecules depends on the conditions of BSA-Cef complexation in the solution.

Figure 8.19 shows the spectra of FL emission and FL excitation of BSA-Cef

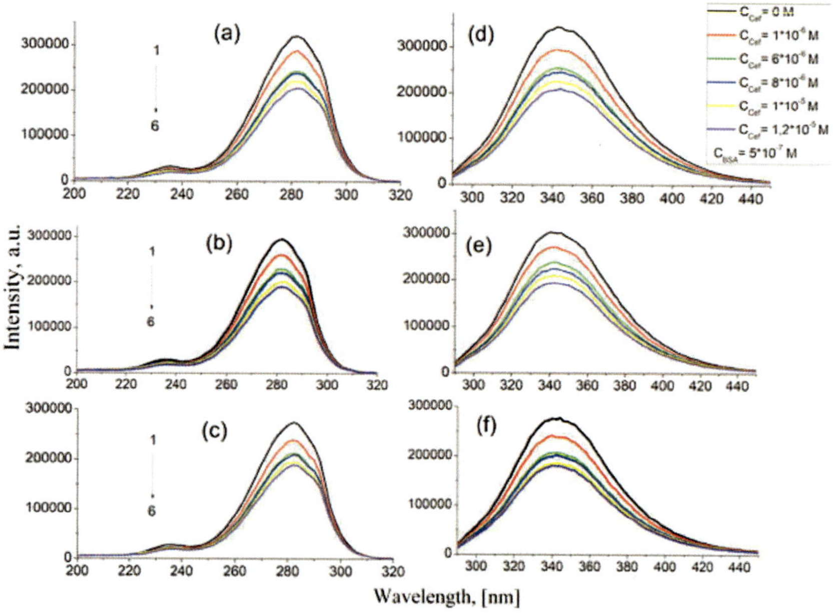


**Fig. 8.19** Excitation (a, b, c) and emission (g, d, e) spectra of fluorescence for aqueous solutions of the BSA-Cef mixture at the concentration of BSA molecules $C_{BSA} = 5 \times 10^{-7}$ M = const and the content of ceftriaxone molecules $C_{Cef} = 0$ M(1); $1 \times 10^{-6}$ M(2), $6 \times 10^{-6}$ M(3), $8 \times 10^{-6}$ M(4), $10 \times 10^{-6}$ M(5), $12 \times 10^{-6}$ M(6) at temperatures $T = 296$ K(a, d), 303 K(b, e), 313 K(c, f)($\lambda_{em} = 345$ nm, $\lambda_{ex} = 280$ nm)

aqueous solutions at different temperatures. The FL emission bands for all solutions and temperatures remain wide without an appreciable shift of the maxima ($\lambda_{em}$ = 345 nm). In the FL excitation spectra, an intense absorption peak $\lambda_{ex}$ = 280 nm is observed. After adding of the antibiotic to the aqueous solution of BSA, the maximum of the absorption band is shifted towards greater wavelengths.

Figure 8.20 demonstrates the experimental values of $F_0/F$, taken from Fig. 8.19, as a function of the concentration $C_{Cef}$ of ceftriaxone in an aqueous solution of BSA and linear approximation of these dependences at different temperatures using the Stern-Volmer Eq. (8.4).

The constants $K_q$ and $K_{SV} = K_q\tau_0$ were determined from the slope of the obtained approximation lines. It can be seen that at all temperatures the FL quenching rate $K_q$ exceeds its maximum value $K_{q,max} = 2 \times 10^{10}\,M^{-1}s^{-1}$ (see Table 8.4) which is the characteristic of the dynamic mechanism of molecular interaction i.e. their collisions in the diffusion process. This fact indicates the implementation of the static mechanism of molecular interaction in the BSA-Cef system with formation of complexes which were analyzed before in the molecular docking study.

The quenching of the FL spectra allowed us to establish the values of equilibrium binding constants $K_A$ and number $n$ of binding sites of the ligand molecule (Cef)

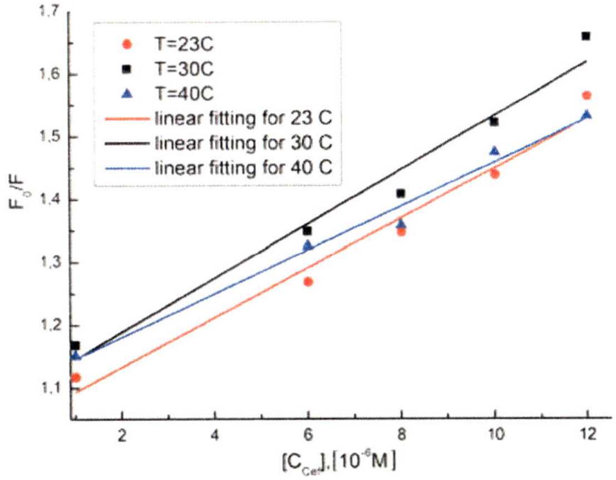


**Fig. 8.20** Dependences of relative intensities of FL emission on concentrations of ceftriaxone (Stern-Volmer dependence) in the aqueous solutions of BSA—Cef ($C_{BSA} = 5 \times 10^{-7}$ M = const) at temperatures 296 K (1), 303 K (2), 313 K (3) (●, ▲, ■)—experimental points, 1, 2, 3—Stern-Volmer fitting lines for temperatures 296, 303, 313 K, respectively

**Table 8.4** FL quenching rate constant $K_q$ and $K_{SV}$ in aqueous solutions of BSA-Cef at different temperatures

| $T$, K | Quenching parameters | | $R$ |
|---|---|---|---|
| | Quenching rate constant $K_q\,(M^{-1}s^{-1})$ | Stern–Volmer constants $K_{SV}\,(M^{-1})$ | |
| 296 | $(3.95 \pm 0.37) \times 10^{12}$ | $(3.95 \pm 0.37) \times 10^4$ | 0.9858 |
| 303 | $(4.29 \pm 0.42) \times 10^{12}$ | $(4.29 \pm 0.42) \times 10^4$ | 0.9868 |
| 313 | $(3.47 \pm 0.24) \times 10^{12}$ | $(3.47 \pm 0.24) \times 10^4$ | 0.9928 |

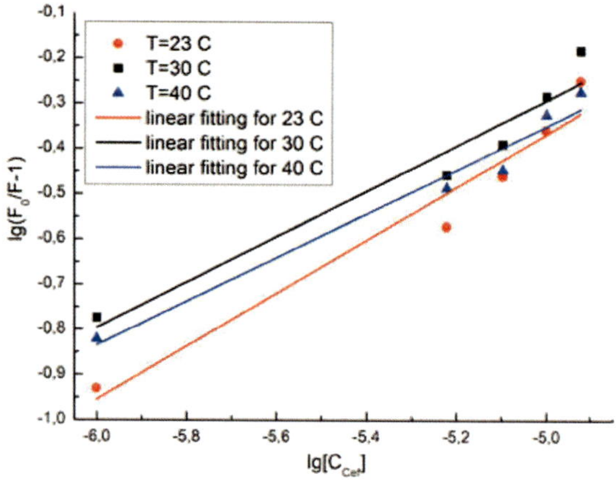


**Fig. 8.21** Hill's dependences of FL relative intensities on the concentration of Cef in the aqueous solutions of BSA-Cef ($C_{BSA} = 5 \times 10^{-7}$ M = const) at temperatures of 296 K (1), 303 K (2), 313 K (3) (●, ▲, ■—experimental points, 1, 2, 3—fitting Hill lines for temperatures 296, 303, 313 K, respectively)

**Table 8.5** Binding parameters in the aqueous solutions of BSA-Cef mixture at different temperatures

| T, K | Binding parameters | | $R$ |
|---|---|---|---|
| | $\lg K_A$ | $n$ | |
| 296 | $2.57 \pm 0.19$ | $0.59 \pm 0.07$ | 0.9769 |
| 303 | $2.22 \pm 0.16$ | $0.5 \pm 0.068$ | 0.9731 |
| 313 | $2.07 \pm 0.06$ | $0.49 \pm 0.05$ | 0.9850 |

to the protein macromolecule. These parameters of complexation in the BSA-Cef conjugates, similarly to the case with DOX, were calculated according to Hill's theory (see Eq. 8.5).

Figure 8.21 represents experimental values of $\lg\left(\frac{F_0}{F} - 1\right)$ for different concentrations $\left[C_{Cef}\right]$ at different temperatures, as well as their linear approximation in accordance with Hill's equations. The obtained values of $K_A$ and $n$ are given in Table 8.5.

Unlike other molecular systems based on BSA, values of the binding constant $K_A$ and $n$ are much smaller. With the rise of temperature, these values decrease monotonically. The number of binding sites $n$ is close to 0.5 at all the temperatures.

### 8.4.5 *Mechanisms of Intermolecular Interaction in the Aqueous Solutions of BSA-Cef*

According to van Hoff's theory [48], the values of binding constant $K_A$ yield the changes of the thermodynamic parameters of an intermolecular interaction: enthalpy $\Delta H$, free enthalpy $\Delta G$, entropy $\Delta S$, which allow to establish the nature of the interaction forces. These changes are caused by the formation of heteroassociations

**Table 8.6** Thermodynamic characteristics for the BSA-Cef aqueous solutions at different temperatures

| T, K | Thermodynamic characteristics | | |
|---|---|---|---|
| | $\Delta G(\times 10^3$ J/mol) | $\Delta H(\times 10^3$ J / mol) | $\Delta S$(J × mol$^{-1}$ K$^{-1}$) |
| 296 | $-14.55 \pm 1.20$ | $-49.66 \pm 5.50$ | $-118.63 \pm 18.20$ |
| 303 | $-12.93 \pm 1.12$ | | $-121.22 \pm 15.30$ |
| 313 | $-12.45 \pm 1.09$ | | $-118.87 \pm 12.80$ |

between BSA and Cef molecules. The thermodynamic parameters for the system BSA-Cef are given in Table 8.6.

The found values of $\Delta H$ and $\Delta S$ for BSA-Cef complexes are negative, which indicates the emergence of hydrophobic interactions in the heteroassociates. The values of $\Delta G$ are negative, which also indicates the spontaneous character of the interaction in the complexes. However, changes of thermodynamic parameters are insignificant which is associated with small values of the interaction constant $K_A$.

## 8.5 Conclusions

From the quantum chemical modeling of electronic structure and localization of molecular orbitals, it was found out that a doxorubicin molecule has optical transitions in UV and visible optical regions, which is well consistent with the positions of the maximum of the electronic transition band around 481 nm. The calculated and experimental spectra of Raman scattering and IR transmission were obtained, which result in the presence of intense bands and breathing modes of the six-element rings of the molecule and the sugar residue. Intense FL bands with maxima around 650 nm were observed for DOX molecules. Taking into account the the FL of BSA molecules at about 345 nm, the parameters of Föster's theory of the resonance transfer of excitation energy from protein to DOX molecules were calculated yielding the distance 2.81 nm, which is consistent with molecular docking results for the position of DOX molecule with respect to the tryptophan amino acid residues. The closest environment of amino acid residues around DOX molecules was established and the interaction energy between these residues and the molecule $E = -3.5 \times 10^4$ J/mol was calculated.

Analyzing the FL quenching of BSA aqueous solution with the addition of DOX at different temperatures, the presence of a static mechanism of intermolecular interaction in the BSA-DOX system with formation of complexes war established. Using Hill's theory, the values of the binding constants $K_A$ and the number of binding sites $n$ in the BSA-DOX conjugate were determined. The value of the $K_A$ monotonically fell from $1.1 \times 10^6$ to $1.86 \times 10^3$ M$^{-1}$ with temperature increase from 293 to 313 K. The value of $n$ remains close to 1. The values of the thermodynamic parameters obtained for BSA-DOX conjugates indicates the predominant van der Waals forces and the presence of hydrogen bonds in the intermolecular interaction. The addition of DOX to the BSA aqueous solution results in significant conformational rearrangement of the polymer biomolecules. At the same time the change of contributions of separate components of protein secondary structure was underlined.

The quantum chemical calculation of the vibration spectrum for Cef molecule showed several regions that can be associated with the vibrations of the following functional groups: β-lactam, triazipyl and thiazine cycles, methyl and methylene, thiomethyl. That is in good agreement with the experimentally obtained spectrum, the exception is for the high-frequency region associated with N-H, O-H groups. The molecular docking simulation demonstrated that the Cef molecule in the BSA-Cef complex is located in a certain environment of amino acid residues at distances 1.18 and 1.78 nm to the tryptophan amino acid residues Trp-212 and Trp-134, respectively. The interaction energy $E$ in the BSA-Cef complex is $-2.1 \times 10^4$ J/mol.

From Forster's theory it was established that in BSA-Cef heteroassociations nonradiative energy transfer is realized at distances $r = 3.5$ nm and $R_0 = 2.2$ nm with energy transfer efficiency $E = 0.114$; overlap integral of FL spectra of donor and absorption spectra of acceptor $J = 0.88 \times 10^{-16}$ cm$^3$M$^{-1}$. Using Stern-Volmer's and Hill's theories, it was shown that the interaction in BSA-Cef complexes occurs due to a static mechanism. The values of binding constant $K_A$ at temperature range 296–313 K were quite small (approximately 200 M$^{-1}$) in comparison with other complexes of BSA with medicines. The interaction in the BSA-Cef heteroassociations has spontaneous hydrophobic character with insignificant energy, which results in small values of $K_A$.

## References

1. Keizer HG, Pinedo HM, Schuurhuis GJ, Joenje H (1990) Doxorubicin (adriamycin): a critical review of free radical-dependent mechanisms of cytotoxicity. Pharmacol Ther 47:219–231. https://doi.org/10.1016/0163-7258(90)90088-J
2. Tritton TR (1991) Cell surface actions of adriamycin. Pharmacol Ther 49:293–309. https://doi.org/10.1016/0163-7258(91)90060-Y
3. Dezhenkova LG, Tsvetkov VB, Shtil AA (2014) Topoisomerase I and II inhibitors: chemical structure, mechanisms of action and role in cancer chemotherapy. Russ Chem Rev 83:82. https://doi.org/10.1070/RC2014v083n01ABEH004363
4. Bijie H, Kulpradist S, Manalaysay M, Soebandrio A (2005) In vitro activity, pharmacokinetics, clinical efficacy, safety and pharmacoeconomics of ceftriaxone compared with third and fourth generation cephalosporins: review. J Chemothwe 17:3–24
5. Duncan CJ, Barr DA, Seaton RA (2012) Outpatient parenteral antimicrobial therapy with ceftriaxone, a review. Int J Chin Drug Invest 34:410–417. https://doi.org/10.1007/s11096-012-9637-z
6. Pozur VV, Skivka LM, Potebnya GP (2008) Staphylococcus aureus peptidoglycan and its immunobiological properties. Biopolymers Cell 24:3–13. https://doi.org/10.7124/bc.00078A
7. Anacona JR, Rodriguez A (2005) Synthesis and antibacterial activity of ceftriaxone metal complexes. Trans Met Chem 30:897–901. https://doi.org/10.1007/s11243-005-6219-0
8. Ingle AP, Duran N, Rai M (2014) Broad-spectrum bioactivities of silver nanoparticles: the emerging trends and future prospects. Appl Microbiol Biotechnol 98:1001–1009. https://doi.org/10.1007/s00253-013-5473-x

9. Lukhin AO, Novikoa GV, Kuzubov AA, Staloverova NA, Sarmatova NI, Varganov SA, Krasnov PO (2014) A complex of ceftriaxone with Pb (II): synthesis, characterization, and antibacterial activity study. Russ J Coord Chem 67(16):2783–2794
10. Studer AM, Limbach LK, Van Duc L, Krumeich F, Athanassiou EK, Gerber LC, Moch H, Stark WJ (2010) Nanoparticle cytotoxicity depends on intracellular solubility: comparison of stabilized copper metal and degradable copper oxide nanoparticles. Toxicol Lett 197:169–174. https://doi.org/10.1016/j.toxlet.2010.05.012
11. Fayaz AM, Girilal M, Mandy SA, Somsundar SS, Venkatesan R, Kalaichelvan PT (2011) Vancomycin bound biogenic gold nanoparticles: a different perspective for development of anti VRSA agents. Process Biochem 46:636–641. https://doi.org/10.1016/j.procbio.2010.11.001
12. Singh R, Smitha MS, Singh SP (2014) The role of nanotechnology in combating multi-drug resistant bacteria. J Nanosci Nanotechnol 14:4745–4756
13. Shah MR, Ali S, Ateeq M, Perveen S, Ahmed S, Ali M, Bertino MF (2014) Morphological analysis of the antimicrobial action of silver and gold New nanoparticles stabilized with ceftriaxone on Escherichia coli using atomic force microscopy. J Chem 38:5633–5640. https://doi.org/10.1039/c4nj00751d
14. Simonov PV, Kovalova DO, Busko TO, Dmitrenko OP, Kulish MP, Zabolotny MA, Kuzmenko MO, Chekman IS (2016) An interaction of the copper nanoparticles-ceftriaxone conjugate with protein, lipid, and carbohydrate components of bacterial cell envelope. Rep NAS Ukraine 5:125–131
15. Tevyashova AN (2014) Creation of prodrugs based on anthracycline antibiotics. Bull MITHT 6:11–25
16. Vyas D, Lopez-Hisijos N, Sulakshana Gandhi M, El-Dakdouki MD, Basson MF, Walsh XH, Vyas AK, Chaturvedi LS (2015) Doxorubicin-hyaluronan conjugated super-paramagnetic iron oxide nanoparticles (DOX-HA-SPION) enhanced cytoplasmic uptake of doxorubicin and modulated apoptosis, IL-6 release and NF-kappaB activity in human MDA-MB-231 breast cancer cells. J Nanosci Nanotechnol 15:6413. https://doi.org/10.1166/jnn.2015.10834
17. Andrýsková P, Šišková KM, Michetschlägerová S, Jiráková K, Kubala M, Jirák D (2020) The effect of fatty acids and BSA purity on synthesis and properties of fluorescent gold nanoclusters. Nanomaterials 10(2):343. https://doi.org/10.3390/nano10020343
18. Goncharenko NA, Dmytrenko OP, Pavlenko OL, Kulish MP, Pundyk IP, Lesyuk AI, Busko TO, Lopatynskyy AM, Chegel VI, Lytvyn VK, Kaniuk MI (2020) Complexation peculiarities in “doxorubicin–bovine serum albumin–gold nanoparticles” heterosystem. The fluorescence study. Ukr J Phys 65:468–475. https://doi.org/10.15407/ujpe65.6.468
19. Vuignier K, Schappler J, Veuthey JL, Carrupt PA, Martel S (2010) Drug–protein binding: a critical review of analytical tools. Anal Bioanal Chem 398:53–66. https://doi.org/10.1007/s00216-010-3737-1
20. Goncharenko NA, Pavlenko OL, Dmytrenko OP, Kulish MP, Lopatynskyi AM, Chegel VI (2019) Gold nanoparticles as a factor of influence on doxorubicin–bovine serum albumin complex. Appl Nanosci (Switzerland) 9:825–833. https://doi.org/10.1007/s13204-018-0748-2
21. Bulavin LA, Goncharenko NA, Dmytrenko OP, Pavlenko OL, Kulish MP, Goncharova O, Demydov PV, Lopatynskyi AM, Chegel VI (2019) Heteroassociation of antitumor agent doxorubicin with bovine serum albumin in the presence of gold nanoparticles. J Mol Liq 284:633–638. https://doi.org/10.1016/j.molliq.2019.04.017
22. Bi S, Sun Y, Qiao C, Zhang H, Liu C (2009) Binding of several anti-tumor drugs to bovine serum albumin: fluorescence study. J Lumin 129(5):541–547. https://doi.org/10.1016/j.jlumin.2008.12.010
23. Daniel A, Philippe B, Julie B, Gervais B, Éric A, Heidar-Ali T-R (2012) Probing the binding sites of antibiotic drugs doxorubicin and N-(trifluoroacetyl) doxorubicin with human and bovine serum albumins. Public Libr Sci 7(8):e43814. https://doi.org/10.1371/journal.pone.0043814
24. Yue Q, Shen T, Wang C, Gao C, Lui J (2012) Study on the interaction of bovine serum albumin with ceftriaxone and the inhibition effect of zinc (II). Int J Spectros 284173:1–9. https://doi.org/10.1155/2012/284173

25. Pon J, Yc Z, Cai X, Wang L, Cao Z (2012) Biophysical study on the interaction of ceftriaxone sodium with bovine serum albumin using spectroscopic methods. J Biochem Mol Toxicol 26(12):487–492. https://doi.org/10.1002/jbt.21446
26. Wu J, Wei R, Wang H, Li T, Ren W (2013) Underlying the mechanism of vancomycin and human serum albumin interaction: a biophysical study. J Biochim Mol Toxial 27(10):463–70. https://doi.org/10.1002/jbt.21511
27. Abu Teir M, Ghithan JH, Abu-Taha MI, Darwish SM, Abu-hadid MM (2014) Spectroscopic approach of the interaction study of ceftriaxone and human serum albumin. J Biophys Struct Biol 6(1):1–12. https://doi.org/10.5897/JBSB2013.0045
28. Naik PN, Chimatadar SA, Nandibewoor ST (2009) Study on the interaction between antibacterial drug and bovine serum albumin: a spectroscopic approach. Spectrochimica acta. Part A Mol Biomol Spectrosc 73(5):841–845. https://doi.org/10.1016/j.saa.2009.04.018
29. Naik PN, Chimatadar SA, Nandibewoor ST (2010) Interaction between a potent corticosteroid drug - dexamethasone with bovine serum albumin and human serum albumin: a fluorescence quenching and Fourier transformation infrared spectroscopy study. J Photochem Photobiol B 100(3):147–159. https://doi.org/10.1016/j.jphotobiol.2010.05.014
30. Bani-Yaseen AD (2011) Spectrofluorimetric study on the interaction between antimicrobial drug sulfamethazine and bovine serum albumin. J Lumin 131:1042–1047. https://doi.org/10.1016/j.jlumin.2011.01.019
31. Egorova AV, Maltsev GV, Skripinets YuV, Antonovich VP (2018) Spectroscopic study of the interaction of iglmeacin and daclatasvir dihydrochloride with human serum albumin. Bull Odessa Natl Univ 23(1):96–108
32. Goncharenko NA, Dmytrenko OP, Kulish MP, Pavlenko OL, Lesiuk AI, Busko TO, Pundyk IP, Pinchuk-Rugal TM, Chegel VI, Lopatynskyi AM, Kanyuk MI, Denis LV (2020) Mechanisms of the interaction of bovine serum albumin with anticancer drug gemcitabine. Mol Cryst Liq Cryst 701(1):59–71. https://doi.org/10.1080/15421406.2020.1732563
33. Meti MD, Nandibewoor ST, Joshi SD, More UA, Chimatadar SA (2015) Multi-spectroscopic investigation of the binding interaction of fosfomycin with bovine serum albumin. J Pharm Anal 5(4):249–255. https://doi.org/10.1016/j.jpha.2015.01.004
34. Durgannavar AK, Patgar MB, Nandibewoor ST, Chimatadar SA (2016) Fluorescent bovine serum albumin interacting with the antitussive quencher dextromethorphan: a spectroscopic insight . Luminescence 31(3):843–50. https://doi.org/10.1002/bio.3040
35. Magdum PA, Gokavi NM, Nandibewoor ST (2017) Study on the interaction between anti-tuberculosis drug ethambutol and bovine serum albumin: multispectroscopic and cyclic voltammetric approaches. Luminescence 32(2):206–216. https://doi.org/10.1002/bio.3169
36. Trott O, Olson AJ (2010) AutoDock Vina: improving the speed and accuracy of docking with a new scoring function, efficient optimization, and multithreading. J Comput Chem 31:455–461. https://doi.org/10.1002/jcc.21334
37. Sturgeon RJ, Schulman SG (1977) Electronic absorption spectra and protolytic equilibria of doxorubicin: direct spectrophotometric determination of microconstants. J Pharm Sci 66(7):958. https://doi.org/10.1002/jps.2600660714
38. Das G, Nicastri A, Coluccio ML, Gentile F, Candeloro P, Cojoc G, Liberale C, De Angelis F, Di Fabrizio E (2010) FT-IR, Raman, RRS measurements and DFT calculation for doxorubicin microsc. Res Tech 73:991–995. https://doi.org/10.1002/jemt.20849
39. Nabiev IR, Morjani H, Manfait M (1991) Selective analysis of antitumor drug interaction with living cancer cells as probed by surface-enhanced Raman spectroscopy. Eur Biophys J 19(6):311–316. https://doi.org/10.1007/BF00183320
40. Strekal N, German A, Gachko G, Maskevich A, Maskevich S (2001) The study of the doxorubicin adsorbed onto chemically modified silver films by surface-enhanced spectroscopy. J Mol Struct 563–564:183–191. https://doi.org/10.1016/S0022-2860(01)00512-9
41. Eliasson C, Lorén A, Murty KVGK, Josefson M, Käll M, Abrahamsson J, Abrahamsson K (2001) Multivariate evaluation of doxorubicin surface-enhanced Raman spectra. Spectrochim Acta - Part A Mol Biomol Spectrosc 57(9):1907–1915. https://doi.org/10.1016/S1386-1425(01)00453-X

42. Guo J, Cai W, Du B, Qian M, Sun Z (2009) Raman spectroscopic investigation on the interaction of malignanthepatocytes with doxorubicin. Biophys Chem 140(1–3):57–61. https://doi.org/10.1016/j.bpc.2008.11.005
43. Das G, Nicastri A, Coluccio ML, Gentile F, Candeloro P, Cojoc G, Liberale C, Angelis FD, Fabrizio ED (2010) Microsc Res Tech 73(10):991–995. https://doi.org/10.1002/jemt.20849
44. Kayal S, Ramanujan RV (2010) Doxorubicin loaded PVA coated iron oxide nanoparticles for targeted drug delivery. Mater Sci Eng C 30(3):484–490. https://doi.org/10.1016/j.msec.2010.01.006
45. Zhou XM, Lü WJ, Su L, Shan ZJ, Chen XG (2012) Binding of phthalate plasticizers to human serum albumin in vitro: a multispectroscopic approach and molecular modeling. Agric J Food Chem 60:1135. https://doi.org/10.1021/jf204380r
46. Shen H, Gu Z, Jian K, Qi J (2013) In vitro study on the binding of gemcitabine to bovine serum albumin. J Pharm Biomed Anal 75:86–93. https://doi.org/10.1016/j.jpba.2012.11.021
47. Bujacz A, Bujacz G (2012) Crystal structure of bovine serum albumin. Acta Crystallogr D Biol Crystallogr 68:1278–1289. https://doi.org/10.1107/S0907444912027047. PDB ID: 4F5S
48. Markarian SA, Aznauryan MG (2012) Study on the interaction between isoniazid and bovine serum albumin by fluorescence spectroscopy: the effect of dimethylsulfoxide. Mol Biol Rep 39:7559. https://doi.org/10.1007/s11033-012-1590-3

# Chapter 9
# Manifestation of Structural Isomerism in the Raman Spectra of Platinum Drugs

**Leonid Bulavin, Victor Gubanov, Dmytro Gryn, and Antonina Naumenko**

**Abstract** This study reports a symmetry group analysis and constructing of the normal vibration forms of two stereoisomorphs of diamminechloridoplatinum (II) cis-[Pt($NH_3$)$_2$$Cl_2$] (cisplatin) and trans-[Pt($NH_3$)$_2$$Cl_2$] (transplatin). The Raman spectra of these compounds have been studied under 488 and 785 $cm^{-1}$ lazer excitation at room temperature. It is proved that the structural isomerism of platinum drugs, namely cisplatin and transplatin, is indeed observed in Raman spectra. The assignments of the obtained bands have been done.

## 9.1 Introduction

Today, chemotherapy remains one of the main, and sometimes the only method of treating patients with malignant tumors. The potential of metal complexes in cancer therapy is of great interest because metals exhibit unique characteristics such as redox activity, variable coordination regimes, and reactivity to an organic substrate. These properties are very attractive in the construction of metal complexes that selectively bind to a biomolecular target with the resulting change in the cellular mechanism of proliferation. Metal complexes and metal-based compounds have the ability to coordinate ligands in a three-dimensional configuration, thereby allowing the creation of groups that will correspond to a particular target molecule. The anticancer action of cisplatin (cis-[Pt(II)($NH_3$)$_2$$Cl_2$]) was discovered in 1960 [1], and in 1978 the substance was approved for use as an antitumor drug [2], thus paving the way for the search for new metal-containing compounds for the treatment of cancer. Even today, platinum drugs are the most commonly used drug in chemotherapy [3, 4], and the actual standard for evaluating the effectiveness of other drugs [5].

L. Bulavin · V. Gubanov · D. Gryn · A. Naumenko (✉)
Faculty of Physics, Taras Shevchenko National University, 64/13, Volodymyrs'ka str. , Kiev 01601, Ukraine
e-mail: a_naumenko@univ.kiev.ua

D. Gryn
e-mail: glizz@ukr.net



L. Bulavin and N. Lebovka (eds.), *Soft Matter Systems for Biomedical Applications*, Springer Proceedings in Physics 266, https://doi.org/10.1007/978-3-030-80924-9_9

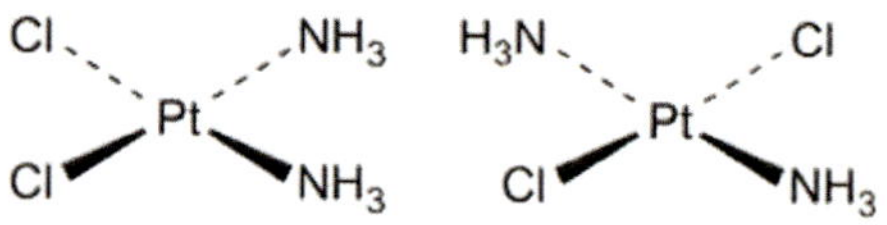


**Fig. 9.1** The structural formulas of cisplatin (left) and transplatin (right)

At the same time, cisplatin has an interesting structural feature–a structural trans-isomer trans-[Pt(II)$(NH_3)_2Cl_2$] [6]. For a long time, transplatin, unlike cisplatin, was considered an ineffective antitumor drug, although it has some advantages over the former, such as good water solubility, or enhances the effect of other drugs used in chemotherapy [7]. Therefore, against the background of numerous experimental and theoretical studies of cisplatin, its transisomer transplatin looks insufficiently studied.

The goal of the present study is to discover the vibrational properties of two isomorphs of Pt(II) cisplatin and transplatin using powerful methods of the symmetry groups and Raman spectroscopy.

## 9.2 Experimental Details

The chemical compounds $Pt(NH_3)_2Cl_2$ – cisplatin and its isomer transplatin, the structural formulas of which are shown in Fig. 9.1, have been investigated. These chemicals are powders of grade A (European Pharmacopoeia (EP) Reference Standard for cisplatin and United States Pharmacopeia (USP) Reference Standard for transplatinum) supplied by Sigma Aldrich.

The Raman measurements were carried out in a quasi-backscattering geometry at room temperature using the Horiba Jobin-Yvon T64000 triple spectrometer with integrated micro-Raman setup–Olympus BX-41 microscope equipped with a motorized XYZ stage and Peltier-cooled CCD detector. Confocal micro-Raman spectroscopy is a very useful technique allowing nondestructive studies of the structure and electronic properties. Samples were excited by $Ar^+$ laser (488 nm, 100 mW) and wavelength stabilized iBeam smart WS diode laser (785 nm, 12 mW) in the region 100–3400 $cm^{-1}$.

## 9.3 Structure and Vibrational Analysis

As is well known, the compounds of the divalent Pt are characterized by the planar structure of the complexes, regardless of their composition. Structural studies have shown that the square coordination is preserved in complete "purity" and in the crystalline state [8].

The crystal structure of transplatin firstly was reported by Porai-Koshits [9]. The determined Hermann-Mauguin symmetry space group was P1 21/a 1. There are two molecules in a monoclinic unit cell with the platinum atoms at 0,0,0 and ½,½, ½. Later, these data were comfirmed in [10] at 120 K and in [11] at 300 and 120 K.

Since the first report on the compound in 1964 there was recognized that identification of the cisplatin's structure depends strongly on availability of high-quality single crystals of this material [12]. Nevertheless, as for now there are two well identified polymorphs of cisplatin known as α and β phases. Until 2010 only the α-cisplatin was thoroughly studied [10] while no structural information has been published on the β–form [13]. In [13] "by the powerful combination of single-crystal X-ray diffraction and neutron powder diffraction" the atom positions and H–Cl distances in α and β cisplatin have been determined at temperatures 100 K and 220 K respectively. The β-phase is dominant at 300 K and converts to α-phase in the temperature range 210–310 K, and at temperatures below 100 K the α-phase is dominant and contains only a single site for Pt center [11]. The crystallographic molecules of these polymorphs differ mainly in relations to the position of the amino H atoms. Crystal of cis-isomer is triclinic with two molecules in the unit cell for α phase and four molecules for β [13]. The planar molecules of cis-isomer are parallel to one another and stacked along the a-axis.

Taking into account such difficulties in determining the structure, and, consequently, difficulties in interpreting experimental vibrational spectra, many attempts have been made to carry out quantum-chemical calculations of platinum (II) compounds: more than 100 calculations of vibrational frequencies for cis- and transplatin have been published over past 25 years [11].

Naturally, both in experimental and theoretical studies, vibrational analysis of cisplatin and transplatin molecules is carried out. Authors of [14] analyzed the vibrations of free cisplatin molecule, hydrogen bonded dimer, α and β crystalline polymorphs with 2 and 4 molecules in the unit cell, respectively, using density functional theory (DFT) method and compared the calculations with experimental IR and Raman spectra at room temperature. The excellent agreement between the simulated and experimental spectra has been achieved for β-form.

Based on the available data, now let us analyze the vibrations of cisplatin and transplatin taking their symmetry into account.

The free cisplatin molecule (Fig. 9.2a) belongs to $C_{2v}$ point group of symmetry and its vibration representations are[1]:

$$\Gamma_{vib} = 9A_1 + 5A_2 + 5B_1 + 8B_2$$

The highest possible symmetry of free transplatin molecule (Fig. 9.2b) is $C_{2v}$ or $D_{2h}$ depending on the positions of $NH_3$ groups: the hydrogen atoms are eclipsed or stagerred, respectively [16]. For the case of $C_{2v}$ symmetry of transplatin molecule the vibration representations are:

$$\Gamma_{vib} = 9A_1 + 5A_2 + 6B_1 + 7B_2$$

[1] Hereinafter, all designations are given accordingly [15].

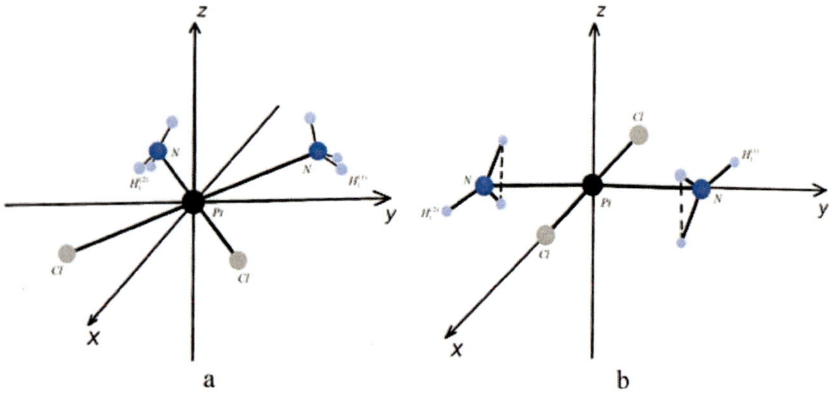


**Fig. 9.2** The molecular structures of cisplatin **a** and transplatin **b**: atom Pt–black, N atoms–blue, Cl atoms–gray, H atoms–light blue

All these 27 vibrational modes are Raman active.

In this research, we present an analysis of the experimentally recorded Raman spectra of two platinum isomorphs, based on an analysis of the normal vibration modes of the skeletons of these molecules cis-$[Pt(II)N_2\ Cl_2]^{6-}$ and trans-$[Pt(II)N_2\ Cl_2]^{6-}$, where the structural differences should be most pronounced. Vibrations of the $NH_3$ groups were analyzed by us earlier [17].

## 9.4 Symmetry, Prenormal Coordinates and Normal Vibrations of Cis- and Trans-Isomer Skeletons

The correct representations of the forms of normal vibrations of molecules and ions allow to give the qualitative estimates of vibrational frequencies and facilitate the interpretation of infrared (IR) and Raman spectra in more complex molecules and in crystals, which include these molecules or ions as structural elements with stronger interatomic bonds. The vibration assignments of two Pt(II)-based isomorphs given in the modern literature have some inaccuracies that arise due to non-compliance with the requirements of orthogonality of eigen vibrational functions. These functions are linear combinations of vibrational symmetrized coordinates (or symmetric displacements), which are the basic vectors in displacement spaces and are transformed by irreducible representations of the symmetry group.

Taking these requirements into account when calculating the normal vibrations of $[Pt(II)N_2\ Cl_2]^{6-}$ ions has also been shown to lead to unambiguous analytical expressions for vibrational modes.

We have constructed normal vibration forms from symmetrized displacements for the cisplatin and transplatin skeletons using well-known projective operator

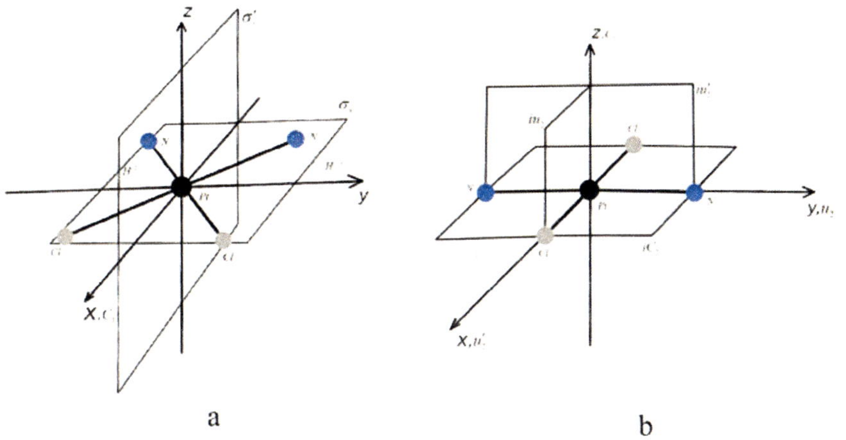

Fig. 9.3 The elements of symmetry of cis-$[Pt(II)\ Cl_2]^{6-}$ **a** and trans-$[Pt(II)\ Cl_2]^{6-}$ **b**

technique method [18]. Application of the projection operator to irreducible representations of the $C_{2v}$ or $D_{2h}$ symmetry groups of cisplatin and transplatin respectively allowed determining symmetrized displacements, which are the same as the symmetry coordinates. Thus, the invariant vector functions, that have been transformed according to the irreducible representations of the $C_{2v}$ or $D_{2h}$ groups could be also determined. The number of linearly independent symmetrized displacements is identical to the number of vibrational modes for each symmetry type. At the next stage, an orthogonal system of the linear combinations of symmetrized displacements is constructed for each symmetry type, and just this system describes a pattern of normal vibrations. We denoted the vectors of the generative basis by the letters $u$ (along the bond stretching direction) and $v$ (ortogonal to $u$) (see Fig. 9.4).

Figure 9.3 shows the elements of symmetry of $[Pt(II)N_2\ Cl_2]^{6-}$ ions (skeletons) for cis-form ($C_{2v}$ point group of symmetry) and trans-form ($D_{2h}$ point group of symmetry).

The symmetry of cisplatin skeleton (cis-$[Pt(II)\ Cl_2]^{6-}$ ion) is described by a point group $C_{2v}$. The representation of all displacements of atoms of the ion in the displacement space is decomposed by irreducible representations of the group as follows:

$\Gamma_{dis} = 5A_1 + 5A_2 + 2B_1 + 3B_2,$

$$\begin{aligned} &\text{including for translations:} && \Gamma_{tr} = A_1 + A_2 + B_2, \\ &\text{for rotations:} && \Gamma_{rot} = A_2 + B_1 + B_2, \\ &\text{and for internal vibrations:} && \Gamma_{vib} = 4A_1 + 3A_2 + B_1 + B_2. \end{aligned}$$

Of the nine internal vibrational modes, all nine ($4A_1 + 3A_2 + B_1 + B_2$) are active in Raman spectra and eight vibrations ($4A_1 + 3A_2 + B_2$) are active in IR absorption. The symmetry of transplatin skeleton (trans-$[Pt(II)\ Cl_2]^{6-}$ ion) is described by a

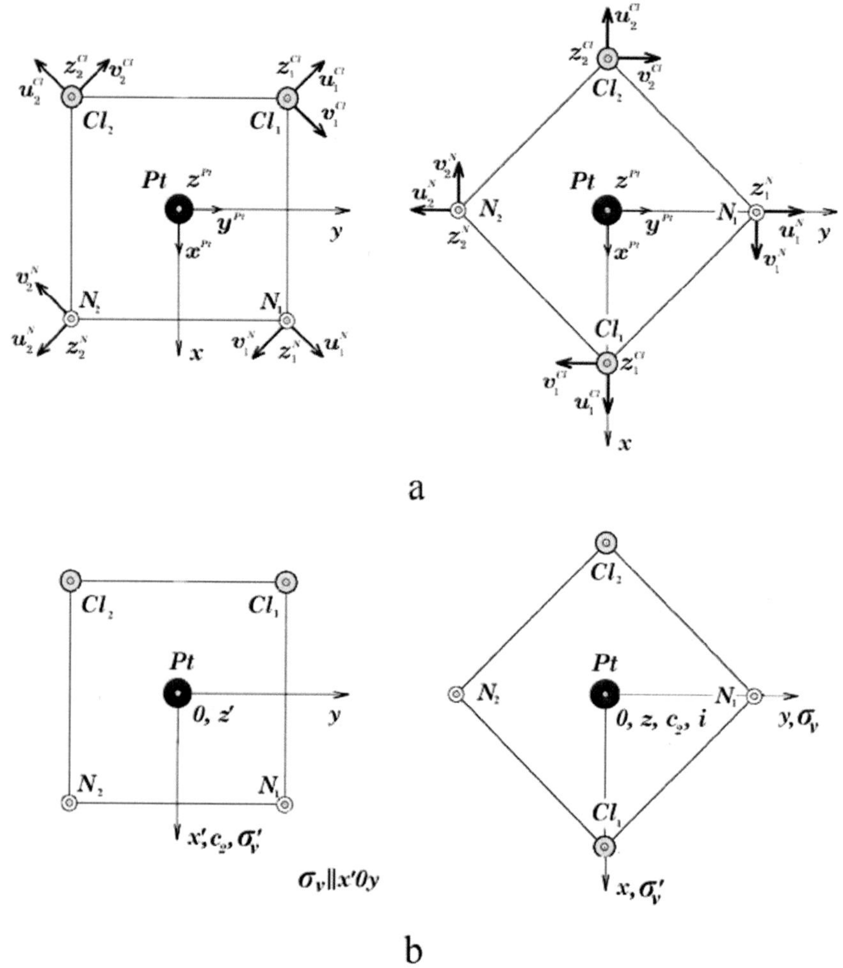


**Fig. 9.4.** **a** The schematic structure and the vectors ($u_i$ and $v_i$) of generating basis of cisplatin skeleton (left) and transplatin skeleton ( ● – Pt, ◎– Cl, ◎– N); **b** the orientation of symmetry elements

point group $D_{2h}$. The representation of all displacements of atoms of ion in the displacement space is decomposed by irreducible representations of the group as follows:

$$\Gamma_{dis} = 2A_1^+ + A_2^+ + 3A_2^- + 2B_1^+ + 3B_1^- + B_2^+ + 3B_2^-,$$

including for translations : $\Gamma_{tr} = A_2^- + B_1^- + B_2^-$

for rotations : $\Gamma_{rot} = A_2^+ + B_1^+ + B_2^+$

and for internal vibrations : $\Gamma_{vib} = 2A_1^+ + 2A_2^- + B_1^+ + 2B_1^- + 2B_2^-$.

Of the nine internal vibrational modes, only three ($\Gamma_{vib} = 2A_1^+ + B_1^+$) are active in Raman spectra and six vibrations ($\Gamma_{vib} = 2A_2^- + 2B_1^- + 2B_2^-$) are active in IR absorption.

Figure 9.4 shows the ions (skeletons) of cis- and trans- isomers of $[Pt(II)N_2Cl_2]^{6-}$ and shows the orientation chosen for calculations relative to the rectangular Cartesian coordinate system, and the orientation of the symmetry elements of the group corresponding to the chosen orientation of the molecule.

Using the matrix of irreducible representations of the groups $C_{2v}$ (for cisplatin) or $D_{2h}$ (for transplatin), the functions that are transformed into generators under the action of the symmetry elements and the standard procedure for the application of projective operators (see the construction of symmetry coordinates [18]), we find the symmetrical displacements of the skeletons. Superscript etter indices of the displacement vectors correspond to atoms they belong to and subscript indices correspond to the actual atom of this type in the chemical structure as indicated in Fig. 9.4.

For cisplatin skeleton these displacements, distributed by types of symmetry in the same way as vibrational modes, are:

for symmetry $A_1$:

$$S_1^{A_1} = x^{Pt};$$
$$S_2^{A_1} = \frac{1}{\sqrt{2}}\left(u_1^{Cl} + u_2^{Cl}\right),\ S_3^{A_1} = \frac{1}{\sqrt{2}}\left(v_1^{Cl} - v_2^{Cl}\right);$$
$$S_4^{A_1} = \frac{1}{\sqrt{2}}\left(u_1^{N} + u_2^{N}\right),\ S_5^{A_1} = \frac{1}{\sqrt{2}}\left(v_1^{N} - v_2^{N}\right)$$

for symmetry $A_2$:

$$S_1^{A_2} = y^{Pt}$$
$$S_2^{A_2} = \frac{1}{\sqrt{2}}\left(u_1^{Cl} - u_2^{Cl}\right),\ S_3^{A_2} = \frac{1}{\sqrt{2}}\left(v_1^{Cl} + v_2^{Cl}\right);$$
$$S_4^{A_2} = \frac{1}{\sqrt{2}}\left(u_1^{N} - u_2^{N}\right),\ S_5^{A_2} = \frac{1}{\sqrt{2}}\left(v_1^{N} + v_2^{N}\right)$$

for symmetry $B_1$:

$$S_1^{B_1} = \frac{2}{\sqrt{2}}\left(z_1^{Cl} - z_2^{Cl}\right),\ S_2^{B_1} = \frac{2}{\sqrt{2}}\left(z_1^{N} - z_2^{N}\right)$$

or symmetry $B_2$:

$$S_1^{B_2} = z^{Pt}$$
$$S_2^{B_2} = \frac{2}{\sqrt{2}}\left(z_1^{Cl} + z_2^{Cl}\right),\ S_3^{B_2} = \frac{2}{\sqrt{2}}\left(z_1^{N} + z_2^{N}\right).$$

Normal vibrations of molecules of each type of symmetry are orthogonal linear combinations of symmetrized displacements of the same type of symmetry for this form of irreducible representations. Establishing the form of these linear combinations is the solution of the problem of finding the forms of normal vibrations. In this case, the form of linear combinations of symmetrized displacements, of course, depends on the form of irreducible representations that determine the displacement, and the very forms of normal vibrations do not depend on the forms of representations.

The construction of forms of normal vibrations $(Q_\mu)_i$ is convenient to begin with the construction of translational and rotational modes because these forms are well known, and linear combinations of symmetrized displacements are obvious. The next step is in finding of the corresponding orthogonal modes, which are divided into quasi-banding and quasi-deformation.

All normal vibrations of the cisplatin molecule skeleton, both exact and approximate, expressed through symmetrical displacements, have the following forms:

for symmetry $A_1$

$$Q_1^{A_1}(x-tr) = \frac{1}{\sqrt{3}}\left(S_1^{A_1} + \frac{1}{\sqrt{2}}\left(-S_2^{A_1} + S_3^{A_1}\right) + \frac{1}{\sqrt{2}}\left(S_4^{A_1} + S_5^{A_1}\right)\right),$$

$$Q_2^{A_1} = \frac{1}{\sqrt{2}}\left(S_2^{A_1} + S_4^{A_1}\right),$$

$$Q_3^{A_1} = \frac{1}{\sqrt{2}}\left(S_2^{A_1} - S_4^{A_1}\right),$$

$$Q_4^{A_1} = \frac{1}{\sqrt{2}}\left(-S_3^{A_1} + S_5^{A_1}\right);$$

$$Q_5^{A_1} = \frac{1}{\sqrt{3}}\left(-S_1^{A_1} + S_3^{A_1} + S_5^{A_1}\right);$$

for symmetry $A_2$

$$Q_1^{A_2}(y-tr) = \frac{1}{\sqrt{3}}\left(S_1^{A_2} + \frac{1}{\sqrt{2}}\left(S_2^{A_2} + S_3^{A_2}\right) + \frac{1}{\sqrt{2}}\left(S_4^{A_2} + S_5^{A_2}\right)\right),$$

$$Q_2^{A_2}(z-\mathrm{rot}) = \frac{1}{\sqrt{2}}\left(S_3^{A_2} + S_5^{A_2}\right)$$

$$Q_3^{A_2} = \frac{1}{\sqrt{3}}\left(-S_1^{A_2} + S_2^{A_2} + S_4^{A_2}\right);$$

$$Q_4^{A_2} = \frac{1}{\sqrt{2}}\left(S_2^{A_2} - S_4^{A_2}\right);$$

$$Q_5^{A_2} = \frac{1}{\sqrt{3}}\left(S_1^{A_2} - S_3^{A_2} + S_5^{A_2}\right);$$

for symmetry $B_1$

$$Q_1^{B_1}(x - rot) = \frac{1}{\sqrt{2}}\left(S_1^{B_1} + S_2^{B_1}\right)$$

$$Q_2^{B_1} = \frac{1}{\sqrt{2}}\left(S_1^{B_1} - S_2^{B_1}\right);$$

for symmetry $B_2$

$$Q_1^{B_2}(z - tr) = \frac{1}{\sqrt{3}}\left(S_1^{B_2} + S_2^{B_2} + S_3^{B_2}\right),$$

$$Q_2^{B_2}(y - rot) = \frac{1}{\sqrt{2}}\left(S_2^{B_2} - S_3^{B_2}\right)$$

$$Q_3^{B_2} = \frac{1}{\sqrt{3}}\left(-S_1^{B_2} + S_2^{B_2} + S_3^{B_2}\right)$$

The forms of both exact and approximate normal vibrations of the cisplatin skeleton are shown in Fig. 9.5.

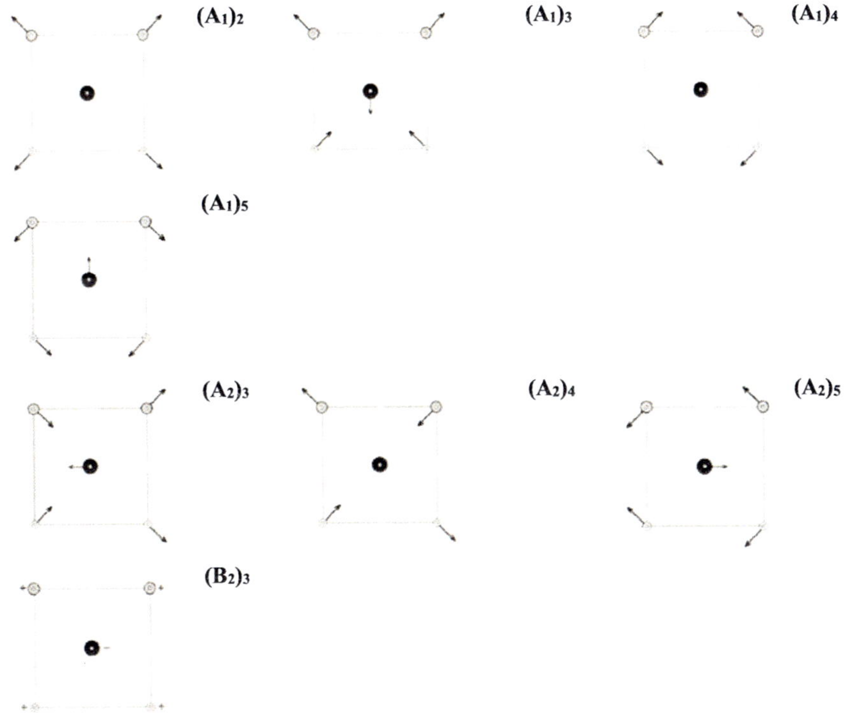


**Fig. 9.5** Normal vibrations of cisplatin skeleton. The translation and rotation modes are not shown. ( – Pt, – Cl, – N)

Similarly, one can find symmetrized displacements and construct forms of normal vibrations of the cisplatin ion, only taking into account that the symmetry of the skeleton is $D_{2h}$. In the latter case $D_{2h}$ symmetry of the transplatin skeleton must be used during construction of the forms.

For transplatin skeleton the symmetrized displacements, distributed by types of symmetry in the same way as vibrational modes, are:

for symmetry $A_1^+$:

$$S_1^{A_1^+} = \frac{1}{\sqrt{2}}\left(u_1^{Cl} + u_2^{Cl}\right);$$

$$S_2^{A_1^+} = \frac{1}{\sqrt{2}}\left(u_1^{N} + u_2^{N}\right);$$

for symmetry $A_2^+$:

$$S_1^{A_2^+} = \frac{4}{\sqrt{2}}\left(z_1^{N} - z_2^{N}\right)$$

for symmetry $A_2^-$:

$$S_1^{A_2^-} = x^{Pt};$$

$$S_2^{A_2^-} = \frac{1}{\sqrt{2}}\left(u_1^{Cl} - u_2^{Cl}\right);$$

$$S_2^{A_2^-} = \frac{1}{\sqrt{2}}\left(v_2^{N} - v_1^{N}\right);$$

for symmetry $B_1^+$:

$$S_1^{B_1^+} = \frac{1}{\sqrt{2}}\left(v_1^{Cl} + v_2^{Cl}\right);$$

$$S_1^{B_1^+} = \frac{1}{\sqrt{2}}\left(v_1^{N} + v_2^{N}\right);$$

for symmetry $B_1^-$:

$$S_1^{B_1^-} = z^{Pt};$$

$$S_2^{B_1^-} = \frac{1}{\sqrt{2}}\left(z_1^{Cl} + z_2^{Cl}\right);$$

$$S_3^{B_1^-} = \frac{1}{\sqrt{2}}\left(z_1^{N} + z_2^{N}\right)$$

for symmetry $B_2^+$:
$S_1^{B_2^+} = \frac{4}{\sqrt{2}}\left(z_1^{Cl} - z_2^{Cl}\right)$;
for symmetry $B_2^-$:

$$S_1^{B_2^-} = y^{Pt};$$
$$S_2^{B_2^-} = \frac{1}{\sqrt{2}}\left(-v_1^{Cl} + v_2^{Cl}\right);$$
$$S_3^{B_2^-} = \frac{1}{\sqrt{2}}\left(u_1^{N} - u_2^{N}\right).$$

All normal vibrations of the transplatin molecule skeleton, both exact and approximate, expressed through symmetrical displacements, have the following forms:
for symmetry $A_1^+$

$$Q_1^{A_1^+} = \frac{1}{\sqrt{2}}\left(S_2^{A_1^+} + S_2^{A_1^+}\right),$$
$$Q_2^{A_1^+} = \frac{1}{\sqrt{2}}\left(S_2^{A_1^+} - S_2^{A_1^+}\right);$$

for symmetry $A_2^+$

$$Q_1^{A_2^+}(x - rot) = S_1^{A_2^+};$$

for symmetry $A_2^-$

$$Q_1^{A_2^-}(x - tr) = \frac{1}{\sqrt{3}}\left(S_1^{A_2^-} + S_2^{A_2^-} + S_3^{A_2^-}\right),$$
$$Q_2^{A_2^-} = \frac{1}{\sqrt{3}}\left(-S_1^{A_2^-} + S_2^{A_2^-} + S_3^{A_2^-}\right),$$
$$Q_2^{A_2^-} = \frac{1}{\sqrt{2}}\left(S_2^{A_2^-} - S_3^{A_2^-}\right);$$

for symmetry $B_1^+$

$$Q_1^{B_1^+}(z - rot) = \frac{1}{\sqrt{2}}\left(S_1^{B_1^+} + S_2^{B_1^+}\right),$$
$$Q_2^{B_1^+} = \frac{1}{\sqrt{2}}\left(S_1^{B_1^+} - S_2^{B_1^+}\right);$$

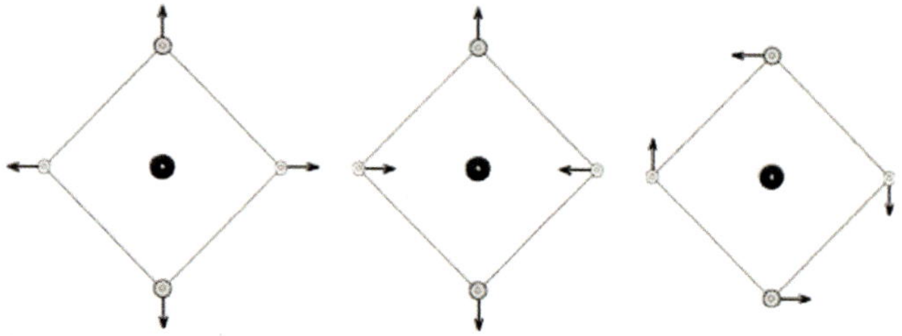

**Fig. 9.6** Normal vibrations of transplatin skeleton. The translation and rotation modes are not shown. Only Raman-active modes are shown. (● – Pt, ◎ – Cl, ◎ – N)

for symmetry $B_1^-$

$$Q_1^{B_1^-}(z - tr) = \frac{1}{\sqrt{3}}\left(S_1^{B_1^-} + S_2^{B_1^-} + S_3^{B_1^-}\right)$$
$$Q_2^{B_1^-} = \frac{1}{\sqrt{3}}\left(S_1^{B_1^-} - S_2^{B_1^-} - S_3^{B_1^-}\right);$$

for symmetry $B_2^+$:

$$Q_1^{B_2^+}(y - rot) = S_1^{B_2^+};$$

for symmetry $B_2^-$:

$$Q_1^{B_2^-} = \frac{1}{\sqrt{3}}\left(S_1^{B_2^-} + S_2^{B_2^-} + S_3^{B_2^-}\right),$$
$$Q_2^{B_2^-} = \frac{1}{\sqrt{3}}\left(-S_1^{B_2^-} + S_2^{B_2^-} + S_3^{B_2^-}\right),$$
$$Q_3^{B_2^-} = \frac{1}{\sqrt{3}}\left(S_1^{B_2^-} - S_2^{B_2^-} + S_3^{B_2^-}\right)$$

The forms of both exact and approximate normal vibrations of the skeleton of the transplatin molecule are shown in Fig. 9.6.

## 9.5 Raman Spectra of Cisplatin and Transplatin

Figures 9.7a and b show the Raman spectra of cis- and transplatin, respectively, in the frequency range of 100–3500 cm$^{-1}$. Each of the figures shows two spectra recorded by two types of detectors–Si detector with the maximum of sensitivity in the visible region and InGaSe detector with the maximum of sensitivity in the IR range. The same excitation wavelength of 785 nm has been applied in both cases.

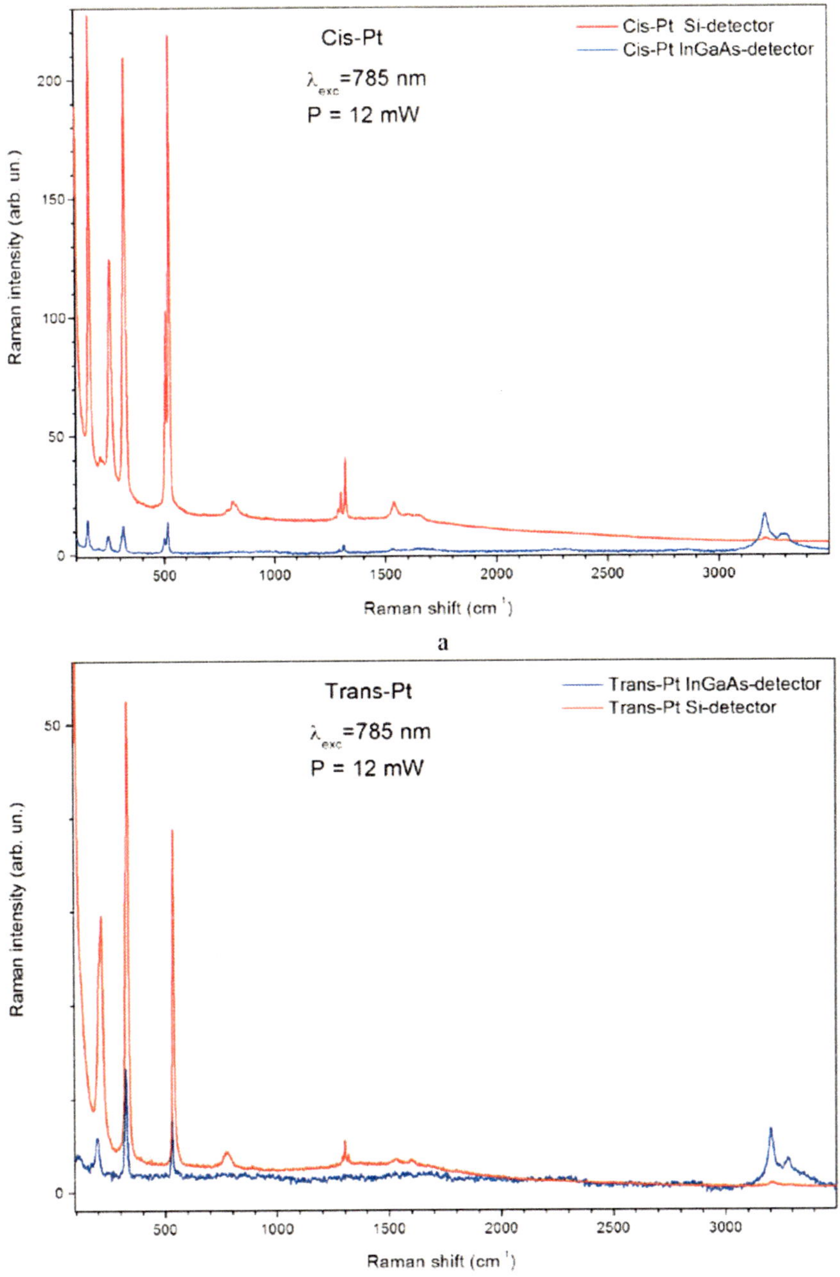


b

**Fig. 9.7** The Raman spectra at 300 K of cis-[Pt(II)$(NH_3)_2Cl_2$] **a** and trans-[Pt(II)$(NH_3)_2Cl_2$] **b** ($\lambda_{ex}$ = 785 nm), recorded using different types of detector: Si (red line) and InGaAs (blue line)

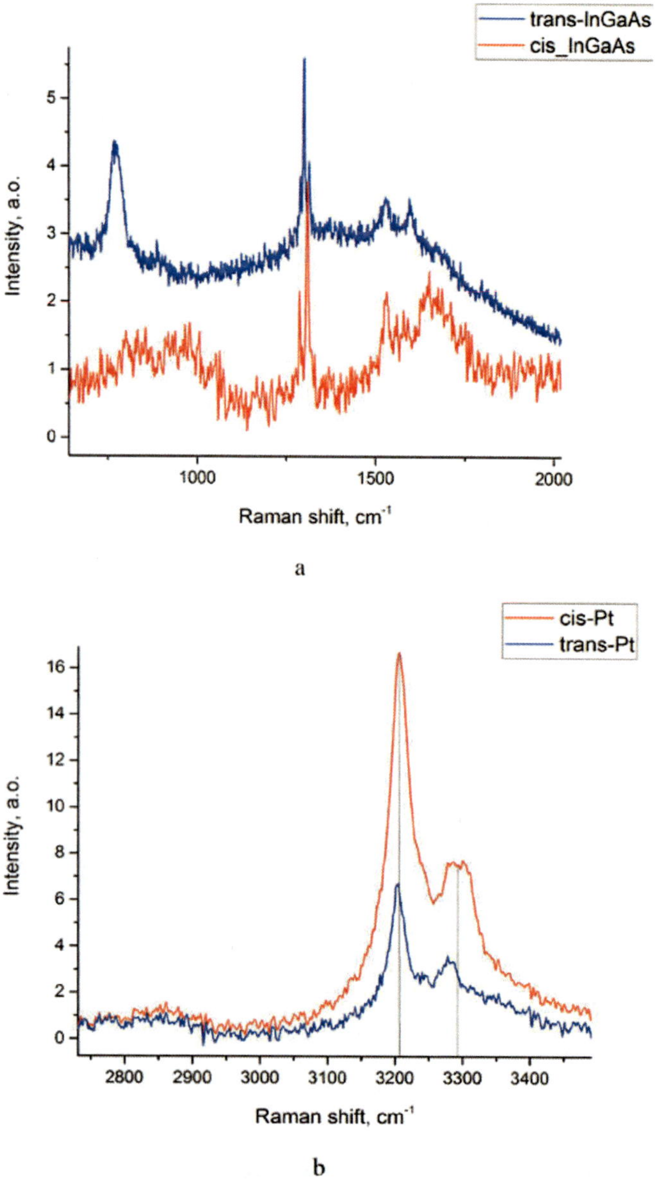


**Fig. 9.8** Raman spectra of cis- and trans- forms of Pt(II) compounds in the region of the vibrations of ammine groups ($\lambda_{ex} = 785$ nm) at room temperature

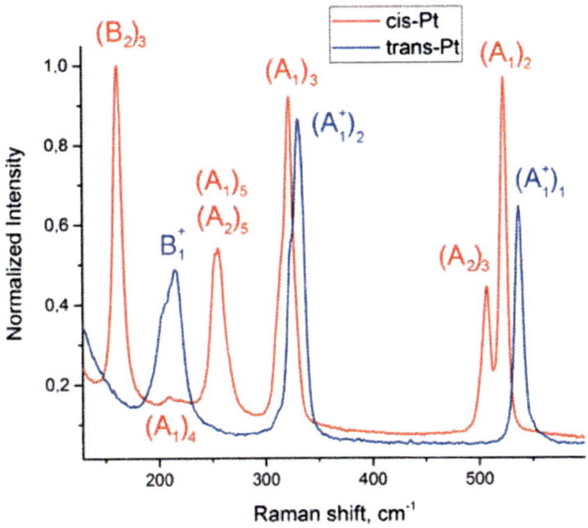


**Fig. 9.9** Raman spectra of cisplatin (red line) and transplatin (blue line) powders under 785 nm laser excitation at T = 300 K in the region of stretching and bending vibrations

We excited the Raman spectra of these two compounds by 488 nm line of $Ar^+$ laser too, but the obtained spectra were less perfect and, in addition, the strong luminescence of transplatin makes it impossible to compare the spectra in the region of vibrations of amino groups.

Figures 9.8a, b show this region, recorded at 785 nm laser excitation at room temperature. As one can see, the spectra have insignificant differences in the position of the maxima of the Raman bands but in the high-frequency region, where the $NH_3$ vibrational modes registered the intensities are rather different, that is caused by their different mutual arrangement in the molecule.

The most interesting and informative for structure determination of two isomorphs of Pt(II) is the region of the spectrum 100 – 600 $cm^{-1}$ (Fig. 9.9), where the banding and stretching vibrations of cis- and trans-isomers $[Pt_2(NH_3)_2Cl_2]$ are located. As expected from the symmetry group analysis, the transplatin molecule, which has a higher symmetry, shows fewer vibrations than transplatin.

Raman spectra of cis- and trans- isomers Pt(II) are shown in Fig. 9.9. As it is seen from Fig. 9.9 all vibrational bands predicted by group-theoretical analyses for cisplatin and transplatin are presented. Conditionally the spectrum of cisplatin can be divided into three parts: (1) stretching vibrations of Pt–N with frequencies 507 and 522 $cm^{-1}$, (2) stretching vibrations of Pt–Cl with asymmetric line shape at 322 $cm^{-1}$, and (3) bending vibrations of the skeletons at 161, 210, and 255 $cm^{-1}$. The spectrum of transplatin demonstrates three bands at 534, 344, and 215 $cm^{-1}$, i.e. these lines are red-shifted in comparison to the corresponding cisplatin lines. While our assignment of the stretching modes of Pt–Cl, Pt–N bonds and of the bending mode of the N–Pt–N unit at 255 $cm^{-1}$ corresponds to the one given by many authors [11–14], the assignment of vibrations at 161 and 210 cm $^{-1}$ is not obvious.

We have no doubt about the interpretation of the stretching modes given by previous authors. We agree that the vibration at 255 $cm^{-1}$ is the bending N–Pt–N vibration and most likely two bands with symmetries $(A_1)_4$ and $(A_1)_5$ (see Fig. 9.5) overlap here. Regarding the band at 210 cm $^{-1}$ we assigned it to N–Pt–Cl normal vibration $(A_2)_5$. The reason for this conclusion is a comparison of the forms of normal vibrations $(A_2)_5$ of cisplatin and $B_1{}^+$ of transplatin, shown in Figs. 9.5 and 9.6. In addition, their Raman shifts are very close: 210 cm$^{-1}$ (cisplatin) and 215 cm$^{-1}$ (transplatin). The band at 161 cm$^{-1}$ we assigned to out-of-plane mode $(B_2)_3$. There is no doubt about the interpretation of the vibrations of transplatin skeleton. All these conclusions are shown in Fig. 9.9, where the symmetry of vibrations is shown in accordance with the forms given in Figs. 9.5 and 9.6.

## 9.6 Conclusions

The Raman spectra of two isomorphs forms of diamminechloridoplatinum (II) cis-$[Pt(NH_3)_2Cl_2]$ (cisplatin) and trans-$[Pt(NH_3)_2Cl_2]$ (transplatin) have been studied at 488 and 785 cm$^{-1}$. It is proved that the structural isomerism of platinum drugs, namely cisplatin and transplatin, is indeed observed in Raman spectra. It is manifested in changes in the vibration frequencies of "skeletons" of isomers formed by platinum, chlorine and nitrogen atoms, in changes in the number of oscillatory bands in the low-frequency part of the Raman spectra (100–600 cm$^{-1}$), and in pumping the intensity of doublet vibrational modes. It was found that the wavelength of the excitation does not affect the frequency position of the bands, however under excitation 785 cm$^{-1}$ the quality of the spectrum improves; the lines become narrower and the doublet in the region of stretching vibrations of cisplatin is clearly resolved.

Forms of normal vibrations of the cis- and trans-isomers are founded and analytical expressions for them are given. The resulting forms of normal vibrations have been used to assignments of the bands that have been controversial and have been shown in Fig. 9.9.

**Acknowledgements** The authors express their gratitude to Dr. Sc., Prof. V. Strelchuk, PhD. O. Kolomys and PhD A. Nikolenko (V.E. Lashkaryov Institute of Semiconductor Physics NAS of Ukraine, Kyiv) for the possibility to obtain Raman spectra

## References

1. Wiltshaw BE (1979) Cisplatin in the treatment of cancer. Platinum Metals Rev 23:90–98
2. Prestayko B, VenCamp L (1979) The successful regression of large solid sarcoma 180 tumors by platinum compounds. Cancer Res 30:1799–1802
3. Abu-Surrah AS, Kettunen M (2006) Platinum group antitumor chemistry: design and development of new anticancer drugs complementary to cisplatin. Curr Med Chem 13:1337–1357
4. Johnstone TC, Lippard SJ (2015) Third row transition metals for the treatment of cancer. Phyl Trans A 373:20140185

5. Hannon MJ (2007) Metal-based anticancer drugs: From a past anchored in platinum chemistry to a post-genomic future of diverse chemistry and biology. Pure Appl Chem 79(12):2243–2261
6. Kauffman GB, Cowan DO, Slusarczuk G, Kirschner S (1963) Cis- and trans-Dichlorodiammineplatinum(II). Inorg Synth 7:239–245
7. Nakata B, Yamagata S, Kanehara I, Shirasaka T, Hirakawa K (2006) Transplatin, a Cisplatin Trans-Isomer, may enhance the anticancer effect of 5-fluorouracil. J Exp Clin Cancer Res 25(2):195–200
8. Porai-Koshits M (1960) Crystal chemical data on the stereochemistry of complex compounds of some divalent transition metals (Fe, Co, Ni, Cu, Pd и Pt), Dokl Acad Nauk SSSR 134:1104–1107
9. Porai-Koshits M. (1954) The structure of the crystals of salt platinum trans-dichlorodiamine. Trudy Inst Krist Acad Nauk SSSR 9:229–238
10. Milburn GHW, Truter MR (1966) The crystal structures of cis- and trans-dochlorodiamineplatinum (II). J Chem Soc A 11:1609–1616
11. Bélanger-Desmarais N, Schütz M, Reber C (2019) Remarcably intricate Raman spectra of Platinum(II)-ligand skeletal modes in diamminedihalido complexes. J Phys Chem A 123:5574–5579
12. Poulet H, Delorme P, Mathieu JP (1964) Spectres de vibration et frequencies fondamentales de composes ammines du platine. Spectrochim Acta 20:1855–1863. https://doi.org/10.1016/0371-1951(64)80190-9
13. Ting VP, Schmidtmann M, Wilson CC, Weller MT (2010) Cisplatin: polymorphism and structural insights into an important chemotherapeutic drug. Angew Chem Int Ed 49:9408–9411
14. Georgieva I, Trendafilova N, Dodoff N, Kovacheva D (2017) DFT study of the molecular and crystal structure and vibrational analysis of cisplatin. Spectrochim Acta Part A Mol Biomol Spectrosc 176:58–66
15. Bir LG, Pikus GE (1974) Symmetry and strain-induced effects in semiconductors. Wiley, New York
16. Nakamoto K, McCarthy PJ, Fujita J, Comndrate RA, Behnke GT (1965) Infrared studies of ligand-ligand interaction in dihalogenodiammineplatinum(II) complexes. Inorg Chem 4:36–43
17. Gubanov VO, Naumenko AP (1997) Normal vibration forms of pyramidal molecules and ions XY3. Kiev Univ Bull 4:342–351
18. Poulet H, Matieu JP (1970) Spectres de vibration et symétrie des cristaux. Gordon and Breach, Paris-London-New York

# Chapter 10
# Approaches for a Closer Look at Problems of Liquid Membranes with Amyloid-Beta Peptides

**Tomáš Kondela, Pavol Hrubovčák, Dmitry Soloviov, Dina Badreeva, Tatiana Murugova, Vadim Skoi, Alexander Kuklin, Oleksandr Ivankov, and Norbert Kučerka**

**Abstract** Alzheimer's disease is a neurodegenerative brain disease, where pathological hallmarks are senile plaques consisting primarily of amyloid-beta peptides. The investigations, however, point out not only the importance of physico-chemical properties of peptides themselves, but the membranes as the targeting environment as well. Commensurately, the interrogation approaches are required to focus on

T. Kondela · P. Hrubovčák · D. Soloviov · T. Murugova · V. Skoi · A. Kuklin · O. Ivankov · N. Kučerka (✉)
Frank Laboratory of Neutron Physics, Joint Institute for Nuclear Research, Dubna, Russia
e-mail: kucerka@nf.jinr.ru

T. Kondela
e-mail: kondela@jinr.ru

D. Soloviov
e-mail: dmitrysoloviov@jinr.ru

T. Murugova
e-mail: murugova@jinr.ru

V. Skoi
e-mail: skoi.vv@phystech.edu

A. Kuklin
e-mail: kuklin@nf.jinr.ru

O. Ivankov
e-mail: ivankov@jinr.ru

T. Kondela
Department of Nuclear Physics and Biophysics, Comenius University in Bratislava, Bratislava, Slovakia

P. Hrubovčák
Department of Condensed Matter Physics, University of P. J. Šafárik in Košice, Košice, Slovakia

D. Soloviov · V. Skoi · A. Kuklin
Moscow Institute of Physics and Technology, Dolgoprudny, Russia

D. Soloviov · O. Ivankov
Institute for Safety Problems of Nuclear Power Plants NAS of Ukraine, Kyiv, Ukraine

 
L. Bulavin and N. Lebovka (eds.), *Soft Matter Systems for Biomedical Applications*, Springer Proceedings in Physics 266, https://doi.org/10.1007/978-3-030-80924-9_10

these systems at various length scales ranging from the submolecular and molecular levels relevant to the structures of peptides and/or membrane lipids locally, to the supermolecular levels covering the aggregates and large structures ensuing from the peptide-membrane interactions. In this chapter, we are summarizing the previous studies that look at complex model and biological membranes using a variety of experimental and theoretical methods based on, or complementing, the scattering techniques. We attempt to examine the effects modulated by the presence of Aβ peptides, and more importantly to determine the modes of interaction between membranes and peptides.

## List of Abbreviations and Chemical Substances

### *Abbreviations*

| | |
|---|---|
| Aβ | amyloid-beta |
| AD | Alzheimer's disease |
| APP | amyloid precursor protein |
| INS | inelastic neutron scattering |
| IXS | inelastic X-ray scattering |
| MD | molecular dynamics |
| NMR | nuclear magnetic resonance |
| NR | neutron reflectometry |
| NSLD | neutron scattering length density |
| PUL | polyunsaturated lipids |
| REMD | replica-exchange molecular dynamics |
| ROS | reactive oxygen species |
| SAND | small angle neutron diffraction |
| SANS | small angle neutron scattering |
| ULV | unilamellar vesicle |

### *Chemical Substances*

| | |
|---|---|
| DMPC | 1,2-dimyristoyl-sn-glycero-3-phosphatidylcholine |

D. Badreeva
Laboratory of Information Technologies, Joint Institute for Nuclear Research, Dubna, Russia
e-mail: badreeva@jinr.ru

N. Kučerka
Department of Physical Chemistry of Drugs, Faculty of Pharmacy, Comenius University in Bratislava, Bratislava, Slovakia

| | |
|---|---|
| DOPC | 1,2-dioleoyl-sn-glycero-3-phosphatidylcholine |
| DOPS | 1,2-dioleoyl-sn-glycero-3-phosphatidylserine |
| DPPC | 1,2-dipalmitoyl-sn-glycero-3-phosphatidylcholine |
| POPC | 1-palmitoyl-2-oleoyl-glycero-3-phosphatidylcholine |
| POPE | 1-palmitoyl-2-oleoyl-sn-glycero-3-phosphatidylethanolamine |
| POPS | 1-palmitoyl-2-oleoyl-sn-glycero-3-phosphatidylserine |

## 10.1 Introduction

Biological membrane is a compositionally complex and fairly dynamic structure fundamental to the function of cells and organelles. The membrane contains three main types of components: lipids, proteins, and carbohydrates. Actual amounts of these components however differ between various biomembranes. For example, the major lipid components of a plasma membrane are 21 mol% phosphatidylcholine, 29 mol% phosphatidylethanolamine and phosphatidylserine, 21 mol% sphingomyelin, and 26 mol% cholesterol [1]. All the phospholipids are polar molecules, which in the presence of polar liquid form specific structures. Hydrophobic parts made up of long hydrocarbon chains move away from polar liquid, while lipid heads, in contrary, move towards the liquid in inter- and outer-cellular environments. This ensues forming the bilayered membrane. It is now generally accepted that active functions such as transport and signaling that take place at membranes are provided by the membrane proteins. Their proper function in turn depends on the physical properties of environment the proteins reside in, the underlying matrix made of lipids [2].

In general, the different membranes serve different functions and consequently have different structural properties. Indeed, the myriad processes taking place in these membranes are reflected in the lipidome's size and diversity [3]. For example, the thermodynamic phase of lipids plays one of the starring roles in determining the membrane's structural properties [4]. At the assistance of increasing temperature, the crystalline phase passes from a highly ordered structure to the liquid-crystalline phase typical of high disorder [5]. It is a lipid membrane in this fluid phase that is biologically most attractive due to its dynamic structure, in which the lipids almost freely diffuse yet embrace their overall bilayered structure of liquid membrane [6]. In addition to the temperature factor and participation of given lipids themselves, the membrane thermodynamic phase is determined by the presence of other membrane components. This is documented well in the case of cholesterol, whose concentration variation across the organelles gives the liquid membranes control over their structure and dynamics [7].

The changes of membrane functionality most likely originate in the structural properties of membrane, that are known to be accompanied by the changes in membrane physico-chemical properties. The cholesterol, for instance, increases the order of lipid hydrocarbon chains and increases the stiffness of the membrane. On the other hand, melatonin was found to increase the fluidity of the membrane and counteracted the effect of cholesterol [8]. Interestingly, decreasing levels of melatonin in the brain tissue were correlated with the aging similarly yet in a counteracting way as increasing levels of cholesterol [9]. The role of membrane properties, its chemical composition, and additives is thus not surprisingly discussed in a connection not only to the membrane function but the disfunction too [10]. This concerns many various diseases, and those related to aging (amyloid toxicity and conformational diseases) in particular [11]. The crucial role in these processes is thought to be imparted by peptide-membrane interactions, which modulated by a membrane composition may cause a peptide to misfold and bring about irreversible conformational changes to other peptides and proteins.

Among conformational diseases, Alzheimer's disease (AD) is a neurodegenerative brain disease manifested by the nerve cell loss, which is the most common cause of dementia. Dementia is a clinical syndrome that can be characterized through the following symptoms: memory and speech disorders, later impairment of cognitive, intellectual, and physical abilities. AD affects 6% of the population over the age of 65 and its incidence increases with increasing age. It is estimated that 47 million people around the world suffer from dementia. Unfortunately, with a ubiquitously increasing aging of the population, this number will increase further, projected to reach 131 million in 2050 [12]. It was discovered in 1906 by Dr. Alois Alzheimer, who first observed morphological and histological changes in the brain of impaired patients [13]. He described amyloid plaques and the neurofibrillary tangles based on the autopsy observations. These two findings are still considered to be the main sequelae of the disease that are visible on the brain. The root cause of the disease, however, eludes explanations to the present days.

There have been several hypotheses suggested before 2019 [14] that try to explain the origin of AD (Fig. 10.1). Metal ions that play various functional roles in organisms, for instance, have been reported to participate in amyloid-beta (Aβ) peptide

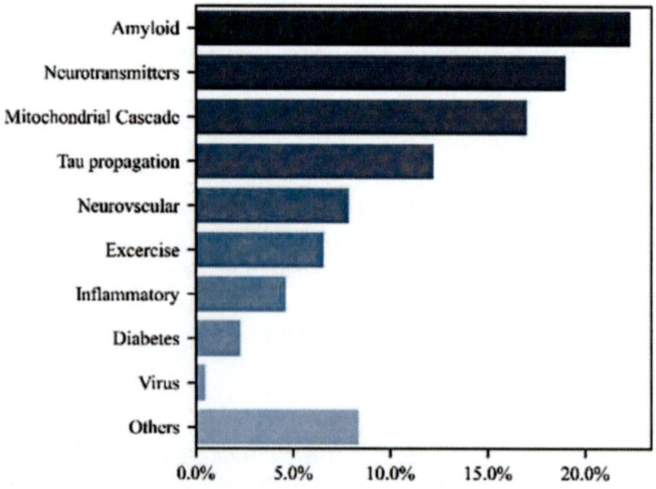


**Fig. 10.1** Distribution of clinical trials performed with regards to the various hypotheses of AD up to the year 2019

aggregation and toxicity [15–17]. The most intriguing, however, are the cholinergic or neurotransmitters hypothesis [18], tau hypothesis [19], and Aβ hypothesis [20]. The cholinergic hypothesis is the oldest, based on the assumption that the cause of the disease is an insufficient synthesis of acetylcholine in the body, which serves as a neurotransmitter. Unfortunately, treatment with drugs that increase acetylcholine production has been shown ineffective [21].

The tau hypothesis identifies the tau protein situated inside the cell as a main contributor to the disease. The normal function of this protein is securing the division and orientation of cellular organelles in the underlying cytoplasm by creating a cytoskeleton system. These microtubes are then disrupted and gradually disintegrated as a result of the protein aggregation, leading eventually to the collapse of the cytoskeleton and subsequent disruption of the neuronal transport system, ultimately resulting in cell death. An anticipated trigger for the aggregation is Aβ peptides, though the details of this process are yet to be known [22–24].

Aβ peptides are proposed to be a key factor also in another hypothesis. It was first postulated in 1991 and it presumes that extracellular Aβ peptides form plaques that are a major cause of AD [20]. Aβ is a peptide that is most often composed of 40–42 amino acids. Its toxicity is associated with the fact that it aggregates into soluble oligomers, which may exist in several forms. It was suggested that a specifically misfolded oligomer may trigger other oligomers to take its form, generating thus a chain reaction and the formation of plaques and fibrils, which may supposedly be a cause of the disease [22, 25]. Recent studies are however more focusing on a monomeric Aβ and small oligomers as an origin of the AD [26, 27].

Aβ peptide is formed by enzymatic cleavage of amyloid precursor protein (APP) in two isoforms 1–40 and 1–42. APP is cleaved in two ways, the so-called non-amyloidogenic and amyloidogenic pathways [28, 29] (Fig. 10.2). The natural function of Aβ is not yet fully understood, although animal studies have not shown any significant physiological function [30]. Nevertheless, some potential activities such as regulation of cholesterol transport [31], enzyme kinase activation [32], or antimicrobial properties [33] have been discovered. As mentioned above, Aβ is most commonly associated with AD where it creates so-called plaques. It has been shown that in the early stages, the non-aggregated form of Aβ can penetrate the cell membrane and initiate intracellular toxicity mechanisms resulting in programmed cell death [34, 35]. As mentioned above, AD is a neurodegenerative brain disease, where about 50% of the brain's dry weight is supplied by lipids that support various structural or metabolic functions [36]. However, the lipids of different shapes and sizes do not affect the static structure only, but their different chemical composition modulates also the elasto-mechanical properties of the membrane. The liquid nature of membrane reflected in its fluidity is one of the examples that has been a topic of a vast number of studies [37–42]. Small molecules such as cholesterol and melatonin may alter the properties of liquid membrane further and in turn affect, in both diseases aiding or normal function protecting ways, amyloid fibril formation and toxicity. The intramembrane interactions become thus obviously a key factor in these hypotheses. One of the parameters proven to play an intriguing role in the interactions of Aβ peptide with membranes was concluded to be connected to

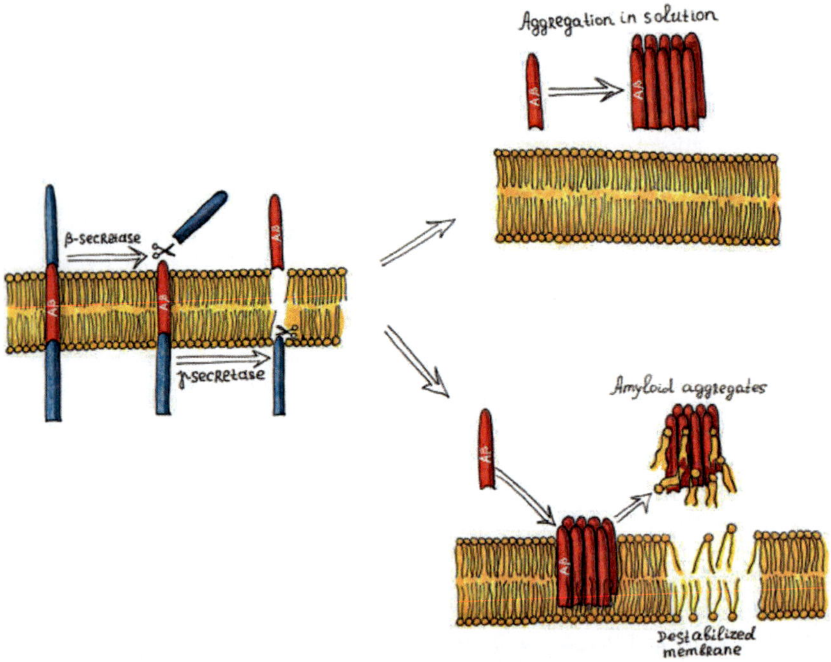


**Fig. 10.2** The sketch of the amyloid-beta hypothesis of the Alzheimer's disease onset. The inter-membrane amyloid precursor protein is cleaved in the amyloidogenic pathway, where extramembranous β-secretase and intramembranous γ-secretase cut APP to form $A\beta_{40}$ or $A\beta_{42}$ peptides. Created peptides may start to aggregate inside or outside the membrane in the case of misfolding conditions

the thermodynamic phase of lipid membrane. Apparently, $A\beta_{1\text{-}40}$ interact with the membrane at temperatures below the main phase transition due to the smoothness of tightly packed gel-phase lipids [43].

Other parameters of importance are the charge and pH. In mostly a zwitterionic lipid bilayer, the peptide will interact with the membrane interface while moving towards its hydrophobic core and thus it will prevent interactions with other peptides that can potentially trigger a change in its secondary structure. In the case of an anionic bilayer, the peptide will be more prone to interact with other peptides as it will be located closer to the membrane surface and exposed to the surrounding solution. The latter is more likely to contribute to oligomerization [44]. Also, a lower pH (pH ~ 5) can induce a more rapid fibrilization of amyloid-beta peptide [45], which means that a local decrease of pH near the anionic lipid bilayer can also contribute to the peptide aggregation. Controversially however, the aggregation of $A\beta_{1-42}$ was proposed not to occur at pH < 5 and pH > 9.5, while it started rather between pH > 6 and <9.5 [46].

A further factor which can be an important element in the aggregation of Aβ is oxidative stress. It has been linked to AD and other age-related neurodegenerative diseases connected to dementia. Interestingly, the Aβ protein itself has been shown to further advocate oxidative stress [47, 48], as the AD infected brain exhibits more oxidative damage than a normal brain [49], and the areas of a high concentration of Aβ have displayed an increased level of protein oxidation [50]. As has been mentioned before, the pathological hallmarks of AD are senile Aβ plaques aggregated together with metal ions like copper, iron, or zinc. These redox-active metal ions together with Aβ can, in turn, produce the reactive oxygen species (ROS). The ROS as highly reactive chemical molecules then contribute to oxidative damage of the nearby molecules of lipids, proteins, and Aβ peptides [51]. All these effects subsequently introduce some characteristic changes to the structural properties of the membrane, and can thus be revealed by the methods of structural investigation. Neutron scattering is one of such influential techniques whose access to the membrane physical properties (thickness, shape, localization, and others) can answer some questions related to the oxidation stress, the impact of charge and pH, and possibly many other factors setting on the AD.

The study of membranes started with the discovery of a microscope in the seventeenth century. It allowed recognizing its bilayered structure and permeable character [52]. The most detailed insights into the membrane structure however became available in the twentieth century with advancement of scientific research based on scattering approaches [11, 41, 53]. The neutron scattering techniques extend these possibilities further owing to the ability of studying the inhomogeneities in membranes, which can be composed of not only different lipids but also proteins and peptides. These inhomogeneities can be in both lateral and normal directions, and can be presented by lipid rafts or the aggregates of membrane-embedded peptides and proteins. The rafts contain lipids with specific properties that are different from the rest of the lipid pool, most often related to the different thermodynamic phases. The structural information about rafts and lateral inhomogeneities in general, can be obtained by appropriate deuteration of lipids and/or adjusting the $H_2O/D_2O$ ratio in the solvent [11, 54–56].

The parameters relevant to Aβ related investigations point out not only the importance of physico-chemical properties of peptides themselves, but the membranes as the targeting environment as well. Commensurately, the interrogation approaches are required to focus on these systems at various length scales ranging from the submolecular and molecular levels relevant to the structures of peptides and/or membrane lipids locally, to the supermolecular levels covering the aggregates and large structures ensuing from the peptide-membrane interactions. There are a few methods that allow to look closely at the membrane and interrogate these systems, that we will discuss in further details.

In this chapter, we are summarizing the previous studies that looked at complex model and biological membranes using a variety of experimental methods based on, or complementing, the scattering techniques. We pay special attention to those researching the possibility to regulate the membrane fluidity by regulating membrane chemical composition. Finally, we attempt to examine further how these effects are modulated by the presence of Aβ peptides, and more importantly how these effects determine the interactions between membranes and peptides.

## 10.2 Membrane Phase Transitions

Biological activities of the membrane surface-active compounds are known to depend on the chemical composition of lipid hydrocarbon chains, polar headgroups, and their capacity to attract water, and on the combination of lipids, cholesterol, melatonin, proteins etc. Indeed, as a function is often tightly coupled to the structure, the myriad specific functions occurring in these membranes are reflected in its compositional diversity. The multicomponent lipid mixtures allow the membrane, in the first place, to take control over its local thermodynamic phase.

A thermodynamic phase is a homogeneous property of a system, that is, a body defined by the same thermodynamic properties at its every point and without the dependence on the amount of matter. The different phases though may coexist while being separated from each other by interfaces. Phase transitions are common for all types of matter and are relatively well described theoretically. When external conditions such as temperature or pressure change, the phases can change from one to another [57]. This process is called a phase transformation, or a phase transition. A characteristic feature of the phase transition is a sharp change in the properties of the substance–for lipid structures, this means for example a change in the lateral diffusion rate by a factor of hundreds or even thousands. In turn, this is of key importance in life sciences. Indeed, the rate of exchange between the cell's comparative and external regulators of the multicellular organism will depend on the rate of diffusion.

The diffusion rates in membrane systems correlate closely with their thermodynamic state due to the balance of intra-membrane forces that in turn depends on thermodynamic conditions. This delicate balance of forces is responsible for minimizing the system's total energy and includes both headgroup and hydrocarbon chain interactions manifested mostly by the attractive hydrophobic forces within the hydrocarbon chain region and headgroup dipolar interactions, and the repulsive steric and entropic effects. When following the consequences of temperature changes, there is an additional factor giving rise to another essential interaction within the lipid membrane, namely trans-gauche isomerization. The probability of trans-gauche isomerization in acyl chains increases with increasing temperature and has the opposite effect to that of van der Waals force. Most importantly however, it introduces the disorder within the hydrocarbon chain region and induces the membrane softening. This has proven a determinant feature in many phenomena in the vicinity of chain-melting phase transition [58].

By the early 1970s, as a result of studies mainly carried out by the method of X-ray structural analysis, it was found that water–lipid systems form a wide range of liquid crystal phases [59]. It was found that lamellar lipid phases, to which the phases of real cell membranes belong, are usually smectic according to Friedel's classification. These phases include the so-called $L_\beta$ gel phase and $L_\alpha$ fluid phase, corresponding to the thermodynamic state of lipids in a living cell. The phase transition from gel to fluid is called the main phase transition. Its study in model membranes of various compositions was considered to be an important task, particularly in the studies of lipid membranes doped with Aβ peptides, since the phase state of the lipid membrane can also influence the localization of peptides [43].

The lipid membranes are characterized by lyotropic mesomorphism (dependence of the state on lipid headgroup hydration) and thermotropic mesomorphism (dependence of the structure on temperature) [60]. Both properties are related. The main phase transition of lipids occurs at a characteristic temperature $T_m$, the value of which depends on the water content in the system. Temperature of the phase transition reaches a minimum as soon as the total water content exceeds the amount that can bind lipid structures. At the same time, lipids with a lack of water can be in an ordered state even at temperatures above $T_m$.

For the studies of systems with lipids and Aβ peptides, it is important to mention the traditional classification of the phase transitions of the first and second order. According to Landau's theory [61], during the phase transitions of the first kind, the distribution function of the system must be bimodal in the vicinity of the phase transition point, that is, it must have two maxima. The highest maximum corresponds to the most stable state of the system, and the second maximum corresponds to a less favorable, metastable state. At the very transition point, the heights of the maxima become the same, and the system can simultaneously coexist in both states. In second-order phase transitions, the distribution function has always only one maximum, which changes discretely at the transition point. Accordingly, in the case of second-order transitions, metastable states do not exist.

One of the characteristics of the lipid bilayer, that allows determining the transition from one phase to another is its mass density $\rho$. In the case of water–lipid systems, $\rho$ means the macroscopic density of the system, whose changes depend on the system volume $V$ only. The coefficient of its thermal expansion (at constant pressure $p$) can be expressed in terms of the volume change ($dV$) due to the change of temperature ($dT$) [62]:

$$\alpha_p = \frac{1}{V}\left(\frac{\partial V}{\partial T}\right)_p \tag{10.1}$$

During the first-order phase transitions, $\alpha_p$ turns to be infinite in the vicinity of phase transition temperature and during the second-order phase transitions, it experiences a jump. Conveniently thus, one can determine the system phase transition by the volume dependence on temperature. This allows the densitometry method to be used to determine the thermodynamic phase of lipids and the characteristics of their main phase transition [63].

One of the modern methods for determining the density of liquid and gaseous samples is densitometry based on the Pulsed Excitation Method [64]. The principle is based on determining the period of forced vibrations of a tube filled with a sample. In this case, the oscillation period $\tau$ is related to the density ρ of the sample filling the tube by the simple relation:

$$\rho = A\tau^2 - B \tag{10.2}$$

where $A$ and $B$ are instrumental constants. Among the existing measuring tube configurations, U-shaped tubes are often used for highly accurate determination of the density of liquid samples with 5-digit accuracy (e.g., DMA densitometers by Anton-Paar, Graz, Austria). Oscillations of such a system occur in a direction perpendicular to the plane of the tube, allowing also for a high precision temperature control. Some disadvantages of the described method on the other hand, include the relatively large sample volume required (about 1 ml), as well as the possibility of its macroscopic separation during long-term measurements. The latter can be avoided by the periodic re-homogenization of the sample [65].

In addition to temperature and pressure, the phase state of lipid membranes can be influenced by the presence of ions, or low and high molecular weight compounds. The above described instrumental method has been utilized recently for example in determining the temperature and width of the main phase transition of lipid bilayers with the addition of cholesterol and melatonin [66]. These can be determined readily from the graph of first derivatives $d\rho/dT$ (see Fig. 10.3). The results have shown that the presence of melatonin and cholesterol dramatically shifts the position and width of the peak of the density first derivative, which turns out to be in accordance with their known influence on the membrane thermodynamic phase.

The fluidizing effect of melatonin can be concluded based on shifting the membrane thermodynamics towards the fluid phase. The mechanism of the effect has been speculated to be linked to the incorporation of melatonin in the head group region, introducing thus the defects in the hydrocarbon region. On the other hand, the effect of cholesterol turned out to be in removing the gel-to-fluid phase transition detected for the neat lipid system previously. This is in good agreement with the

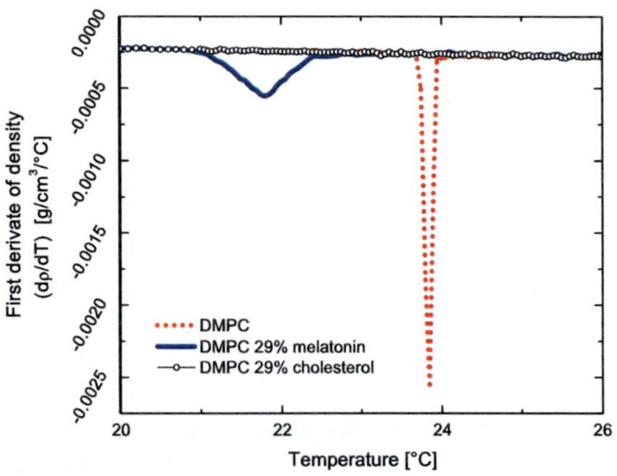


**Fig. 10.3** The dependencies of the first derivative of mass density vs temperature for 1,2-dimyristoyl-sn-glycero-3-phosphocholine (DMPC) multilayer vesicles aqueous dispersion (error bars are smaller than the symbol size). A sharp peak corresponding to the main gel-to-liquid phase transition is observed for pure DMPC (red). For DMPC with the addition of melatonin, the peak position and shape change significantly (blue). For DMPC with the addition of cholesterol, the peak vanishes completely (black circles)

formation of cholesterol-induced liquid ordered (Lo) phase, in which the membrane is fluid even below the phase transition temperature of lipid. In an aim to understand the systems at high cholesterol concentrations, the presumed cholesterol's ordering effect on the lipid chains and the overall membrane structure has to be scrutinized further by other experimental techniques, such as those based on scattering principles.

## 10.3 Overall Membrane Thickness

Small angle neutron scattering (SANS) is a powerful technique for studying the structure of membranes, such as lipidic model membranes and biological membranes isolated from alive systems. Its main advantage comes from the possibility to measure samples in biologically relevant conditions, and in retrieving structural characteristics on several length scales. SANS gives some essential information about the overall size of objects on the level of thousands of angstroms, at the same time with the thickness, curvature, inhomogeneity, and asymmetry of the membrane on the level of tens of angstroms.

The membrane-containing samples for SANS mainly present liquid dispersions, in which the solution is typically made deuterium rich for enhancing the neutron scattering contrast between the membrane and solution. While biological membranes need to be extracted according to the particular protocols required for their isolation, it is desirable to use the heavy water buffers in the final steps. If the membranes are formed by synthetic lipids, the lipid powders or films are dispersed in $D_2O$-based buffers initially. For multicomponent model membranes, all components are pre-dissolved in an organic solvent (chloroform, methanol, trifluoro-ethanol, or their mixtures). That allows one to achieve a homogeneous mixing of all components, at the same time as it enforces the efficacy of the incorporation of various transmembrane components, such as proteins and peptides. The protocols for embedding Aβ peptide into model membranes have their specifics that can be found elsewhere [11, 67]. All components (in desired ratios and after receiving required pre-treatments) are thus mixed in glass tubes, and the solvent is evaporated by a nitrogen stream and subsequent evacuation. The resulting film that forms on the walls of the glass tube is then hydrated with $D_2O$ or $D_2O/H_2O$ mixture to obtain the membrane concentration of about 1 wt% that is required by SANS measurements.

In the case of most commonly used lipids mixed with water, they tend to spontaneously form large multilayered vesicles. These objects result in strong diffraction patterns when examined by neutron (and/or X-ray) scattering. Although the diffraction patterns have their advantages and are successfully employed in various structural studies, they overwhelm the smooth scattering signal due to the membrane inner structure [68]. Therefore, the multilamellar vesicles are often extruded through polycarbonate filters with pores of 300–1000 Å to break the multilayers into unilamellar vesicles (ULVs), and to unify them in size. The prepared samples are typically poured into quartz cuvettes and placed in a specialized thermo-holder [69].

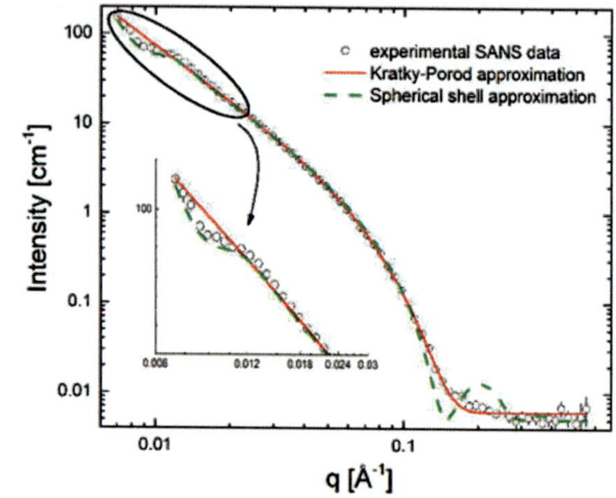


**Fig. 10.4** SANS curve for lipid ULVs with embedded A$\beta_{25\text{–}35}$ peptide. The Kratky-Porod approximation is superior for extracting membrane thickness as seen by its close fitting to the curve in the high *q* region, while the spherical shell approximation improves the fitting in the low *q* region that corresponds to the overall vesicle parameters as demonstrated in the insert

The experimentally measured SANS curve in the case of non-interacting ULVs is typically smooth and almost featureless. It remains true after the addition of peptides if their fraction in the membrane is small enough to prevent aggregation or other specific inter-lamellar interactions. The analysis of the experimental data can then be carried out in a manner that is used for lipid membrane systems. The example of the experimental SANS curve and its theoretical approximations are presented in Fig. 10.4.

The simplest way to calculate membrane thickness is according to Kratky-Porod approximation for a region of medium scattering vectors ($0.02 < q < 0.1$ Å$^{-1}$):

$$I(q) = \frac{A}{q^2} \cdot e^{-q^2 \cdot R_t^2} + b \tag{10.3}$$

where $q$ is a scattering vector defined as $4\pi\ \sin\theta/\lambda$, $\theta$ is the angle of incident and $\lambda$ is the neutron wavelength; $A$ is a constant depending on the average scattering length density, volume, and quantity of membranes in an experimental sample; $b$ is a scattering background; and $R_t$ is a radius of gyration along the thickness of a membrane. The latter relates to the membrane thickness as $d = \sqrt{12} \cdot R_t$. In this case, the membrane is supposed to have a homogeneous distribution of scattering length density. The function (Eq. 10.3) is simple and unambiguous. The shape of the theoretical scattering curve depends only on value $R_t$ and does not depend on the size and overall shape of the membrane. The example of the approximation is presented in Fig. 10.4. This approach is widely used to study the effect of various factors on membrane thickness [70], including the presence of Aβ peptides [66].

There is however additional information readily obtainable from SANS experiments that relates to the overall form and size of objects. Since model and biological membranes in solution tend to form vesicles, their scattering curves can be approximated by function for spherical shells:

$$I(q) = \frac{A}{q} \cdot [(R+T)^2 j_1(q(R+T)) - R^2 j_1(qR)]^2 + b, \qquad (10.4)$$

where $j_1(x) = \frac{\sin x}{x^2} - \frac{\cos x}{x}$ is the first-order spherical Bessel function, $R$–is the inner radius of a shell, $T$ is the thickness of a membrane, $A$ is a constant depending on membrane concentration and contrast between membrane and solvent.

The function assumes that the scattering length density of the membrane is uniform, which however, can be improved easily by introducing more sophisticated models [71]. In addition, Eq. 10.4 has to be convoluted further with the size distribution function depending on radius $R$, due to the size polydispersity of real ULV samples [71].

Among the other studies, SANS allowed us to study the effect of cholesterol and melatonin on the membrane parameters in the presence of Aβ peptide, and the regime of peptide interactions with the membrane [66]. In the case of single lipid membranes, cholesterol makes the membrane thicker due to its "condensing effect" that increases the ordering and consequently the length of lipid tails. Melatonin, on the other hand, reduces the membrane thickness because of its disordering effect on lipid tails. Thus, cholesterol and melatonin can regulate the membrane fluidity in opposite directions: cholesterol decreasing, and melatonin increasing, as suggested already by the results of densitometry.

Importantly, both cholesterol and melatonin are present in the brain where they are believed to regulate the binding of Aβ peptide with membrane, and to affect the amyloid fibril formation [72, 73]. Our further SANS experiments have indeed shown the incorporation of Aβ peptide into the phospholipid membrane [66]. In the case of a pure lipid membrane, the peptide induces the thinning of the membrane. This effect can indicate that the peptide binds to the lipid-water interface of the bilayer, thereby increasing disorder in the bilayer and shortening the lipid tails, similarly to the case of the addition of melatonin. The same thickness behavior was observed upon the addition of Aβ peptide into the cholesterol containing membranes. In the case of melatonin containing membranes, the thickness change was negligibly small. That can be explained by the peptide incorporation into the hydrocarbon chain region as outlined in Fig. 10.5. It may be noted however, that the speculated results are at the limit of SANS experiments, and the approaches providing higher resolution are in place for further confirmation.

## 10.4 Detailed Membrane Structure

While the above discussed experimental approach of SANS covers several length scale regimes, it is nevertheless a low-resolution method with the limitation on the level of tens of angstroms (corresponding roughly to $2\pi/q_{max}$, where a typical $q_{max}$ ~ 0.2–0.3 $Å^{-1}$). This satisfies the information related to the overall membrane properties. It is however not possible to look more closely at the membrane local structure and even submolecular arrangements. The extension of the high-resolution limit is

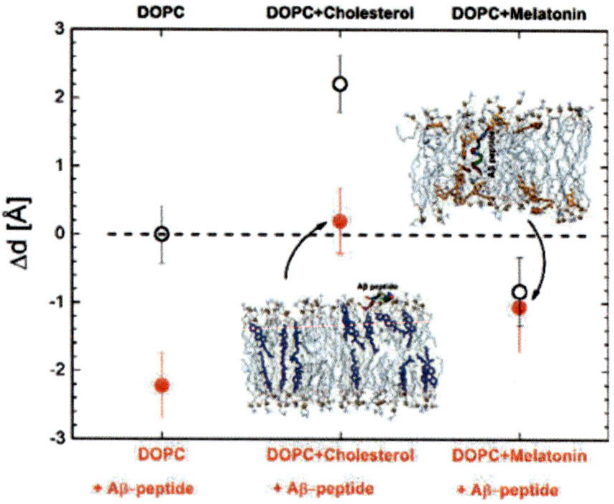


**Fig. 10.5** The changes in the thickness of DOPC bilayer Δd induced by the incorporation of Aβ peptide at the presence of cholesterol or melatonin [66]. The thickness of pure DOPC membrane was taken as a reference point (dashed line). The black empty points and top axis correspond to DOPC bilayer without Aβ peptide, and the red points and bottom axis correspond to the DOPC bilayer with 3 mol% of Aβ peptide. The inserts contain snapshots of molecular dynamic simulations of corresponding systems, corroborating the peptide's location near the lipid-water interface in the case of pure DOPC and that with the addition of cholesterol, while it is localized in the hydrocarbon region in the case of melatonin addition

often possible thanks to the changing the geometry of samples from unilamellar to multilamellar membranes, and from unoriented spherical vesicles to the parallelly oriented arrays [68]. As a result, several diffraction orders that become achievable experimentally may provide a spatial resolution of several angstroms (typical $q_{max} \sim 0.6\text{–}0.8$ Å$^{-1}$).

Similarly to SANS, small angle neutron diffraction (SAND) measurements are performed in an inverse $q$-space followed by the reconstruction of neutron scattering length density (NSLD) profiles in a real space. This is achieved straightforwardly through the Fourier transform of diffraction form factors $F_h$ (calculated as a square root of integral diffraction intensity) [74]:

$$NLSD(z) = \frac{2}{D}\sum_{h=1}^{h_{\max}} F_h \cos\left(\frac{2\pi h z}{D}\right) \tag{10.5}$$

where $h$ is a peak order, and $D$ is a lattice spacing calculated from the Bragg equation. In this, neutron diffraction in particular provides an unprecedented possibility to solve the infamous scattering phase problem directly in the experiment. The phase signs (+ or – for centro-symmetric systems in the case of most membranes) of diffraction form factors can be deduced from the systematic substitution of $H_2O$ by $D_2O$ in the hydrating solution that must yield a linear change as a function of $D_2O$ fraction [74].

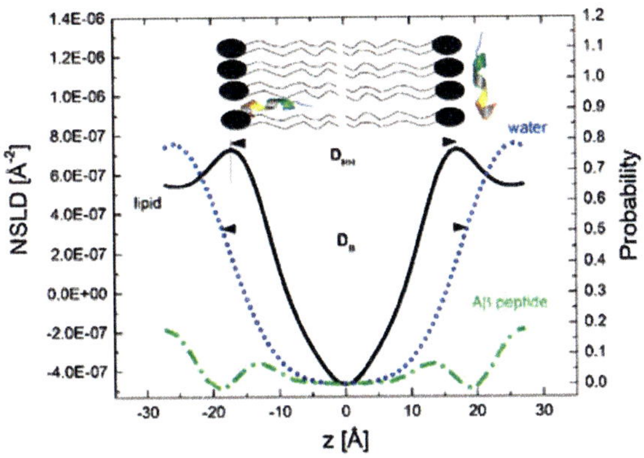


**Fig. 10.6** The NSLD profile reconstructed from SAND experiment. The solid line scaled to the left-hand axis shows the total NSLD profile of the entire bilayer hydrated with 8% D2O. The broken lines scaled according to the right-hand axis show the probability distributions of water (blue color) and 3 mol% Aβ peptide (green color). The insert on top of the graph depicts schematically the plausible distributions of Aβ peptides within the bilayer

The NSLD profiles obtained at various $D_2O$ concentrations are then utilized in determining several membrane structural properties. The profiles obtained at 8% $D_2O$ solution, for example, show only the membrane not obscured by the scattering from solvent (see Fig. 10.6). This is because the neutron scattering length for H is –3.739 fm and that for D is 6.671 fm (together with 5.803 fm for O) [75], making the contribution of 8% $D_2O$ water solution to the NSLD practically zero. These profiles are thus often used to extract lipid head-to-head thickness $D_{HH}$ from peak to peak distance across the bilayer. The $D_{HH}$ variations resulting from the compositional changes are then rudimentary parameters when looking for correlations between structure and function. The evaluation of $D_{HH}$ allowed for example to postulate the condensing effect of cholesterol and fluidizing effect of melatonin [8], as discussed in the previous section on SANS and densitometry.

In order to better understand how various components affect the interactions of biomolecules with the membrane, it is necessary to determine also their location within the membrane. The SAND experiments allow just that by providing quantitative data on the distribution of structural moieties, their sizes, shapes, and correlation lengths. At first, these can be determined following the assumption that the given component does not disturb the overall structure of the membrane and rather changes its features locally. In fact, this approach can be easily applied not only to neutron techniques, but to the X-ray diffraction as well, assuming the additions are made in small quantities. For example, 1.5 mol% of Aβ peptide was localized either inside the membrane core or at its surface depending whether shortened or full length peptide was added, respectively [76]. Our recent results demonstrate this approach in Fig. 10.6.

Another important result attainable from SAND experiment while utilizing contrast varied NSLD profiles is the water probability profile (see Fig. 10.6). The

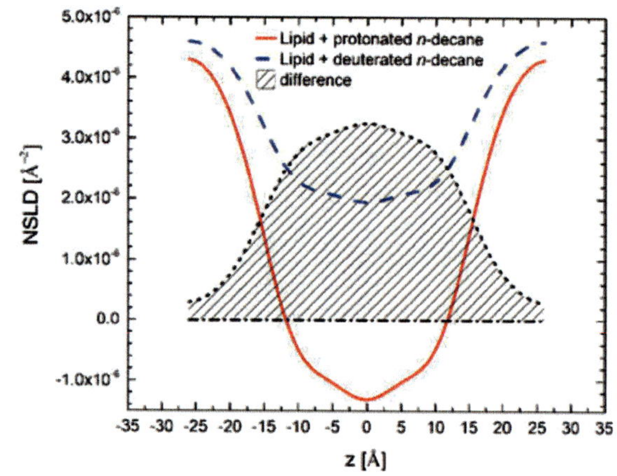


**Fig. 10.7** The example of the NSLD profile difference of labeled (blue) and unlabeled (red) samples. The difference reveals the label distribution along the profile, whose integrated area under the curve is directly proportional to the amount of the label

general NSLD profile consists of the contribution due to lipid membrane (and its substances as melatonin, cholesterol etc.) and water. The subtraction of NSLD profiles measured at any two different $D_2O$ contents provides thus the probability distribution of water only, as the unchanged membrane contributions cancel out each other [74]. The mean position of water-bilayer interface present in the water distribution (often obtained by error function fitting) determines then the bilayer thickness $D_B$ (Fig. 10.6).

Finally, the most intriguing advantage of neutron diffraction and the contrast variation technique in particular, is the labeling approach. The sensitivity of neutrons to various membrane components can be changed similarly to the case of $D_2O/H_2O$ exchange discussed above. The subtraction of the two NSLD profiles yields a distribution of label and thus a given moiety in question [77]. An example of this approach is illustrated in Fig. 10.7, where protonated vs. deuterated n-decane was added to the lipid membrane [78]. The differences are visible clearly when the labeled components are contained at high concentrations (up to 2:1 ratio of n-decane:lipid in the work mentioned). Unfortunately, often employed minute amounts of labeled components preclude the practicality of this approach in SANS and SAND methods. Its applicability may nevertheless benefit from even higher resolution achievable in reflectometry as we will discuss further.

## 10.5 Single Planar Membrane

Neutron reflectometry (NR) turned out to be a feasible tool for the investigation of model lipid membranes with the ever-progressing advancement of neutron sources. Particular attention has been paid to the application of NR for the study of interactions between the membranes and various proteins. Currently, there is a great expansion in its application also to the research of membranes with Aβ peptides, driven mainly by two factors. The first is the wide evidence on the Aβ peptides aggregation at the cell membranes that is correlated with the pathology of many serious disorders including Alzheimer's, Parkinson's, and prion disease or type II diabetes [79, 80]. The second factor is a great effort behind the optimization of the sample preparation methods for the NR studies of biological processes. As a result, NR allows scrutinizing the submolecular details of single planar membranes. Despite the availability of other well-developed experimental methods with comparable resolution (e.g. SANS and SAND), the application of NR can provide structural information on the systems with a specific architecture.

The suitability of NR follows from the fundamental principles of reflectometry as an optical method designed for the examination of thin layers. At a certain level of generalization, a single lipid membrane can be treated as a system of parallel layers, hence appropriate for scrutinizing by the NR. In addition, the crucial difference with respect to other methods is that NR allows inspecting a lipid bilayer in its individual planar form. This is in a contrast to the diffraction experiments where adjacent bilayers in the stack may intercept or hinder the access of solvent (and additives) in-between the lamellae. Although SANS enables experiments with lipid bilayers in a single form, ULVs of spherical shape and significant curvature may also exhibit some limitations with respect to the concept of the cell membrane model. The drawbacks that have to be considered are the difference of lateral pressure in inner and outer bilayer leaflets as well as the actual thickness of the lamella.

From the fundamental point of view, NR probes the structure of thin homogeneous layers and their interfaces with a possible resolution down to a few angstroms. The lipid bilayer deposited onto an optically flat surface can be regarded as a system of such layers with corresponding refraction indices. Impinging neutron beam penetrates the sample and it is reflected at the interfaces according to the laws of optics. The neutron reflectivity, *R*, refers to the ratio of the incoming and exiting neutron beams (the ratio of elastically and specularly scattered neutrons to incident neutrons) and it is measured as a function of momentum transfer $q$ (see the definition in 'Overall membrane thickness' section). *R(q)* is given by the Fresnel's law and Born approximation as the Fourier transform of *NSLD(z)* distribution in the direction perpendicular to the interface, similar to that in previous cases of SANS and SAND [81]:

$$R(q) = \frac{16\pi^2}{q^2}\left|\int NSLD(z)e^{-iqz}dz\right|^2 \tag{10.6}$$

Therefore, the intensity of the reflected beam depends on the thickness, density, and interface roughness of the sample.

Figure 10.8 shows reflectivity curves obtained for the single lipid bilayer systems deposited on a planar solid substrate. The influence of cholesterol (29 mol%), melatonin (29 mol%), and both compounds together (29 mol% + 29 mol%) on the structural properties of DOPC bilayer was investigated. The systems were hydrated by immersing the bilayer face of the substrate to heavy water ($D_2O$). Characteristic features of the *R(q)* curves are fringes with their magnitudes and local minima positions with respect to $q$. They relate to the layer's NSLDs and thicknesses in each system.

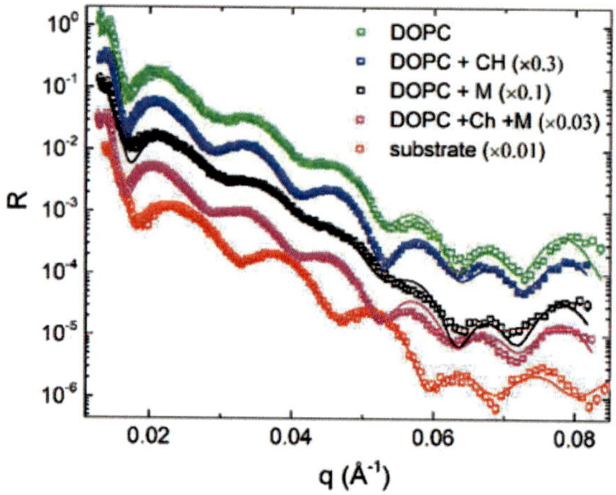


**Fig. 10.8** NR data obtained for the DOPC bilayer (green), and systems composed of DOPC with the addition of 29 mol% cholesterol (blue) and melatonin (black) separately or together (29 mol% + 29 mol%, magenta). Solid lines represent the best model fits to the experimental data. The data are scaled according to the factors shown in the legend for enhancing the clarity of presentation

In order to obtain the NSLD profile of the system and hence to determine its composition along the normal to the surface, reflectivity curves are modeled and subsequently fitted to the experimental data. The models employ the approach where the *NSLD(z)* is decomposed in a finite series of layers, each of which is defined by three parameters: NSLD, thickness, and roughness. Roughness refers to the interface of two adjacent layers and it is usually expressed by error function. Some advanced models account for an extra parameter that reflects less than the complete coverage of the substrate by the membrane. The data shown in Fig. 10.8 were fitted by the use of a model that assumes a lipid bilayer composed of three homogenous slabs comprising the lipid heads, lipid tails, and another side lipid heads. The results obtained by NR for the examined single planar membranes were [107] found to corroborate the data reported above for the membranes in the form of multilayers or ULVs obtained using SAND and SANS techniques, respectively [8, 66]. The counteracting effect of cholesterol and melatonin on membrane thickness (increase vs decrease, respectively) has been reconfirmed. More intriguingly, the submolecular resolution of NR allows extending the research by identifying the positions of the two components and their effects on the incorporation of other bio-related membrane components.

In order to address the tremendous variability of lipid membranes and the biological processes they facilitate, several sample preparation methods for the examination of protein adsorption by different biomimetic model membranes have been developed and employed in NR. Currently, three general approaches are utilized most frequently, depending on specific conditions that are required to maintain in a given examined system. The most common types of biomimetic membranes they are expected to model are (i) phospholipid monolayer at air/water interface, (ii) phospholipid monolayer on solid supports, and (iii) fluid lipid bilayers on solid support. Monolayers at the air/water interface (i) provide advantages in their simplicity, stability, ease of preparation and controlling of their properties. Proteins are adsorbed to the monolayer from the aqueous phase, while the lipid packing density of the phospholipid monolayer formed using the Langmuir trough can be precisely monitored by controlling the applied surface pressure. This ability is available even during the experiment and it is absent in the majority of other artificial membrane models. Its unique application potential stems from the fact that lipid packing density affects the protein or peptide interactions with membranes. It has been shown with the aid of NR that the amount of proteins associated with lipid monolayers decreases with the increase of surface pressure [82].

Instead of water, monolayers can be also spread on solid supports (ii). Since phospholipid monolayer mimics one leaflet of a biological membrane, it can be employed in the studies of the interactions and adsorption of peripheral proteins [83]. Finally, fluid lipid bilayers on a solid support (iii) represent the group of the most popular artificial membrane models. Unlike the lipid monolayers, they mimic the plasma membrane more accurately, hence being physiologically more relevant. In the wide range of their applications in NR, these systems have shown their suitability in the studies of protein-membrane interactions where the effect of bilayer environment has also been taken into account.

The foremost investigated environmental effect in membrane research is that determining the membrane thermodynamic state, in which the concept of lipid rafts is attracting a lot of attention. For example, the interaction of Aβ peptides with a single asymmetric complex membrane mimicking the lipid raft in its composition is suggested important for Aβ peptides settling and seeding [84]. NR with the aid of solvent contrast variation method, and contrast enhancement between deuterated lipids and hydrogen rich peptides was able to localize the place of action of the Aβ peptide structures and assess their penetration depth. Apparently, the structured oligomers of $A\beta_{1-42}$, considered as the most membrane-active state, are penetrating the outer leaflet of the membrane only, where they can serve as seeds for further Aβ aggregation. On the other hand, the unstructured $A\beta_{1-42}$ early-oligomers were localized even in the inner leaflet. Based on the peculiar NSLD profile, Aβ monomers are hypothesized to organize themselves within the membrane environment to form the pores.

A different approach is employed in the case of polymer cushioned membranes that are deposited on a soft polymeric layer. The layer provides an environment that attempts to mimic the cytoskeleton or extracellular matrix. It facilitates the reconstruction of membrane domains morphology and allows for mobility of the transmembrane proteins. In this way, the role of $A\beta_{1-42}$ peptides on polymer cushioned membranes with composition mimicking the neural membrane could has been studied at physiological conditions [79]. By monitoring the kinetics of the bilayer formation followed by the injection of Aβ peptides, the NR curves and consequently reconstructed NSLD profiles revealed the penetration of Aβ peptides to the membrane. This could have been correlated nicely with the concomitant increase of membrane thickness and softening. Although the membrane thickness is certainly one of the primary parameters achievable in NR (as well as SANS and SAND), the softening is a parameter related to the membrane dynamics, and is therefore best researched by other methods, including the inelastic scattering.

## 10.6 Membrane Dynamics

Inelastic X-ray and neutron scattering techniques (IXS and INS) are widely used to study the dynamic properties of various materials. The method is based on measuring the energy and momentum changes of photons (neutrons), which are determined by excitations of atoms and/or molecules (known as phonons) inside the sample. IXS method requires incident photons with energies $(10 \div 25)$ keV with transfer energies during scattering on a sample within $(1 \div 100)$ meV [85].

IXS and INS represent two of the few techniques capable to detect rotational and vibrational excitations of lipid molecules [86, 87], diffusion processes inside the membrane [88, 89], and to determine energies of intermolecular (lipid-lipid and lipid-protein) interactions in picosecond-nanometer time–space window [90, 91] in particular. IXS and INS are complementary techniques, with their advantages and disadvantages. For example, INS is nondestructive for the sample and provides excellent energy resolution, but has a limited dynamic range due to kinematic constraints. In contrast, IXS has no superior energy resolution (comparing to INS) and the lipid sample may be damaged by radiation during the experiment. At the same time however, IXS scattering cross-section is inherently coherent, so the deuteration of lipid tails (often performed for experiments with neutrons), which changes lipid's physical properties, is not required. This makes IXS method a bit more preferable for studying the dynamical properties of lipid membranes.

A typical spectrum of in-plane inelastic X-ray scattering on oriented lipid multilayers is shown in Fig. 10.9. In addition to the central elastic peak, which corresponds to a zero-energy transfer of incident photons, the IXS curve exhibits few inelastic peaks whose energies are determined by the energies of the corresponding propagating phonon modes. Positive and negative values of the transferred energy correspond to the creation and annihilation of phonon, respectively. The number of inelastic peaks is determined by the number of unique phonon modes. The dependences of transferred energies on the scattering vector $q$ are called the dispersion curves. The range for the $q$ vector is determined by the size of quasi-Brillouin zone, which can be understood as the mean distance between the hydrocarbon chains in the case of lipid membranes.

The IXS technique used recently for investigating the diffusion and relaxation processes occurring in the lipid membrane has provided experimental evidence for a dispersive transverse acoustic mode in both $L_\beta$ and $L_\alpha$ phase of a pure dipalmitoylphosphatidylcholine (DPPC) that has not been reported before in the literature [90]. The transverse mode exhibited a low-frequency phonon gap when the DPPC transitions into the $L_\alpha$ phase. It was argued that such band gaps are directly related to spontaneous short-lived (on the picosecond timescale) local lipid clustering and, consequently, the formation of volume voids in the membrane. The void formation process then in turn appears to underlie the passive transport of small solutes through the cell membrane.

Further, the investigation of collective vibrations in DPPC-cholesterol binary mixtures has revealed the existence of an optical phonon, which emerges due to

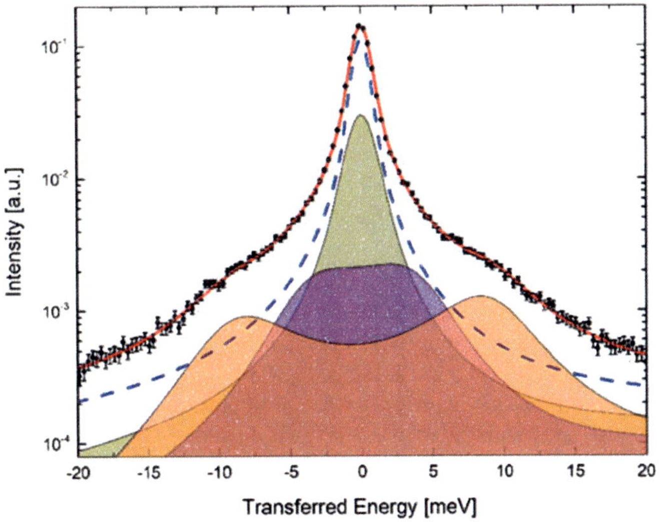


**Fig. 10.9** A typical spectrum of in-plane inelastic X-Ray scattering on oriented lipid membranes (black points). The central elastic peak is shown by the blue dashed line, and several inelastic peaks are shown by lines with color filled regions

out-of-phase vibrations of coupled DPPC and cholesterol molecules [91]. The effect of cholesterol molar fraction on the phonon band gap opening and evolution has been determined. The opening of the optical phononic gap with increased cholesterol concentration is direct evidence of the nanoscale domains in Lo phase and their relationship to the mechanical properties of the cell membrane (such as surface tension, or bending rigidity), which in turn is related to various cell functions, like passive transport, cell curvature generation and sustainability.

Considering the recent interest in membrane complexes with cholesterol, melatonin, and Aβ peptides, it will be interesting to employ the above discussed approach in determining the effect of cholesterol and melatonin – separately as well as together – on the phonon band gap opening in model phospholipid membrane with Aβ peptide. Such a system may ensue a formation of rafts as already mentioned in previous sections. Conveniently, the formation of lateral inhomogeneities creates defects at the interfaces of these structures that can be well detected and characterized by inelastic scattering. The integration of peptides amid lipid molecules would also have a characteristic fingerprint in experimental data. The concentration and temperature studies of the phonon band gap evolution will shed a light on the role of cholesterol and melatonin in the considerable changes to the membrane structural properties. This will provide the base information for furthering our understanding of the processes like Aβ peptide insertion and lateral distribution, direct transport of small molecules, and energy transfers in cell membranes.

## 10.7 Aβ Peptide Secondary Structure

One of the main interests in the Aβ peptide related investigations is the mechanism of peptide incorporation into the biological membrane and/or its withdrawal from the membrane. This mechanism is expected to be influenced by both the membrane elasto-mechanical properties and structural properties of Aβ peptides. Although neither is understood yet fully, the quaternary structure of amyloid fibrils points the attention on the tertiary and secondary structures of Aβ peptides (not to discuss the importance of primary structure that is documented by the toxicity of Aβ peptides and their amino acids 25–35 in particular).

One of the methods for determining the secondary structure of peptides is circular dichroism. This method has been used since the very first investigation of the $A\beta_{25-35}$ peptide aggregation. The results showed a balance between the α-helical, β-sheet and random coil structures with the changes of various conditions [92]. The balance was shown to be pH dependent with the concentration driven transition of the random coil to β-sheet happening in sodium acetate buffer at pH 4 and 5.5, and it shifted the distribution towards the β-sheets with the increasing of pH up to 7.4. The addition of negatively charged lipid vesicles to the system comprising $A\beta_{25-35}$ also caused the distribution balance to shift in the direction of β-sheet formation, which could be obtained alternatively by increasing the $A\beta_{25-35}$ concentration almost 10 times at the condition without the lipids.

The same structural changes were observed in full-length $A\beta_{1-40}$ peptide. Investigations of the addition of lipid membranes to $A\beta_{1-42}$ and $A\beta_{1-40}$ under various pH and salt concentrations, however, concluded the peptide secondary structure being not the only factor in the membrane disrupting activity [93]. The additional NMR study of Aβ peptide binding to the lipid membranes showed no changes in lipid headgroup region conformation, and in the lipid chain flexibility and ordering [94]. The peptides therefore could appear to exist with lipid membranes in a non-interacting, or at least non-disturbing way at some conditions, while a severe disturbance may occur in other circumstances.

The questions of Aβ peptide interactions with membranes evolved in the necessity to differentiate between the incorporation and adsorption of peptides to the lipid membrane. The investigations of conformational changes in the Aβ structure over the time period of 18 days have shown that incorporated peptides avoid the conformational changes, while the surface associated Aβ peptides show dramatic changes by going from disordered structures to predominantly β-sheet structures after several weeks [95]. The mechanisms of these changes are however yet to be understood. A great deal of detailed information needed for shedding more light on the problem may perhaps come from theoretical studies based on molecular dynamics (MD) simulations.

In view of the continuous development of high-performance heterogeneous computing systems, the efficiency of MD methods for modeling the biological structures, such as lipid membranes and proteins interacting with them, is steadily increasing. MD simulations help to interpret and corroborate experimental data obtained from SANS, SAND, NR, and INS discussed earlier as well as many other experimental techniques. In turn, the experimental results are an essential part in the verification and further improvements of MD force fields. Nowadays, one of the popular areas of the joined MD-experimental approaches is the study of the interaction of the Aβ peptide with model lipid membranes. The principal interest in this direction, registering particularly with MD methodology, is the problem of searching conformations and location of the Aβ peptide in membrane-water systems.

The replica–exchange molecular dynamics (REMD) and the umbrella sampling simulation methods are often used to solve the above problems [96–98]. The idea of the replica exchange method is that a set of replicas is created, each of which is assigned a certain temperature value, and those replicas are simulated further in parallel with a periodic exchange of conformational states. Umbrella sampling is one of the accelerated sampling methods, the idea of which is to insert an additional potential reducing potential barriers, that allows the system to bypass the conformational space more easily, and then the effect of this potential is subtracted from the obtained free energy profile. Well-tempered metadynamics is also a method of enhanced sampling and calculating the free energy landscape, which is accelerated by the bias potential depending on collective variables.

The first step towards investigating the mutual influence of Aβ peptides and lipid membranes is modeling systems with peptide monomers. Currently, a number of MD simulations have been carried out. They are aimed at studying the Aβ interaction with model membranes depending on the length and conformation of the peptide, as well as the composition of the bilayer (e.g., saturated or unsaturated lipid chains, zwitterionic or anionic lipid headgroups, cholesterol or melatonin addition, etc.). A general pattern of the Aβ interaction with lipid membranes is then justified by the consistency of results.

Simulations of the $A\beta_{25-35}$ monomer in an implicit water-membrane environment and explicit DMPC membrane with water models by REMD have shown for example that the peptide can locate in the membrane in two states – a surface-bound state and a less stable inserted state, between which Aβ can often be interconverted (see Fig. 10.10 for the illustration) [99, 100]. It is encouraging to recall similar conclusions achieved based on the SAND experiments [76], though the capability of MD expands

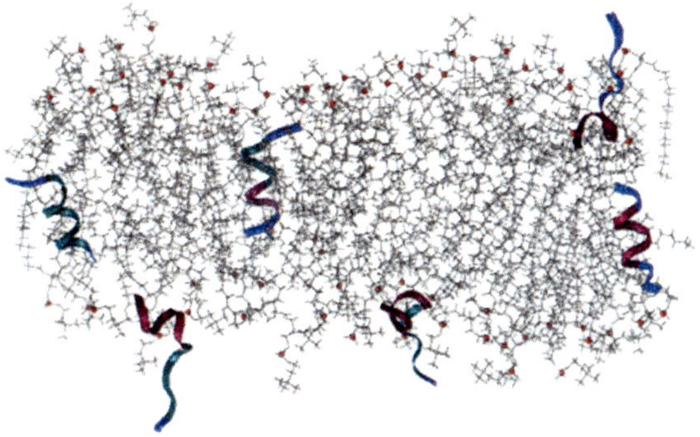

**Fig. 10.10** Snapshots of $A\beta_{25-35}$ conformations (helix shown by purple, turn by cyan, and coil by blue colors) and positions in phospholipid membrane (phosphorus atoms shown by red points, rest of the lipids shown by gray color; water was omitted for the clarification of presentation)

the results further. Accordingly, the peptide has highly extended conformations in the surface-bound state. In the second state, the C-terminus has a stable helix shape and is located deep in the membrane, while the N-terminus is more disordered with lower helix propensities and closer to hydrophilic heads. In addition, the peptide binding slightly decreases membrane thickness and does not cause strong lipid disordering.

The REMD simulations of A$\beta_{10-40}$ monomer with an explicit DMPC bilayer demonstrate the existence of one stable peptide state in the membrane, in which the C-terminus penetrates the hydrophobic part of phospholipids and forms a stable helix, while the N-terminus interacts mainly with the bilayer surface [101]. At the same time, the membrane thickness and area per lipid are significantly reduced, though the increase in the disorder of lipid tails is insignificant due to the shallow insertion of A$\beta_{10-40}$.

The results obtained by umbrella sampling simulations of the A$\beta_{1-42}$ peptide in zwitterionic DPPC and anionic DOPS bilayers, and molecular dynamics simulations of the A$\beta_{1-40}$ in zwitterionic POPC, anionic POPS, and mixed POPC/POPE bilayers suggest that the anionic bilayers may stabilize the peptide secondary structure to a greater extent than the zwitterionic ones. Besides, the C-terminus is embedded into the membrane hydrophobic core in all but the case of anionic membranes, while the N-terminus is located closer to the water-membrane interface and more strongly exposed to a solvent [44, 102]. The main A$\beta_{1-40}$ conformations are helices and random coils, although β-hairpins are also observed, mainly in N-terminus peptide region.

The composition of lipid bilayers significantly influences the insertion, orientation, and aggregation of Aβ peptides, regulating Aβ-membrane interactions. Polyunsaturated lipids (PUL) cause, apparently, stronger adsorption of Aβ on the membrane and lead to a weaker binding between peptides in aggregates [9]. In the presence of cholesterol in the lipid membrane, that has been discussed above to ensure a stiffened membrane as opposed to the highly fluid PUL membranes, the A$\beta_{25-35}$ peptide does not permeate into the bilayer hydrophobic core due to the high order of lipids. This leads to the peptide aggregation on the membrane surface and a slight decrease of the bilayer thickness [103, 104]. In the case of the longer A$\beta_{1-40}$ and A$\beta_{1-42}$ peptides, A$\beta_{1-40}$ also ends up preferably at the surface of cholesterol-depleted membranes, though it exhibits a partially or completely embedded state in cholesterol-enriched membranes. In turn, A$\beta_{1-42}$ peptide inserts partially or completely into cholesterol-depleted membranes and completely inserts into cholesterol-enriched membranes [105].

## 10.8 Concluding Remarks

Alzheimer's disease was first described in 1906 based on its clinical symptoms and microscopic changes in the brain structure [106]. For many years then to come, microscopy was the main method for recognizing the disease at the autopsy of patients. With its increasing social impact, however, increased also an interest in its identification. Over more than a century, it has seen a huge development in the research approaches available more or less readily for its interrogation.

The research in general and structural biophysics in particular, have proven to benefit from the development of dedicated X-ray sources. Starting with the fixed tube, coincidently around the same time as the AD was identified, all the way to the free electron lasers, their brilliance has advanced by impressive 16 orders of magnitude. The golden era of neutrons following its discovery in 1932, brought other expectations and advancements. The peculiar properties of neutrons allowed for building sample holders capable to reproduce biologically relevant conditions during experiments. The *in situ* approaches expanded to include contrast variation and deuterium labeling techniques as the most powerful tools. The increase of computational power, doubling every 2 years, earned theoretical approaches their niche in structural biology most recently. MD simulations, following the sample dynamics for microseconds now, provide an unprecedented level and amount of details. We have entered a century of big data, built on promises to combine all the information into one model.

The research on AD may not be at the end just yet, but we certainly have a lot of knowledge by now. Owing to the advancements in research approaches, there is a final number of hypotheses describing the origin of the disease. There is a final amount of models depicting the interactions leading to the onset of the disease. Most importantly though, there is finally a limited number of proposals for combating the disease. Some of the discussed problems certainly find their place at the membrane, either as the place of origin, development, or even action of Aβ peptides. The closer and closer looks at these problems will undoubtedly provide an understanding of relevant mechanisms and perhaps will be a part of the final solution.

**Acknowledgements** This work has been supported by the Russian Science Foundation under grant 19-72-20186.

## References

1. Lodish H, Arnold B, Zipursky L, Matsudaira P, Baltimore D, Darnell J (1999) Molecular Cell Biology, 4th edn. W. H. Freeman, New York
2. Lee AG (2004) How lipids affect the activities of integral membrane proteins. Biochim Biophys Acta - Biomembr 1666:62–87
3. Van Meer G, Voelker DR, Feigenson GW (2008) Membrane lipids: where they are and how they behave. Nat Rev Mol Cell Biol 9:112–124
4. Koynova R, Caffrey M (1998) Phases and phase transitions of the phosphatidylcholines. Biochim. Biophys Acta - Rev Biomembr 1376:91–145
5. Tardieu A, Luzzati V, Reman FC (1973) Structure and polymorphism of the hydrocarbon chains of lipids: a study of lecithin-water phases. J Mol Biol 75:711–733
6. Chapman D (1975) Phase transitions and fluidity characteristics of lipids and cell membranes. Q Rev Biophys 8:185–235
7. Bretscher MS, Munro S (1993) Cholesterol and the Golgi apparatus. Science 261:1280–1281

8. Drolle E et al (2013) Effect of melatonin and cholesterol on the structure of DOPC and DPPC membranes. Biochim Biophys Acta - Biomembr 1828:2247–2254
9. Ntarakas N, Ermilova I, Lyubartsev A (2019) Effects of lipid saturation on amyloid-beta peptides partitioning and aggregation in neuronal membranes: molecular dynamics simulations. Eur Biophys J 48:813–824
10. Martel A et al (2017) Membrane permeation versus amyloidogenicity: a multitechnique study of islet amyloid polypeptide interaction with model membranes. J Am Chem Soc 139:137–148
11. Dante S, Hauß T, Brandt A, Dencher NA (2008) Membrane fusogenic activity of the Alzheimer's peptide Aβ(1–42) demonstrated by small-angle neutron scattering. J Mol Biol 376:393–404
12. Prince M, Comas-Herrera A, Knapp M, Guerchet M, Karagiannidou M (2016) World Alzheimer report 2016: improving healthcare for people living with dementia: coverage, quality and costs now and in the future
13. Hippius H, Neundörfer G (2003) The discovery of Alzheimer's disease. Dialogues Clin Neurosci 5:101–108
14. Liu PP, Xie Y, Meng XY, Kang JS (2019) History and progress of hypotheses and clinical trials for Alzheimer's disease. Signal Transduct Target Ther 4:29
15. Bush AI et al (1994) Rapid induction of Alzheimer Aβ amyloid formation by zinc. Science 265:1464–1467
16. Spinello A, Bonsignore R, Barone G, K. Keppler B, Terenzi A (2016) Metal ions and ietal complexes in Alzheimer's Disease. Curr Pharm Des 22:3996–4010
17. Clements A, Allsop D, Walsh DM, Williams CH (1996) Aggregation and metal-binding properties of mutant forms of the amyloid Aβ peptide of Alzheimer's disease. J Neurochem 66:740–747
18. Francis PT, Palmer AM, Snape M, Wilcock GK (1999) The cholinergic hypothesis of Alzheimer's disease: a review of progress. J Neurol Neurosurg Psychiatry 66:137–147
19. Mudher A, Lovestone S (2002) Alzheimer's disease – do tauists and baptists finally shake hands? Trends Neurosci 25:22–26
20. Hardy J, Allsop D (1991) Amyloid deposition as the central event in the aetiology of Alzheimer's disease. Trends Pharmacol Sci 12:383–388
21. Martorana A, Esposito Z, Koch G (2010) Beyond the cholinergic hypothesis: do current drugs work in Alzheimer's disease? CNS Neurosci Ther 16:235–245
22. Nussbaum JM, Seward ME, Bloom GS (2013) Alzheimer disease. Prion 7:14–19
23. Mohandas E, Rajmohan V, Raghunath B (2009) Neurobiology of Alzheimer's disease. Indian J Psychiatry 51:55–61
24. Di J, Cohen LS, Corbo CP, Phillips GR, El Idrissi A, Alonso AD (2016) Abnormal tau induces cognitive impairment through two different mechanisms: Synaptic dysfunction and neuronal loss. Sci Rep 6:1–12
25. Hardy J, Selkoe DJ (2002) The amyloid hypothesis of Alzheimer's disease: progress and problems on the road to therapeutics. Science 297:353–356
26. Vestergaard M, Hamada T, Takagi M (2008) Using model membranes for the study of amyloid beta:lipid interactions and neurotoxicity. Biotechnol Bioeng 99:753–763
27. Buchsteiner A, Hauß T, Dante S, Dencher NA (2010) Alzheimer's disease amyloid-B peptide analogue alters the PS-dynamics of phospholipid membranes. Biochim Biophys Acta - Biomembr 1798:1969–1976
28. Ehehalt R, Keller P, Haass C, Thiele C, Simons K (2003) Amyloidogenic processing of the Alzheimer beta-amyloid precursor protein depends on lipid rafts. J Cell Biol 160:113–123
29. Hattori C et al (2006) BACE1 interacts with lipid raft proteins. J Neurosci Res 84:912–917
30. Sadigh-Eteghad S, Talebi M, Farhoudi M, Golzari SEJ, Sabermarouf B, Mahmoudi J (2014) Beta-amyloid exhibits antagonistic effects on alpha 7 nicotinic acetylcholine receptors in orchestrated manner. J Med Hypotheses Ideas 8:49–52
31. Yao ZX, Papadopoulos V (2002) Function of beta-amyloid in cholesterol transport: a lead to neurotoxicity. FASEB J 16:1677–1679

32. Bogoyevitch MA, Boehm I, Oakley A, Ketterman AJ, Barr RK (2004) Targeting the JNK MAPK cascade for inhibition: basic science and therapeutic potential. In: Biochimica et Biophysica Acta - Proteins and Proteomics. Elsevier, pp 89–101
33. Soscia SJ, et al (2010) The Alzheimer's disease-associated amyloid β-Protein is an antimicrobial peptide. PLoS One 5:e9505
34. Dahlgren KN, Manelli AM, Stine WB, Baker LK, Krafft GA, LaDu MJ (2002) Oligomeric and fibrillar species of amyloid-β peptides differentially affect neuronal viability. J Biol Chem 277:32046–32053
35. Kim H-S et al (2002) Amyloid beta peptide induces cytochrome C release from isolated mitochondria. NeuroReport 13:1989–1993
36. Lim L, Wenk MR (2009) Neuronal membrane lipids–their role in the synaptic vesicle cycle. Handbook of neurochemistry and molecular neurobiology. Springer, US, Boston, pp 223–238
37. Chu N, Kučerka N, Liu Y, Tristram-Nagle S, Nagle JF (2005) Anomalous swelling of lipid bilayer stacks is caused by softening of the bending modulus. Phys Rev E 71:41904
38. Ermilova I, Lyubartsev AP (2019) Cholesterol in phospholipid bilayers: positions and orientations inside membranes with different unsaturation degrees. Soft Matter 15:78–93
39. Kučerka N, Marquardt D, Harroun TA, Nieh M-P, Wassall SR, Katsaras J (2009) The functional significance of lipid diversity: orientation of cholesterol in bilayers is determined by lipid species. J Am Chem Soc 131:16358–16359
40. Martinez-Seara H, Róg T, Pasenkiewicz-Gierula M, Vattulainen I, Karttunen M, Reigada R (2008) Interplay of unsaturated phospholipids and cholesterol in membranes: effect of the double-bond position. Biophys J 95:3295–3305
41. Nagle JF, Tristram-Nagle S (2000) Structure of lipid bilayers. Biochim Biophys Acta - Rev Biomembr 1469:159–195
42. Petrache HI, Gouliaev N, Tristram-Nagle S, Zhang R, Suter RM, Nagle JF (1998) Interbilayer interactions from high-resolution x-ray scattering. Phys Rev E 57:7014–7024
43. Yoda M, Miura T, Takeuchi H (2008) Non-electrostatic binding and self-association of amyloid β-peptide on the surface of tightly packed phosphatidylcholine membranes. Biochem Biophys Res Commun 376:56–59
44. Davis CH, Berkowitz ML (2009) Interaction between amyloid-β (1–42) peptide and phospholipid bilayers: A molecular dynamics study. Biophys J 96:785–797
45. Kirkitadze MD, Condron MM, Teplow DB (2001) Identification and characterization of key kinetic intermediates in amyloid β-protein fibrillogenesis. J Mol Biol 312:1103–1119
46. Kobayashi S, Tanaka Y, Kiyono M, Chino M, Chikuma T, Hoshi K, Ikeshima H (2015) Dependence pH and proposed mechanism for aggregation of Alzheimer's disease-related amyloid-β(1–42) protein. J Mol Struct 1094:109–117
47. Butterfield DA, Lauderback CM (2002) Lipid peroxidation and protein oxidation in Alzheimer's disease brain: potential causes and consequences involving amyloid β-peptide-associated free radical oxidative stress. Free Radic Biol Med 32:1050–1060
48. Butterfield DA, Boyd-Kimball D (2004) Amyloid β-peptide(1–42) contributes to the oxidative stress and neurodegeneration found in Alzheimer disease brain. Brain Pathol Int Soc Neuropathol 14:426–432
49. Marcus D et al (1998) Increased peroxidation and reduced antioxidant enzyme activity in Alzheimer's disease. Exp Neurol 150:40–44
50. Hensley K et al (1995) Brain regional correspondence between Alzheimer's disease histopathology and biomarkers of protein oxidation. J Neurochem 65:2146–2156
51. Cheignon C, Tomas M, Bonnefont-Rousselot D, Faller P, Hureau C, Collin F (2018) Oxidative stress and the amyloid beta peptide in Alzheimer's disease. Redox Biol 14:450–464
52. Quincke G (1888) Ueber periodische Ausbreitung an Flüssigkeitsoberflächen und dadurch hervorgerufene Bewegungserscheinungen. Ann der Phys und Chemie 271:580–642
53. Pabst G, Kučerka N, Nieh MP, Rheinstädter MC, Katsaras J (2010) Applications of neutron and X-ray scattering to the study of biologically relevant model membranes. Chem Phys Lipids 163:460–479

54. Marquardt D, Heberle FA, Nickels JD, Pabst G, Katsaras J (2015) On scattered waves and lipid domains: detecting membrane rafts with X-rays and neutrons. Soft Matter 11:9055–9072
55. Knoll W, Haas J, Stuhrmann HB, Füldner H-H, Vogel H, Sackmann E (1981) Small-angle neutron scattering of aqueous dispersions of lipids and lipid mixtures. A contrast variation study. J Appl Crystallogr 14:191–202
56. Murugova TN et al (2011) Potentials of small-angle neutron scattering for studies of the structure of "live" mitochondria. Neutron News 22:11–14
57. Armstrong CL et al (2012) Co-existence of gel and fluid lipid domains in single-component phospholipid membranes. Soft Matter 8:4687–4694
58. Kuklin A et al (2020) On the origin of the anomalous behavior of lipid membrane properties in the vicinity of the chain-melting phase transition. Sci Rep 10:1–8
59. Luzzati V, Tardieu A (1974) Lipid phases: structure and structural transitions. Annu Rev Phys Chem 25:79–94
60. Lewis R, McElhaney R (2011) The mesomorphic phase behavior of lipid bilayers. In: The structure of biological membranes, 3rd edn, pp 19–89. CRC Press, Cambridge
61. Tolédano JC, Tolédano P (1987) The Landau theory of magnetic phase transitions. World Sci Lecture Notes Phys C 3:307–373
62. Kittel C, Kroemer H, Scott HL (1998) Thermal physics, 2nd ed. Am J Phys 66:164–167
63. Nagle JF (1973) Lipid bilayer phase transition: density measurements and theory. Proc Natl Acad Sci U S A 70:3443–3444
64. Kratky O, Leopold H, Stabinger H (1973) [5] The determination of the partial specific volume of proteins by the mechanical oscillator technique. Methods Enzymol 27:98–110
65. Murugova TN, Balgavý P (2014) Molecular volumes of DOPC and DOPS in mixed bilayers of multilamellar vesicles. Phys Chem Chem Phys 16:18211–18216
66. Murugova T et al (2020) Structural changes introduced by cholesterol and melatonin to the model membranes mimicking preclinical conformational diseases. Gen Physiol Biophys 39:135–144
67. Lau TL et al (2006) Amyloid-β peptide disruption of lipid membranes and the effect of metal ions. J Mol Biol 356:759–770
68. Kučerka N, Liu Y, Chu N, Petrache HI, Tristram-Nagle S, Nagle JF (2005) Structure of fully hydrated fluid phase DMPC and DLPC lipid bilayers using X-Ray scattering from oriented multilamellar arrays and from unilamellar vesicles. Biophys J 88:2626–2637
69. Kuklin AI, et al (2017) Neutronographic investigations of supramolecular structures on upgraded small-angle spectrometer YuMO. J Phys Conf Ser 848:012010. Institute of Physics Publishing
70. Sreij R et al (2018) DMPC vesicle structure and dynamics in the presence of low amounts of the saponin aescin. Phys Chem Chem Phys 20:9070–9083
71. Kučerka N, Nagle JF, Feller SE, Balgavý P (2004) Models to analyze small-angle neutron scattering from unilamellar lipid vesicles. Phys Rev E - Stat Physics, Plasmas, Fluids, Relat Interdiscip Top 69:9
72. Puglielli L, Tanzi RE, Kovacs DM (2003) Alzheimer's disease: the cholesterol connection
73. Karasek M (2004) Melatonin, human aging, and age-related diseases. Exp Gerontol 39:1723–1729
74. Kučerka N, Nieh M, Pencer J, Sachs J, Katsaras J (2009) What determines the thickness of a biological membrane. Gen Physiol Biophys 28:117–125
75. Sears VF (1992) Neutron scattering lengths and cross sections. Neutron News 3:26–37
76. Barrett M, Alsop R, Hauß T, Rheinstädter M (2015) The position of Aβ22-40 and Aβ1-42 in anionic lipid membranes containing cholesterol. Membranes (Basel) 5:824–843
77. Dante S, Hauss T, Dencher NA (2002) β-amyloid 25 to 35 is intercalated in anionic and zwitterionic lipid membranes to different extents. Biophys J 83:2610–2616
78. Hrubovčák P et al (2018) Location of the general anesthetic n-decane in model membranes. J Mol Liq 276:624–629

79. Dante S, Hauß T, Steitz R, Canale C, Dencher NA (2011) Nanoscale structural and mechanical effects of beta-amyloid (1–42) on polymer cushioned membranes: a combined study by neutron reflectometry and AFM Force Spectroscopy. Biochim Biophys Acta - Biomembr 1808:2646–2655
80. Hellstrand E et al (2013) Adsorption of α-synuclein to supported lipid bilayers: Positioning and role of electrostatics. ACS Chem Neurosci 4:1339–1351
81. Jackson AJ (2008) Introduction to small-angle neutron scattering and neutron reflectometry. NIST Center for Neutron Research. https://ftp.ncnr.nist.gov/summerschool/ss10/pdf/SANS_NR_Intro.pdf
82. Maierhofer AP, Bucknall DG, Bayerl TM (2000) Modulation of cytochrome C coupling to anionic lipid monolayers by a change of the phase state: a combined neutron and infrared reflection study. Biophys J 79:1428–1437
83. Lu JR, Murphy EF, Su TJ, Lewis AL, Stratford PW, Satija SK (2001) Reduced protein adsorption on the surface of a chemically grafted phospholipid monolayer. Langmuir 17:3382–3389
84. Rondelli V et al (2016) Amyloid β Peptides in interaction with raft-mime model membranes: a neutron reflectivity insight. Sci Rep 6:20997
85. Baron Alfred QR (2016) High-Resolution Inelastic X-Ray scattering I: Context, spectrometers, samples, and superconductors. In: Jaeschke Eberhard J, Khan Shaukat, Schneider Jochen R, Hastings Jerome B (eds) Synchrotron light sources and free-electron lasers. Springer, Cham, pp 1643–1719. https://doi.org/10.1007/978-3-319-14394-1_41
86. Bonn M, Bakker HJ, Ghosh A, Yamamoto S, Sovago M, Campen RK (2010) Structural inhomogeneity of interfacial water at lipid monolayers revealed by surface-specific vibrational pump-probe spectroscopy. J Am Chem Soc 132:14971–14978
87. Mashaghi A et al (2012) Interfacial water facilitates energy transfer by inducing extended vibrations in membrane lipids. J Phys Chem B 116:6455–6460
88. Busch S, Smuda C, Pardo LC, Unruh T (2010) Molecular mechanism of long-range diffusion in phospholipid membranes studied by quasielastic neutron scattering. J Am Chem Soc 132:3232–3233
89. Pronk S, Lindahl E, Kasson PM (2015) Coupled diffusion in lipid bilayers upon close approach. J Am Chem Soc 137:708–714
90. Zhernenkov M et al (2016) Revealing the mechanism of passive transport in lipid bilayers via phonon-mediated nanometre-scale density fluctuations. Nat Commun 7:11575–11575
91. Soloviov D et al (2020) Functional lipid pairs as building blocks of phase-separated membranes. Proc Natl Acad Sci U S A 117:4749–4757
92. Terzi E, Seelig J, Hölzemann G (1994) Alzheimer β-Amyloid peptide 25–35: electrostatic interactions with phospholipid membranes. Biochemistry 33:7434–7441
93. McLaurin JA, Chakrabartty A (1997) Characterization of the interactions of Alzheimer β-amyloid peptides with phospholipid membranes. Eur J Biochem 245:355–363
94. Terzi E, Hölzemann G, Seelig J (1997) Interaction of Alzheimer β-amyloid peptide(1–40) with lipid membranes. Biochemistry 36:14845–14852
95. Bokvist M, Lindström F, Watts A, Gröbner G (2004) Two types of Alzheimer's β-Amyloid (1–40) peptide membrane interactions: aggregation preventing transmembrane anchoring versus accelerated surface fibril formation. J Mol Biol 335:1039–1049
96. Sugita Y, Okamoto Y (1999) Replica-exchange molecular dynamics method for protein folding. Chem Phys Lett 314:141–151
97. Torrie GM, Valleau JP (1977) Nonphysical sampling distributions in Monte Carlo free-energy estimation: umbrella sampling. J Comput Phys 23:187–199
98. Barducci A, Bussi G, Parrinello M (2008) Well-tempered metadynamics: A smoothly converging and tunable free-energy method. Phys Rev Lett 100:020603
99. Smith AK, Klimov DK (2018) Binding of cytotoxic Aβ25-35 peptide to the dimyristoylphosphatidylcholine lipid bilayer. J Chem Inf Model 58:1053–1065
100. Tsai HHG, Bin LJ, Tseng SS, Pan XA, Shih YC (2010) Folding and membrane insertion of amyloid-beta (25–35) peptide and its mutants: Implications for aggregation and neurotoxicity. Proteins Struct Funct Bioinforma 78:1909–1925

101. Lockhart C, Klimov DK (2014) Alzheimer's Aβ10-40 peptide binds and penetrates DMPC bilayer: an isobaric-isothermal replica exchange molecular dynamics study. J Phys Chem B 118:2638–2648
102. Lemkul JA, Bevan DR (2011) Lipid composition influences the release of Alzheimer's amyloid β-peptide from membranes. Protein Sci 20:1530–1545
103. Ermilova I, Lyubartsev AP (2020) Modelling of interactions between Aβ(25–35) peptide and phospholipid bilayers: effects of cholesterol and lipid saturation. RSC Adv 10:3902–3915
104. Ivankov OI et al (2020) Interactions in the model membranes mimicking preclinical conformational diseases. Adv Biomembr Lipid Self-Assembly 31:185–214
105. Qiu L, Buie C, Reay A, Vaughn MW, Cheng KH (2011) Molecular dynamics simulations reveal the protective role of cholesterol in β-amyloid protein-induced membrane disruptions in neuronal membrane mimics. J Phys Chem B 115:9795–9812
106. Alzheimer A (1907) Über eine eigenartige Erkrankung der Hirnrinde [article in German]. Allg Z Psych Psych-gerich Med 64:146–148
107. Pavol H, Ermuhammad D, Tomáš K, Oleksandr T, Kholmirzo K, Norbert K Reflectometry and molecular dynamics study of the impact of cholesterol and melatonin on model lipid membranes. Eur Biophys J. https://doi.org/10.1007/s00249-021-01564-y

# Chapter 11
# Analysis of Natural and Engineered Amyloid Aggregates by Spectroscopic and Scattering Techniques

**T. R. Heyn, V. M. Garamus, Anja Steffen-Heins, K. Schwarz, and J. K. Keppler**

**Abstract** The increasing knowledge about natural functional fibrils has triggered the interest in synthetic or engineered fibrils. Naturally occurring amyloid fibrils (functional and pathogenic) have been analyzed for many years at different structural levels. Engineered fibrils are structurally similar to natural fibrils and the main sub-structural feature of amyloids is characterized by cross-beta structure stabilizing the fibril formation. However, a number of peculiarities exist comparing natural and engineered fibrils that may affect their analysis, especially in spectroscopic and scattering methods. For this reason, several methods that are commonly used for natural fibril analysis are presented and particularities for their application in the characterization of engineered fibrils are described. In addition, the understanding about structure–function relation of fibrils studied in the different research areas may mutually improve when using the same analytical approaches for natural and engineered fibrils.

T. R. Heyn · A. Steffen-Heins · K. Schwarz
Institute of Human Nutrition and Food Science, Division of Food Technology, Kiel University, Kiel, Germany
e-mail: theyn@foodtech.uni-kiel.de

A. Steffen-Heins
e-mail: aheins@foodtech.uni-kiel.de

K. Schwarz
e-mail: kschwarz-2@foodtech.uni-kiel.de

V. M. Garamus (✉)
Helmholtz-Zentrum Hereon, Geesthacht, Germany
e-mail: vasyl.haramus@hereon.de

J. K. Keppler
Food Process Engineering Laboratory, Agrotechnology and Food Sciences Group, Wageningen University, Wageningen, The Netherlands
e-mail: julia.keppler@wur.nl

 
L. Bulavin and N. Lebovka (eds.), *Soft Matter Systems for Biomedical Applications*, Springer Proceedings in Physics 266, https://doi.org/10.1007/978-3-030-80924-9_11

## List of Abbreviations, Symbols, and Chemical Substances

### Abbreviations

| | |
|---|---|
| AFM | Atomic force microscopy |
| ATR | Attenuated total reflection |
| COSMiCS | Complex objective structural analysis of multi component system |
| CW EPR | Continuous wave EPR |
| DEER | Double electron–electron resonance |
| DLS | Dynamic light scattering |
| EPR | Electron paramagnetic resonance |
| FTIR | Fourier-transform infrared |
| SAXS | Small angle X-ray scattering |
| SANS | Small angle neutron scattering |
| SEM | Scanning electron microscopy |
| SDSL | Site directed spin labelling |
| ssNMR | Solid state nuclear magnetic resonance |
| TEM | Transmission electron microscopy |
| Thio-T | Thioflavin-T |
| WAXS | Wide angle X-ray scattering |

### Symbols

| | |
|---|---|
| Rg | Radius of gyration |

### Chemical Substances

| | |
|---|---|
| α-B-Cry | α-B-crystallin |
| BLG | Whey protein beta-lactoglobulin |
| BSA | Bovine serum albumin |
| EGCG | Epigallocatechin gallate |
| IPSL | N-(1-oxyl-2,2,5,5-tetramethyl-3-pyrrolidinyl) iodoacetamide |
| MTSSL | (1-Oxyl-2,2,5,5-tetramethylpyrroline-3-methyl) methanethiosulfonate |
| PGG | Pentagalloyl glucose |

## 11.1 Introduction

Amyloid fibrils are ordered aggregates of normally soluble peptides or proteins [1]. The amyloid fibril structure is associated with more than 30 neurodegenerative diseases, for example with Alzheimer's disease (Abeta peptide amyloids) or Parkinson, (α-synuclein amyloids) [2, 3].

In addition to human ***pathogenic fibrils***, there are also so-called ***functional fibrils*** that fulfil a plethora of biological functions. Such functional structures can be found above all where surface adhesion and structural stability are required, e.g., in the biofilm of bacteria, fungal attachments or in the eggshell of silk moths [4–6]. The main difference between the two ***natural fibril*** manifestations is that the human ***pathogenic amyloids*** can be considered as protein misfolding. They exist in a wide variety of polymorphs and phenotypes which intermediate oligomeric products are found as toxic [7]. They have no intrinsic mechanism to degrade them, and thus to control their accumulation.

***Functional fibrils***, in contrast, have highly conserved structural motifs to ensure their vital function in addition to a rapid fibrillization process, thereby omitting toxic intermediate products [8].

The highly ordered arrangement of proteins in functional amyloid fibrils has raised considerable interest and the targeted production of such structures leads to many novel and promising applications in life-, environmental-, and material science [9]. Because these structures consist of proteins that do not exhibit an intrinsic amyloid tendency in their natural environment but require specific processing conditions, they are referred to below as synthetic ***engineered fibrils***.

The analysis of the amyloid formation kinetics, morphology, and structure is an important part of recent research because it gives feedback on how the aggregates are affected by changing environments, which in turn is an important part of understanding their formation and properties.

For ***natural amyloids,*** there are a number of well-established analytical techniques, all covering different aspects of their structure, morphology and formation kinetics. For example, crystallization of smaller segments has been conducted to study the protein stacking, solid state nuclear magnetic resonance spectroscopy (ssNMR) allows insights into the local structure arrangement and contacts between atoms, and the overall length and morphology are followed by different microscopy techniques (e.g., atomic force microscopy (AFM), transmission electron microscopy (TEM)).

The present chapter will focus only on the broader structural analysis which is of particular interest in order to draw conclusions regarding the functional properties of **engineered fibrils** in biomaterials. The most common and simple analysis of amyloids is the colorimetric Thioflavin-T (Thio-T) test, which basically reacts to the characteristic cross-β structure [10] but also cross-α [11] motif itself. However, this test does not give further insights into structural and morphological features. Thus, often a set of experiments is conducted to give additional information on the different levels (secondary to quaternary structure) of fibril formation. In an interdisciplinary approach, many of the techniques for elucidating the ***natural fibril*** structure have been used in the analysis of ***engineered fibrils.*** The analysis of fibril engineering requires process monitoring and a higher sample throughput, thus many methods for ***engineered fibrils*** have to be more practical. In addition, there are some analytical restrictions and/or modifications required when transferring the methodology. Therefore, the methodological and analytical aspects of three very different spectroscopic and scattering analysis techniques that can be used for ***natural*** and ***engineered amyloids*** in combination with the Thio-T test are described in the following subchapters.

The three techniques were selected to cover a most broad representation of different set-ups (measurement of functional groups in structural elements (FTIR, cf Sect. 11.3.1), measurement of specific labeled amino acids (EPR, cf Sect. 11.3.2), measurement of X-ray scattering intensity (SAXS, cf Sect. 11.3.3)) to sketch an overall picture of common aspects to consider when analyzing ***engineered fibrils***. In order to further limit the complexity of the topic and due to its wide dissemination, the following chapters are focusing on fibrils from whey protein beta-lactoglobulin (BLG) as reference for ***engineered fibrils***. BLG fibrils are among the most abundantly characterized engineered amyloid aggregates so far and their formation is typical for many other engineered fibrils.

## 11.2 Differences in the Origin of Natural Functional and Engineered Amyloids

Before discussing the methods to analyze ***natural*** and ***engineered fibrils***, it is important to give an overview on the main structural differences to be expected. Differences between ***natural*** and ***engineered fibrils*** are related to their formation conditions that are relevant in terms of their analysis (structure, morphology, kinetics).

Amyloid fibrils can be seen as a structural superfamily that shares a common substructure, irrespective of the precursor proteins [12]. Most amyloid aggregates have a characteristic cross-β motif, which consists of multiple β-strands that are stacked perpendicularly to the long axis of the fibrils and are linked by hydrogen bonds. β-Strands with the same primary structure can either align themselves evenly, resulting in a parallel, in-register arrangement, or they can have a shifted alignment, which is found in a parallel but out-register arrangement (Fig. 11.1A). Multiple variation in arrangement of β-sheets along the fibril axis or in the building blocks that form the fibril, such as specific peptides sequences or entire small proteins result in a diverse morphology and functionality of amyloids. The β-sheet motifs can often form micrometer long and even intertwined fibers of few to several nanometers thickness (Fig. 11.1B) which are straight or curly.

***Natural functional*** amyloids are triggered by defined biochemical pathways and require highly coordinated system of amyloid subunits, molecular chaperones, and transport systems. Well known functional fibrils are formed by bacteria. They

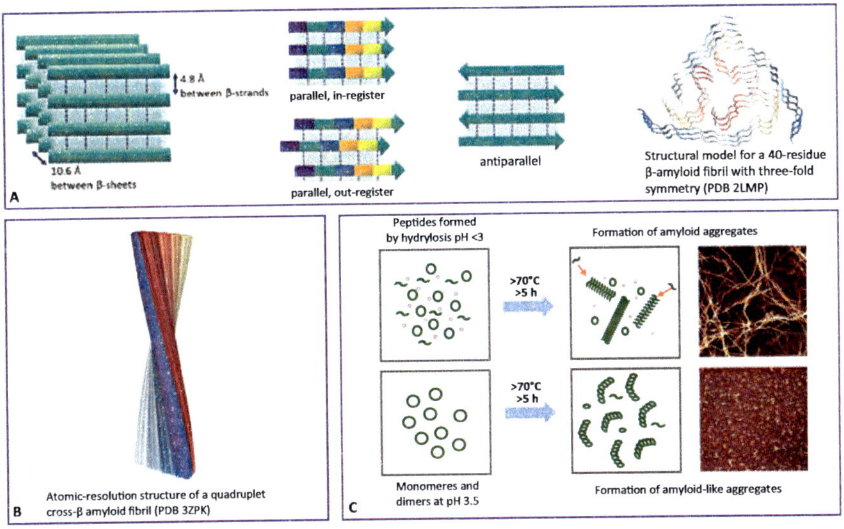


**Fig. 11.1** The variety of fibrils formed by their registry of β-sheets and protein building blocks differ in morphology, functionality and pathogenicity. **A** Left: Scheme of β-sheet alignment in amyloid aggregates, parallel in-register, out-register or antiparallel. Right: Structural motif of β-amyloid fibrils with three-folded symmetry (PDB 2LMP). **B** Structure of a quadruplet cross-beta amyloid fibril as an example of a pathogenic fibril (PDB 3ZPK). **C** Examples of engineered BLG-fibrils as a function of their building blocks after heat treatment. At pH 2 and elevated temperature, aggregates are formed from hydrolyzed BLG (peptides as building blocks), while at higher pH values, the aggregation initially occurs with non-hydrolyzed BLG (whole protein as building blocks)

are found extracellular and mediate the initial attachment of bacteria to surfaces and provide stability and functionality to mature biofilms. The rapid formation of *natural functional fibrils* is particular remarkable for those covering the surface of the frog's tongue and cause flies to stick to its surface. Those fibrils are formed by the millisecond friction of the moving tongue [13].

In contrast, *engineered amyloids* are produced in non-living systems usually under extreme conditions (e.g., by high temperature, solvents, urea, extreme pH value) [14, 15], as they are practically forced into these aggregate structures. For BLG, amyloid fibrils of several micrometer length and few nanometers thickness are typically produced in acidic conditions < pH 3, after prolonged heating > 70 °C for at least several hours. A combination of denaturation, acid hydrolysis and partial reassociation of certain peptides with high amyloidal propensity results in a conversion rate of approximately 35% (Fig. 11.1C) [16–19].

These fibrils consist of individual protofilaments of which several are intertwined to thicker structures [20, 21]. The low conversion rate in combination with hydrolysis results in a very polydisperse sample which composition strongly changes as a function of incubation time. Different polymorphs exist in dependence of the incubation conditions, they can range from differences in the number of intertwined protofilaments to their morphology [22]. For example, at pH 3.5, acid hydrolysis is restricted and thus whole BLG assembles into worm-like structures that are considered as amyloid-like and have a higher conversion rate (Fig. 11.1C) [20]. This shows that various morphologies and structures can be produced in engineered fibrils in dependence of the applied conditions. However, despite the often low conversion rate, engineered fibrils are present in higher concentration than most natural fibrils and do not require sophisticated isolation procedures in order to inhibit structural changes or interference with other compounds during the analysis [23] compared to amyloids formed in living organisms.

## 11.3 Spectroscopic and Scattering Techniques for Fibril Analysis

### *11.3.1 Infrared Spectroscopy*

*Background:* Circular dichroism spectroscopy and Fourier transform infrared spectroscopy (FTIR) are two common techniques for the analysis of secondary up to tertiary structures of proteins. FTIR is especially sensitive to beta-sheet conformations, which are an important aspect of fibrils. Various set-ups are available for FTIR, which allow the analysis of liquid, turbid, gel, or even powder samples. This also opens the possibility of *in-situ* analyses of aggregation.

IR uses the interaction of molecular bonds with electromagnetic radiation (light) in the infrared wavelength range to obtain information about the type of bond and thus about the structure of the molecule [24, 25]. For the analysis of proteins or protein aggregates, oscillations specific for amino acid residues are of relevance, e.g., the stretching vibrations of the $C = O$ bonds, which absorb at a wave number of 1600 to 1700 $cm^{-1}$ (amide I band) [26]. The $C = O$ bonds, involved in the protein secondary structure, absorb at different wave numbers depending on their location within the conformation: e.g., α-helix at wavenumber 1655 to 1658 $cm^{-1}$, β-sheets at 1633 $cm^{-1}$ and 1688 $cm^{-1}$, β-turns at 1672 $cm^{-1}$ and random coil elements at 1644 $cm^{-1}$ [25, 27].

During amyloid aggregation, the concentration of alpha-helix and intramolecular (parallel and antiparallel) beta-sheets decreases, while the concentration of random coil elements increases in the course of protein denaturation. This is reflected by a lower intensity at 1655 $cm^{-1}$ and 1633 $cm^{-1}$ as well as a higher intensity at 1644 $cm^{-1}$ in Fig. 11.2A. The formation of amyloid structures results in the arrangement of large and extremely well-ordered intermolecular beta-sheet structures. In this case, the vibration is distributed even more strongly over the surrounding fiber backbone than in intramolecular beta-sheets (increased normal mode vibration). Therefore, an increase in the intensity at lower wavenumbers 1618 to 1630 $cm^{-1}$ occurs, which can result in the formation of a maximum (see Fig. 11.2B) [19, 20, 28, 29]. The exact position of the maximum depends on the measurement set-up, the individual protein and the microenvironment (pH value, ion concentration etc.) (see Table 11.1) [28,

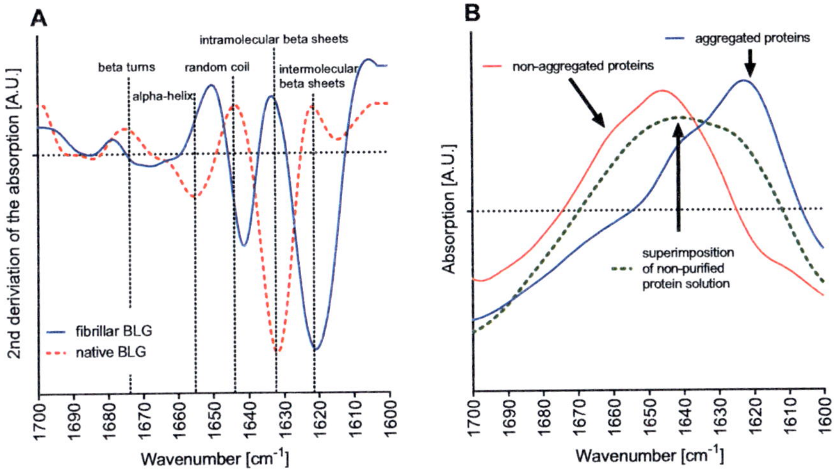


**Fig. 11.2** A) 2$^{nd}$ derivative of amide I FTIR spectra of fibrillar (blue) and native (red) BLG. Dotted vertical lines indicate the different structural elements and their typical wavenumbers in ATR set-up B) Amide I band spectra of non-purified BLG fibril solution (green) after 5 h at 90 °C, as well as the isolated aggregate (amyloids, blue) and non-aggregate fraction (non-amyloid material, red) at pH 3.5 [20]

**Table 11.1** Position (wavenumber) of amide I maxima in the FTIR spectra of various engineered and natural fibrils

| Protein | Wavenumber |
|---|---|
| **Engineered fibrils** | |
| BLG (pH2) [20, 43] | 1622 $cm^{-1}$, 1624 $cm^{-1}$ |
| BLG (pH3.5) [20] | 1619 $cm^{-1}$ |
| BSA [25] | 1620 $cm^{-1}$ |
| **Pathogenic fibrils** | |
| Alpha-synuclein [23] | 1624 $cm^{-1}$ |
| Alpha–beta (1–42) peptide [45] | 1630 $cm^{-1}$ |
| Human Lysozyme (pH2) [33] | 1622 $cm^{-1}$ |
| HET's prion fibers [46] | 1620 $cm^{-1}$ |
| Beta2-microglobulin [47] | 1620 $cm^{-1}$ |
| Transthyretin [48] | 1625 $cm^{-1}$ |
| Myoglobulin [49] | 1623 $cm^{-1}$ |
| Insulin (bovine) [34] | 1628 $cm^{-1}$ |
| **Functional fibrils** | |
| FabC (***P. Aeruginosa*** biofilms) [39] | 1625 $cm^{-1}$ |
| CsgA (***E. coli*** curli) [40] | 1623 $cm^{-1}$ |

BSA, bovine serum albumin; BLG, beta-Lactoglobulin.

30]. However, many other studies described this maximum in the formation of non-amyloid protein aggregates as it only indicates intermolecular β-sheet association [31]. Therefore, FTIR should always be used in combination with another analysis methods to unequivocally confirm the presence of amyloids [30].

*Natural amyloid aggregates:* A sufficiently high protein concentration is necessary to assure a low signal-to-noise ratio. This is of particular relevance for proteins that are only available in low concentration such as recombinantly produced or those extracted from organisms.

The necessary concentration depends on the IR spectrometer and the measuring method (e.g., liquid vs. dry). The challenge of an adequate signal-to-noise ratio is based on the high absorption of $H_2O$ around 1650 $cm^{-1}$, which must be background corrected from the absorption spectrum [32].

For human lysozyme, adequate signals were reached, using a 0.7 mM protein solution and $H_2O$ background subtraction. Depending on the pH-value, the aggregate characteristic shoulder can be shifted by a few wave numbers [33]. An alternative is the use of $D_2O$ as a solvent, as demonstrated by Bouchard et al. (2000) [34] for the formation of amyloid fibrils from bovine insulin. However, influences on the formation of aggregates are possible [35, 36]. Using liquid samples gives the opportunity of *in-situ* kinetics measurements of the fibril formation but can be limited due to a low signal-to-noise ratio.

Dry samples can be measured using the attenuated total reflection (ATR) technique in combination with FTIR or by pressing the dried protein powder with KBr into pellets. Both methods deliver very strong protein signals [37–40]. However, it must be taken into account that during the drying or pressing process, the structure of the fibrils can be affected.

The formation of ***natural pathogenic*** fibrils can be characterized by intermediate products, which superimpose with the fibrils in the spectra if the conversion is not yet completed. In the case of alpha-synuclein, for example, these can be alpha-helix intermediates, which pass through a helix to β-sheet transition relatively late in the process [41]. Therefore, removal of nonfibrillar protein by centrifugation was performed before measurement [38].

*Engineered amyloid aggregates:* A substantial advantage in the analysis of engineered amyloid aggregates is the usually good available protein material (e.g., whey or egg protein) (*cf* 11.2). A high protein concentration results in a low number of required scans (i.e., analysis speed is enhanced) due to the low signal-to-noise ratio for aqueous protein solutions [19, 42, 43]. Lower scan numbers can also help to capture the aggregation kinetics during *in-situ* experiments with a high temporal resolution. When synthesizing hydrolysis-induced engineered amyloid aggregates directly on the FTIR crystal, however, the chemical and physical stability of the set-up (e.g., ATR-unit) must be considered. For example, the extreme conditions usually applied for the fibril formation of BLG (pH 2 and 90 °C) are outside the limits for many FTIR devices. Similar to the procedure described for natural fibrils, $D_2O$ can be used as solvent for engineered fibrils to prevent an intersecting absorbance. ATR-FTIR analyses of engineered amyloids from BLG in $D_2O$ indicated the same characteristics as measurements with $H_2O$ [28, 29, 31]. Shifts of the exact maximum wavenumber might be caused by slightly different proton activities [25].

The comparison of diluted to ATR-crystal dried engineered fibrils from whey protein indicated that drying also leads to particularly clean spectra with similar bands [17]. Although the drying process requires additional time and can affect the fibrils, it may be necessary if the protein concentration is too low. For engineered fibrils, this could occur for example after several purification steps or addition of expensive reactants [44].

In contrast to most natural fibrils, for many engineered fibrils it is strongly recommended to separate fibrils and non-fibrillar material because of the strong polydispersity (*cf* Sect. 11.2). Ultrafiltration can be used (as with SAXS (*cf* Sect. 11.3.3)) to separate the samples into the different fractions in order to obtain the infrared spectrum of isolated aggregates. The FTIR spectra of purified fibrils (Fig. 11.2) demonstrate how strongly the signals can vary for BLG fibrils formed at pH 2 (semi-flexible fibers, building blocks are peptides): The isolated fibrils now give evidence of a much stronger signal at the 1619 $cm^{-1}$ band, which was clearly less visible in the unpurified sample. Likewise, the spectra of the aggregate morphologies obtained at pH 3.5 (worm-like amyloid-like aggregates, building blocks are whole proteins, see Fig. 11.1C and Sect. 11.2 for clarification) also show the formation of the typical shoulder at 1619 $cm^{-1}$ after isolation. However, due to the lower degree of hydrolysis at pH 3.5 and the overall higher conversion rate, less pronounced superposition effects were observed for these aggregates than for pH 2 fibrils [20]. Therefore, the comparison between the non-ultrafiltered and the purified aggregate fraction for these morphologies does not show any noticeable difference. Additionally, a higher degree of random coil was observed in the spectra. This confirms that the BLG aggregates formed at pH 3.5 are built from unhydrolyzed BLG [18] and thus integrate a high number of unstructured sidechains in their aggregate backbone, interrupting and reducing the purity of intermolecular beta-sheets. Overall, the FTIR is a powerful tool to derive the morphological differences of amyloid aggregates based on the specific structural characteristics, if the sample-specific problems (low protein concentration, polydispersity, device damaging effects) are counteracted by appropriate measures.

### 11.3.2 Electron Paramagnetic Resonance Spectroscopy

*Background:* EPR spectroscopy is a very sensitive approach to monitor the chemical microenvironment of spins, *i.e.*, unpaired electrons as found in radicals or transition metals. The spectroscopic approach is based on the magnetic moment of an electron, so that electrons align themselves parallel or anti-parallel in a strong external magnetic field (Zeeman effect). This alignment leads to splitting into a high ($M_s = +½$) and a low energy level ($M_s = -½$), whereas its energy difference (ΔE) relates proportionally with the outer magnetic field strength (B). Electromagnetic wave frequencies for EPR spectroscopy are in the range of microwave radiation, where among others the X-band (9–10 GHz) is most commonly used for spatially resolved protein analysis. If the energy difference at a certain magnetic field strength corresponds to the energy of the irradiated microwave frequency (resonance conditions), the electrons absorb the microwave. The resulting EPR spectrum, which represents the absorption intensity as a function of the magnetic field (B), is typically the first derivative of this energy absorption function. During spectral acquisition, the microwave (resonance) frequency is usually kept constant (so-called continuous-wave or cw EPR), while the magnetic field varies. However, also pulsed microwaves, such as short triggered frequencies in the range of 1–100 ns with high power, which approximate the resonance conditions, can be used at constant magnetic fields. These so-called pulsed EPR approaches provide additional information with neighboring nuclei and electron spins such as intramolecular or intermolecular distances ranging from ≈8 to 80 A° between the spins and their spatial distribution *e.g.*, by double electron–electron resonance (DEER) [50, 51].

As a biophysical technique, EPR spectroscopy is used in protein analysis to answer relevant structural and dynamic questions related to biological systems [52]. Because of lower number of spins in proteins, nitroxide radicals are frequently used as spin labels, the most common ones being *e.g.*, methanothiolsulfonate spin label (MTSSL) and iodoacetamido proxyl spin label (IPSL) [53]. These can bind covalently to specific cysteine-thiol residues and thus generate a stable EPR-active spin-label side chain [53], which is called site-directed spin labelling (SDSL). Among other factors, the small size of the labels is an important prerequisite that only minimally affects the structure and function of the protein [54, 55]. At the same time, however, labels are very sensitive to the local environment as well as structure and dynamics of the protein backbone, so that complexation, aggregation, adhesion, and fibrillization of proteins can be studied. The evaluation of EPR spectra is based on line- shape analysis, such as peak width, height, and shape. This encodes the interactions of the nitroxides towards the near environment, such as micro viscosity, micro polarity, nitroxide side-chain mobility, accessibility of *e.g.*, water, oxygen, and reducing agents. It is important to note that nitroxides with different spatial distribution and therefore different environments (spectral domains) have individual line shapes, which all superimpose each other in the acquired EPR spectra (Fig. 11.3).

However, these individual spectra can be separated by simulation assuming different biophysical models, which result in the characteristic parameters of the respective environmental conditions of the individual nitroxide compounds [44, 53, 56, 57] (Fig. 11.3). Thus, the EPR data analysis for a series of spin-labeled proteins, particularly at multiples sites of proteins, allows modeling of the protein structures with a spatial resolution at the level of the backbone fold [52].

*Natural amyloid aggregates:* SDSL was fundamentally pushed forward by Hubble and colleagues since the 1990$^{th}$ for protein complexes and membrane proteins [58], and became later likewise relevant for enzymes and small fibril-forming proteins [52]. For the site-directed binding of nitroxides to proteins, specific mutagenesis is performed to introduce a unique cysteine residue at the desired protein sidechain, e.g., as in α-synuclein [50]. Alternatively, naturally occurring cysteine residues are removed and replaced by amino acids such as alanine or serine. These modifications

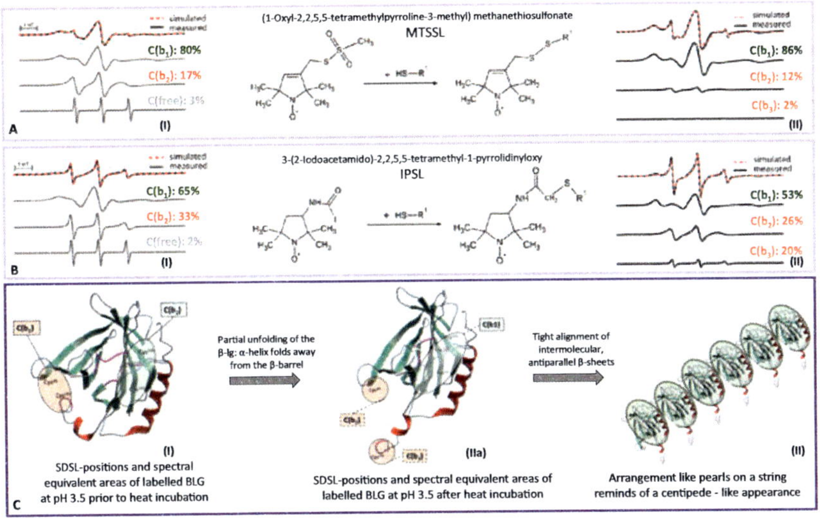


**Fig. 11.3** Proposed aggregation mechanism of amyloid-like aggregates from BLG (pH 3.5, 90 °C for 5 h) described by SDSL-EPR and mass spectrometry in Lux et al. (2021) [44]. For comparison purpose, two spin labels MTSSL (A) and IPSL (B) were used. Measured and simulated EPR spectra of MTSSL (A(I)) and IPSL (B(I)) bound to BLG before heat treatment result in two bound compounds (C($b_1$) and C($b_2$) and one free compound C(free). Measured and simulated EPR spectra of MTSSL (A(II)) and IPSL (B(II)) bound to BLG after thermal aggregation showed three individual bound compounds (C (IIa)). Schematic representation of the formation of centipede-like aggregates (C(II)). The structural model of BLG (PDB 3NPO) was created using the ICM browser from molsoft L.L.C. (La Jolla, California)

are realized in a variety of mutagenesis approaches reviewed *e.g.*, by Schmidt et al. (2015) [77].

SDSL-EPR has therefore been used for some time in various areas of structural biology and medicine investigating mostly pathogenic amyloids [58–62]. Fundamental insights obtained from combination of CW and pulsed EPR approaches concerning pathogenic amyloids are clarifying the nature of the intermolecular contacts and the structural stacking of the molecules along the fibril axis [52]. The main objectives are to determine the exact location of fibril forming core regions and to investigate local secondary structures within such core regions.

This allows the precise identification of sequences in the protein/peptide that form and stabilize the cross-β core of the fibrils, as opposed to those sections that are flexible and accessible to the solvent. The arrangement and the registry of β-strands in the β-sheets lead to different models of fibril formation and are determined by the intermolecular hydrogen bonds formed between the amyloidogenic sequence segments. Molecular modelling studies have shown that different amounts of energy are required for these alignments [63], where parallel in-register alignment requires the least, so that stackings of the same residue type are formed along the fibril axis [58, 63]. This is particularly the case for amyloids that contain an extensive core region of 20 amino acids or more can be explained by energy reasons and by maximizing the hydrophobic contact surface. In contrast, very small peptides (< 20 amino acids in the fibril core) β-strands mainly arrange antiparallel [58, 63].

Because of the proteins and peptides proned to fibrillization contain core regions with more than 20 amino acids, this parallel in-register structure is probably the most common form of natural amyloids.

*Engineered amyloid aggregates:* In contrast to natural (pathogenic) fibrils, the application of SDSL-EPR for engineered amyloids is completely new. So far, there exists only one pilot study evaluating the suitability of SDSL-EPR to assess the aggregation mechanism of bovine β-lactoglobulin [44] as a function of *e.g.*, the pH-value (cf. Sect. 11.2, Fig. 11.1C). Alterations in amyloid aggregation by labeled BLG could be neglected [64]. It is important to emphasize that this labeling study differs from those of pathogenic amyloids in that no mutagenesis of BLG was performed. This study aimed to investigate the aggregation mechanism of native BLG during thermal and acidic treatment, using the five naturally present cysteine residues for labeling. The two spin labels MTSSL and IPSL were bound to BLG binding to all cysteine residues, but MTSSL and IPSL preferred different but reproducible binding positions (Fig. 11.3A & B (I)). Those binding positions could be assigned to two spectral components of the labels bound to native BLG (Fig. 11.3C(I)). C(b1) indicated by a non-polar environment outside of the β-barrel of BLG (Cys106, Cys119, and Cys121) and thus describes the microenvironment of the β-sheets. C(b2) was the superposition of labeled Cys66 embedded in strand D and Cys160 contained in the random coil. Both binding positions were located in the DMSO/water environment and were in close proximity to the tertiary protein structure. During amyloid-like aggregation at pH 3.5 (worm-like amyloid-like aggregates, building blocks are whole proteins), these previously assigned three different binding positions of the spin labels can be more clearly distinguished after possible rearrangement when compared to the natural BLG monomer. Therefore, it is likely that the C-terminus C(b3) will fold outward, as evidenced by high accessibility to the aqueous phaseand fast rotational motion (Fig. 11.3A & B (II)). This rearrangement allows the α-helix to be shifted out of the autonomous folding domain during partial unfolding, which subsequently leads to greater accessibility of the β-sheets to the aqueous environment (Fig. 11.3C(IIa)). Meanwhile, the β-sheets are most likely compressed by the accumulation of intermolecular β-sheets, which subsequently support the assumption that amyloid-like structures can be formed from the building blocks of the entire primary BLG structure (Fig. 11.3C(II)). Furthermore, pulsed EPR technique confirmed the spatial distribution of the labels within the amyloid-like fraction in a one-dimensional arrangement of the entire protein aggregates, which was similar to a string of pearls (Fig. 11.3C(II)). These findings were also in accordance with SAXS measurements (cf. Sect. 11.3.3). Therefore, SDSL proteins containing several cysteine residues can be successfully used to gain deep insights into the aggregation mechanism of engineered amyloids if the binding positions are known.

### 11.3.3 Small Angle X-ray Scattering

*Background:* SAXS is the scattering of photons on electrons of studied objects. Usually, the scattered intensity is represented as a function of the scattering vector that relates to the scattering angle and wavelength of photons. It is an elastic scattering approximation, which means that there is no exchange of energy between photons and electrons. Additionally, only the coherent part of such scattering includes structural information and is therefore further analyzed. SAXS is detected at low values of the scattering vector and provides information on low resolution which is sensitive not to the location of single atoms but to the electron density (more correct, the scattering length density of electrons on length scale of few atoms). For separate particles in solution, the scattering intensity is determined by the scattering contrast i.e., the difference between the scattering densities of the particle and the solvent. Amplitude of the scattering by the particle is the Fourier transformation of the scattering contrast within the particle, which also depends on its shape and size. For a population of particles in solution, the scattering intensity ($I(q)$), which is observed by SAXS experiments, is the function of a number of particles, their interaction and modulus of scattering amplitude of each single particle (scattering contrast, shape, and size of particle). SAXS is the structural method of low resolution and different structures can provide quite similar scattering signals. As result, the obtained structure is not unique, and its proper choice requires additional information from other methods.

Small angle X-ray scattering is a powerful technique to follow the amyloid aggregation kinetics while statistically relevant information in mesoscopic length scale 1 to 1000 nm on the morphology of aggregates with regard to overall size and shape (Fig. 11.4), flexibility and stiffness, cross section parameters and proteins (peptides) density can be deduced [65]. However, for samples with high polydispersity, each protein, peptide, or their aggregate structure in the sample adds to the scattering pattern and thus, only an averaged scattering pattern of the heterogeneous system is

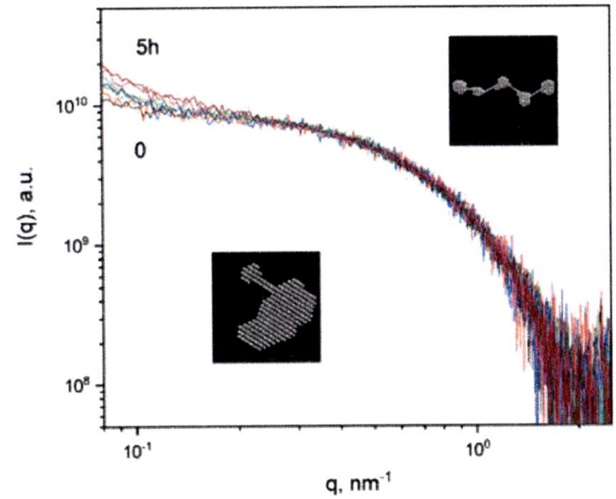


**Fig. 11.4** SAXS intensities of BLG solutions with moderate growth and corresponding 3D models (for initial state and after 5 h). The formation of small globular like objects which transform to elongated ones is followed by *ab initio* analysis [76] and represents the average state within the system

detected [65]. The most promising way is to apply SAXS in combination with small angle neutron scattering (SANS) and imaging methods that helps to obtain the real picture of the structures formed. The application of different types of probes (i.e., X-rays and neutrons) gives the possibility to study the effect of magnetic nanoparticles on amyloids structure [67, 68]. In comparison to dynamic light scattering (DLS), it gives so-called "dry" parameters of aggregates without hydration water.

By using additivity properties of a SAXS signal, which is valid for systems of low concentration and small number of different types of aggregates, it is possible to decompose intensities and obtained scattering signals from single components for further analysis [66]. The situation becomes more complex if initial and final states are not pure or systems consist of more components than number of independent scattering curves. For the latter case, the further development of SAXS analysis has been performed such as COSMiCS (Complex objective structural analysis of multi-component systems) [69]. It does not require the information from other experiments and works with SAXS data in different representations. This approach has been successfully applied for fibrillogenic forms of insulin and the familial mutant E46K of α-synuclein.

*Natural amyloid aggregates:* Pathogenic natural D187N mutant of gelsolin (actin-severing protein) has been intensively studied by SAXS [70] for comparison with wild type at physiological conditions and at low pH 2. At physiological conditions both proteins show similar globular particles with radius of gyration (Rg, the radius of inertion of scattering length density contrast within the particle) ~ 3.3 nm and no sign of additional aggregation. The pair-distance distribution function, which is a histogram over distances between pair of points within the particle, was calculated with maximal diameter for both gelsolins of ~ 9.1 nm. Obtained molecular mass was much closed to monomeric proteins and obtained shape matches quite good with crystal structure. The storage of proteins at low pH 2 up to 72 h leads to dramatically increase in scattering intensities especially for mutant protein and the formation of higher order aggregation effected the species, which are very polydisperse. SAXS together with other methods (electron and atomic force microscopy, mass spectrometric analysis, atomistic and replica exchange molecular dynamics simulations) suggests the alternative way of secondary oligomerization-dependent aggregation of gelsolin amyloidosis without the specific proteolytic step which can be used for improvement of therapeutic strategies.

SAXS has been applied for the determination of the structure and amyloidogenic feature of mutation R69C and D109H in human αB-crystallin (αB-Cry) [71]. It was observed that oligomers of D109H and R69C mutants are significantly larger than those of Wt α-B-Cry but keep a similar globular form as Wt α-B-Cry. The SAXS data were fitted with a model of several simple homogeneous bodies, assuming these samples to be monodisperse. The mutations enhance the amyloidogenic propensity of human αB-Cry and diminish the chaperone-like activity of protein.

Suppression of the growth of amyloids that form in the extracellular matrix of the bacterial biofilm as bacteria protection against environmental stress and antibiotics has been achieved by addition of polyphenols. The effect of polyphenols epigallocatechin gallate (EGCG), penta-O-galloyl-β-d-glucose (PGG), baicalein, oleuropein, and procyanidin B2 on the amyloid growth in a system of the highly fibrillation-prone protein FapC has been evaluated by SAXS [72]. The solutions with all polyphenols showed similar SAXS scattering patterns and corresponding pair distance distribution functions were determined. The main feature between 0 and 20 nm indicated that all aggregates had a somewhat compact structure. It was possible to fit the data with a model which consists of a compact prolate ellipsoidal core and a shell of flexible proteins. It can be concluded that polyphenols inhibit amyloid formation by redirecting the aggregation of FapC monomers into oligomeric species, which according to analysis of SAXS data are the core–shell complexes of short axis diameters ~ 25 nm consisting of less than 10 monomers. It was possible to correlate the biophysical observation (key features from SAXS) that the extent of amyloid inhibition is connected with the ability to reduce the biofilm formation, which shows a strong perspective for therapeutic application against biofilm formation.

*Engineered amyloid aggregates:* Peptide A1H1 of squid sucker ring teeth family has been shown to undergo β-strand-driven elongation into amyloid-like fibrils with a rich polymorphism. The structure of the fibrils was investigated by SAXS and wide-angle X-ray scattering (WAXS) and AFM. SAXS data were fitted by a model of polydisperse core two-shell cylinders. This model explains the presence of hydrophilic and hydrophobic segments in the peptide sequence. The obtained parameters are in agreement with molecular simulations. The most probable distance between fibrils has been also obtained from SAXS and confirmed by some anisotropy of 2D spectra [73].

In our work [20] the engineered fibrils formed by BLG at different pH (2 and 3.5, see Sect. 11.2 and Fig. 11.1 C) have been characterized by SAXS. At pH 2 (semi-flexible fibers, building blocks are peptides), no growth of fibrils in the first 0.5 h of incubation has been detected. In this initial phase the aggregates are in size of few nanometers (most probably monomers of proteins, Rg ~ 1.8 nm). In further time points (0.5 to 5 h) an increase in the scattering intensity and its power law behavior of $I(q) \sim q^{-1}$ support the occurrence of larger (longer) and rod-like objects, as expected for BLG fibrils. The integral parameters of the SAXS signal, such as Rg and scattering at "zero angle" follow a well-known sigmoidal-form and the different phases in the formation of amyloid fibrils (i.e., lag phase, steep elongation phase and depletion of building blocks) can be determined. The further parameters of fibrils′ cross section and flexibility have been obtained via modelling of SAXS curves. At pH 3.5 (worm-like aggregates, building blocks are whole proteins, see Fig. 11.1C and Sect. 11.2), the scattering intensity dramatically increases already at 0.5 h of incubation time and in the SAXS data one can observe the scattering from flexible wormlike fibrils. An increase of the apparent Rg from 2.6 nm (consistent with previous findings by SAXS for BLG dimers [74] to a Rg of 38 nm reached after 2 h was observed. These findings are consistent with AFM measurements, which described long semiflexible fibrils at pH 2 [75] and flexible wormlike aggregates at pH 3.5 [20]. As mentioned in the chapter of FTIR (cr. Sect. 11.3.1), the flexibility can be explained by high amount of random coil elements, which were also recognized by ESR as side arms of the centipede-like structure (cr. Sect. 11.3.2).

The SAXS method is used for the characterization of pH dependent formation kinetics of natural and engineered amyloid aggregates. The similarity of the formed structures in mesoscopic range gives the possibility to apply the same models especially at initial stage of aggregate formation. The extraction of more detailed information will require some special treatment of studied solution to separate aggregates of different morphology and/or complementary information from other methods (DLS, AFM, TEM, SANS).

## 11.4 Conclusion/Outlook

The three scattering and spectroscopic techniques discussed in this chapter are powerful tools to elucidate different structural elements of natural and synthetic engineered aggregates alike.

However, some restrictions and modifications become evident when transferring the methods for the different fibril classes. Especially the polydispersity of the engineered amyloids provides some significant effects in FTIR (cf. Sect. 11.3.1) and SAXS (cf. Sect. 11.3.3). In both cases, a sample fractionation into amyloid and non-amyloid material is required to achieve detailed structural information. Due to the various analysis options of FTIR (from liquid to powder), the method can also easily be adapted to different sample concentrations and environments, even allowing *in-situ* analyses with certain set-ups. However, without a second reference analysis (e.g., Thioflavin-T or microscopy), the information provided is not sufficient to unequivocally identify the presence of amyloid or fibril morphologies in samples. This holds also true for SAXS, which is otherwise a very versatile tool to follow the aggregation kinetics, as well as to determine the shape and flexibility of natural and engineered structures. Here, several data treatment approaches have been published that also allow for some corrections and extrapolation of the scattering signal, especially for polydisperse samples of the engineered fibrils. SDSL-EPR (cf. Sect. 11.3.3) has so far been used almost exclusively for the investigation of pathogenic amyloids aimed at identifying those peptide sequences that are particularly prone to fibril formation by evaluating the registry, distance, and arrangement of β-sheets. This requires the recombinant addition of a spin labeling site. For engineered fibrils, this method is not yet widely used, as it may be less relevant for them to know the exact registry. However, it is interesting to make an adaptation and label natural cysteine residues of a protein, which allows, for example, to follow the aggregation fate of edible proteins in food processes. However, this only applies if the labeling rate per naturally present cysteine residue is known.

Overall, it is very possible to transfer these methods established for natural fibrils to the analysis of functional amyloids. For some techniques, this transfer has already been done years before (e.g., FTIR), because the analysis is simple to adapt, while SAXS is more complicated in relation to having access to measuring facilities and the application of the data modeling. For EPR, the novel approach without introducing cysteines could be a compromise to the different level of information needed for engineered fibrils. Using the same analytical approaches for natural and engineered fibrils may mutually inspire the understanding about structure–function relation of fibrils studied in the different research areas in the future.

**Acknowledgements** TRH, ASH, KS and JKK would like to thank the German Research Foundation (DFG), project number 315456892, within the priority program SPP1934.

## References

1. Nilsson MR (2004) Techniques to study amyloid fibril formation in vitro. Methods 34(1):151–160
2. Araki K, Yagi N, Aoyama K et al (2019) Parkinson's disease is a type of amyloidosis featuring accumulation of amyloid fibrils of α-synuclein. Proc Natl Acad Sci U S A 116(36):17963–17969
3. Kollmer M, Close W, Funk L et al (2019) Cryo-EM structure and polymorphism of Aβ amyloid fibrils purified from Alzheimer's brain tissue. Nat Commun 10(1):4760
4. Iconomidou VA, Cordopatis P, Hoenger A et al (2011) The silkmoth eggshell as a natural amyloid shield for the safe development of insect oocyte and embryo: Insights from studies of silkmoth chorion protein peptide-analogues of the B family. Biopolymers 96(6):723–733
5. Shanmugam N, Baker MODG, Ball SR et al (2019) Microbial functional amyloids serve diverse purposes for structure, adhesion and defence. Biophys Rev 11(3):287–302
6. Zeng G, Vad BS, Dueholm MS et al (2015) Functional bacterial amyloid increases pseudomonas biofilm hydrophobicity and stiffness. Front Microbiol 6:1099
7. Sengupta U, Nilson AN, Kayed R (2016) The role of amyloid-β oligomers in toxicity, propagation, and immunotherapy. EBioMedicine 6:42–49
8. Deshmukh M, Evans ML, Chapman MR (2018) Amyloid by design: intrinsic regulation of microbial amyloid assembly. J Mol Biol 430(20):3631–3641
9. Knowles TPJ, Mezzenga R (2016) Amyloid fibrils as building blocks for natural and artificial functional materials. Adv Mater 28(31):6546–6561
10. Biancalana M, Koide S (2010) Molecular mechanism of thioflavin-t binding to amyloid fibrils. Biochim Biophys Acta 1804(7):1405–1412
11. Tayeb-Fligelman E, Salinas N, Tabachnikov O et al (2020) Staphylococcus aureus PSMα3 Cross-α fibril polymorphism and determinants of cytotoxicity. Structure 28(3):301-313.e6
12. Sunde M, Serpell LC, Bartlam M et al (1997) Common core structure of amyloid fibrils by synchrotron X-ray diffraction. J Mol Biol 273(3):729–739
13. Kleinteich T, Gorb SN (2015) Frog tongue acts as muscle-powered adhesive tape. R Soc Open Sci 2(9):150333
14. Kayser JJ, Arnold P, Steffen-Heins A et al (2020) Functional ethanol-induced fibrils: Influence of solvents and temperature on amyloid-like aggregation of beta-lactoglobulin. J Food Eng 270:109764
15. Yoshida K, Vogtt K, Izaola Z et al (2012) Alcohol induced structural and dynamic changes in β-lactoglobulin in aqueous solution: a neutron scattering study. Biochim Biophys Acta 1824(3):502–510
16. Arosio P, Beeg M, Nicoud L et al (2012) Time evolution of amyloid fibril length distribution described by a population balance model. Chem Eng Sci 78:21–32
17. Ye X, Hedenqvist MS, Langton M et al (2018) On the role of peptide hydrolysis for fibrillation kinetics and amyloid fibril morphology. RSC Adv 8(13):6915–6924
18. Keppler JK, Heyn TR, Meissner PM et al (2019) Protein oxidation during temperature-induced amyloid aggregation of beta-lactoglobulin. Food Chem 289:223–231

19. Heyn TR, Mayer J, Neumann HR et al (2020) The threshold of amyloid aggregation of beta-lactoglobulin: Relevant factor combinations. J Food Eng 283:110005
20. Heyn TR, Garamus VM, Neumann HR, et al (2019) Influence of the polydispersity of pH 2 and pH 3.5 beta-lactoglobulin amyloid fibril solutions on analytical methods. Eur Polym J 120:109211
21. Akkermans C, Venema P, van der Goot AJ et al (2008) Peptides are building blocks of heat-induced fibrillar protein aggregates of beta-lactoglobulin formed at pH 2. Biomacromol 9(5):1474–1479
22. Loveday SM, Anema SG, Singh H (2017) β-Lactoglobulin nanofibrils: the long and the short of it. Int Dairy J 67:35–45
23. Eymsh B, Drobny A, Heyn TR et al (2020) Toxic metamorphosis-how changes from lysosomal to cytosolic pH modify the alpha-synuclein aggregation pattern. Biomacromol 21(12):4673–4684
24. Yadav LDS (2005) Infrared (IR) Spectroscopy. In: Yadav LDS (ed) Organic Spectroscopy. Springer, Dordrecht. https://doi.org/10.1007/978-1-4020-2575-4_3
25. Barth A (2007) Infrared spectroscopy of proteins. Biochim Biophys Acta 1767(9):1073–1101
26. Militello V, Casarino C, Emanuele A et al (2004) Aggregation kinetics of bovine serum albumin studied by FTIR spectroscopy and light scattering. Biophys Chem 107(2):175–187
27. Kavanagh GM, Clark AH, Ross-Murphy SB (2000) Heat-induced gelation of globular proteins: part 3. molecular studies on low pH β-lactoglobulin gels. Int J Biol Macromol 28(1):41–50
28. Moran SD, Zanni MT (2014) How to Get Insight into amyloid structure and formation from infrared spectroscopy. J Phys Chem Lett 5(11):1984–1993
29. Baldassarre M, Bennett M, Barth A (2016) Simultaneous acquisition of infrared, fluorescence and light scattering spectra of proteins: direct evidence for pre-fibrillar species in amyloid fibril formation. Analyst 141(3):963–973
30. Hahn S, Kim S-S, Lee C et al (2005) Characteristic two-dimensional IR spectroscopic features of antiparallel and parallel beta-sheet polypeptides: simulation studies. J Chem Phys 123(8):84905
31. Allain A-F, Paquin P, Subirade M (1999) Relationships between conformation of β-lactoglobulin in solution and gel states as revealed by attenuated total reflection Fourier transform infrared spectroscopy. Int J Biol Macromol 26(5):337–344
32. Venyaminov S, Prendergast FG (1997) Water (H2O and D2O) molar absorptivity in the 1000–4000 cm-1 range and quantitative infrared spectroscopy of aqueous solutions. Anal Biochem 248(2):234–245
33. Mossuto MF, Dhulesia A, Devlin G et al (2010) The non-core regions of human lysozyme amyloid fibrils influence cytotoxicity. J Mol Biol 402(5):783–796
34. Bouchard M, Zurdo J, Nettleton EJ et al (2000) Formation of insulin amyloid fibrils followed by FTIR simultaneously with CD and electron microscopy. Protein Sci 9(10):1960–1967
35. Petrenko VI, Avdeev MV, Garamus VM et al (2014) Structure of amyloid aggregates of lysozyme from small-angle X-ray scattering data. Phys Solid State 56(1):129–133
36. Avdeev MV, Aksenov VL, Gazová Z et al (2013) On the determination of the helical structure parameters of amyloid protofilaments by small-angle neutron scattering and atomic force microscopy. J Appl Crystallogr 46(1):224–233
37. Cerf E, Sarroukh R, Tamamizu-Kato S et al (2009) Antiparallel beta-sheet: a signature structure of the oligomeric amyloid beta-peptide. Biochem J 421(3):415–423
38. Conway KA, Harper JD, Lansbury PT (2000) Fibrils formed in vitro from alpha-synuclein and two mutant forms linked to parkinson's disease are typical amyloid. Biochemistry 39(10):2552–2563
39. Bleem A, Christiansen G, Madsen DJ et al (2018) Protein engineering reveals mechanisms of functional amyloid formation in pseudomonas aeruginosa biofilms. J Mol Biol 430(20):3751–3763
40. Dueholm MS, Nielsen SB, Hein KL et al (2011) Fibrillation of the major curli subunit CsgA under a wide range of conditions implies a robust design of aggregation. Biochemistry 50(39):8281–8290

41. Ghosh D, Singh PK, Sahay S et al (2015) Structure based aggregation studies reveal the presence of helix-rich intermediate during α-Synuclein aggregation. Sci Rep 5:9228
42. Krebs MRH, Devlin GL, Donald AM (2009) Amyloid fibril-like structure underlies the aggregate structure across the pH range for beta-lactoglobulin. Biophys J 96(12):5013–5019
43. Oboroceanu D, Wang L, Brodkorb A et al (2010) Characterization of beta-lactoglobulin fibrillar assembly using atomic force microscopy, polyacrylamide gel electrophoresis, and in situ fourier transform infrared spectroscopy. J Agric Food Chem 58(6):3667–3673
44. Lux J, Azarkh M, Fitzner L et al (2021) Amyloid aggregation of spin-labeled β-lactoglobulin. Part ii: identification of spin-labeled protein and peptide sequences after amyloid aggregation. Food Hydrocolloids 112:106174
45. Sarroukh R, Goormaghtigh E, Ruysschaert J-M et al (2013) ATR-FTIR: A "rejuvenated" tool to investigate amyloid proteins. Biochim Biophys Acta 1828(10):2328–2338
46. Berthelot K, Ta HP, Géan J et al (2011) In vivo and in vitro analyses of toxic mutants of HET-s: FTIR antiparallel signature correlates with amyloid toxicity. J Mol Biol 412(1):137–152
47. Fabian H, Gast K, Laue M et al (2008) Early stages of misfolding and association of beta2-microglobulin: Insights from infrared spectroscopy and dynamic light scattering. Biochemistry 47(26):6895–6906
48. Cordeiro Y, Kraineva J, Suarez MC et al (2006) Fourier transform infrared spectroscopy provides a fingerprint for the tetramer and for the aggregates of transthyretin. Biophys J 91(3):957–967
49. Dong A, Randolph TW, Carpenter JF (2000) Entrapping intermediates of thermal aggregation in alpha-helical proteins with low concentration of guanidine hydrochloride. J Biol Chem 275(36):27689–27693
50. Drescher M, Jeschke G, Bordignon E (eds) (2012) EPR Spectroscopy: Applications in Chemistry and Biology. Topics in current chemistry, vol 321. Springer, Berlin
51. Jeschke G, Polyhach Y (2007) Distance measurements on spin-labelled biomacromolecules by pulsed electron paramagnetic resonance. Phys Chem Chem Phys 9(16):1895–1910
52. Klare JP (2013) Site-directed spin labeling EPR spectroscopy in protein research. Biol Chem 394(10):1281–1300
53. Berliner LJ (1976) Spin Labeling: Theory and Applications. Elsevier Science, Burlington
54. Berliner LJ, Grunwald J, Hankovszky HO et al (1982) A novel reversible thiol-specific spin label: papain active site labeling and inhibition. Anal Biochem 119(2):450–455
55. Mchaourab HS, Lietzow MA, Hideg K et al (1996) Motion of spin-labeled side chains in T4 lysozyme. Correlation with protein structure and dynamics. Biochemistry 35(24):7692–7704
56. Strancar J, Koklic T, Arsov Z et al (2005) Spin label EPR-based characterization of biosystem complexity. J Chem Inf Model 45(2):394–406
57. Strancar J, Kavalenka A, Urbancic I et al (2010) SDSL-ESR-based protein structure characterization. Eur Biophys J EBJ 39(4):499–511
58. Margittai M, Langen R (2008) Fibrils with parallel in-register structure constitute a major class of amyloid fibrils: Molecular insights from electron paramagnetic resonance spectroscopy. Q Rev Biophys 41(3–4):265–297
59. Török M, Milton S, Kayed R et al (2002) Structural and dynamic features of Alzheimer's Abeta peptide in amyloid fibrils studied by site-directed spin labeling. J Biol Chem 277(43):40810–40815
60. Gu L, Tran J, Jiang L et al (2016) A new structural model of Alzheimer's Aβ42 fibrils based on electron paramagnetic resonance data and Rosetta modeling. J Struct Biol 194(1):61–67
61. Sepkhanova I, Drescher M, Meeuwenoord NJ et al (2009) Monitoring Alzheimer amyloid peptide aggregation by EPR. Appl Magn Reson 36(2–4):209–222
62. Chen M, Margittai M, Chen J et al (2007) Investigation of alpha-synuclein fibril structure by site-directed spin labeling. J Biol Chem 282(34):24970–24979
63. Trovato A, Chiti F, Maritan A et al (2006) Insight into the structure of amyloid fibrils from the analysis of globular proteins. PLoS Comput Biol 2(12):e170
64. Lux J, Heyn TR, Kampen I et al (2021) Amyloid aggregation of spin-labeled β-lactoglobulin. Part I: Influence of spin labeling on amyloid aggregation. Food Hydrocolloids 112:106178

65. Kikhney AG, Svergun DI (2015) A practical guide to small angle X-ray scattering (SAXS) of flexible and intrinsically disordered proteins. FEBS Lett 589(19 Pt A): 2570–2577
66. Vestergaard B, Groenning M, Roessle M et al (2007) A helical structural nucleus is the primary elongating unit of insulin amyloid fibrils. PLoS Biol 5(5):e134
67. Bolisetty S, Vallooran JJ, Adamcik J et al (2013) Magnetic-responsive hybrids of Fe3O4 nanoparticles with β-lactoglobulin amyloid fibrils and nanoclusters. ACS Nano 7(7):6146–6155
68. Majorosova J, Petrenko VI, Siposova K et al (2016) On the adsorption of magnetite nanoparticles on lysozyme amyloid fibrils. Colloids Surf B Biointerfaces 146:794–800
69. Herranz-Trillo F, Groenning M, van Maarschalkerweerd A et al (2017) Structural analysis of multi-component amyloid systems by chemometric SAXS data decomposition. Structure 25(1):5–15
70. Srivastava A, Singh J, Singh Yadav SP et al (2018) The gelsolin pathogenic D187N mutant exhibits altered conformational stability and forms amyloidogenic oligomers. Biochemistry 57(16): 2359–2372
71. Ghahramani M, Yousefi R, Krivandin A et al (2020) Structural and functional characterization of D109H and R69C mutant versions of human αB-crystallin: THE BIOCHemical pathomechanism underlying cataract and myopathy development. Int J Biol Macromol 146:1142–1160
72. Najarzadeh Z, Mohammad-Beigi H, Nedergaard Pedersen J et al (2019) Plant polyphenols inhibit functional amyloid and biofilm formation in pseudomonas strains by directing monomers to off-pathway oligomers. Biomolecules 9(11):659 (2019)
73. Sánchez-Ferrer A, Adamcik J, Handschin S et al (2018) Controlling supramolecular chiral nanostructures by self-assembly of a biomimetic β-sheet-rich amyloidogenic peptide. ACS Nano 12(9):9152–9161
74. Keppler JK, Martin D, Garamus VM et al (2015) Differences in binding behavior of (-)-epigallocatechin gallate to β-lactoglobulin heterodimers (AB) compared to homodimers (A) and (B). J Mol Recognit 28(11):656–666
75. Adamcik J, Jung J-M, Flakowski J et al (2010) Understanding amyloid aggregation by statistical analysis of atomic force microscopy images. Nat Nanotechnol 5(6):423–428
76. Svergun DI (1999) Restoring low resolution structure of biological macromolecules from solution scattering using simulated annealing. Biophys J 76(6):2879–2886
77. Schmidt MJ, Fedoseev A, Summerer D, Drescher M (2015) Chapter eighteen – genetically encoded spin labels for in vitro and in-cell EPR studies of native proteins. Methods Enzymol 563:483–502

# Part III Magnetic Field Effects

# Chapter 12
# Liquid Biosystems in Gradient Magnetic Fields: Electrokinetic, Magnetophoretic and Orientation Effects

**Yu. I. Gorobets, S. V. Gorobets, and O. Yu. Gorobets**

**Abstract** This chapter describes the state of the art in investigation of gradient magnetic field influence on liquid biosystems including electrokinetic, magnetophoretic and orientation effects. The gradient magnetic force acting on bioobjects under moderate magnetic field strength is analyzed depending on their effective magnetic susceptibility and size of the objects. The attention of the Reader is also attracted to the existence of not only artificial but also biogenic sources of gradient magnetic fields such as biogenic magnetic nanoparticles. The hypothesis is discussed that the chains of biogenic magnetic nanoparticles are important functional elements of the cells and represent magnetic organelle because high-gradient magnetic fields of the biogenic magnetic nanoparticles influence on vesicles, granules and other components. The liquid–liquid phase separation is demonstrated experimentally as an example of the effect of high-gradient magnetic fields on colloids. Besides, influence of strong magnetic fields on the orientation of cells with anisotropic diamagnetic susceptibility in liquids is described.

## List of Abbreviations and Symbols

### *Abbreviations*

| | |
|---|---|
| BMN | biogenic magnetic nanoparticles |
| EMF | electromotive force |
| MPP | magnetophoresis potential |

Yu. I. Gorobets · O. Yu. Gorobets (✉)
Institute of Magnetism of NAS and MES of Ukraine, 36b Acad. Vernadskoho Blvd., Kyiv 03142, Ukraine

S. V. Gorobets
National Technical University of Ukraine «Igor Sikorsky Kyiv Polytechnic Institute», 37 Peremohy Ave., Kyiv 03056, Ukraine
e-mail: pitbm@ntu-kpi.kiev.ua



L. Bulavin and N. Lebovka (eds.), *Soft Matter Systems for Biomedical Applications*, Springer Proceedings in Physics 266, https://doi.org/10.1007/978-3-030-80924-9_12

## Symbols

| | |
|---|---|
| $a,b$ | semi-major and semi-minor axis of an ellipsoid |
| $c$ | $2.9979 \times 10^{10}$ cm·s$^{-1}$, speed of light in vacuum |
| $c(\vec{r})$ | concentration of *magnions* in a point $\vec{r}$ |
| $E_{ef}$ | average energy of randomizing motion |
| $e$ | $1.602176634 \times 10^{-19}$ C, charge of electron |
| $\vec{E}_{mPh}$ | electric field of the MPP |
| $F$ | Henry function |
| $\vec{F}_m$ | gradient magnetic force |
| $\vec{F}_{\nabla \vec{H}^2}$ | gradient magnetic force density |
| $g_L$ | Lande $g$-factor |
| $\vec{H}$ | magnetic field strength |
| $\vec{H}_0$ | external homogeneous magnetic field strength |
| $\vec{i}$ | current density |
| $\vec{j}$ | flux density of *magnions* |
| $\vec{J}_i$ | generalized flow, $i = 1, 2$ |
| $k_B$ | $1.381 \times 10^{-23}$ J/K, Boltzmann constant |
| $m$ | $9.109 \times 10^{-28}$ g, mass of the electron |
| $\vec{r}$ | radius vector |
| $R_0$ | characteristic size of a *magnion* |
| $T$ | absolute temperature |
| $U$ | energy of interaction a magnetic moment with magnetic field |
| $\vec{v}_m$ | velocity of a *magnion* |
| $V_p$ | particle (impurity) volume or volume of *magnion* |
| $\vec{X}_i$ | generalized force, $i = 1, 2$ |
| $Ze$ | charge of a *magnion* |

## Greek

| | |
|---|---|
| $\Delta\varphi_N(\vec{r})$ | Nernst type voltage in a point $\vec{r}$ |
| $\varepsilon$ | ratio of magnetic and thermal energy |
| $\eta$ | dynamic viscosity of liquid |
| $\theta$ | angle between the greater semi-axis of an ellipsoid and direction of magnetic field |
| $\kappa$ | Debye–Huckel parameter |
| $\mu$ | electrophoretic mobility of a *magnion* |
| $\chi$ | effective magnetic susceptibility |
| $\chi_{\parallel}$ | longitudinal component of $\chi$ |
| $\chi_{\perp}$ | transversal component of $\chi$ |
| $\Delta\chi$ | $\lvert\chi_{\parallel}-\chi_{\perp}\rvert$ |
| $\chi_p$ | magnetic susceptibility of a paramagnetic particle or *magnion* |

| | |
|---|---|
| $\chi_l$ | magnetic susceptibility of liquid |
| $\mu_0$ | $0.927 \times 10^{-20}$ erg/G, Borh magneton |
| $\phi$ | volume fraction of *magnions* in the electrolyte |
| $\phi_{mPh}(\vec{r})$ | MPP at a point $\vec{r}$ |
| $\Delta\varphi(\vec{r})$ | superposition of the potential of Nernst type and MPP |

## 12.1 Magnetic Field Sources of Artificial and Biogenic Origin Acting on Biosystems

To date, an extensive literature is devoted to the study of the effect of external magnetic fields on living organisms at different levels of their organization in organisms as a whole, individual organs and tissues, cellular and molecular [1–5]. The results of the latest research on "life on a magnet" are focused on the effect of artificial sources of constant magnetic field on the submicron components of the internal environment of organisms, and thus on the life and death of a cell [2]. "Magnetic tweezers" were described also as devices in which external inhomogeneous magnetic fields were used to separate liquid biological media [6].

At the same time, it is well known that magnetic fields that accompany electrical processes in organisms are extremely small, even compared to the magnitude of the Earth's magnetic field. Therefore, until recently the own magnetic fields of living organisms have been considered mainly in connection with the development of diagnostic tools, rather than to study their effects on metabolism. This stereotype arose because of small value of intrinsic magnetic fields in organisms, and a very weak response of tissues and organisms in general to magnetic fields [7, 8]. However, there are the sources of strong gradient magnetic fields of biogenic origin, which go beyond the above-described stereotype, namely, magnetostatic stray fields of so-called biogenic magnetic nanoparticles (BMN). Blackmore discovered the synthesis of such particles in unicellular organisms in 1975 [9], and later such nanoparticles, which are natural permanent nanomagnets, were found in multicellular organisms, including humans [10, 11]. The physical mechanisms of its influence on metabolism is based on the existence of its strong gradient magnetic fields of biogenic origin.

## 12.2 Magnetophoretic Effects in Liquid Biosystems

The term "magnetophoresis" has been proposed to describe the particle movement through a viscous liquid under the influence of an external gradient magnetic field. The gradient magnetic field influence on electrochemical reactions in aqueous solutions and on liquid biosystems on the basis of magnetophoresis.

### 12.2.1 Magnetic Tweezers for Biomedical Applications

For many years magnetic tweezers have been used for the investigation of the local forces in biological tissues, for stretching and manipulating of DNA, for transport of ferrofluids, and cells probing [12–14]. The devices based on micromagnets or microconductors, known as chips, that form local high gradient magnetic fields in the space, were also used for trapping and manipulation of atoms [15–17]. For these purposes, the most promising way is the creation of natural medium with high magnetic characteristics, which elements are capable to provide high gradient magnetic fields and can be easily moved by applying a weak external field. The domain structures of garnet-ferrite films were discussed in [6, 18] for this purpose. The method of the micron and submicron para- and ferromagnetic particles transportation using a potential barrier of an isolated magnetic domain on the surface of garnet-ferrite films with perpendicular anisotropy was proposed in [6] (Fig. 12.1).

Let us consider the force acting on a particle in a magnetic trap of the stripe domain. If the particle is made from paramagnetic or superparamagnetic material, it cannot considerably affect magnetization distribution in the domain due to its small magnetic moment. To calculate the force acting on a paramagnetic particle with magnetic susceptibility $\chi_p$ let us consider a small volume $dV_p$ inside the particle. The potential energy of this volume $dU$ under an external field strength $\vec{H}$ and the force $d\vec{F}_m$ acting on it is determined by

$$dU = -\frac{\chi_p}{2}\vec{H}^2 dV_p \tag{12.1}$$

$$d\vec{F}_m = -\nabla(dU) = \frac{\chi_p}{2}\nabla\vec{H}^2 dV_p \tag{12.2}$$

The force acting on the particle can be obtained by integration of this equation over the particle volume and taking into account the Gauss' theorem

$$\vec{F}_m = \frac{\chi_p}{2}\oint dS(\vec{n}\cdot\vec{e}_i)\vec{H}^2 \tag{12.3}$$

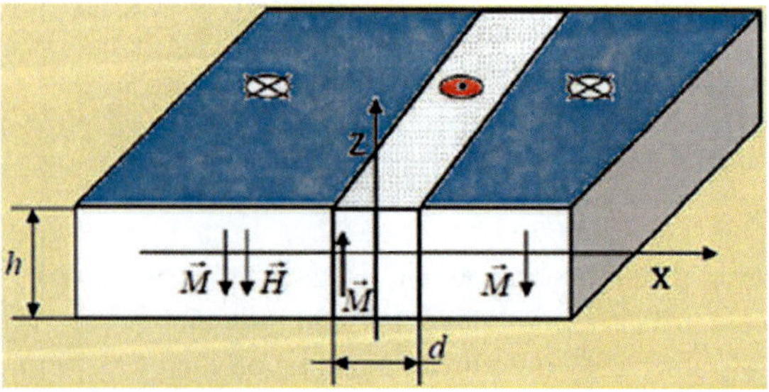


**Fig. 12.1** The schematic image of the garnet-ferrite film with a stripe domain. Here $\vec{M}$ is the magnetization of film material, $\vec{H}$ is the external magnetic field, *h* is the thickness of the film, and *d* is the width of the domain

where $\vec{n}$ is the vector, which is normal to the particle surface and $\vec{e}_i$ are orts. The integration has been carried out over the particle surface. This force is called the gradient magnetic force.

It is clear from the last equation that $\frac{\chi_p}{2}\vec{H}^2 = P$, $P$ is just pressure acting on the surface of the particle. If the external field is homogeneous, the resulting force is equal to zero. The maximum force value can be reached in those areas where the magnetic field has the highest gradients. The value of $\vec{H}^2$ decreases practically twice for the distance of about of domain wall width away from domain. The successive stages of microparticle transportation by an isolated domain loop on the surface of garnet-ferrite film is presented in Fig. 12.2 a)-d). This control scheme allows moving the magnetic particles in two dimensions.

The gradient magnetic force causes magnetophoresis of particles. It is used for high gradient magnetic separation. High gradient magnetic separation is widely used in the processes of extracting magnetic target objects from liquids and for purification of wastewaters, removal of radioactive and ferromagnetic impurities from the system of cooling of heat and nuclear power plants, etc. [19–24]. Lately, high gradient magnetic separation has been extensively used in biology and medicine. Using it one conducts highly specific removal of cells, magnetic separation of proteins, and nonpolar organic impurities from water and liquid biological media [19]. Extraction of paramagnetic and effectively paramagnetic target objects from liquids was

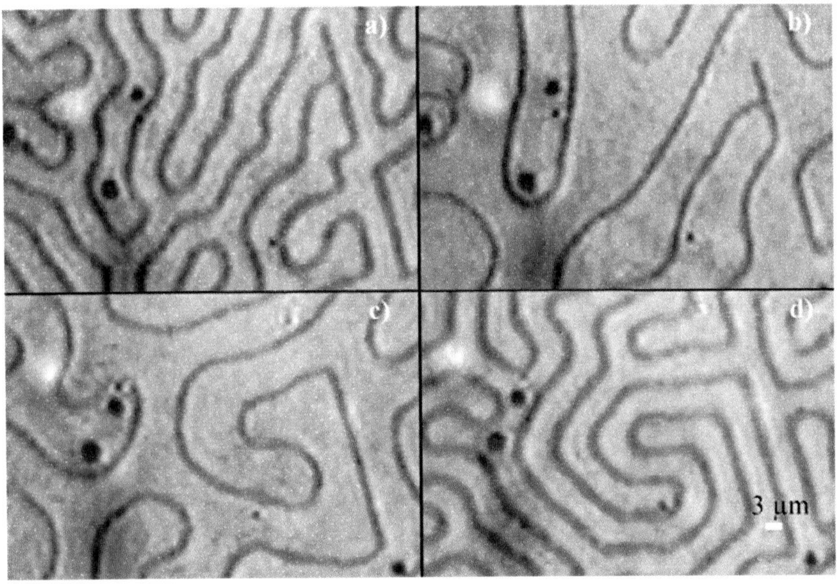


**Fig. 12.2** Sequential displacement of two Ni particles by the stripe domain in a film of epitaxial iron-yttrium garnet, $(YSmLuCa)_3(FeGe)_5O_{12}$ (diameters of the transported particles are about 3 micron); Images (a), (b), (c), (d) show subsequent decreasing of the distance between these particles caused by changing of the stripe domain structure

carried out by means of magnetic separators containing high gradient ferromagnetic matrices [19, 20]. Particles (impurities) may be removed from the flow of liquid by a magnetized high gradient ferromagnetic matrix in the external magnetic field. The basic equation of high gradient magnetic separation is

$$\vec{F}_m + \vec{F}_a = -\vec{F}_d, \tag{12.4}$$

where $\vec{F}_a$ is the force producing an active movement of a particle relative to liquid of different origin including randomizing Brownian motion, motion due to chemical or biochemical processes, motion caused by external forces not connected with gradient magnetic field. The force $\vec{F}_m$ is a gradient magnetic force acting on a particle in a stray magnetic field of a high gradient ferromagnetic matrix elements, $\vec{F}_d$ is the hydrodynamic force, described by the Stokes law for spherical particle of radius $r_p$

$$\vec{F}_m = \frac{1}{2}\hat{\chi} V_p grad(\vec{H}^2), \tag{12.5}$$

$$\vec{F}_d = 6\pi r_p \eta \vec{v} \tag{12.6}$$

where $\hat{\chi} = \hat{\chi}_p - \hat{\chi}_l$ ($\hat{\chi}$ is effective magnetic susceptibility tensor, $\hat{\chi}_p$ is tensor of magnetic susceptibility of an impurity, $\hat{\chi}_l$ is tensor of magnetic susceptibility of liquid); $V_p$ is particle (impurity) volume. $\vec{H}$ is superposition of strength of the external magnetic field and heterogeneous magnetic field created by a separate element of high gradient ferromagnetic matrix, $\eta$ is dynamic viscosity of liquid; $\vec{v}$ is liquid velocity.

First, let us consider the case when the force of an active movement $\vec{F}_a$ is zero and the tensors of magnetic susceptibilities $\hat{\chi}_p$ and $\hat{\chi}_l$ are isotropic $\hat{\chi}_p = \begin{pmatrix} \chi_p & 0 & 0 \\ 0 & \chi_p & 0 \\ 0 & 0 & \chi_p \end{pmatrix}$, $\hat{\chi}_l = \begin{pmatrix} \chi_l & 0 & 0 \\ 0 & \chi_l & 0 \\ 0 & 0 & \chi_l \end{pmatrix}$. In this case, the effective magnetic susceptibility tensor has the form

$$\hat{\chi} = \begin{pmatrix} \chi & 0 & 0 \\ 0 & \chi & 0 \\ 0 & 0 & \chi \end{pmatrix}, \tag{12.7}$$

where $\chi = \chi_p - \chi_l$.

The particle behavior under influence of a gradient magnetic field is similar at the solid state surface (see the example of magnetic tweezers Fig. 12.1, Fig. 12.2, Eqs. (12.1)–(12.7)) and in a liquid. The only difference is that the effective magnetic susceptibility $\chi$ is a key parameter governing the magnetophoresis of a particle in liquid instead of its magnetic susceptibility at the solid state surface. The effective magnetic susceptibility of a particle in liquid is equal to the difference between the magnetic susceptibilities of the material inside of the particle $\chi_p$ and the liquid $\chi_l$. For example, the presence of the particle with positive effective magnetic susceptibility leads to attraction of the particle in liquid to the region of higher magnetic field strength in order to minimize free energy similarly to the behavior of a particle with the positive magnetic susceptibility in a nonuniform magnetic field. While the particle with negative effective magnetic susceptibility is repulsing into the region of lesser magnetic field strength in a liquid.

Particles in liquids with positive or negative magnetic susceptibilities can represent ions, molecules, or the region of continuous medium (in solid, liquid or gaseous state of matter). Let us introduce the term *magnion* denoting the mentioned above region of continuous medium, which is characterized by magnetic susceptibility different from the magnetic susceptibility of liquid. *Magnion* behavior under gradient magnetic field is characterized (as it was illustrated above) by the effective magnetic susceptibility which is equal to the difference between the magnetic susceptibility of *magnion* and liquid. It allows considering the behavior of this *magnion* in thermodynamic approximation where the effective magnetic susceptibility plays the major role. Such effective magnetic susceptibility means the magnetic susceptibility of a *magnion* relative to the medium [25]. For example, the presence of the *magnion* with positive effective magnetic susceptibility leads to the repulsing of the *magnion* into the region of higher magnetic field strength in order to minimize the free energy. The notion of effective magnetic susceptibility cannot be introduced for the single ion or molecule because it is essentially thermodynamic parameter.

Impact of magnetic field on an ion, a molecule or a *magnion* can be estimated using the ratio of magnetic and thermal energy

$$\varepsilon \frac{\chi V_P \overline{H}^2}{2k_B T}, \tag{12.8}$$

where $k_B = 1.381 \times 10^{-23}$ J/K is the Boltzmann constant and $T$ is the absolute temperature [26].

Typically, the organisms, organs, tissues, biological cells, cellular compartments, lipids and other organic compounds have diamagnetic susceptibilities $\chi$ that are very close to that of water $\chi_{\text{water}} = -0.719 \times 10^{-6}$ (CGS units) [7, 27–29]. Particularly, $\chi_{\text{p}} = -0.774 \times 10^{-6}$ for cellular proteins, $\chi_{\text{lip}} = -0.669 \times 10^{-6}$ for cellular lipids [30, 31], $\chi_{\text{BSA}} = -0.826 \times 10^{-6}$ for BSA [29], and $\chi_{\text{ch}} = -0.735 \times 10^{-6}$ for cholesterol [28, 30] (all in CGS units). The theoretical molar magnetic susceptibility of a peptide bond is $-5.36 \times 10^{-6}$ emu (CGS units) [32, 33].

The different diamagnetic ions (for example, $H^+$, $Ca^{+2}$, $Zn^{+2}$) or molecules, and paramagnetic ions such as iron, copper, or manganese can be naturally present in metalloenzymes and in several proteins with redox function [34]. Besides, naturally occurring diamagnetic ions in metalloproteins are often replaced with their paramagnetic counterparts (for example, oxovanadium(IV), manganese(II), cobalt(II), nickel(II) and copper(II) for naturally occurring zinc(II) or trivalent lanthanoid ions for naturally occurring calcium(II)) [34]. Lanthanoid(III) ions (for example, $Ce^{3+}$, $Pr^{3+}$, $Nd^{3+}$, $Sm^{3+}$, $Eu^{3+}$, $Tb^{3+}$, $Dy^{3+}$, $Tm^{3+}$, $Yb^{3+}$, $Er^{3+}$, $Ho^{3+}$) are applied in the development of contrast agents for magnetic resonance imaging, drugs and in structural biology [34]. Metal-based paramagnetic center can be chemically inserted even in proteins without any metal binding sites using special probes [34]. This method is used also in drug discovery applications [34].

For diamagnetic species the ratio $\varepsilon$ can be evaluated using Langevin formula

$$\varepsilon = e^2 \sum_{i=1}^{Z} \langle r_i^2 \rangle \vec{H}^2 / \left(12mc^2 k_B T\right).$$

Here $m$ and $e$ are the mass and the charge of the electron, correspondingly, $c$ is the light speed in vacuum, $\langle r_i^2 \rangle \approx 10^{-16}$ cm$^2$ is the averaged square radius of the $i$-th electron orbit projection on a plane perpendicular to the magnetic field (the summation is over all $Z$ electrons of the ion) [26].

For paramagnetic species the ratio $\varepsilon$ can be evaluated using Brillouin formula

$$\varepsilon = \frac{(g_L \mu_0)^2 J(J+1)\vec{H}^2}{6(k_B T)^2}.$$

Here $g_L$ is Lande g-factor of the ion, $J$ is the total angular momentum of the ion, $\mu_0 = 0.927 \times 10^{-20}$ erg/G is the Borh magneton.

Typically, at moderate magnetic fields $H \approx 1$–10 kOe the ratios $\varepsilon$ are very small ($\varepsilon << 1$) for both the diamagnetic and paramagnetic species and therefore the impact of magnetic fields on these species can be neglected.

Table 12.1 illustrates the different examples of ratios $\varepsilon$ at moderate magnetic fields ($H \approx 1$–10 kOe) for combinations of diamagnetic/paramagnetic *magnions* dispersed in diamagnetic/paramagnetic liquids.

For large diamagnetic *magnions* (lipid droplets, starch granules, liposomes, and vesicles) with size exceeding of several hundred of nm, the effects of gradient magnetic fields may be significant even for moderate values of $H$. The large diamagnetic species are typical for biological systems. For example, diameter of intracellular transport vesicles is $d \approx 30$–100 nm [35], of vesicles within a clathrin cage is $d \approx$ 40–100 nm [35], of transport vesicles associated with the Golgi is $d \approx 60$ nm [36], of synaptic vesicles in neurons is $d \approx 40$ nm [37]. Outside of these ranges, there are larger, non-spherical carriers such as tubulovesicular recycling vesicles, while the lower limit for a vesicle is estimated to be $d \approx 20$ nm [38]. The size range of the lysosomes is $d \approx 100$ nm-1 μm [39]. A unilamellar liposome/vesicle (spherical vesicle, bounded by a single lipid bilayer) are classified into the small ($d \approx 20$–100 nm), large ($d \approx 100$–1000 nm) and giant ($d \approx 1$–200 μm) [40, 41]. The effects of magnetic fields was confirmed experimentally for giant unilamellar vesicles or cells into various microstructures with spatially coded configurations and spatialized cascade biochemical reactions using a stainless steel mesh [42].

**Table 12.1** Ratio of magnetic and thermal energy $\varepsilon$ (Eq. 12.8) at moderate magnetic fields ($H \approx$ 1–10 kOe) for *magnions* of different sizes and magnetic properties in diamagnetic (water, cytoplasm), $\chi_1 = -0.719 \times 10^{-6}$ (CGS units) and paramagnetic ($\chi_1$ = $-10^{-5} - 10^{-4}$ (CGS units) [45, 46]) liquids. Here, $\chi_p$ is a magnetic susceptibility of a *magnion,* $|\chi|$ is a difference between the magnetic susceptibilities of magnion $\chi_p$ and the liquid $\chi_l$, $R_0$ is a characteristic size of a *magnion,* and a volume of *magnion* is calculated as $V_p = 4\pi R_0{}^3/3$

| *Magnion* system | $\chi_p$ | $\lvert\chi\rvert$ | $\lvert R_0\rvert$,nm | ε |
|---|---|---|---|---|
| 1. Diamagnetic *magnion* in diamagnetic liquids | | | | |
| Model *magnion* | $10^{-6} - 10^{-5}$ [47] | $10^{-6} - 10^{-5}$ | 10 | $10^{-4} - 10^{-1}$ |
| | | | 100 | $10^{-1} - 10^{2}$ |
| Lipid droplet | $-0.669 \times 10^{-6}$ [30, 31], | $1{\cdot}10^{-7}$ | $10 - 10^5$ [48] | $5{\cdot}10^{-6} - 5{\cdot}10^{8}$ |
| Protein | $-0.774 \times 10^{-6}$ [30, 31], | $0.5{\cdot}10^{-7}$ | 1 – 10 [49] | $3{\cdot}10^{-9} - 3{\cdot}10^{-4}$ |
| Cholesterol | $-0.735 \times 10^{-6}$ [28, 30], | $0.1{\cdot}10^{-7}$ | 10 –20 [50] | $5{\cdot}10^{-7} - 4{\cdot}10^{-4}$ |
| Starch granules | $- 0.87 \times 10^{-6}$ [51], | $0.15{\cdot}10^{-6}$ | $500–5{\cdot}10^4$ [52] | $0.9 - 9{\cdot}10^{7}$ |
| Amyloplasts | $-8 \times 10^{-7}$ [53] | $8 \times 10^{-8}$ [53] | 1000 | $4 - 4{\cdot}10^{2}$ |
| Mitochondria | $-0.30 \times 10^{-8}$ [54] | $0.7 \times 10^{-6}$ | 400 [55] | $2 - 2{\cdot}10^{2}$ |
| Lymphocytes in water | $-(0.37–0.46) \times 10^{-8}$ [54], | $0.7 \times 10^{-6}$ | 7000 | $1{\cdot}10^{4} - 1{\cdot}10^{6}$ |
| 2. Paramagnetic *magnion* in a diamagnetic liquids | | | | |
| Mitochondria aggregates | $+ 50.0 \times 10^{-8}$ [54] | $1.2 \times 10^{-6}$ | 400 [55] | $3.8 - 4{\cdot}10^{2}$ |
| Methemoglobin containing erythrocytes | $+ (22–35) \times 10^{-8}$ [54] | $(0.9–4.2) \times 10^{-6}$ | 5000 [56] | $6{\cdot}10^{3} - 6{\cdot}10^{5}$ |
| Ferritin | $+ (3.1–2400) \times 10^{-8}$ [54], | $(0.75–24.7) \times 10^{-6}$ | 12 [57] | $6{\cdot}10^{-5} - 0.2$ |
| 3. Diamagnetic *magnion* in a paramagnetic liquid | | | | |
| Model *magnion* | $10^{-6}$ | $10^{-5} - 10^{-4}$ | 10 | $10^{-3} - 1$ |
| | | | 100 | $1 - 10^{3}$ |
| 4. Paramagnetic *magnion* in a paramagnetic liquid | | | | |
| Model *magnion* | $10^{-5} - 10^{-4}$ [44] | $10^{-5} - 10^{-4}$ | 10 | $10^{-3} - 1$ |
| | | | 100 | $1 - 10^{3}$ |

For large paramagnetic *magnions* with sizes 10–100 nm, the effects of magnetic fields may be also significant (Table 12.1). Note that paramagnetic *magnions* do not exist in naturally occurring form in bio-organisms with several exceptions including, mitochondria aggregates, met-hemoglobin containing erythrocytes, and ferritin (Table 12.1). As for the artificial paramagnetic *magnions*, a substantial number of reviews has been published on magnetic drug delivery [43]. A promising drug delivery approach is magnetic drug targeting which can be realized if a drug delivery vehicle possesses a strong magnetic moment [43, 44]. For example, estimated specific magnetic moment of a magnetic nanoliposome is about 0.2 emu/g at $H = 6$ kOe at room temperature and magnetic susceptibility is $\chi_P \approx 10^{-5}$ [44].

Note that, typically, the sizes of living cells are much larger than the sizes of granules/liposomes/vesicles/cellular compartments and they can significantly influenced by external magnetic fields. For example, magnetophoretic motion of both nonlabeled HeLa tumor cells ($\chi_{\mathrm{p}} \approx -0.5136 \times 10^{-6}$ (CGS units)) and labeled yeast, liver, and carcinoma cells ($\chi_{\mathrm{p}} \approx (1.0-1.6) \times 10^{-6}$ (CGS units) in a nonhomogeneous magnetic field has been observed [29].

However, very limited information on magnetic susceptibility for the biological samples at their physiological conditions is available [33]. For example, the measured of mass magnetic susceptibility of a transplanted hepatoma is $-(0.688 \pm 0.0046) \times 10^{-6}$ emu/g at 263 K [58]. For the liver tissues of a tumor bearing rat and normal rat these values are $-(0.670 \pm 0.0012) \times 10^{-6}$ emu/g and $-(0.637 \pm 0.0059) \times 10^{-6}$ emu/g, respectively. However, the magnetic susceptibility information of these tissues at 37 C (310 K) is unknown [33].

### 12.2.2 Phase Separation of Liquid Media Under Gradient Magnetic Fields

Liquid–liquid phase separation was revealed during electrochemical reactions in aqueous solutions under influence of the gradient magnetic field [26]. These investigations illustrate mechanisms of the gradient magnetic field influence on transport processes and biochemical reactions. The study of the influence of the gradient magnetic field on the form of the interface of liquid–liquid phase separation was carried out in [26] using the installation, schematically shown in Fig. 12.3. The time evolution of the shape of the areas with high concentration of *magnions* during the deposition of copper on the surface of a steel sphere is shown in Fig. 12.4 [26].

The shape of the interface of liquid–liquid phase separation (Fig. 12.4) is described by model of a hydrostatic equilibrium on the interface taking on account the osmotic pressure and the magnetic pressures of *magnion* solution, as well as the hydrostatic pressure due to the Earth's gravity field [21, 26].

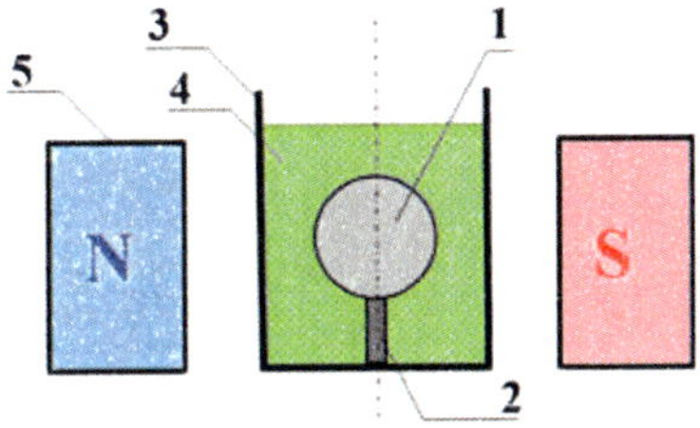


**Fig. 12.3** The schematic image of the experimental device used in [26] for investigation of the effects of the gradient magnetic field on the form of the interface of liquid–liquid phase separation. Here 1 is a steel sphere which is the source of the gradient magnetic field when magnetized, 2 is a glass or polymer holder, 3 is a glass cell, 4 is a liquid (an aqueous solution of $CuSO_4$), 5 is a magnetic system (electromagnet) creating a homogeneous constant magnetic field between the pole tips "N" and "S" (blue and red colors correspond to the south pole "S" and the north pole "N" of the magnet respectively, opposite poles of the magnet are located opposite each other). The vector of the gravitational acceleration $\vec{g}$ is directed vertically downwards. The scale is not respected

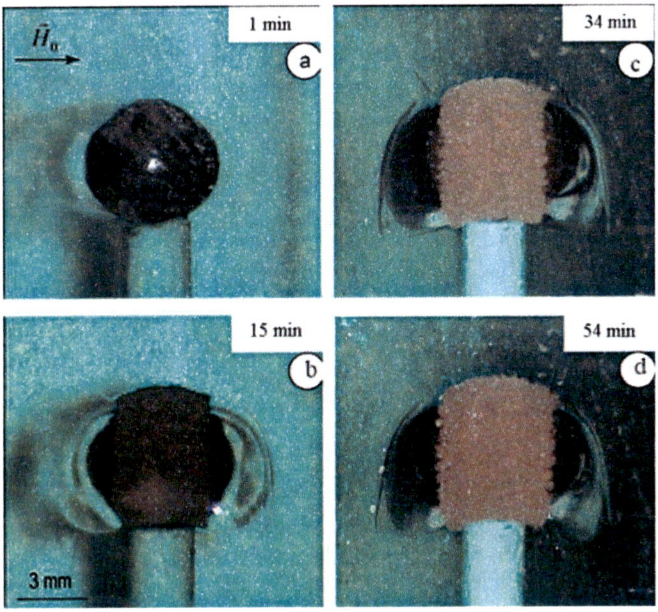


**Fig. 12.4** The time evolution of the shape of the areas with high concentration of magnions during the deposition of copper from the solution of $CuSO_4$ on the surface of a steel sphere. The initial radius of steel sphere was $R_0 = 3$ mm. The external magnetic field $H_0 = 3$ kOe was applied in the horizontal direction

### 12.2.3 Magnetophoresis of the Internal Components of the Body. Biogenic Magnetic Nanoparticles as Magnetic Organelle

Recently an interdisciplinary direction of research area emerged about the influence of gradient magnetic fields on metabolism associated with gradient scattering fields of BMN. BMNs are nanosized crystals of ferrites, namely magnetite, maghemite, greigite, etc. with characteristic sizes from tens to hundreds of nanometers [59]. As a rule, ferrite BMNs are permanent nanomagnets because they have a residual magnetization and they are single domain. Therefore BMNs create in their vicinity a gradient magnetic field about 4 orders of magnitude greater than the Earth's magnetic field. Since ferrites are minerals (i.e. inorganic substances with a crystalline structure), the biosynthesis of these compounds has been called the biomineralization of BMN. BMNs became the subject of intensive research since 1975, when they were first found in magnetotactic bacteria [9]. Since then, BMNs have been found in a number of organisms belonging to all three domains: prokaryotes, archaea, and eukaryotes, from bacteria [9], algae, and protozoa to mammals and humans [9, 59, 60]. Under normal conditions BMN are in the human body in the walls of capillaries [61] of the brain, heart, liver, spleen and other organs [11, 60]. The general genetic mechanism of biomineralization of BMN in all multicellular organisms, including man [3, 4], and in a number of unicellular was discovered by bioinformatics methods in the papers [62, 63]. It is shown on the example of some unicellular organisms that the amount of BMN is regulated by such factors as oxygen and iron ion concentrations [62], as well as the value of the external magnetic field [64]. Experimental data are also known that iron ions catalyze the Fenton reaction of production of reactive oxygen species [65], so the biosynthesis of BMN in bacteria reduces the concentration of Fe-ions in the cytosol, and, consequently, reduces the quantity of reactive oxygen species [65]. It is known that such diseases as cardiovascular, atherosclerosis, neurodegenerative, cancer, infectious are accompanied by hypoxia and it has been proven that a number of oncological and neurodegenerative diseases are characterized by a significant increase in the concentration of BMN [66–68].

The BMN can have common functions in cells of both prokaryotes and eukaryotes. The possible functions of BMN are the additional transport driven forces near the cell membrane due to at least two different mechanisms. The first mechanism is magnetophoretic one [21–24] and it can result in high gradient magnetic accumulation, concentration and separation of internal components of the body of *magnion*-type under influence of inhomogeneous magnetic field of BMN. The gradient magnetic force can lead also to creation of quasi-stationary liquid phases with different magnetic susceptibilities in diamagnetic aqueous solutions as it was illustrated above [21, 26] (Fig. 12.4). Gradient magnetic capture of internal components of the body of *magnion*-type can increase significantly the probability of fusion reactions between *magnions* of different types. The second mechanism is based on generation of electromotive force (EMF) of concentration circuit induced by the inhomogeneous magnetic field [69] of BMN at the cell membrane as the secondary magnetochemical effect of the first mechanism. The second mechanism can give rise to the additional electrophoretic driven force acting on ions, charged molecules and other charged components near the cell membrane. Besides, the magnetohydrodynamic stirring effect [70, 71] can take place in the combined magnetic and electric fields, generated by BMN in the vicinity of the cell membrane. The stirring can lead not only to the change of transport processes but it can change the rates of chemical and biochemical reactions with diffusion or mixed kinetics [72]. Thus, the BMN can represent the biogenic nano-device for regulation of transport processes in cells.

Let's write the Boltzmann distribution of concentration *magnions* $c(\vec{r})$ in an aqueous solution in the vicinity of a magnetized sample (see Fig. 12.4 for illustration) [3]:

$$c(\vec{r}) = c_0 \exp(\varepsilon) \tag{12.9}$$

where $c_0$ is a constant concentration. Then the exponent $\varepsilon$ in the Eq. (12.8) is about $10^{-6}$ for the separate paramagnetic ion in aqueous solution (see Table 12.1) [21, 26], when the stray magnetic field strength of a magnetite BMN is estimated as $\left|\vec{H}\right| \approx 4\pi M_0$, $M_0 = 447$ G is magnetization saturation for magnetite. Indeed, the gradient magnetic field of BMN cannot change significantly distribution of concentration of separate paramagnetic ions or molecules in a cell. However, the exponent $\varepsilon$ can be greater than 1 for sufficiently large volume of effectively para- or diamagnetic *magnion* of volumes $V_p$, which are equivalent to the number of para- or diamagnetic ions $n_g \cong 10^6 \div 10^8$ correspondingly (Table 12.1). Thus, the gradient magnetic field of BMN can render significant influence on distribution of concentration of *magnions* consisting of a large number of paramagnetic ions or molecules [21, 26]. The nanobubbles, stabilized by the paramagnetic ions in aqueous solution, dispersive colloid nanoparticles, bubbles of gases, vesicles, granules and cells can behave as *magnions* especially for biosystems at certain conditions such as gene and drug delivery, transport of vesicles and cells in a blood system [3].

Let us discuss the second mechanism for possible metabolic functions of BMN. It follows from the general equation for the electric cell voltage [73] that the EMF of concentration circuit is induced in inhomogeneous magnetic field of BMN between the points with radius vectors $\vec{r}_1$ and $\vec{r}_2$ in the vicinity of BMN. The Nernst type voltage can be defined by the expression taking on account the difference of chemical potentials of *magnions* in the inhomogeneous magnetic field $\vec{H}$ in the case of only one type of *magnions* [69]:

$$\Delta\varphi_N(\vec{r}) = \varphi_N(\vec{r}_1) - \varphi_N(\vec{r}_2) = \frac{k_B T}{Ze} \ln \frac{c(\vec{r}_2)}{c(\vec{r}_1)}, \tag{12.10}$$

where $Ze$ is the charge of a *magnion*, $c(\vec{r}_1)$, $c(\vec{r}_2)$ are the concentrations of *magnions* at the points with radius vectors $\vec{r}_1$ and $\vec{r}_2$ correspondingly, $\vec{r} = \vec{r}_2 - \vec{r}_1$. It is possible to transform the expression (12.10) [69, 74]:

$$\Delta\varphi_N(\vec{r}) = \frac{\chi V_p}{2Ze}\left[\vec{H}^2(\vec{r}_1) - \vec{H}^2(\vec{r}_2)\right]. \tag{12.11}$$

The formula (12.11) describes the distribution of electric potential $\Delta\varphi_N(\vec{r})$ at the cell membrane for membrane-bound BMNs and it shows that the nonuniform distribution of the magnetic field of BMN generates the electric field in a cell. Usually, the charge of a *magnion* $Ze << n_g \cdot e$ where $e$ is the charge of electron. For example, the charge of the nanobubbles $Ze$ can be estimated as $10^2 \cdot e$ [69] and the voltage $\Delta\varphi_N(\vec{r})$ between magnetic pole and equator of magnetized particle can be about $(3 \cdot 10^{-4} - 3 \cdot 10^{-2})$ emu in Gauss units which correspond to approximately (0.1 – 10 mV) in SI units for $n_g$ from $10^6$ to $10^8$ [69]. The electric current arises in aqueous solution under voltage according to the Eq. (12.11) [69]. The current density and the induced Lorentz force were calculated in [74, 75]. The last results in rotation of aqueous solution [69, 74, 75].

The modeling experiments [26] (see Fig. 12.4 for illustration) show that the effects of inhomogeneous magnetic field on transport processes and electrochemical reaction in aqueous solutions become apparent at wide-ranging of scales starting from meso-, micro- and to nanoscales. It is in accordance with theoretical calculations of scale independent distribution of magnetic pressure and the Eq. (12.11) for voltage as the function of the dimensionless coordinates in the units of magnetized particle radius [69, 74, 75].

As it was mentioned above, the fundamental genetically programmed mechanism of BMN biosynthesis in all living organisms was revealed in the paper [3]. This genetic mechanism is unique to representatives of unicellular and multicellular organisms, including humans. The general mechanism of BMN biosynthesis is one of the main mechanisms that form the basis of the functioning of living things, as it arose before the emergence of multicellular organisms. According to this concept, the BMN chain is an important functional element of the cell (magnetic organelle), which uses the effect of high-gradient magnetic fields of the BMN on vesicles, granules and other components of the cell [3, 61].

## 12.3 Electrokinetic Effects Under Gradient Magnetic Fields

The electric potential arises under influence of the nonuniform magnetostatic field on *magnions* as colloid particles [69]. This electric potential consists of the Nernst-type potential (Eq. (12.11)) of inhomogeneous concentration distribution of *magnions* caused by gradient magnetic field [74] and magnetophoresis potential (MPP) at movement of *magnions* under inhomogeneous magnetic field. For calculation of the MPP, it is assumed only the case when the charge and *magnion* transport in liquid takes place without appearance of hydrodynamic flows. Let's use Onsager's phenomenological equations for relation between the generalized flows and the generalized thermodynamic forces in the case of the presence of a magnetic field. The consideration is restricted to the case of two cross-processes: the electric current and the mass transport of *magnions*. It is known that Onsager's kinetic coefficient tensor can be represented conveniently as a sum of symmetric and antisymmetric parts in the presence of a magnetic field [76]. Thus, the symmetric (even) part of this tensor does not change the sign during inversion of the magnetic field, but the antisymmetric part is an odd function of the magnetic field in the medium. However, for electrolytes in the range of moderate magnetic fields up to $10^4$ G ($10^4$ G = 1 T) the antisymmetric part of the kinetic coefficient's tensor is much less than its symmetric part. This takes place due to the low Hall conductivity of electrolytes [77]. In the case of an isotropic medium, the relationship between the generalized flows and generalized forces – after neglecting the antisymmetric part of the kinetic coefficients tensor – can be expressed through just the even part of the kinetic coefficients tensor with respect to the inversion of the magnetic field direction [78]:

$$\begin{aligned} \vec{J}_1 &= L_{11}\vec{X}_1 + L_{12}\vec{X}_2, \\ \vec{J}_2 &= L_{21}\vec{X}_1 + L_{22}\vec{X}_2 \end{aligned} \tag{12.12}$$

where the matrix of phenomenological coefficients $L = \begin{pmatrix} L_{11} & L_{12} \\ L_{21} & L_{22} \end{pmatrix}$, the generalized forces $\vec{X}_1$, $\vec{X}_2$ and generalized flows $\vec{J}_1$, $\vec{J}_2$ are introduced. Let us choose the electric current density in the electrolyte $\vec{J}_1 = \vec{i}$ and the flux density of matter $\vec{J}_2 = \vec{j}$ as the thermodynamic fluxes, and the electric field $\vec{X}_1 = \vec{E}$ as the first generalized force. Basing on the expression for the rate of entropy production $S$[79, 80]:

$$\dot{S} = \frac{1}{T}\sum_{i=1}^{2}\vec{X}_i\vec{J}_i, \tag{12.13}$$

where $T$ is the absolute temperature, in this case we can get a second generalized force in the form $\vec{X}_2 = \vec{F}_{\nabla\vec{H}^2}$ where $\vec{F}_{\nabla\vec{H}^2}$ is the gradient magnetic force acting on a unit volume of a *magnion* as follows from the Eqs. (12.5) and (12.7) in the case of isotropic effective magnetic susceptibility tensor:

$$\vec{F}_{\nabla\vec{H}^2} = \frac{\chi}{2}\nabla\vec{H}^2, \tag{12.14}$$

where $\vec{H}$ is the magnetic field strength in the electrolyte.

There is a relationship between the phenomenological coefficients according to Onsager-Casimir relations

$$L_{12} = L_{21}, \tag{12.15}$$

here the fact that in the above approximation kinetic coefficients do not depend on the direction of the magnetic field is taken into account.

It can be written from the relations (12.12)–(12.15) that

$$\left(\left|\vec{i}\right| \times \left|\vec{F}_{\nabla \vec{H}^2}\right|^{-1}\right)_{\vec{E}=0} = \left(\left|\vec{j}\right| \times \left|\vec{E}\right|^{-1}\right)_{\vec{F}_{\nabla \vec{H}^2}=0}. \quad (12.16)$$

The ratio of the right side of (12.16) corresponds to the electrophoretic effects and the ratio of the left side corresponds to effects associated with the influence of the gradient magnetic force on *magnions.*

Let us consider in the left side of the relation (12.16) the fact that current $\vec{i}$ is related to the electric field $\vec{E}_{mPh}$ of the MPP in the following way:

$$\vec{i} = K_\infty \vec{E}_{mPh}, \quad (12.17)$$

where $K_\infty$ is the conductivity of pure solution without *magnions.*

The expression for the flux density of *magnions* is used in the right side of the formula (12.16).

$$\vec{j} = \vec{v}_m n, \quad (12.18)$$

where $n$ is the concentration of *magnions* and the velocity of a *magnion* equals to.

$$\vec{v}_m = \mu \vec{E}, \quad (12.19)$$

where $\mu$ is the electrophoretic mobility of a *magnion* as a colloid particle.

The expression for the electric field created by the magnetic field gradient can be obtained by substituting the Eqs. (12.17)–(12.19) to the formula (12.16)

$$\vec{E}_{mPh} = \frac{\phi \mu \chi}{2K_\infty} \nabla \vec{H}^2, \quad (12.20)$$

where $\phi = nV_p$ is volume fraction of *magnions* in the electrolyte.

Let us write MPP difference between the points of the radius vectors $\vec{r}_1$ and $\vec{r}_2$ that may be obtained by integrating of the expression (12.20):

$$\varphi_{mPh}(\vec{r}_1) - \varphi_{mPh}(\vec{r}_2) = \frac{\phi \mu \chi}{2K_\infty}\left[\vec{H}^2(\vec{r}_1) - \vec{H}^2(\vec{r}_2)\right]. \quad (12.21)$$

The expressions (12.9) and (12.10) are valid for dilute solutions $\phi << 1$. However, the generalized expression for the electric field strength created by the MPP difference can be obtained similarly to the calculation of sedimentation potential in the paper [81]

$$\vec{E}_{mPh} = \frac{\phi(1-\phi)\mu \chi}{2(1+\phi/2)K_\infty} \nabla \vec{H}^2, \quad (12.22)$$

and for the MPP difference

$$\varphi_{mPh}(\vec{r}_1) - \varphi_{mPh}(\vec{r}_2) = \frac{\phi(1-\phi)\mu\chi}{2(1+\phi/2)K_\infty}\left[\vec{H}^2(\vec{r}_1) - \vec{H}^2(\vec{r}_2)\right] \tag{12.23}$$

of concentrated solution of *magnions.*

It can be seen that the functional expression for the MPP difference differs from the expression for the EMF of the concentration circuit between two points of the electrode surface [82] obtained from the Nernst Eq. (12.11)

$$\Delta\varphi_N(\vec{r}) = \varphi_N(\vec{r}_1) - \varphi_N(\vec{r}_2) = \frac{\chi V_p}{2Ze}\left[\vec{H}^2(\vec{r}_1) - \vec{H}^2(\vec{r}_2)\right] \tag{12.24}$$

only by the factor. Thus, the superposition of the potential of Nernst type $\Delta\varphi_N(\vec{r})$ and MPP $\Delta\varphi_{mPh}(\vec{r})$ may take place:

$$\Delta\varphi(\vec{r}) = \frac{\chi V_p}{2Ze}\left(1 + \frac{\phi(1-\phi)\mu}{(1+\phi/2)K_\infty}\frac{Ze}{V_p}\right)\left[\vec{H}^2(\vec{r}_1) - \vec{H}^2(\vec{r}_2)\right]. \tag{12.25}$$

Let's use the relationship between the electrophoretic mobility of *magnions* and zeta potential similar to the one obtained during the calculation of sedimentation potential [83]. It allows estimating the relative contribution of MPP of moving *magnions* to the general electrical potential (12.25). The breaking of the diffusion layer symmetry around the *magnion* during the free movement of phases in opposite directions (relaxing effect) [84] is neglected as well as the surface and convective electrical conductivities of the double layer. The dynamic mobility $\mu$ at the motion of a *magnion* with the radius $R_0$ and zeta-potential $\zeta$ in the electrolyte with the dynamic viscosity $\eta$ exposed to an electric field can be represented using the generalized Helmholtz-Smoluchowski equation [84, 85]

$$\mu = \frac{\tilde{\varepsilon}\zeta}{4\pi\eta} f(\kappa R_0), \tag{12.26}$$

where $\tilde{\varepsilon}$ is the dielectric constant of the medium, $\kappa$ is the Debye–Huckel parameter, the dimensionless function $f(\kappa R_0)$ is the Henry function [85] that has the order of unity ($f(\kappa R_0) = 2/3$ when the thickness of the diffusion layer is much larger than the particle size (i.e. *magnion* size) and $f(\kappa R_0) = 1$ in the opposite limiting case).

Let's estimate the ratio of the MPP difference to the Nernst potential difference according to formulas (12.23) and (12.24) using the formula (12.26) for the dynamic mobility of *magnions.*

$$\gamma = \frac{\varphi_{mPh}(\vec{r}_1) - \varphi_{mPh}(\vec{r}_2)}{\varphi_N(\vec{r}_1) - \varphi_N(\vec{r}_2)} = \frac{\phi\mu Ze}{K_\infty V_p}, \tag{12.27}$$

which is independent on the magnetic field and the effective magnetic susceptibility of *magnion* substance.

Magnetophoretic effects have found their broad application, in particular, in microbiology and biomedicine in filtering and separation of solutions and mixtures containing magnetic particles [86], cell bioseparation [87], drug delivery, labeling and manipulation of biomaterials and others. Therefore, the results of theoretical modeling of this work depending on the physical characteristics of *magnions* and the conducting medium can be used both for creating functional materials by magneto-electrolysis methods and for modelling of the impact of biogenic magnetic nanoparticles on transport processes and biochemical reactions in the cells of living organisms [69].

## 12.4 Orientation Effects in Strong Magnetic Fields

If the force producing an active movement of a *magnion* relative to liquid is not zero and tensor of an effective magnetic susceptibility of a *magnion* is anisotropic than directed movement of *magnions* arise under influence of even uniform external magnetic field. As it was described above, there are magnetotactic bacteria able to move along geomagnetic field or low enough external magnetic field (about 1 G) direction due to existence of active movement forces of flagellum and anisotropy of magnetic susceptibility formed in presence of chain of BMNs along the long axis of bacterium. However, the similar directed movement of bacteria with anisotropic diamagnetic susceptibility can arise under influence of strong uniform external magnetic field about $10^4$ - $10^5$G.

The statistical model describing orientation of trajectories of such unicellular diamagnetic organisms in a magnetic field was proposed in the reference [8]. The statistical parameter such as the effective energy was calculated on basis of this model. The resulting effective energy is the statistical characteristics of trajectories of diamagnetic microorganisms in a magnetic field connected with their metabolism. The statistical model is applicable for the case when the energy of the thermal motion of bacteria is negligible in comparison with their energy in a magnetic field and the bacteria manifest the significant "active random movement", i.e. there is the randomizing motion of the bacteria of nonthermal nature, for example, movement of bacteria by means of flagellum. The energy of the randomizing active self-motion of bacteria is characterized by the new statistical parameter for biological objects. The parameter replaces the energy of the randomizing thermal motion in calculation of the statistical distribution.

The existence of a large number of moving bacteria and large number of their trajectories allows to obtain the statistical laws of their movement which as it is well known give the new information about the important macroscopic parameters of the system, in particular, the average energy of randomizing motion $E_{ef}$ (in this case the randomizing motion is not of thermal nature). The energy of interaction of an external homogeneous magnetic field $\vec{H}_0$ with a magnetic moment of diamagnetic microorganism in the form of ellipsoid can be written as:

$$U = \frac{\Delta\chi H_0^2}{2}\cos^2\theta - \frac{H_0^2}{2}\chi_\perp V_p, \tag{12.28}$$

where $\Delta\chi$ is the absolute value of the difference between the longitudinal $\chi_{II}$ (along the greater semi-axis of the ellipsoid $a$) and transversal $\chi_\perp$ (along the shorter semi-axis $b$) components of diamagnetic susceptibility $\Delta\chi = |\chi_\perp - \chi_{II}|V_p$ is about 6,7 $\times$ $10^{-17}$ cm$^3$ because $a = 100$ μ, $b = 20$ μ and the difference $|\chi_\perp - \chi_{II}|$ is about $10^{-3}|\chi_\perp|$ [88], $\theta$ is the angle between the greater semi-axis of diamagnetic microorganism and direction of an external magnetic field and $|\chi_{II}| > |\chi_\perp|$, in the case when the minimum of magnetic energy (Eq. (12.28)) is realized for the perpendicular orientation of the greater semi-axis of microorganism with respect to the external magnetic field direction. Another case is realized when $|\chi_\perp| > |\chi_{II}|$ and the energy minimum takes place for parallel orientation of the greater axis of microorganism and an external magnetic field. In such a case, the energy has the following form:

$$U = -\frac{\Delta\chi H_0^2}{2}\cos^2\theta - \frac{H_0^2}{2}\chi_\perp V_p. \tag{12.29}$$

The last case was observed experimentally in the paper [88], it is unlike parallel movement of magnetotactic bacteria along a magnetic field direction.

The Boltzmann distribution for polar and azimuth angles $\theta$ and $\varphi$ of microorganism orientation in a magnetic field can be written in the following form [8]:

$$dw = \begin{cases} b\exp(-\alpha x^2)dxd\varphi, |\chi_{II}| > |\chi_\perp| \\ b\exp(\alpha x^2)dxd\varphi, |\chi_\perp| > |\chi_{II}| \end{cases}, \tag{12.30}$$

where $x = \cos\theta$, $\alpha = \frac{\Delta\chi H_0^2}{2E_{ef}}$,

$$b = \begin{cases} \dfrac{1}{4\pi\int\limits_0^1 \exp(-\alpha x^2)dx}, |\chi_{II}| > |\chi_\perp| \\ \dfrac{1}{4\pi\int\limits_0^1 \exp(\alpha x^2)dx}, |\chi_\perp| > |\chi_{II}| \end{cases} \tag{12.31}$$

and the angle $\theta$ can vary within the range from 0 to $\pi$ (i.e. $x$ from $-1$ to 1), and the angle $\varphi$ from 0 to $2\pi$. $E_{ef}$ is an effective energy which can be considered as temperature in Boltzmann distribution and it characterizes the intrinsic randomly oriented movement of diamagnetic microorganisms connected with their metabolism in the investigated medium [8].

The quantitative fitting of the experimental [88] and theoretically calculated dependence of the order parameter average magnetization on a magnetic field strength can be obtained numerically for value of the effective energy of diamagnetic unicellular organism $E_{ef} = 1.3 \times 10^{-8}$ g cm$^2$/s$^2$ [8]. The statistical model was proposed for description of orientation of trajectories of unicellular diamagnetic organisms in a magnetic field [8]. The statistical parameter such as the effective energy $E_{ef}$ is calculated on basis of this model. The resulting effective energy $E_{ef}$ is the statistical characteristics of trajectories of diamagnetic microorganisms in a magnetic field connected with their metabolism. Summarizing, it is worthy to note that this research is aimed to statistical modeling the case when

1) the energy of the thermal motion of bacteria is negligible in comparison with their energy in a magnetic field and
2) the bacteria manifest the significant "active random movement", i.e. there is the randomizing motion of the bacteria of non-thermal nature, for example, movement of bacteria by means of flagellum.

The energy of the randomizing active self-motion of bacteria is characterized by the new statistical parameter $E_{ef}$ for biological objects. The parameter $E_{ef}$ replaces the energy of the randomizing thermal motion in calculation of the statistical distribution. The procedure of derivation of this statistical function is similar to the calculation of the classical example of Langevin function for paramagnetic molecules and superparamagnetic nanoparticles. The statistical approach resulting in the formular (12.30) was tested on the experiments [88] but it is applicable to the wide spectrum of bacteria able to move, for example, by means of flagellum (including magnetotactic bacteria and bacteria with magnetically sensitive inclusions). The measurement of the statistical parameter $E_{ef}$ can be exploited as a one more quantitative characterization of behavior of concrete species and strains of bacteria, namely the average energy of the active randomizing movement of swimming populations of unicellular organisms in a magnetic field. The averaging for $E_{ef}$ calculation can be carried out both on the ensemble of trajectories of different bacteria in the population or for the individual bacterium for a long enough trajectory in a constant magnetic field. In the last case, the parameter $E_{ef}$ would characterize the energy of the active randomizing movement of the separate swimming unicellular organism in a magnetic field. In this connection, the statistical modeling of the experimental data gives the important new information about the metabolic processes. The parameter $E_{ef}$ represents the new biological characteristics of organisms. The parameter $E_{ef}$ is specific for the species, strains etc. and it can correlate with other biological characteristics of the organisms. Besides, this parameter $E_{ef}$ can provide information about the change of state of organisms under the influence of different external physical factors and changes of the environment. It is possible to introduce the similar parameter $E_{ef}$ for any bacteria with flagellum and with anisotropic tensor of magnetic susceptibility.

## 12.5 Final Remarks and Future Perspectives

Gradient magnetic fields and gradient magnetic forces are widely used in biotechnology and medicine for manipulation of a variety of bioobjects in liquids from biomolecules to magnetically labelled cells. The physical model of magnetophoretic motion is well understood and the extensive literature is devoted to development of modern applications. However, the physical models of gradient magnetic field influence on electrochemical reactions, liquid–liquid phase separation are far from comprehensive. The influence of the stray gradient magnetic field of biogenic magnetic nanoparticles on charge and mass transport through the cell membrane belong to open problems at the intersection of molecular biology, physics of magnetism and biochemistry. In this connection, the presented material sheds light on metabolic functions of biogenic magnetic nanoparticles.

**Acknowledgements** This work was not supported by any funding sources.

## References

1. Bingi VN, Savin AV (2003) Effects of weak magnetic fields on biological systems: physical aspects. Phys Usp 46:259–291. https://doi.org/10.1070/PU2003v046n03ABEH001283
2. Zablotskii V, Dejneka A, Kubinová Š, Le-Roy D, Dumas-Bouchiat F, Givord D, Dempsey NM, Syková E (2013) Life on magnets: stem cell networking on micro-magnet arrays. PLoS ONE 8:e70416. https://doi.org/10.1371/journal.pone.0070416
3. Gorobets OY, Gorobets SV, Gorobets YI (2014) Biogenic Magnetic Nanoparticles. Biomineralization in Prokaryotes and Eukaryotes. Dekker Encycl. Nanosci. Nanotechnology, 3rd edn. CRC Press, New York, pp 300–308
4. Gorobets O, Gorobets S, Koralewski M (2017) Physiological origin of biogenic magnetic nanoparticles in health and disease: from bacteria to humans. Int J Nanomed 12:4371–4395. https://doi.org/10.2147/IJN.S130565
5. Gorobets OY (2015) Biomagnetism and biogenic magnetic nanoparticles. Visn Nac Akad Nauk Ukrai'ni, 53–64 . https://doi.org/10.15407/visn2015.07.053
6. Gorobets YI, Dzhezherya YI, Melnichuk IA, Cherepov SV, Kuz' AP, (2010) Application of domain structures elements of ferrite-garnet films for transport of magnetic microparticles. J Appl Phys 108:123902. https://doi.org/10.1063/1.3517111
7. Pavlovich NV (1985) The Magnetic Susceptibility of Organisms. Nauka, Minsk
8. Gorobets YI, Gorobets OY (2015) Statistical characteristics of trajectories of diamagnetic unicellular organisms in a magnetic field. Prog Biophys Mol Biol 117:125–128. https://doi.org/10.1016/j.pbiomolbio.2012.06.001
9. Blakemore R (1975) Magnetotactic bacteria. Science 80(190):377–379. https://doi.org/10.1126/science.170679
10. Schultheiss-Grassi PP, Dobson J (1999) Magnetic analysis of human brain tissue. Biometals 12:67–72. https://doi.org/10.1023/A:1009271111083
11. Grassi-Schultheiss PP, Heller F, Dobson J (1997) Analysis of magnetic material in the human heart, spleen and liver. Biometals 12:67–72. https://doi.org/10.1023/A:1018340920329
12. Smith SB, Finzi L, Bustamante C (1992) Direct mechanical measurements of the elasticity of single DNA molecules by using magnetic beads. Science 258:1122–1126. https://doi.org/10.1126/science.1439819

13. De Vlaminck I, Dekker C (2012) Recent advances in magnetic tweezers. Annu Rev Biophys 41:453–472. https://doi.org/10.1146/annurev-biophys-122311-100544
14. Sarkar R, Rybenkov VV (2016) A guide to magnetic tweezers and their applications. Front Phys 4:1–20. https://doi.org/10.3389/fphy.2016.00048
15. Denschlag J, Schmiedmayer J (1998) A neutral atom and a thin wire: on the road to mesoscopic atom optics. In: Technical Digest. Summaries of Papers Presented at the International Quantum Electronics Conference. Conference Edition. 1998 Technical Digest Series, vol 7 (IEEE Cat. No.98CH36236). Opt. Soc. America, pp 234–235
16. Bartenstein M, Cassettari D, Calarco T, Chenet A, Folman R, Brugger K, Haase A, Hartungen E, Hessmo B, Kasper A, Kruger P, Maier T, Payr F, Schneider S, Schmiedmayer J (2000) Atoms and wires: toward atom chips. IEEE J Quantum Electron 36:1364–1377. https://doi.org/10.1109/3.892555
17. Leung VYF, Pijn DRM, Schlatter H, Torralbo-Campo L, La Rooij AL, Mulder GB, Naber J, Soudijn ML, Tauschinsky A, Abarbanel C, Hadad B, Golan E, Folman R, Spreeuw RJC (2014) Magnetic-film atom chip with 10 μm period lattices of microtraps for quantum information science with Rydberg atoms. Rev Sci Instrum 85:053102. https://doi.org/10.1063/1.4874005
18. Helseth LE, Fischer TM, Johansen TH (2003) Paramagnetic beads surfing on domain walls. Phys Rev E 67:042401. https://doi.org/10.1103/PhysRevE.67.042401
19. Iranmanesh M, Hulliger J (2017) Magnetic separation: its application in mining, waste purification, medicine, biochemistry and chemistry. Chem Soc Rev 46:5925–5934. https://doi.org/10.1039/c7cs00230k
20. Gorobets SV, Mikhailenko NA (2014) High-gradient ferromagnetic matrices for purification of wastewaters by the method of magnitoelectrolysis. J Water Chem Technol 36:153–159. https://doi.org/10.3103/S1063455X14040018
21. Gorobets OY, Gorobets YI, Bondar IA, Legenkiy YA (2013) Quasi-stationary heterogeneous states of electrolyte at electrodeposition and etching process in a gradient magnetic field of a magnetized ferromagnetic ball. J Magn Magn Mater 330:76–80. https://doi.org/10.1016/j.jmmm.2012.10.015
22. Friedlaender F, Gerber R, Kurz W, Birss R (1981) Particle motion near and capture on single spheres in HGMS. IEEE Trans Magn 17:2801–2803. https://doi.org/10.1109/TMAG.1981.1061683
23. Moyer C, Natenapit M, Arajs S (1986) Diffusive capture of ultrafine particles by spheres. J Magn Magn Mater 54–57:1475–1477. https://doi.org/10.1016/0304-8853(86)90891-7
24. Yellen BB, Forbes ZG, Halverson DS, Fridman G, Barbee KA, Chorny M, Levy R, Friedman G (2005) Targeted drug delivery to magnetic implants for therapeutic applications. J Magn Magn Mater 293:647–654. https://doi.org/10.1016/j.jmmm.2005.01.083
25. Oberteuffer J (1973) High gradient magnetic separation. IEEE Trans Magn 9:303–306. https://doi.org/10.1109/TMAG.1973.1067673
26. Gorobets OYu, Gorobets YuI, Rospotniuk VP, Kyba AA, Legenkiy YuA (2015) Liquid-liquid phase separation occurring under the influence of inhomogeneous magnetic field in the process of the metal deposition and etching of the magnetized ferromagnetic ball. J Solid State Electrochem 19(10):3001–3012. https://doi.org/10.1007/s10008-015-2904-x
27. Schenck JF (1996) The role of magnetic susceptibility in magnetic resonance imaging: MRI magnetic compatibility of the first and second kinds. Med Phys 23:815–850. https://doi.org/10.1118/1.597854
28. Zablotskii V, Polyakova T, Lunov O, Dejneka A (2016) How a high-gradient magnetic field could affect cell life. Sci Rep 6:37407. https://doi.org/10.1038/srep37407
29. Yaman S, Tekin HC (2020) Magnetic susceptibility-based protein detection using magnetic levitation. Anal Chem 92:12556–12563. https://doi.org/10.1021/acs.analchem.0c02479
30. He X, Yablonskiy DA (2009) Biophysical mechanisms of phase contrast in gradient echo MRI. Proc Natl Acad Sci U S A 106:13558–13563. https://doi.org/10.1073/pnas.0904899106
31. Zablotskii V, Polyakova T, Dejneka A (2018) Cells in the non-uniform magnetic world: how cells respond to high-gradient magnetic fields. BioEssays 40:1800017. https://doi.org/10.1002/bies.201800017

32. Pauling L (1979) Diamagnetic anisotropy of the peptide group. Proc Natl Acad Sci 76:2293–2294. https://doi.org/10.1073/pnas.76.5.2293
33. Tao Q, Zhang L, Han X, Chen H, Ji X, Zhang X (2019) Magnetic susceptibility difference-induced nucleus positioning in gradient ultrahigh magnetic field. Biophys J. https://doi.org/10.1016/j.bpj.2019.12.020
34. Ravera E, Takis PG, Fragai M, Parigi G, Luchinat C (2018) NMR spectroscopy and metal ions in life sciences. Eur J Inorg Chem 2018:4752–4770. https://doi.org/10.1002/ejic.201800875
35. Vigers GP, Crowther RA, Pearse BM (1986) Location of the 100 kd-50 kd accessory proteins in clathrin coats. EMBO J 5:2079–2085. https://doi.org/10.1002/j.1460-2075.1986.tb04469.x
36. Cosson P, Amherdt M, Rothman JE, Orci L (2002) A resident Golgi protein is excluded from peri-Golgi vesicles in NRK cells. Proc Natl Acad Sci U S A 99:12831–12834. https://doi.org/10.1073/pnas.192460999
37. Takamori S, Holt M, Stenius K, Lemke EA, Grønborg M, Riedel D, Urlaub H, Schenck S, Brügger B, Ringler P, Müller SA, Rammner B, Gräter F, Hub JS, De Groot BL, Mieskes G, Moriyama Y, Klingauf J, Grubmüller H, Heuser J, Wieland F, Jahn R (2006) Molecular anatomy of a trafficking organelle. Cell 127:831–846. https://doi.org/10.1016/j.cell.2006.10.030
38. Huang C-H (1969) Studies on phosphatidylcholine vesicles. Form Phys Characteristics Biochem 8:344–352. https://doi.org/10.1021/bi00829a048
39. Su QP, Du W, Ji Q, Xue B, Jiang D, Zhu Y, Ren H, Zhang C, Lou J, Yu L, Sun Y (2016) Vesicle size regulates nanotube formation in the cell. Sci Rep 6:24002. https://doi.org/10.1038/srep24002
40. Rideau E, Dimova R, Schwille P, Wurm FR, Landfester K (2018) Liposomes and polymersomes: a comparative review towards cell mimicking. Chem Soc Rev 47:8572–8610. https://doi.org/10.1039/C8CS00162F
41. Olga W, Krystyna M, Jadwiga M, Andrzej BH (2009) Giant unilamellar vesicles - a perfect tool to visualize phase separation and lipid rafts in model systems. Acta Biochimica Polonica 56(1). https://doi.org/10.18388/abp.2009_2514
42. Li Q, Li S, Zhang X, Xu W, Han X (2020) Programmed magnetic manipulation of vesicles into spatially coded prototissue architectures arrays. Nat Commun 11:232. https://doi.org/10.1038/s41467-019-14141-x
43. Price PM, Mahmoud WE, Al-Ghamdi AA, Bronstein LM (2018) Magnetic drug delivery: where the field is going. Front Chem 6:1–7. https://doi.org/10.3389/fchem.2018.00619
44. Liu Y, Yang F, Yuan C, Li M, Wang T, Chen B, Jin J, Zhao P, Tong J, Luo S, Gu N (2017) Magnetic nanoliposomes as in situ microbubble bombers for multimodality image-guided cancer theranostics. ACS Nano 11:1509–1519. https://doi.org/10.1021/acsnano.6b06815
45. Magnetic susceptibility of the elements and inorganic compounds, the source The European School on Magnetism. http://www.fizika.si/magnetism/MagSusceptibilities.pdf
46. Bwambok DK, Thuo MM, Atkinson MBJ, Mirica KA, Shapiro ND, Whitesides GM (2013) Paramagnetic ionic liquids for measurements of density using magnetic levitation. Anal Chem 85:8442–8447. https://doi.org/10.1021/ac401899u
47. Kuchel PW, Chapman BE, Bubb WA, Hansen PE, Durrant CJ, Hertzberg MP (2003) Magnetic susceptibility: solutions, emulsions, and cells. Concepts Magn Reson 18A:56–71. https://doi.org/10.1002/cmr.a.10066
48. Suzuki M, Shinohara Y, Ohsaki Y, Fujimoto T (2011) Lipid droplets: size matters. Microscopy 60:S101–S116. https://doi.org/10.1093/jmicro/dfr016
49. Erickson HP (2009) Size and shape of protein molecules at the nanometer level determined by sedimentation, gel filtration, and electron microscopy. Biol Proced Online
50. Suh S, Park H-D, Kim SW, Bae JC, Tan AH-K, Chung HS, Hur KY, Kim JH, Kim K-W, Lee M-K (2011) Smaller mean LDL particle size and higher proportion of small dense LDL in Korean type 2 diabetic patients. Diab Metab J 35:536. https://doi.org/10.4093/dmj.2011.35.5.536
51. Kashevskii BE, Kashevskii SB, Prokhorov IV, Aleksandrova EN, Istomin YP (2006) Magnetophoresis and the magnetic susceptibility of HeLa tumor cells. Biophysics (Oxf). https://doi.org/10.1134/S0006350906060091

52. Cabálková J, Pribyl J, Skládal P, Kulich P, Chmelík J (2008) Size, shape and surface morphology of starch granules from Norway spruce needles revealed by transmission electron microscopy and atomic force microscopy: effects of elevated CO(2) concentration. Tree Physiol 28:1593–1599. https://doi.org/10.1093/treephys/28.10.1593
53. Kuznetsov OA, Hasenstein KH (1996) Intracellular magnetophoresis of amyloplasts and induction of root curvature. Planta 198:87–94. https://doi.org/10.1007/BF00197590
54. Norina SB, Park SH, Soh KS (2005) Diamagnetic microparticle movement in high gradient magnetic separation analyses. J Korean Phys Soc 47:297–305
55. Su X-T, Singh K, Rozmus W, Backhouse C, Capjack C (2009) Light scattering characterization of mitochondrial aggregation in single cells. Opt Express 17:13381. https://doi.org/10.1364/OE.17.013381
56. Turgeon ML (2012) Clinical Hematology Theory & Procedures, Wolters Kl
57. Gupta R, Alamrani NA, Greenway GM, Pamme N, Goddard NJ (2019) Method for determining average iron content of ferritin by measuring its optical dispersion. Anal Chem 91:7366–7372. https://doi.org/10.1021/acs.analchem.9b01231
58. Sentle FE, Thorpe A (1961) Magnetic susceptibility of normal liver and transplantable hepatoma tissue. Nature 190:410–413. https://doi.org/10.1038/190410a0
59. Faivre D, Godec TU (2015) From bacteria to mollusks: the principles underlying the biomineralization of iron oxide materials. Angew Chemie Int Ed 54:4728–4747. https://doi.org/10.1002/anie.201408900
60. Robin Baker R, Mather JG, Kennaugh JH (1983) Magnetic bones in human sinuses. Nature 301:78–80. https://doi.org/10.1038/301078a0
61. Gorobets S, Gorobets O, Gorobets Y, Bulaievska M (2018) Ferrimagnetic organelles in multicellular organisms. ArXiv ID 181106717
62. Schubbe S, Wurdemann C, Peplies J, Heyen U, Wawer C, Gluckner FO, Schuler D (2006) Transcriptional organization and regulation of magnetosome operons in magnetospirillum gryphiswaldense. Appl Environ Microbiol 72:5757–5765. https://doi.org/10.1128/AEM.00201-06
63. Elfick A, Rischitor G, Mouras R, Azfer A, Lungaro L, Uhlarz M, Herrmannsdörfer T, Lucocq J, Gamal W, Bagnaninchi P, Semple S, Salter DM (2017) Biosynthesis of magnetic nanoparticles by human mesenchymal stem cells following transfection with the magnetotactic bacterial gene mms6. Sci Rep 7:39755. https://doi.org/10.1038/srep39755
64. Wang X, Liang L (2009) Effects of static magnetic field on magnetosome formation and expression of mam A, mms 13, mms 6 and mag A in Magnetospirillum magneticum AMB-1. Bioelectromagnetics 30:313–321. https://doi.org/10.1002/bem.20469
65. Guo FF, Yang W, Jiang W, Geng S, Peng T, Li JL (2012) Magnetosomes eliminate intracellular reactive oxygen species in Magnetospirillum gryphiswaldense MSR-1. Environ Microbiol 14:1722–1729. https://doi.org/10.1111/j.1462-2920.2012.02707.x
66. Hautot D, Pankhurst QA, Khan N, Dobson J (2003) Preliminary evaluation of nanoscale biogenic magnetite in Alzheimer's disease brain tissue. Proc R Soc London Ser B Biol Sci 270:S62–S64. https://doi.org/10.1098/rsbl.2003.0012
67. Brem F, Hirt AM, Winklhofer M, Frei K, Yonekawa Y, Wieser H-G, Dobson J (2006) Magnetic iron compounds in the human brain: a comparison of tumour and hippocampal tissue. J R Soc Interface 3:833–841. https://doi.org/10.1098/rsif.2006.0133
68. Kobayashi A, Yamamoto N, Kirschvink J (1997) Studies of inorganic crystals in biological tissue: magnetic in human tumor. J Japan Soc Powder Powder Metall 44:294–300. https://doi.org/10.2497/jjspm.44.294
69. Gorobets OY, Gorobets YI, Rospotniuk VP (2015) Magnetophoretic potential at the movement of cluster products of electrochemical reactions in an inhomogeneous magnetic field. J Appl Phys 118:073902. https://doi.org/10.1063/1.4928671
70. Gorobets YI, Gorobets SV (2000) Stationary flows of electrolytes in the vicinity of ferromagnetic particles in a constant magnetic field. Bull Kherson State Tech Univ 3:276–281
71. Gorobets SV, Gorobets OY, Reshetnyak SA (2003) Electrolyte vortex flows induced by a steady-state magnetic field in the vicinity of a steel wire used as an accelerator of the chemical reaction rate. Magnetohydrodynamics 39:211–214

72. Levich VG, Tobias CW (1963) Physicochemical hydrodynamics. J Electrochem Soc 110:251C. https://doi.org/10.1149/1.2425619
73. Antropov LI (1972) Theoretical Electrochemistry. Mir, Moscow
74. Gorobets G, Rospotniuk L (2014) Electric cell voltage at etching and deposition of metals under an inhomogeneous constant magnetic field. Condens Matter Phys 17:43401. https://doi.org/10.5488/CMP.17.43401
75. Gorobets OY, Gorobets YI, Rospotnyuk VP (2012) Electromotive force under etching of uniformly magnetized steel ball in electrolyte. Metallofiz i Noveĭshie Tekhnologii 34:895–906
76. de Groot SR, Mazur P (1962) Non-Equilibrium Thermodynamics. Wiley, New York
77. Conway BE (1999) The Effect of Magnetic Fields on Electrochemical Processes. Plenum Pub, New York
78. de Groot SR (1952) Nonequilibrium thermodynamics of the sedimentation potential and electrophoresis. J Chem Phys 20:1825. https://doi.org/10.1063/1.1700322
79. Onsager L (1931) Reciprocal Relations in Irreversible Processes. II. Phys Rev 38:2265–2279. https://doi.org/10.1103/PhysRev.38.2265
80. Prigogine I (1947) Etude Thermodynamique des Processus Irreversibles. Dunod, Paris
81. Ohshima H (2000) Cell model calculation for electrokinetic phenomena in concentrated suspensions: an Onsager relation between sedimentation potential and electrophoretic mobility. Adv Colloid Interface Sci 88:1–18. https://doi.org/10.1016/S0001-8686(00)00038-5
82. Gorobets OY, Gorobets YI, Rospotniuk VP (2014) Movement of electrolyte at metal etching and deposition under a non-uniform steady magnetic field. Magnetohydrodynamics 50:317–332 . https://doi.org/10.22364/mhd.50.3.9
83. Booth F (1954) Sedimentation potential and velocity of solid spherical particles. J Chem Phys 22:1956–1968. https://doi.org/10.1063/1.1739975
84. Masliyah JH, Bhattacharjee S (2006) Electrokinetic and Colloid Transport Phenomena. John Wiley, New Jersey
85. Henry DC (1948) The electrophoresis of suspended particles. IV. The surface conductivity effect. Trans Faraday Soc 44:1021–1026. https://doi.org/10.1039/tf9484401021
86. Pamme N, Eijkel JCT, Manz A (2006) On-chip free-flow magnetophoresis: separation and detection of mixtures of magnetic particles in continuous flow. J Magn Magn Mater 307:237–244. https://doi.org/10.1016/j.jmmm.2006.04.008
87. Furlani EP (2007) Magnetophoretic separation of blood cells at the microscale. J Phys D Appl Phys 40:1313–1319. https://doi.org/10.1088/0022-3727/40/5/001
88. Guevorkian K, Valles JM (2006) Aligning paramecium caudatum with static magnetic fields. Biophys J 90:3004–3011. https://doi.org/10.1529/biophysj.105.071704

# Chapter 13
# Magnetic Fluids in Biosciences, Biotechnology and Environmental Technology

**Ivo Safarik and Kristyna Pospiskova**

**Abstract** This chapter provides a brief overview of the literature about the applicability of magnetic fluids in various areas of biosciences, biotechnology and environmental technology. Magnetic fluids were used for magnetic modification of diamagnetic materials and synthesis of magnetically responsive materials. Both free ferrofluids and ferrofluid modified materials were employed for separation and removal of biologically active compounds, pollutants and cells, immobilization of target molecules and cells, magnetic solid phase extraction and related (bio)analytical procedures. Ferrofluid modified materials also exhibit peroxidase-like catalytic activity. The discussed materials and processes have a great application potential.

**Keywords** Magnetic fluids · Ferrofluids · Magnetic modification · Magnetic separation · Peroxidase-like activity

## List of Abbreviations, Symbols, and Chemical Substances

### *Abbreviations*

| | |
|---|---|
| CNTs | carbon nanotubes |
| DC | direct current |

I. Safarik (✉) · K. Pospiskova (✉)
Department of Nanobiotechnology, Biology Centre, ISB, CAS, Na Sadkach 7, 370 05 Ceske Budejovice, Czech Republic
e-mail: ivosaf@yahoo.com

Regional Centre of Advanced Technologies and Materials, Palacky University, Slechtitelu 27, 783 71 Olomouc, Czech Republic
e-mail: kristyna.pospiskova@upol.cz

I. Safarik
Department of Magnetism, Institute of Experimental Physics, SAS, Watsonova 47, 040 01 Kosice, Slovakia



L. Bulavin and N. Lebovka (eds.), *Soft Matter Systems for Biomedical Applications*, Springer Proceedings in Physics 266, https://doi.org/10.1007/978-3-030-80924-9_13

| | |
|---|---|
| FFs | ferrofluids |
| LMF-DMMLE | low-density magnetofluid dispersive liquid–liquid microextraction |
| MFs | magnetic fluids |
| MCNTs | magnetically modified carbon nanotubes |
| MRI | magnetic resonance imaging |
| MSPE | magnetic solid phase extraction |
| NPs | nanoparticles |
| P-L | peroxidase-like |
| SEM | scanning electron microscopy |
| SPE | solid-phase extraction |
| TEM | transmission electron microscopy |

## *Symbols*

| | |
|---|---|
| $Q_{max}$ | maximum adsorption capacity (mg/g) |

## *Chemical Substances*

| | |
|---|---|
| CGTase | cyclodextrin glucanotransferase |
| HRP | horseradish peroxidase |
| NADH | nicotinamide adenine dinucleotide, reduced form |
| PVA | polyvinyl alcohol |

## 13.1 Introduction

Due to their unique magnetic properties magnetically responsive nano- and microparticles have found great potentials in a variety of biological applications, especially for the isolation, separation, purification, detection, determination, immobilization and modification of target biologically active compounds, analytes, pollutants, subcellular structures and cells. Typical magnetic materials consist of magnetite ($Fe_3O_4$), maghemite ($\gamma$-$Fe_2O_3$) or ferrites with the particle size ranging from nanometer to micrometer size. Magnetically responsive materials exhibit several types of responses to external magnetic field, namely the possibility of their selective separation, targeting and localization, heat generation (which is caused by magnetic particles subjected to a high-frequency alternating magnetic field), increase of a negative T2 contrast during magnetic resonance imaging (MRI) or great increase of apparent viscosity of magnetorheological fluids when subjected to a magnetic field. In addition, magnetic iron oxide(s) nanoparticles (NPs) exhibit peroxidase-like activity [1, 2].

Magnetic fluids (MFs), also named ferrofluids (FFs) are typically colloidal suspensions of very fine (ca 10 nm) surface stabilized superparamagnetic nanoparticles dispersed in an appropriate carrier liquid [3]; they are thus a special group of nanofluids [4]. Both polar and nonpolar carrier liquid can be employed for FF synthesis. The first real FFs were synthesized in the nineteen sixties, and since that time their research and applications are steadily growing [3].

Ferrofluids are stable colloidal dispersions of nanosized particles of ferro- or ferrimagnetic materials in a diamagnetic carrier liquid. The main characteristic feature of FFs (in comparison with other types of magnetic (nano)suspensions) is their long-term colloidal stability, even in strong and non-uniform magnetic fields. To achieve this property, the overall particle interactions have to be repulsive; it means the attractive van der Waals and magnetic forces have to be balanced by Coulombic, steric or other interactions, in order to keep nanoparticles apart from each other [5, 6].

There are several procedures to prepare stable FFs, where magnetic nanoparticles are stabilized in a proper way, namely [3, 5]:

- Charge (electrostatic, ionic) stabilization (magnetic nanoparticles have to carry the same charge, either positive or negative)
- Steric stabilization of magnetic nanoparticles with either appropriate low molecular weight molecules (often tensides, in the single-layer or double-layer form) or by modification with high molecular weight synthetic polymers (e.g., polyethylene glycol) or biopolymer (e.g., dextran or starch)
- Mixed charge and steric stabilization (e.g., magnetic iron oxide nanoparticles bearing citric acid)

The first heptane-based, oleic acid stabilized FF was prepared by the top-down procedure (mechanical milling) in the early sixties of the 20th century [7]. For most bioapplications, however, water based FFs are necessary. The first bottom-up synthesis of mineral acid-stabilized FF was described by Massart in the years 1980 and 1981 [8, 9], while the first dextran-based MF was prepared by Molday in 1982 [10]. There are several well written review journal papers [4, 5, 11–20], book chapters [6, 21] and books [22–25] devoted to the preparation, characterization and application of various FFs, where the reader can find all necessary information.

Magnetic fluids have already been employed both in many important technological applications and in various areas of biosciences, biotechnology, environmental technology and medicine. In this review chapter we will highlight typical examples of both FFs and FF-modified materials and related processes in various areas of biochemistry, biotechnology, environmental technology and related areas. Because the term magnetic fluid or ferrofluid is sometimes used incorrectly, we will strictly focus only on papers where the "proper" magnetic fluids exhibiting long-term colloidal stability were used. Medical applications of MFs have been summarized in large number of reviews where the reader can find relevant information.

## 13.2 Bioapplications of Magnetic Fluids

The following chapters will present selected examples of the preparation of FF-modified materials and the important (bio)applications of both free FFs and FF-modified materials.

### *13.2.1 Magnetic Modification of Diamagnetic Materials*

Many diamagnetic particulate and high aspect ratio materials (e.g., catalysts, adsorbents, carriers, microbial cells, chromatography materials, waste biological materials etc.) are available. Magnetic modification can substantially increase their application potential. Several procedures enabling the preparation of magnetically responsive materials from diamagnetic precursors have been developed. Magnetic fluid modification is usually caused by the deposition of magnetic nanoparticles within the treated materials pores, on the materials surface or within the polymer gels [26, 27]. Magnetically modified materials can be easily separated using appropriate permanent magnets, electromagnets or magnetic separators. Magnetic fluid modification of selected materials follows.

#### 13.2.1.1 Plant Derived Materials and Biopolymers

Perchloric acid stabilized MF enabled magnetic modification of diamagnetic materials suspended in methanol; during mixing FF-derived magnetic iron oxide nanoparticles firmly precipitated on the particles surface. This simple procedure was employed for magnetic modification of sawdust [28], spent grain [29, 30], spent coffee grounds [31], spent tea leaves [32], or peanut husks (see Fig. 13.1) [33]. These magnetically modified materials were used as adsorbents for the removal of water-soluble organic dyes or heavy metal ions or as carriers for the immobilization of selected enzymes. Other low-cost adsorbents (defective green coffee, coffee silverskin and spent coffee grounds) were magnetically modified with tetramethylammonium hydroxide stabilized MF [34]. Properly performed magnetic modification has not caused substantial decrease of adsorption capacities of modified adsorbents. Alternatively, spent tea leaves [35] and *Posidonia oceanica* [36] were magnetically modified by the direct mixing of the material to be modified with water-based ionic MF stabilized with perchloric acid; after drying, aggregates of magnetic iron oxide nanoparticles were deposited on the treated material.

Also (bio)polymer gels such as bacterial cellulose produced by *Komagataeibacter sucrofermentans* can be efficiently magnetically modified using perchloric acid stabilized MF and subsequently used as a carrier for the immobilization of affinity ligands, enzymes and cells [37] or for the construction of a novel adsorbent for selective and efficient recognition of thymidine nucleoside [38]. Crosslinked, beaded form

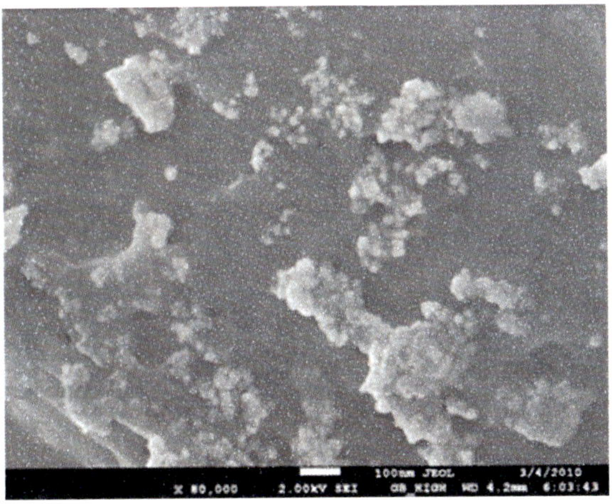

**Fig. 13.1** SEM image of ferrofluid modified peanut husk particle (bar = 100 nm) (reprinted with permission from [33])

of agarose known as Sepharose was packed in a column and water-based FF was pumped through to enable magnetic modification of the beads [39].

#### 13.2.1.2 Microbial and Microalgal Cells

Also microbial and microalgal cells were magnetically modified using MFs [1, 2, 40]. *Saccharomyces cerevisiae* cells were modified with three MFs under different conditions (tetramethylammonium hydroxide stabilized FF in 0.1 M glycine–NaOH buffer, pH 10.6; perchloric acid stabilized FF in 0.1 M acetate buffer, pH 4.6; citrate FF in 0.1 M glycine–HCl buffer, pH 2.2). All procedures enabled to form magnetically responsive yeast cells in a short time due to magnetic iron oxide nanoparticles precipitation on the yeast cell surfaces (see Fig. 13.2). The best results were obtained using tetramethylammonium hydroxide stabilized FF; the modified yeast cells exhibited higher intracellular catalase and invertase activities and that's why this type of whole-cell biocatalyst was used for hydrogen peroxide decomposition and sucrose conversion into glucose and fructose [41]. The different physiological state of the yeast cells results in different magnetic modification; dormant yeast cells were surface-modified while actively growing yeast cells accumulated the magnetic modifier in the periplasmic space [42].

Magnetic modification of dried *Kluyveromyces marxianus* (fodder yeast) and *Chlorella vulgaris* cells required the removal of a substantial portion of soluble macromolecules that otherwise caused spontaneous precipitation of MF; after washing and suspending the cells in acetic acid solution, the addition of perchloric acid–stabilized MF resulted in the formation of magnetically modified yeast [43] and microalgae cells [44]. Also naturally occurring micrometer-sized, high aspect ratio native and autoclaved *Leptothrix* sp. sheaths were efficiently modified with perchloric acid-stabilized MF [45] and subsequently used as an inexpensive adsorbent for crystal

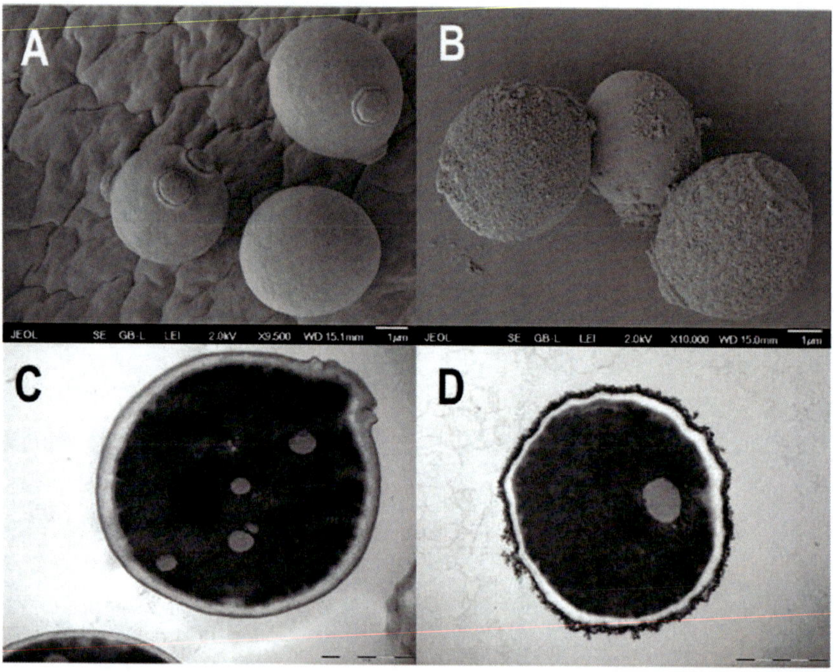


**Fig. 13.2** **A** Scanning electron microscope (SEM) image of native *Saccharomyces cerevisiae* cells. **B** SEM image of ferrofluid-modified *S. cerevisiae* cells, showing attached magnetic nanoparticles and their aggregates on the cell surface. **C** Transmission electron microscope (TEM) image of a native *S. cerevisiae* cell. **D** TEM image of a ferrofluid-modified *S. cerevisiae* cell with attached magnetic iron oxide nanoparticles on the cell wall. Bars correspond to 1 μm. Magnetic modification of dormant yeast cells was performed using perchloric acid-stabilized magnetic fluid as described previously [41] (reprinted with permission from [2])

violet [45] and Amido black 10B [46] removal from aqueous solutions. The same MF was successfully employed for magnetic modification of both native [47] and nanogold-modified [48] diatom cells of *Diadesmis gallica*; magnetic modification was performed both in methanol and by direct mixing.

#### 13.2.1.3 Inorganic and Organic Particulate Materials

Postmagnetization procedures have been also successfully used for magnetic modification of non-biological particulate materials. Direct mixing of water based acid FF with the treated material enabled to prepare magnetically modified montmorillonite [35]. The same procedure enabled to prepare magnetically modified SEPABEADS® EC-HA beads which were subsequently used as a carrier for enzyme immobilization [49].

The natural saponite clay was magnetically modified by contact with citric acid stabilized MF [50]. Preparation of magnetic derivatives of saponite, palygorskite, and spondyle clay was performed after their magnetic modification with oleic acid stabilized MF [51, 52]; specific magnetization of modified clays ranged between 3.9 and 6.3 A $m^2$/kg [51].

#### 13.2.1.4 High Aspect Ratio Materials

Perchloric acid-stabilized MF was employed for magnetic modification of lysozyme amyloid fibrils; binding of magnetic nanoparticles to fibrils was observed. It was found that magnetic nanoparticles altered amyloid structure, leading to the increase of the fibril diameter and decrease of fibril length [53]. Similar experiments dealing with the adsorption of magnetic nanoparticles on lysozyme amyloid fibrils were studied with the aim to clarify previously found disaggregation activity of the nanoparticles in amyloid fibril dispersions and to get deeper insight into interaction processes between amyloids and MNPs [54].

Carbon nanotubes (CNTs) have exceptional mechanical, electrical and optical properties and also an extremely large surface area and varied possibilities for their functionalization. Magnetic CNTs can find many applications in medicine, diagnosis, cell biology, analytical chemistry, and environmental technology. Several strategies can be employed for modification of CNTs with magnetic nanoparticles, including the modification with FFs [55–58]. The filling process is the simplest procedure in order to fabricate magnetic carbon nanotubes. Filling CNTs with a commercial FF can lead to almost 100% magnetic modification and subsequent simple manipulation in external magnetic field. The first example of this strategy was published in 2005 by Korneva et al. [59] who synthesized magnetic CNTs by introducing commercial FFs (containing 10 nm particles of $Fe_3O_4$) into the inner cavity of 300 nm-diameter nanotubes. The preparation procedure consisted of several steps:

(i) synthesis of the CNTs by chemical vapor deposition in the pores of an alumina template (to obtain CNTs with open ends);
(ii) filling the CNTs with a FF suspension containing $Fe_3O_4$ through a magnetic-assisted capillary strategy;
(iii) separation of the FF filled CNTs from the alumina template; and
(iv) drying of the carrier fluid to leave only magnetic nanoparticles within the CNTs.

Transmission electron microscopy study showed a high density of magnetic nanoparticles inside the CNTs. These one-dimensional magnetic nanostructures can find numerous applications in nanotechnology, memory devices, optical transducers for wearable electronics, and in medicine [59]. Hexane-based ferrofluid was employed for similar synthesis of magnetic carbon nanotubes; the measured saturation magnetization 65 emu/g makes this composite material promising for a variety of applications [60].

In an alternative procedure, tetramethylammonium hydroxide stabilized MF was mixed with water-based CNT nanofluid (prepared by the addition of CNTs into water solution of gum arabic) and the mixture was sonicated. The hybrid nanofluid was dried and then the material was resuspended in water. Magnetic CNTs were used to study the effects of both constant and alternating magnetic fields on the laminar forced convective heat transfer of this hybrid nanofluid flowing through a heated tube [61]. In another procedure CNTs were mixed with commercial organic FF; due to the hydrophobic nature of nanotubes they easily interacted with FF. Microscopic examination showed the presence of magnetic iron oxide nanoparticles on the outer surface of the treated CNTs [57]. α-Iron filled multi-walled CNTs were prepared through filling of commercially available CNTs, after oxidative opening, with commercial organic FF. After washing and reduction with hydrogen clean, CNTs filled with α-iron were obtained [62].

Halloysite nanotubes are the tubular forms of halloysite and are chemically similar to kaolin. They are layered aluminosilicates with hollow tubular structure and high aspect ratio. Halloysite can be used for many applications, such as for clay polymer nanocomposite preparation, catalysis and adsorption [63]. In order to explore the possibility of augmentation of magnetic field induced chain formation in dodecanoic acid stabilized water-based MF, the addition of halloysite nanotubes into FF in various concentrations was tested. It was shown that the addition of tube shaped halloysite to FF led to the augmentation of magnetic fluid chains; the augmentation was attributed to the increase in the magnetization of the mixture [64].

The magnetic modification of four different electrospun nanofibrous textiles, based on polyamide, polyvinyl alcohol, polycaprolactone and polyurethane, was performed by the simple spray modification procedure employing two types of MFs. Magnetic modification led to the deposition of magnetic iron oxide nanoparticles on the surface of textile nanofibers. In addition to magnetic response, the modified nanotextile exhibited peroxidase-like activity [65]. Alternatively, electrospinning of appropriate polymer containing MF was performed; in a typical example polyvinyl butyral nanofibers were generated from a polymer solution/mineral oil-based FF mixture. The prepared magnetic nanofibers were tested as a material for magnetic hyperthermia applications [66].

### *13.2.2 Synthesis of Magnetically Responsive Biobased and Related Materials*

Magnetic fluids have been often used during the synthesis of magnetically responsive biocomposites, including magnetic biopolymers, magnetoliposomes, polymersomes, etc. as shown below.

#### 13.2.2.1 Magnetically Responsive (Bio)Polymers

Various biopolymers including chitosan, alginate, agarose, starch, dextran and plant gums in several forms (nano- and microparticles, films, fibers, etc.) were prepared in magnetic form using magnetic fluid during syntheses.

Chitosan is one of the most abundant natural polysaccharide in nature [67]. Its magnetic derivatives can be used as efficient adsorbents or carriers. Ethylenediamine-modified cross-linked magnetic chitosan particles were prepared by the addition of water-based FF into chitosan solution in acetic acid; in the next step glutaraldehyde and ethylenediamine were used to form the modified chitosan particles, which were used for the adsorption of chromium (VI) ions from aqueous solution [68].

In another procedure chitosan dissolved in diluted acetic acid was mixed with MF and the solution was then dropwise poured into paraffin containing Span-80 as an emulsifier. The suspension medium was stirred and then glutaraldehyde was added. After some time, NaOH solution was gradually added to increase the pH and after further mixing the product was filtered and washed consecutively with petroleum, acetone, and distilled water [69].

Magnetic chitosan particles were also formed by complexation of poly(methacrylic acid)-stabilized MF with chitosan; the composite flocculation occurred due to physical cross-linking of the MF stabilized with negatively charged polymer and the positively charged chitosan chains. The precipitated magnetic chitosan composite particles exhibited a great response to external magnetic field [70].

Citrate-stabilized MF was employed for the preparation of spherical magnetic alginate microparticles (25–60 μm in diameter) using the microemulsion system. Water-saturated 1-pentanol was used as the organic phase; the limited solubility of 1-pentanol in water enabled simple removal of the organic solvent from the prepared beads with calcium chloride solution which is a very cheap and safe procedure. The magnetic phase was homogeneously distributed throughout the particle volume [71].

Composite magnetic alginate beads for organic pollutants removal, containing both magnetic nanoparticles from citrate-stabilized FF and activated carbon, were prepared by an extrusion technique and cross-linked with epichlorohydrin. The beads were easily magnetically recovered or manipulated. The bead pollutant adsorption capacity was linked to encapsulated activated carbon and to active sites present both in magnetic nanoparticles and alginate [72].

The citrate stabilized FF was also employed to prepare magnetic alginate beads used for the removal of both cetylpyridinium chloride [73] and Pb(II) ions [74]. In both cases, sodium alginate solution was mixed with FF and the mixture was then introduced into a plastic syringe equipped with a needle; the beads were formed by dropping this mixture in calcium chloride solution. The saturation magnetization value of a single wet bead was 1.15 kA/m for cetylpyridinium chloride adsorbent [73] while the value 32.2 emu/g was observed for lead ions adsorbent [74]. Addition of an extractant (Cyanex 272) into alginate/FF mixture enabled to prepare magnetic beads for Ni(II) ions removal [75].

Magnetic dextran microspheres were prepared by the mixing of alkaline dextran solution with commercial FF followed by emulsification of the mixture in the cotton seed oil/Tween 80 system. The microspheres formed were cross-linked by the addition of cyanogen bromide. Subsequently, the particles were converted into the aminodextran microspheres by the treatment with cyanogen bromide and 1,6-hexanediamine [76].

Magnetic starch was prepared from an aqueous suspension of native corn starch which was stirred and heated with MF. The resulting magnetic gel was modified through three freezing/thawing cycles. Gel was frozen at −18 °C for 24 h and then thawed at room temperature in each cycle. Subsequently, the insoluble magnetic starch gel formed was disintegrated and sieved [77].

Polyethylene glycol-stabilized FF was used to prepare magnetically responsive microbeads composed of artemisia seed gum and chitosan; the mixture was added dropwise to the dispersion medium composed of mineral oil, petroleum ether and emulsifier under mixing. Activation was performed with glutaraldehyde which enabled the immobilization of β-galactosidase [78].

Polyvinyl alcohol magnetic films were prepared by entrapment of MF into polymer solution and subsequently polymer film was formed. The magnetic polymer film prepared by solvent casting method was either kept between two magnetic poles or without the influence of magnetic field. It was shown that the film developed in the absence of magnetic field possesses randomly oriented domains, whereas film developed under the influence of magnetic field showed well aligned flux lines. The number of flux lines developed in the polymer matrix was directly proportional to applied external magnetic field [79]. Biocompatible polyvinyl alcohol particles and hydrogels modified with appropriate MFs were also prepared and studied, especially as drug carriers enabling controlled drug release [80] and smart magnetic polyvinyl alcohol ferrogels and films [81, 82].

Fabrication of flexible agarose magnetic microfibers having tunable size and mechanical properties was developed. The basic principle was based on the magnetic field-driven self-assembly of citrate-stabilized FF containing agarose at a temperature above its gelling point in the oil emulsion. The emulsion was exposed to a DC uniaxial magnetic field. Once the magnetic agarose fibers have formed, the gelification of the agarose was caused by the sample cooling. Subsequent addition of surfactant prevented lateral aggregation of the fibers. Resulting fibers had diameters in the range between a few to tens of micrometers. A phase diagram presenting the influence of surfactant concentration and residence time under magnetic field during the synthesis was constructed [83].

#### 13.2.2.2 Magnetoliposomes and Magnetic Polymersomes

Natural lipids (mostly phospholipids) or their synthetic alternatives can self-assemble into vesicles (spherical bilayer), termed liposomes. Liposomes are a simplified model of a cell membrane without accompanying proteins. Liposomes are highly biocompatible, but they are usually rather unstable and sensitive. Alternative materials have thus been developed employing versatile, easily obtainable and tunable entities, behaving in many aspects like lipids, namely amphiphilic polymers. Self-assembled amphiphilic polymers can form polymeric vesicles called polymersomes [84]. In both cases, magnetic derivatives employing magnetic fluid can be prepared [85, 86].

Combining liposomes with magnetic nanoparticles is an interesting approach to create smart multifunctional vesicles (magnetoliposomes) for biological and medical applications, including controlled drug delivery and diagnostic imaging [86]. First described by De Cuyper and Joniau in 1988 [87], magnetoliposomes have become remarkable hybrids due to the multivalent properties of both the carriers and the triggers. Several methodologies can be used for magnetoliposomes formation, including sonication, spontaneous swelling, and reverse-phase evaporation. Detailed information about magnetoliposomes preparation, characterization and application can be found in several review papers [86, 88, 89].

In a typical example, magnetoliposomes were prepared by lipid film hydration followed by extrusion, employing citrate-stabilized FF as the magnetic label. Cholesterol and 1,2-dioleoyl-sn-glycero-3-phosphocholine dissolved in chloroform were added to a rotary evaporator and evaporated to dryness. Lipids were hydrated with the FF and homogenized by vortexing, which resulted in the formation of a dispersion of multilamellar vesicles. Large unilamellar vesicles were obtained by hand extrusion of the resulting lipid dispersion through 200 nm and then 100 nm polycarbonate filters. Nonentrapped maghemite nanoparticles were eliminated, first by NaCl-induced aggregation, then by size exclusion chromatography; magnetoliposomes were concentrated and sorted by magnetic chromatography [90].

Micromixer technology was an efficient way to manufacture polymeric vesicles (Pluronic®L-121) dual loaded with the anticancer drug camptothecin and commercial MF; successful incorporation of magnetic nanoparticles and a relatively narrow size distribution of the hybrid polymersomes was observed [91].

## *13.2.3 Separation of Biologically Active Compounds*

Isolation and separation of specific molecules, organelles and cells is used in almost all areas of biosciences and biotechnology. Diverse procedures can be used to achieve this goal. Affinity ligand techniques represent currently the most powerful tool available to the downstream processing both in term of their selectivity and recovery. Recently, increased attention has been paid to the development and application of magnetic separation techniques, which employ small magnetic particles or ferrofluid assisted aqueous two-phase systems [92, 93].

### 13.2.3.1 Application of Ferrofluid-Modified Materials

Standard affinity or ion-exchange chromatography material can be efficiently magnetically modified by interaction of the sorbent with water-based ferrofluid. Magnetic particles accumulated within the pores of chromatography adsorbent thus modifying this material into magnetic form [39]. Swollen gel of 5′-AMP-Sepharose 4B was packed in a column and commercial water-based FF was pumped through the column to modify the beads. After washing, the magnetic adsorbent exhibited affinity for beef heart lactate dehydrogenase; the bound enzyme can be eluted with NADH [39]. S-Sepharose Fast flow (ion-exchanger consisting of a cross-linked agarose) was magnetically modified in a batch mode using water-based FF; these magnetic particles were designed for use in magnetically stabilized fluidized beds [94].

Magnetic alginate microparticles employing citrate-stabilized FF were used as magnetic affinity adsorbents for specific purification of α-amylases of various origin (porcine α-amylase from porcine pancreatic acetone powder and α-amylase from *Bacillus amyloliquefaciens* from a technical preparation). Enzyme was eluted by 1.0 M maltose. α-Amylases from *Bacillus amyloliquefaciens* and porcine pancreatic acetone powder were purified 9- and 12-fold with 88 and 96% activity recovery, respectively [71].

Magnetic porous corn starch was prepared as an affinity adsorbent for the efficient and simple scale-up procedure for one-step purification of cyclodextrin glucanotransferase (CGTase) from *Bacillus circulans*. Magnetic affinity separation enabled isolation of CGTase from cultivation media (volumes between 10 and 400 mL) with ca 60–70% recovery after elution with alkaline buffers containing soluble starch; the enzyme purification factor was 19–25 in different batches [77].

Magnetic bacterial cellulose nanofibers were converted to a novel adsorbent for selective and efficient recognition of thymidine nucleoside by the method of molecular imprinting. In this process, magnetic bacterial cellulose nanofibers were silanized and further polymerized with a hydrophilic monomer for templating thymidine via metal chelate coordination. Highly selective adsorption of thymidine at the same concentration of other nucleosides (cytidine and uridine) was observed [38].

#### 13.2.3.2 Ferrofluid Assisted Aqueous Two-Phase Systems

Aqueous two-phase systems have a great potential in the downstream processing of biomolecules [95, 96]. A new technique to speed up the phase separation of aqueous two-phase systems is based on the addition of magnetically susceptible additives (ferrofluids or iron oxide particles). In a magnetic field such additives will induce a faster phase separation [92]. This method seems to be useful when the two phases have very similar densities, the volumetric ratio between the phases is very high or low, or the systems are viscous. Magnetically enhanced phase separation usually increases the speed of phase separation by a factor of about 10 in well-behaved systems, but it may increase by a factor of many thousands in difficult systems. The addition of ferrofluids and/or iron oxide particles was shown to have usually no influence on enzyme partitioning or enzyme activity [92, 97–101].

In one approach, dextran-stabilized ferrofluid was added to an aqueous two-phase system containing polyethylene glycol and dextran. The ferrofluid was totally partitioned to the dextran phase. After mixing of the two-phase system, it was possible to reduce the separation time by a factor of 35 by applying a magnetic field to the system. Both batch and semicontinuous separation of test mixtures of lactate dehydrogenase and ß-galactosidase were carried out in aqueous two-phase systems containing dextran ferrofluid. The recoveries of the enzymes were high (80–100%) in a system containing 2.5% ferrofluid [97].

### *13.2.4 Ferrofluid Modified (Bio)sorbents for Pollutant Removal*

Various types of biological materials including agriculture and food waste and byproducts and other low-cost inorganic and organic materials can be used as biosorbents for the removal of organic and inorganic pollutants and radionuclides from contaminated waters. Magnetic modification of adsorbents by appropriate magnetic fluids enabled to prepare magnetic composites which facilitated efficient magnetic separation [102].

Sawdust modified with perchloric acid stabilized magnetic fluid was used for the removal of model organic dyes from aqueous solutions [28, 103]; subsequently magnetic characterization of this adsorbent was performed [104]. Similar ferrofluid-based modification procedures were employed for the preparation of other plant-derived materials including spent grain [30], spent coffee grains and other coffee derivatives [31, 34, 105, 106], peanut husk [33, 105, 107], or waste tea leaves [35], all of them being used for dye removal for water systems. Alternatively, also microbial and microalgae cells such as *Saccharomyces cerevisiae* (baker's yeast) [108, 109], *Saccharomyces cerevisiae* subsp. *uvarum* (brewer's yeast) [110], *Kluyveromyces fragilis* (fodder yeast) [43], *Leptothrix* sp. [45, 46] and *Chlorella vulgaris* [44] have been prepared and used for the same purpose. Comparison of maximum adsorption capacities $Q_{max}$ (mg/g) of ferrofluid modified biological materials for the tested dyes is presented in Table 13.1.

Microbial cells exhibit high affinity for heavy metal ions and radionuclides. Ferrofluid modified *Saccharomyces cerevisiae* subsp. *uvarum* cells were used for the removal of $Hg^{2+}$, $Cu^{2+}$, $Ni^{2+}$ and $Zn^{2+}$ ions from water solutions; after ion adsorption the magnetic biosorbent was efficiently regenerated with 0.1 M $HNO_3$ [111, 112]. In addition, FF-modified *Kluyveromyces fragilis* cells were used as an adsorbent for the removal of $Sr^{2+}$ ions from aqueous solutions [113], while FF-modified *Rhodotorula glutinis* cells were used for uranium biosorption [114]. Ferrofluid modified sawdust and orange peel were used for Pb(II), Ni(II) and Cd(II) ions removal [115].

Also ferrofluid modified inorganic materials have been tested for pollutant removal. The modified natural saponite was employed for the adsorption of model dyes from water solutions [50]. Magnetic nanocomposite sorbents for disposal of synthetic detergents from wastewater were prepared from clay minerals (saponite, palygorskite, and spondyle clay) after their magnetic modification with oleic acid stabilized magnetic fluid [51]. Similar magnetic adsorbents were used for the removal of malachite green and Congo red [52].

**Table 13.1** Comparison of maximum adsorption capacities $Q_{max}$ (mg/g) of ferrofluid modified biological materials for tested dyes

| Dyes | Colour index number | Maximum adsorption capacities of ferrofluid modified plant derived materials (mg/g) | | | | | | | | | | | |
|---|---|---|---|---|---|---|---|---|---|---|---|---|---|
| | | Spruce sawdust [103] | Spruce sawdust [28] | Peanut husks [33] | Spent coffee grounds [31] | Spent grain [30] | Spent tea leaves [35] | *Posidonia oceanica* [36] | Baker's yeast [108] | Brewer's yeast [110] | Fodder yeast [43] | *Leptothrix* sp. [45, 46] | *Chlorella vulgaris* [44] |
| Acridine orange | 46005 | 24.1 | | 71.4 | 73.4 | | 98.8 | | 82.8 | | 62.2 | | |
| Amido black 10B | 20470 | | | | 1.24 | | | | | 11.6 | 29.9 | 339.2 | |
| Aniline blue | 42755 | | | | | 44.7 | | | 430.2 | 228.0 | | | 257.9 |
| Bismarck brown | 21000 | 52.1 | | 95.3 | 69.2 | 72.4 | 71.7 | | | | 75.7 | | 201.9 |
| Congo red | 22120 | | | | 9.43 | | | | | 93.1 | 49.7 | | 156.7 |
| Crystal violet | 42555 | 52.4 | 51.1 | 80.9 | 68.1 ara> | 40.2 | 100.1 | | 85.9 | 41.7 | 42.9 | 166.6 | 42.9 |
| Malachite green | 42000 | | | | 43.0 | | 66.2 | | 19.6 | | | | |
| Methyl green | 42585 | | | | | | 30.8 | | | | | | |
| Methylene blue | 52015 | | | | | | | 133.3 | | | | | |
| Nile blue A | 51180 | | | | | 64.1 | 87.1 | | | | | | |
| Safranin O | 50240 | 25.0 | | 86.1 | 59.0 | | 84.7 | | 90.3 | 46.6 | 138.2 | | 115.7 |
| Saturn blue LBRR | 34140 | | | | | | | | | | 33.0 | | 24.2 |

### 13.2.5 (Bio)analysis Using Ferrofluids and Ferrofluid-Modified Materials

Solid-phase extraction (SPE) is a standard procedure used to isolate target analytes from a sample using an appropriate adsorbent placed in a small column; the extraction efficacy is influenced by the sample flow-rate and potential presence of particle impurities in the sample.

Several alternative extraction procedures have been developed. Magnetic solid-phase extraction (MSPE) developed by Safarikova and Safarik and described for the first time in 1999 is based on the application of magnetic adsorbents suspended (dispersed) in the analyzed solution or suspension [116]. MSPE can be successfully used for analyte extractions even from difficult-to handle samples including suspensions because majority of accompanying impurities has diamagnetic properties. Magnetic adsorbents can be rapidly, efficiently and selectively separated using appropriate magnetic separator; the analyte is consequently eluted from the recovered adsorbent and analyzed [117]. MSPE is currently one of the most often used preconcentration procedure of wide variety of analytes due the easy operation, high extraction efficiency, and potential reusability of adsorbent [55, 118–124].

Appropriate diamagnetic biosorbent surface modified with magnetic fluid can be employed for MSPE. Spent coffee grounds magnetically modified with perchloric acid stabilized magnetic fluid was employed for MSPE of crystal violet; it was possible to detect 1 μg of dye in 1000 mL of water (1 ppb) using a simple spectrophotometry [31].

Many other magnetic fluid modified materials have been applied for MSPE. Oleic acid stabilized ferrofluid was used for the synthesis of magnetite/silica/poly (methacrylic acid–co-ethylene glycol dimethacrylate) spherical microparticles (diameter ca 1.3 μm); this adsorbent was employed for the extraction of eleven sulfonamides in milk samples [125].

In recent years, the use of magnetically modified carbon nanotubes (MCNTs) as adsorbents for MSPE has become an active area of research [55, 56, 126, 127]. Currently, however, ferrofluid modified carbon nanotubes have not been used for MSPE.

Magnetic fluids based on the water insoluble carrier liquids enabled to develop new liquid–liquid (micro)extraction procedures. A recently described micro-extraction technique named low-density magnetofluid dispersive liquid–liquid microextraction (LMF-DMMLE) enabled analysis without the need of special devices and complicated operations. Dispersion of the organic solvent-based magnetic fluid into the aqueous sample was achieved by using vortex mixing. The extraction magnetic fluid was collected with an external magnet placed outside the extraction container after dispersing. Then, the ferrofluid based magnetic nanoparticles were easily removed by adding a precipitation reagent under the magnetic field. In order to evaluate the validity of this method, ten organochlorine pesticides were chosen as the analytes [128].

Ferrofluid based extraction procedure was successfully applied for the determination of four phenolic compounds (bisphenol-A, bisphenol-AF, tetrabromobisphenol-A and 4-tert-octylphenol) in milks and fruit juices. Different alkyl (C4–C10) alcohols as the carrier liquid were used for the preparation of ferrofluids; an appropriate chain length of alkyl alcohol can improve the extraction efficiency and maintain the integrity of ferrofluids during extraction. 1-Heptanol (C7) was selected as the optimal carrier liquid [129].

### 13.2.6 Immobilization of Biologically Active Compounds and Cells on Ferrofluid-Modified Carriers

Immobilization of enzymes, other low- and high-molecular weight biologically active compounds, affinity ligands, subcellular organelles and cells is a very important technique used in various areas of biochemistry, cell biology, microbiology and biotechnology. Molecules and cells immobilized on magnetic carriers can be removed from the system simply by using an external magnetic field, or can also be targeted to the desired place. Immobilized compounds can be used for the isolation of complementary biologically active compounds, subcellular components and cells, to perform their specific activities (e.g., immobilized enzymes and cells) or be transported to the desired place [93].

Magnetic montmorillonite, prepared by direct mixing of acid FF and diamagnetic precursor, was used as a carrier for the immobilization of lipase and beta-galactosidase; immobilized enzymes showed long-term stability without leaching of enzyme from the support and enabled their repeated use without significant loss of activity [35]. Similar procedure was used for modification of commercially available SEPABEADS® EC-HA which were subsequently employed for the immobilization of diamine oxidase from *Pisum sativum*; this complex was part of a fiber optic biosensor for the determination of biogenic amines [57]. Ferrofluid-modified bacterial cellulose was used as a carrier for the immobilization of an affinity ligand (reactive copper phthalocyanine dye), bovine pancreas trypsin and *Saccharomyces cerevisiae* cells. Immobilized yeast cells retained approximately 90% of their initial activity after 6 repeated cycles of sucrose solution hydrolysis. Trypsin covalently bound to cellulose after periodate activation was very stable during operational stability testing; it could be repeatedly used for ten cycles of low molecular weight substrate hydrolysis without loss of its initial activity [37].

Composite particles formed by complexation of poly(methacrylic acid)-stabilized magnetic fluid with chitosan were used as a magnetic carrier for immobilization of enzymes and cells. *Candida rugosa* lipase was covalently attached after cross-linking and activation of chitosan using glutaraldehyde while *Saccharomyces cerevisiae* cells were incorporated into the chitosan composite during its preparation. Both biocatalysts were active after reaction with appropriate substrates [70].

Also potential of ferrofluid-modified plant derived materials was studied. Ferrofluid-modified spent grain was used as a low-cost, biocompatible and magnetically responsive carrier for the immobilization of *Candida rugosa* lipase. Several immobilization procedures were tested using both native and poly(ethyleneimine)-modified magnetic spent grain. Magnetic spent grain modified with poly(ethyleneimine) bound a smaller amount of active lipase than unmodified magnetic spent grain, but the operational and storage stabilities of enzyme immobilized on poly(ethyleneimine)-modified carrier were very high [29].

*Artemisia* seed gum/chitosan/magnetic fluid microspheres were prepared and subsequently cross-linked and activated using glutaraldehyde. The β-galactosidase was immobilized by covalent binding to the surface of the activated magnetic beads. The pH and thermal stability of the immobilized β-galactosidase were higher than those of the soluble one [78].

### 13.2.7 Other Biotechnology Applications of Ferrofluid Modified Materials

Biodesulfurization employing microbial cells is an interesting biotechnology process for the desulfurization of liquid fuels. *Pseudomonas delafieldii* was immobilized in magnetic polyvinyl alcohol (PVA) beads containing MF and sodium alginate. Beads were formed by extruding the mixture through a syringe into gelling solution of calcium chloride with boric acid. The immobilized cells exhibited sufficient desulfurizing activity; magnetic immobilized cells maintained a high desulfurization activity and remained in a good shape after 7 times of repeated use [130]. In an alternative procedure, *Pseudomonas delafieldii* cells in phosphate buffer were mixed with aqueous solution of PVA and oleic acid stabilized MF and subsequently dropped into liquid nitrogen for quick freezing; the beads had superparamagnetic properties and their saturation magnetization was 8.02 emu/g [131].

### 13.2.8 Peroxidase-Like Activity of Ferrofluid-Derived Materials

In the year 2007 the new property of magnetic iron oxide nanoparticles was described, namely the presence of intrinsic peroxidase-like (P-L) activity similar to horseradish peroxidase (HRP); this was the first time inorganic nanoparticles were considered as an enzyme mimetic material for potential biomedical applications [132]. Their P-L activity has been observed during the typical colorimetric reaction involving hydrogen peroxide and appropriate chromogenic reagents. This P-L activity requires similar optimal conditions as those for HRP, including the optimal temperature at 37–40 °C under the optimal pH (pH 3–6.5) in an acidic buffer [133]. The term "nanozyme" was introduced to define nanomaterials with intrinsic enzyme-like activities [133].

Four different electrospun nanofibrous textiles, based on polyamide, polyvinyl alcohol, polycaprolactone and polyurethane were magnetically modified by spraying employing two types of magnetic fluids, namely perchloric acid stabilized FF and chloroform FF. Magnetically modified nanotextile exhibited peroxidase-like activity and thus it represents a promising composite material applicable in various biochemical, biomedical and biotechnology applications [65]. In addition, poly(carboxymethyl methacrylate-co-ethylene dimethacrylate) particles and their thionine derivative were magnetically modified with a nitric acid-stabilized ferrofluid in methanol; saturation magnetization of 11.1 $Am^2/kg$ was observed. These particles exhibiting high peroxidase-like activity at pH 4–6 and at 37 °C, represent a highly sensitive sensor component potentially useful in enzyme-based immunoassays [134].

## 13.3 Conclusions

Progressive magnetically responsive materials have been applied in many important research and technological fields of biosciences, biotechnologies, environmental technologies and related areas. This review demonstrates promising possibility of magnetic fluid utilization for successful magnetic modification of wide variety of biological and non-biological diamagnetic materials of various origin, shapes and sizes, ranging from particles up to high aspect ratio materials, (bio)polymer gels and (nano)textiles. Derivatization through the ferrofluids leads to the deposition of magnetic iron oxide nanoparticles on the surface or within the treated material pores or inside the structure of the (bio)polymer gels. Magnetically modified materials can be selectively separated even from difficult-to-handle suspensions by means of appropriate magnetic separator. Magnetically modified materials show several types of responses to external magnetic field. According to the published research, ferrofluid modified materials have been successfully used as adsorbents, carriers or whole-cell biocatalysts. These materials have been also studied for their catalytic properties. Peroxidase-like activity, caused mainly by the bound magnetic iron oxide nanoparticles, is very interesting from the point of view of mimicking the enzyme activity of biomolecules.

**Acknowledgements** This research was supported by the ERDF/ESF projects CZ.02.1.01/0.0/0.0/17_048/0007399, CZ.02.1.01/0.0/0.0/17_048/0007323 and CZ.02.1.01/0.0/0.0/16_019/0000754 of the Ministry of Education, Youth and Sports of the Czech Republic. Research was also supported by the project ITMS 313011T548 (Structural Funds of EU, Ministry of Education, Slovakia) and bilateral project NASU-20-04 (Academy of Sciences, Czech Republic). This work was carried out in the frame of the COST Actions CA16215 (European network for the promotion of portable, affordable and simple analytical platforms) and CA18238 (European transdisciplinary networking platform for marine biotechnology).

## References


1. Safarik I, Maderova Z, Pospiskova K, Horska K, Safarikova M (2014) Chapter 10. Magnetic decoration and labeling of prokaryotic and eukaryotic cells. In: Fakhrullin RF, Choi I, Lvov YM (eds) Cell surface engineering: fabrication of functional nanoshells. Royal Society of Chemistry, pp 185–215. https://doi.org/10.1039/9781782628477-00185
2. Safarik I, Pospiskova K, Baldikova E, Maderova Z, Safarikova M (2016) Chapter 5. Magnetic modification of cells. In: Grumezescu A (ed) Applications of NanoBioMaterials, volume II: Engineering of NanoBioMaterials. Elsevier, pp 145–180. https://doi.org/10.1016/B978-0-323-41532-3.00005-1
3. Scherer C, Figueiredo Neto AM (2005) Ferrofluids: properties and applications. Brazil J Phys 35:718–727. https://doi.org/10.1590/S0103-97332005000400018
4. Rocha PAC, Santos R, Lima RJP, da Silva MEV (2019) A review on nanofluids: preparation methods and applications. Period Tche Quimica 16(31):365–380
5. Socoliuc V, Peddis D, Petrenko VI, Avdeev MV, Susan-Resiga D, Szabó T, Turcu R, Tombácz E, Vékás L (2020) Magnetic nanoparticle systems for nanomedicine - a materials science perspective. Magnetochemistry 6(1):2. https://doi.org/10.3390/magnetochemistry6010002
6. Vékás L, Avdeev MV, Bica D (2009) Magnetic nanofluids: Synthesis and structure. In: Shi D (ed) NanoScience in biomedicine. Springer, Heidelberg, pp 650–728. https://doi.org/10.1007/978-3-540-49661-8_25
7. Pappel SS (1963) Low viscosity magnetic fluid obtained by the colloidal suspension of magnetic particles. USA Patent US3215572A
8. Massart R (1980) Preparation of aqueous ferrofluids without using surfactant - behavior as a function of the pH and the counterions. C R Acad Hebd Seances Acad Sci C 291(1):1–3
9. Massart R (1981) Preparation of aqueous magnetic liquids in alkaline and acidic media. IEEE Trans Magn 17(2):1247–1248. https://doi.org/10.1109/TMAG.1981.1061188
10. Molday RS, Mackenzie D (1982) Immunospecific ferromagnetic iron-dextran reagents for the labeling and magnetic separation of cells. J Immunol Methods 52(3):353–367. https://doi.org/10.1016/0022-1759(82)90007-2
11. Shokrollahi H (2013) Structure, synthetic methods, magnetic properties and biomedical applications of ferrofluids. Mat Sci Eng C - Mater 33(5):2476–2487. https://doi.org/10.1016/j.msec.2013.03.028
12. Afifah AN, Syahrullail S, Sidik NAC (2016) Magnetoviscous effect and thermomagnetic convection of magnetic fluid: a review. Renew Sustain Energy Rev 55:1030–1040. https://doi.org/10.1016/j.rser.2015.11.018
13. Alberto N, Domingues MF, Marques C, Andre P, Antunes P (2018) Optical fiber magnetic field sensors based on magnetic fluid: a review. Sensors 18(12):4325. https://doi.org/10.3390/s18124325
14. Ebrahimi M (2016) A short review on ferrofluids surface modification by natural and biocompatible polymers. Nanomed J 3(3):155–158. https://doi.org/10.7508/nmj.2016.03.002
15. Felicia LJ, Vinod S, Philip J (2016) Recent advances in magnetorheology of ferrofluids (magnetic nanofluids) - a critical review. J Nanofluids 5(1):1–22. https://doi.org/10.1166/jon.2016.1203
16. Mehta JS, Kumar R, Kumar H, Garg H (2018) Convective heat transfer enhancement using ferrofluid: a review. J Therm Sci Eng Appl 10(2):020801. https://doi.org/10.1115/1.4037200
17. Novopashin SA, Serebryakova MA, Khmel SY (2015) Methods of magnetic fluid synthesis (review). Thermophys Aeromech 22(4):397–412. https://doi.org/10.1134/S0869864315040010
18. Suriyanto, Ng EYK, Kumar SD (2017) Physical mechanism and modeling of heat generation and transfer in magnetic fluid hyperthermia through Neelian and Brownian relaxation: a review. Biomed Eng Online 16:36. https://doi.org/10.1186/s12938-017-0327-x
19. Verma DK, Chandra P, Sinha P (1998) Theoretical studies in ferrofluid lubrication: a review. Indian J Eng Mater Sci 5(6):396–399. http://hdl.handle.net/123456789/29663

20. Zhao Y, Li H, Lv RQ, Hu HF (2014) Reviews on simulation methods for the microstructure of magnetic fluid with and without applied magnetic field. Int J Appl Electromagn Mech 46(3):593–610. https://doi.org/10.3233/JAE-141958
21. Banerjee S (2020) Basics and applications of ferrofluids. In: Prasanta S (ed) Handbook of research on developments and trends in industrial and materials engineering. IGI Global, Hershey, pp 366–400. https://doi.org/10.4018/978-1-7998-1831-1.ch016
22. Odenbach S (ed) (2009) Colloidal magnetic fluids. Basics, development and application of ferrofluids. Springer, Heidelberg. https://doi.org/10.1007/978-3-540-85387-9
23. Blums E, Cebers A, Maiorov MM (1997) Magnetic fluids. Walter de Gruyter, Berlin, New York. https://doi.org/10.1515/9783110807356
24. Berkovsky BM, Bashtovoy V (1996) Magnetic fluids and applications handbook. Begell House
25. Odenbach S (ed) (2002) Ferrofluids. Magnetically controllable fluids and their applications. Springer, Heidelberg. https://doi.org/10.1007/3-540-45646-5
26. Safarik I, Pospiskova K, Horska K, Safarikova M (2012) Potential of magnetically responsive (nano)biocomposites. Soft Matter 8:5407–5413. https://doi.org/10.1039/C2SM06861C
27. Safarik I, Prochazkova J, Baldikova E, Timko M, Kopcansky P, Rajnak M, Torma N, Pospiskova K (2020) Modification of diamagnetic materials using magnetic fluids. Ukr J Phys 65(9):751–760. https://doi.org/10.15407/ujpe65.9.751
28. Safarik I, Lunackova P, Mosiniewicz-Szablewska E, Weyda F, Safarikova M (2007) Adsorption of water-soluble organic dyes on ferrofluid-modified sawdust. Holzforschung 61(3):247–253. https://doi.org/10.1515/HF.2007.060
29. Pospiskova K, Safarik I (2013) Magnetically modified spent grain as a low-cost, biocompatible and smart carrier for enzyme immobilisation. J Sci Food Agric 93(7):1598–1602. https://doi.org/10.1002/jsfa.5930
30. Safarik I, Horska K, Safarikova M (2011) Magnetically modified spent grain for dye removal. J Cereal Sci 53(1):78–80. https://doi.org/10.1016/j.jcs.2010.09.010
31. Safarik I, Horska K, Svobodova B, Safarikova M (2012) Magnetically modified spent coffee grounds for dyes removal. Eur Food Res Technol 234(2):345–350. https://doi.org/10.1007/s00217-011-1641-3
32. Rozumova L, Seidlerova J, Safarik I, Safaríkova M, Cihlarova M, Gabor R (2014) Magnetically modified tea for lead sorption. Adv Sci Eng Med 6(4):473–476. https://doi.org/10.1166/asem.2014.1527
33. Safarik I, Safarikova M (2010) Magnetic fluid modified peanut husks as an adsorbent for organic dyes removal. Phys Procedia 9:274–278. https://doi.org/10.1016/j.phpro.2010.11.061
34. Zuorro A, Lavecchia R, Natali S (2014) Magnetically modified agro-industrial wastes as efficient and easily recoverable adsorbents for water treatment. Chem Eng Trans 38:349–354. https://doi.org/10.3303/CET1438059
35. Safarik I, Horska K, Pospiskova K, Safarikova M (2012) One-step preparation of magnetically responsive materials from non-magnetic powders. Powder Technol 229:285–289. https://doi.org/10.1016/j.powtec.2012.06.006
36. Safarik I, Ashoura N, Maderova Z, Pospiskova K, Baldikova E, Safarikova M (2016) Magnetically modified *Posidonia oceanica* biomass as an adsorbent for organic dyes removal. Medit Mar Sci 17(2):351–358. https://doi.org/10.12681/mms.1549
37. Baldikova E, Pospiskova K, Ladakis D, Kookos IK, Koutinas AA, Safarikova M, Safarik I (2017) Magnetically modified bacterial cellulose: a promising carrier for immobilization of affinity ligands, enzymes, and cells. Mater Sci Eng C 71:214–221. https://doi.org/10.1016/j.msec.2016.10.009
38. Saylan Y, Gokturk I, Pospiskova K, Safarik I, Denizli A (2020) Magnetic bacterial cellulose nanofibers for nucleoside recognition. Cellulose 27:9479–9492. https://doi.org/10.1007/s10570-020-03425-x
39. Mosbach K, Andersson L (1977) Magnetic ferrofluids for preparation of magnetic polymers and their application in affinity chromatography. Nature 270(5634):259–261. https://doi.org/10.1038/270259a0

40. Safarik I, Safarikova M (2007) Magnetically modified microbial cells: a new type of magnetic adsorbents. China Particuol 5(1–2):19–25. https://doi.org/10.1016/j.cpart.2006.12.003
41. Safarikova M, Maderova Z, Safarik I (2009) Ferrofluid modified *Saccharomyces cerevisiae* cells for biocatalysis. Food Res Int 42(4):521–524. https://doi.org/10.1016/j.foodres.2009.01.001
42. Azevedo RB, Silva LP, Lemos APC, Bao SN, Lacava ZGM, Safarik I, Safarikova M, Morais PC (2003) Morphological study of *Saccharomyces cerevisiae* cells treated with magnetic fluid. IEEE Trans Magn 39(5):2660–2662. https://doi.org/10.1109/TMAG.2003.815547
43. Safarik I, Rego LFT, Borovska M, Mosiniewicz-Szablewska E, Weyda F, Safarikova M (2007) New magnetically responsive yeast-based biosorbent for the efficient removal of water-soluble dyes. Enzyme Microb Technol 40(6):1551–1556. https://doi.org/10.1016/j.enzmictec.2006.10.034
44. Safarikova M, Pona BMR, Mosiniewicz-Szablewska E, Weyda F, Safarik I (2008) Dye adsorption on magnetically modified *Chlorella vulgaris* cells. Fresenius Environ Bull 17(4):486–492
45. Safarik I, Angelova R, Baldikova E, Pospiskova K, Safarikova M (2017) *Leptothrix* sp. sheaths modified with iron oxide particles: magnetically responsive, high aspect ratio functional material. Mater Sci Eng C 71:1342–1346. https://doi.org/10.1016/j.msec.2016.10.056
46. Angelova R, Baldikova E, Pospiskova K, Safarikova M, Safarik I (2017) Magnetically modified sheaths of *Leptothrix* sp. as an adsorbent for Amido black 10B removal. J Magn Magn Mater 427:314–319. https://doi.org/10.1016/j.jmmm.2016.10.094
47. Kratošová G, Schröfel A, Šafařík I, Horská K, Urban M, Rosenbergová K, Šafaříková M, Slabotinský J (2013) Bionanokompozit, způsob jeho výroby a použití (Bionanocomposite, process for its preparation and use). Czech Patent 304046
48. Holišová V, Natšinová M, Kratošová G, Chromčáková Ž, Schröfel A, Vávra I, Životský O, Šafařík I, Obalová L (2019) Magnetically modified nanogold-biosilica composite as an effective catalyst for CO oxidation. Arab J Chem 12(7):1148–1158. https://doi.org/10.1016/j.arabjc.2018.12.002
49. Pospiskova K, Safarik I, Sebela M, Kuncova G (2013) Magnetic particles–based biosensor for biogenic amines using an optical oxygen sensor as a transducer. Microchim Acta 180(3):311–318. https://doi.org/10.1007/s00604-012-0932-0
50. Makarchuk OV, Dontsova TA, Astrelin IM (2016) Magnetic nanocomposites as efficient sorption materials for removing dyes from aqueous solutions. Nanoscale Res Lett 11(1):161. https://doi.org/10.1186/s11671-016-1364-2
51. Makarchuk O, Dontsova T, Perekos A, Skoblik A, Svystunov Y (2017) Magnetic mineral nanocomposite sorbents for wastewater treatment. J Nanomater. https://doi.org/10.1155/2017/8579598.Article no 8579598
52. Dontsova TA, Yanushevskaya EI, Nahirniak SV, Makarchuk OV, Ivanets, AI, Roshchina MY, Kutuzova AS, Kulikov LM (2018) Directional control of the structural adsorption properties of clays by magnetite modification. J Nanomater. https://doi.org/10.1155/2018/6573016.Article no 6573016
53. Gdovinová V, Tomašovičová N, Batko I, Batková M, Balejčíková L, Garamus VM, Petrenko VI, Avdeev MV, Kopčanský P (2017) Interaction of magnetic nanoparticles with lysozyme amyloid fibrils. J Magn Magn Mater 431:8–11. https://doi.org/10.1016/j.jmmm.2016.09.035
54. Majorosova J, Petrenko VI, Siposova K, Timko M, Tomasovicova N, Garamus VM, Koralewski M, Avdeev MV, Leszczynski B, Jurga S, Gazova Z, Hayryan S, Hu C-K, Kopcansky P (2016) On the adsorption of magnetite nanoparticles on lysozyme amyloid fibrils. Colloids Surf B 146:794–800. https://doi.org/10.1016/j.colsurfb.2016.07.024
55. Herrero-Latorre C, Barciela-García J, García-Martín S, Peña-Crecente RM, Otárola-Jiménez J (2015) Magnetic solid-phase extraction using carbon nanotubes as sorbents: a review. Anal Chim Acta 892:10–26. https://doi.org/10.1016/j.aca.2015.07.046
56. Samadishadlou M, Farshbaf M, Annabi N, Kavetskyy T, Khalilov R, Saghfi S, Akbarzadeh A, Mousavi S (2018) Magnetic carbon nanotubes: preparation, physical properties, and applications in biomedicine. Artif Cells Nanomed Biotechnol 46(7):1314–1330. https://doi.org/10.1080/21691401.2017.1389746

57. Samouhos S, McKinley G (2006) Carbon nanotube–magnetite composites, with applications to developing unique magnetorheological fluids. J Fluids Eng 129(4):429–437. https://doi.org/10.1115/1.2436581
58. Masotti A, Caporali A (2013) Preparation of magnetic carbon nanotubes (Mag-CNTs) for biomedical and biotechnological applications. Int J Mol Sci 14(12):24619–24642. https://doi.org/10.3390/ijms141224619
59. Korneva G, Ye H, Gogotsi Y, Halverson D, Friedman G, Bradley J-C, Kornev KG (2005) Carbon nanotubes loaded with magnetic particles. Nano Lett 5(5):879–884. https://doi.org/10.1021/nl0502928
60. Pal S, Chandra S, Phan MH, Mukherjee P, Srikanth H (2009) Carbon nanostraws: nanotubes filled with superparamagnetic nanoparticles. Nanotechnology 20(48):485604. https://doi.org/10.1088/0957-4484/20/48/485604
61. Shahsavar A, Saghafian M, Salimpour MR, Shafii MB (2016) Experimental investigation on laminar forced convective heat transfer of ferrofluid loaded with carbon nanotubes under constant and alternating magnetic fields. Exp Therm Fluid Sci 76:1–11. https://doi.org/10.1016/j.expthermflusci.2016.03.010
62. Jain D, Wilhelm R (2007) An easy way to produce α-iron filled multiwalled carbon nanotubes. Carbon 45(3):602–606. https://doi.org/10.1016/j.carbon.2006.10.012
63. Yuan P, Bergaya F, Thill A (2016) Chapter 1 - General introduction. In: Yuan P, Thill A, Bergaya F (eds) Developments in clay science, vol 7. Elsevier, pp 1–10. https://doi.org/10.1016/B978-0-08-100293-3.00001-7
64. Desai R, Upadhyay RV, Mehta RV (2014) Augmentation of chain formation in a magnetic fluid by the addition of halloysite nanotubes. J Phys D 47(16):165501. https://doi.org/10.1088/0022-3727/47/16/165501
65. Prochazkova J, Pospiskova K, Safarik I (2019) Magnetically modified electrospun nanotextile exhibiting peroxidase-like activity. J Magn Magn Mater 473:335–340. https://doi.org/10.1016/j.jmmm.2018.10.106
66. Molcan M, Safarik I, Pospiskova K, Paulovicova K, Timko M, Kopcansky P, Torma N (2020) Magnetically modified electrospun nanofibers for hyperthermia treatment. Ukr J Phys 65(8):655–661. https://doi.org/10.15407/ujpe65.8.655
67. Tong J, Chen L (2013) Preparation and application of magnetic chitosan derivatives in separation processes. Anal Lett 46(17):2635–2656. https://doi.org/10.1080/00032719.2013.807815
68. Hu X-J, Wang J-S, Liu Y-G, Li X, Zeng G-M, Bao Z-L, Zeng X-X, Chen A-W, Long F (2011) Adsorption of chromium (VI) by ethylenediamine-modified cross-linked magnetic chitosan resin: isotherms, kinetics and thermodynamics. J Hazard Mater 185(1):306–314. https://doi.org/10.1016/j.jhazmat.2010.09.034
69. Jiang D-S, Long S-Y, Huang J, Xiao H-Y, Zhou J-Y (2005) Immobilization of *Pycnoporus sanguineus* laccase on magnetic chitosan microspheres. Biochem Eng J 25(1):15–23. https://doi.org/10.1016/j.bej.2005.03.007
70. Safarik I, Stepanek M, Uchman M, Slouf M, Baldikova E, Nydlova L, Pospiskova K, Safarikova M (2016) Composite particles formed by complexation of poly(methacrylic acid) — stabilized magnetic fluid with chitosan: magnetic material for bioapplications. Mater Sci Eng C 67:486–492. https://doi.org/10.1016/j.msec.2016.05.017
71. Safarikova M, Roy I, Gupta MN, Safarik I (2003) Magnetic alginate microparticles for purification of α-amylases. J Biotechnol 105(3):255–260. https://doi.org/10.1016/j.jbiotec.2003.07.002
72. Rocher V, Bee A, Siaugue JM, Cabuil V (2010) Dye removal from aqueous solution by magnetic alginate beads crosslinked with epichlorohydrin. J Hazard Mater 178(1–3):434–439. https://doi.org/10.1016/j.jhazmat.2010.01.100
73. Obeid L, El Kolli N, Dali N, Talbot D, Abramson S, Welschbillig M, Cabuil V, Bee A (2014) Adsorption of a cationic surfactant by a magsorbent based on magnetic alginate beads. J Colloid Interface Sci 432:182–189. https://doi.org/10.1016/j.jcis.2014.06.027

74. Bee A, Talbot D, Abramson S, Dupuis V (2011) Magnetic alginate beads for Pb(II) ions removal from wastewater. J Colloid Interface Sci 362(2):486–492. https://doi.org/10.1016/j.jcis.2011.06.036
75. Ngomsik A-F, Bee A, Siaugue J-M, Cabuil V, Cote G (2006) Nickel adsorption by magnetic alginate microcapsules containing an extractant. Water Res 40(9):1848–1856. https://doi.org/10.1016/j.watres.2006.02.036
76. Pulfer SK, Gallo JM (1998) Enhanced brain tumor selectivity of cationic magnetic polysaccharide microspheres. J Drug Target 6(3):215–227. https://doi.org/10.3109/10611869808997896
77. Safarikova M, Horska K, Maderova Z, Tonkova A, Ivanova-Pashkoulova V, Safarik I (2012) Magnetic porous corn starch for the affinity purification of cyclodextrin glucanotransferase produced by *Bacillus circulans*. Biocatal Biotransformation 30(1):96–101. https://doi.org/10.3109/10242422.2012.646665
78. Zhang J, Zhang ST, Zeng JY, Wang YP, Wang YF (2006) Immobilized β-galactosidase: preparation and characterization. Polym Polym Compos 14(8):813–823. https://doi.org/10.1177/096739110601400806
79. Jeyasubramanian K, Selvakumar N, Ilakkiya J, Santhoshkumar P, Satish N, Sahoo SK (2013) Magnetic flux alignment studies on entrapped ferrofluid nanoparticles in poly vinyl alcohol matrix. J Mater Sci Technol 29(10):903–908. https://doi.org/10.1016/j.jmst.2013.07.002
80. Bertoglio P, Jacobo SE, Daraio ME (2010) Preparation and characterization of PVA films with magnetic nanoparticles: the effect of particle loading on drug release behavior. J Appl Polym Sci 115(3):1859–1865. https://doi.org/10.1002/app.31315
81. Hernández R, López G, López D, Vázquez M, Mijangos C (2011) Magnetic characterization of polyvinyl alcohol ferrogels and films. J Mater Res 22(8):2211–2216. https://doi.org/10.1557/jmr.2007.0298
82. Albornoz C, Jacobo SE (2006) Preparation of a biocompatible magnetic film from an aqueous ferrofluid. J Magn Magn Mater 305(1):12–15. https://doi.org/10.1016/j.jmmm.2005.11.021
83. Aldana S, Vereda F, Hidalgo-Alvarez R, de Vicente J (2016) Facile synthesis of magnetic agarose microfibers by directed self-assembly in W/O emulsions. Polymer 93:61–64. https://doi.org/10.1016/j.polymer.2016.04.010
84. Rideau E, Dimova R, Schwille P, Wurm FR, Landfester K (2018) Liposomes and polymersomes: a comparative review towards cell mimicking. Chem Soc Rev 47(23):8572–8610. https://doi.org/10.1039/C8CS00162F
85. Liarou E, Varlas S, Skoulas D, Tsimblouli C, Sereti E, Dimas K, Iatrou H (2018) Smart polymersomes and hydrogels from polypeptide-based polymer systems through α-amino acid N-carboxyanhydride ring-opening polymerization. From chemistry to biomedical applications. Prog Polym Sci 83:28–78. https://doi.org/10.1016/j.progpolymsci.2018.05.001
86. Monnier CA, Burnand D, Rothen-Rutishauser B, Lattuada M, Petri-Fink A (2014) Magnetoliposomes: opportunities and challenges. Eur J Nanomed 6(4):201–215. https://doi.org/10.1515/ejnm-2014-0042
87. De Cuyper M, Joniau M (1988) Magnetoliposomes: formation and structural characterization. Eur Biophys J 15(5):311–319. https://doi.org/10.1007/bf00256482
88. Namdari M, Cheraghi M, Negahdari B, Eatemadi A, Daraee H (2017) Recent advances in magnetoliposome for heart drug delivery. Artif Cell Nanomed Biotechnol 45(6):1051–1057. https://doi.org/10.1080/21691401.2017.1299159
89. Fattahi H, Laurent S, Liu FJ, Arsalani N, Elst LV, Muller RN (2011) Magnetoliposomes as multimodal contrast agents for molecular imaging and cancer nanotheragnostics. Nanomedicine 6(3):529–544. https://doi.org/10.2217/nnm.11.14
90. Garnier B, Tan S, Miraux S, Bled E, Brisson AR (2012) Optimized synthesis of 100 nm diameter magnetoliposomes with high content of maghemite particles and high MRI effect. Contrast Media Mol Imaging 7(2):231–239. https://doi.org/10.1002/cmmi.487
91. Bleul R, Thiermann R, Marten GU, House MJ, Pierre TGS, Häfeli UO, Maskos M (2013) Continuously manufactured magnetic polymersomes – a versatile tool (not only) for targeted cancer therapy. Nanoscale 5(23):11385–11393. https://doi.org/10.1039/C3NR02190D

92. Safarik I, Safarikova M (2004) Magnetic techniques for the isolation and purification of proteins and peptides. BioMagn Res Technol 2(1):7. https://doi.org/10.1186/1477-044X-2-7
93. Safarik I, Safarikova M (1997) Overview of magnetic separations used in biochemical and biotechnological applications. In: Hafeli U, Schutt W, Teller J, Zborowski M (eds) Scientific and clinical applications of magnetic carriers. Plenum Press, New York and London, pp 323–340. https://doi.org/10.1007/978-1-4757-6482-6_24
94. Nixon L, Koval CA, Noble RD, Slaff GS (1992) Preparation and characterization of novel magnetite-coated ion-exchange particles. Chem Mater 4(1):117–121. https://doi.org/10.1021/cm00019a025
95. Grilo AL, Raquel Aires-Barros M, Azevedo AM (2016) Partitioning in aqueous two-phase systems: fundamentals, applications and trends. Sep Purif Rev 45(1):68–80. https://doi.org/10.1080/15422119.2014.983128
96. Kee PE, Ng T-C, Lan JC-W, Ng H-S (2020) Recent development of unconventional aqueous biphasic system: characteristics, mechanisms and applications. Crit Rev Biotechnol 40(4):555–569. https://doi.org/10.1080/07388551.2020.1747388
97. Wikström P, Flygare S, Gröndalen A, Larsson P-O (1987) Magnetic aqueous two-phase separation: a new technique to increase rate of phase-separation, using dextran-ferrofluid or larger iron oxide particles. Anal Biochem 167(2):331–339. https://doi.org/10.1016/0003-2697(87)90173-4
98. Larsson P-O (1994) Magnetically enhanced phase separation. Meth Enzymol 228:112–117. https://doi.org/10.1016/0076-6879(94)28012-6
99. Goto M, Nikamoto Y, Nakashio F (1994) Promotion of phase separation in aqueous two-phase systems utilizing magnetic fluid. Kagaku Kogaku Ronbunshu 20(5):728–731. https://doi.org/10.1252/kakoronbunshu.20.728
100. Flygare S, Wikstrom P, Johansson G, Larsson PO (1990) Magnetic aqueous two-phase separation in preparative applications. Enzyme Microb Technol 12(2):95–103. https://doi.org/10.1016/0141-0229(90)90080-A
101. Kamihira M (2000) Affinity partitioning using magnetic two-phase systems. In: Hatti-Kaul R (ed) Aqueous two-phase systems: methods and protocols. Springer, pp 381–390. https://doi.org/10.1385/1-59259-028-4:381
102. Safarik I, Baldikova E, Prochazkova J, Safarikova M, Pospiskova K (2018) Magnetically modified agricultural and food waste: preparation and application. J Agric Food Chem 66(11):2538–2552. https://doi.org/10.1021/acs.jafc.7b06105
103. Safarik I, Safarikova M, Weyda F, Mosiniewicz-Szablewska E, Slawska-Waniewska A (2005) Ferrofluid-modified plant-based materials as adsorbents for batch separation of selected biologically active compounds and xenobiotics. J Magn Magn Mater 293(1):371–376. https://doi.org/10.1016/j.jmmm.2005.02.033
104. Mosiniewicz-Szablewska E, Safarikova M, Safarik I (2007) Magnetic studies of ferrofluid-modified spruce sawdust. J Phys D 40(21):6490–6496. https://doi.org/10.1088/0022-3727/40/21/003
105. Besharati N, Alizadeh N, Shariati S (2018) Removal of cationic dye methylene blue (MB) from aqueous solution by coffee and peanut husk modified with magnetite iron oxide nanoparticles. J Mex Chem Soc 62(3):110–124. https://doi.org/10.29356/jmcs.v62i3.433
106. Zuorro A, Di Battista A, Lavecchia R (2013) Magnetically modified coffee silverskin for the removal of xenobiotics from wastewater. Chem Eng Trans 35:1375–1380. https://doi.org/10.3303/CET1335229
107. Besharati N, Alizadeh N (2018) Adsorption of malachite green dye on different natural absorbents modified with magnetite nanoparticles. J Nanoanalysis 5(3):143–155. https://doi.org/10.22034/jna.2018.542763
108. Safarik I, Ptackova L, Safarikova M (2002) Adsorption of dyes on magnetically labeled baker's yeast cells. Eur Cells Mater 3(Suppl. 2):52–55
109. Wu Q, Shan Z, Shen M, Li S, Chen H (2009) Biosorption of direct scarlet dye on magnetically modified *Saccharomyces cerevisiae* cells. Chin J Biotechnol 25(10):1477–1482

110. Safarikova M, Ptackova L, Kibrikova I, Safarik I (2005) Biosorption of water-soluble dyes on magnetically modified *Saccharomyces cerevisiae* subsp. *uvarum* cells. Chemosphere 59(6):831–835. https://doi.org/10.1016/j.chemosphere.2004.10.062
111. Yavuz H, Denizli A, Gungunes H, Safarikova M, Safarik I (2006) Biosorption of mercury on magnetically modified yeast cells. Sep Purif Technol 52(2):253–260. https://doi.org/10.1016/j.seppur.2006.05.001
112. Uzun L, Saglam N, Safarikova M, Safarik I, Denizli A (2011) Copper biosorption on magnetically modified yeast cells under magnetic field. Sep Sci Technol 46(6):1045–1051. https://doi.org/10.1080/01496395.2010.541400
113. Ji YQ, Hu YT, Tian Q, Shao XZ, Li JY, Safarikova M, Safarik I (2010) Biosorption of strontium ions by magnetically modified yeast cells. Sep Sci Technol 45(10):1499–1504. https://doi.org/10.1080/01496391003705664
114. Bai J, Wu XL, Fan FL, Tian W, Yin XJ, Zhao L, Fan FY, Li Z, Tian LL, Qin Z, Guo JS (2012) Biosorption of uranium by magnetically modified *Rhodotorula glutinis*. Enzyme Microb Technol 51(6–7):382–387. https://doi.org/10.1016/j.enzmictec.2012.08.007
115. Rozumova L, Kus P, Safarik I (2018) Modified biological sorbents from waste for the removal of metal ions from the water system. In: Vilarinho C, Castro F, de Lurdes Lopes M (eds) Wastes - solutions, treatments and opportunities II. CRC Press, pp 377–382
116. Safarikova M, Safarik I (1999) Magnetic solid-phase extraction. J Magn Magn Mater 194(1–3):108–112. https://doi.org/10.1016/S0304-8853(98)00566-6
117. Safarik I, Horska K, Pospiskova K, Safarikova M (2012) Magnetic techniques for the detection and determination of xenobiotics and cells in water. Anal Bioanal Chem 404(4):1257–1273. https://doi.org/10.1007/s00216-012-6056-x
118. Vasconcelos I, Fernandes C (2017) Magnetic solid phase extraction for determination of drugs in biological matrices. TrAC Trends Anal Chem 89:41–52. https://doi.org/10.1016/j.trac.2016.11.011
119. Li N, Jiang HL, Wang XL, Wang X, Xu GJ, Zhang BB, Wang LJ, Zhao RS, Lin JM (2018) Recent advances in graphene-based magnetic composites for magnetic solid-phase extraction. TrAC Trends Anal Chem 102:60–74. https://doi.org/10.1016/j.trac.2018.01.009
120. Maya F, Cabello CP, Frizzarin RM, Estela JM, Palomino GT, Cerda V (2017) Magnetic solid-phase extraction using metal-organic frameworks (MOFs) and their derived carbons. TrAC Trends Anal Chem 90:142–152. https://doi.org/10.1016/j.trac.2017.03.004
121. Tekkeli SEK, Durmus Z (2019) Magnetic solid phase extraction applications combined with analytical methods for determination of drugs in different matrices. Rev J Chil Chem Soc 64(2):4448–4458. https://doi.org/10.4067/S0717-97072019000204448
122. Wierucka M, Biziuk M (2014) Application of magnetic nanoparticles for magnetic solid-phase extraction in preparing biological, environmental and food samples. TrAC Trends Anal Chem 59:50–58. https://doi.org/10.1016/j.trac.2014.04.007
123. Hemmati M, Rajabi M, Asghari A (2018) Magnetic nanoparticle based solid-phase extraction of heavy metal ions: a review on recent advances. Microchim Acta 185(3):160. https://doi.org/10.1007/s00604-018-2670-4
124. Wan Ibrahim WA, Nodeh HR, Aboul-Enein HY, Sanagi MM (2015) Magnetic solid-phase extraction based on modified ferum oxides for enrichment, preconcentration, and isolation of pesticides and selected pollutants. Crit Rev Anal Chem 45(3):270–287. https://doi.org/10.1080/10408347.2014.938148
125. Gao Q, Luo D, Ding J, Feng Y-Q (2010) Rapid magnetic solid-phase extraction based on magnetite/silica/poly(methacrylic acid–co–ethylene glycol dimethacrylate) composite microspheres for the determination of sulfonamide in milk samples. J Chromatogr A 1217(35):5602–5609. https://doi.org/10.1016/j.chroma.2010.06.067
126. Moazzen M, Ahmadkhaniha R, Gorji ME, Yunesian M, Rastkari N (2013) Magnetic solid-phase extraction based on magnetic multi-walled carbon nanotubes for the determination of polycyclic aromatic hydrocarbons in grilled meat samples. Talanta 115:957–965. https://doi.org/10.1016/j.talanta.2013.07.005

127. Rastkari N, Ahmadkhaniha R (2013) Magnetic solid-phase extraction based on magnetic multi-walled carbon nanotubes for the determination of phthalate monoesters in urine samples. J Chromatogr A 1286:22–28. https://doi.org/10.1016/j.chroma.2013.02.070
128. Shen Z, He Z, Wang P, Zhou Z, Sun M, Li J, Liu D (2013) Low-density magnetofluid dispersive liquid–liquid microextraction for the fast determination of organochlorine pesticides in water samples by GC-ECD. Anal Chim Acta 793:37–43. https://doi.org/10.1016/j.aca.2013.07.050
129. Yang D, Li G, Wu L, Yang Y (2018) Ferrofluid-based liquid-phase microextraction: analysis of four phenolic compounds in milks and fruit juices. Food Chem 261:96–102. https://doi.org/10.1016/j.foodchem.2018.04.038
130. Shan G-b, Xing J-m, Luo M-f, Liu H-z, Chen J-y (2003) Immobilization of *Pseudomonas delafieldii* with magnetic polyvinyl alcohol beads and its application in biodesulfurization. Biotechnol Lett 25(23):1977–1981. https://doi.org/10.1023/B:BILE.0000004388.15751.8c
131. Guobin S, Jianmin X, Chen G, Huizhou L, Jiayong C (2005) Biodesulfurization using *Pseudomonas delafieldii* in magnetic polyvinyl alcohol beads. Lett Appl Microbiol 40(1):30–36. https://doi.org/10.1111/j.1472-765X.2004.01617.x
132. Gao L, Zhuang J, Nie L, Zhang J, Zhang Y, Gu N, Wang T, Feng J, Yang D, Perrett S, Yan X (2007) Intrinsic peroxidase-like activity of ferromagnetic nanoparticles. Nature Nanotechnol 2(9):577–583. https://doi.org/10.1038/nnano.2007.260
133. Gao LZ, Fan KL, Yan XY (2017) Iron oxide nanozyme: a multifunctional enzyme mimetic for biomedical applications. Theranostics 7(13):3207–3227. https://doi.org/10.7150/thno.19738
134. Zasonska BA, Salek P, Prochazkova J, Mullerova S, Svoboda J, Petrovsky E, Proks V, Horak D, Safarik I (2019) Peroxidase-like activity of magnetic poly(glycidyl methacrylate-co-ethylene dimethacrylate) particles. Sci Rep 9. https://doi.org/10.1038/s41598-018-38012-5.Article no 1543

# Part IV Nanosystems and Nanomaterials for Biomedical Applications

# Chapter 14
# Smart Polymer-Based Multicomponent Nanosystem for Enhanced Anticancer Photodynamic Therapy

**Nataliya Kutsevol, Yuliia Kuziv, Leonid Bulavin, and Vasyl Chekhun**

**Abstract** Nanotechnology is a promising approach for creation of polymer-based nanocomposite dye for Photodynamic therapy (PDT). Polymer molecules can be preferentially accumulated in tumor die, prevent photosensitizer aggregation. Thermosensitive star-like Dextran-graft-Poly-N-isopropylacrylamide copolymer (D-PNIPAM) was used as matrix for creation of multicomponent nanosystems for PDT. Au nanoparticles (AuNPs) and photosensitizer Chlorin e6 (Ce6) were incorporated into polymer. The behavior of nanosystems was studied by optical absorption, dynamic light scattering in water and in Hank's buffer solution within temperature range of 25–40 °C. No drastic changes accompanied by aggregation process in the nanosystem in Hank's saline solution in comparison with one prepared in water were registered. In vitro examination of the PDT activity of the prepared nanosystem on the culture of MT-4 human malignant lymphocytes revealed the death of 40% lymphocytes preincubated with nanocomposite.

---

N. Kutsevol · Y. Kuziv (✉)
Faculty of Chemistry, Taras Shevchenko National University, Volodymyrska Street 60, Kyiv 01601, Ukraine

N. Kutsevol
e-mail: kutsevol@ukr.net

L. Bulavin
Faculty of Physics, Taras Shevchenko National University, Volodymyrska Street 60, Kyiv 01601, Ukraine

V. Chekhun
R.E. Kavetsky Institute for Experimental Pathology Oncology and Radiobiology, Vasilkivska Street 45, Kyiv 03022, Ukraine
e-mail: chekhun@nas.gov.ua



L. Bulavin and N. Lebovka (eds.), *Soft Matter Systems for Biomedical Applications*, Springer Proceedings in Physics 266, https://doi.org/10.1007/978-3-030-80924-9_14

## List of Abbreviations and Chemical Substances

### *Abbreviations*

| | |
|---|---|
| DLS | Dynamic Light Scattering |
| HPLC | High-performance Liquid Chromatography |
| LCST | Lower critical solution temperature |
| PDT | Photodynamic therapy |
| RPMI | Roswell Park Memorial Institute medium |
| SEC | Size Exclusion Chromatography |
| SPR | Surface plasmon resonance |
| TDA | Triple Detector Array |
| TEM | Transmission Electron Microscopy |
| UV | Ultraviolet |

### *Chemical Substances*

| | |
|---|---|
| AuNPs | Au nanoparticles, gold nanoparticles |
| Ce6 | Chlorin e6 |
| DMSO | Dimethyl sulfoxide |
| D-PNIPAM | Dextran-graft-Poly(N-isopropylacrylamide) |
| NIPAM | N-isopropylacrylamide |
| PNIPAM | Poly(N-isopropylacrylamide) |
| PS | Photosensitizer |

## 14.1 Introduction

Recently, photodynamic therapy (PDT) is considered as an alternative therapy of cancer diseases. PDT based on photochemical reactions with photosensitizing drug (PS). Photosensitizer with subsequent light irradiation at the proper wavelength caused the photochemical reaction: molecular oxygen ($^3O_2$) converts into high reactive singlet oxygen ($^1O_2$) and provokes a damage of malignant cells via apoptosis or necrosis due to high cytotoxicity. Oxidative process causes damage of nucleic acids, proteins and lipids in the cells [1–4]. The classic PDT based on accumulation of a photosensitizer in tumor tissue. Then, photosensitizer is activated by illumination with light of the certain wave length and leads to the cell death. For efficient PDT photosensitizer should have high absorption in the range of 650–850 nm when the tissue penetration is high; PS should have high singlet oxygen quantum yield and have low dark toxicity; and should not form aggregates in water medium. Nowadays, despite several clinically approved PS, the highly efficient and selective PS agent is still not found [1].

Nanotechnology is a promising approach for creation of highly efficient PS for PDT treatment. Polymers can be used as nanocarrier for drug delivery, including hydrophobic PS [5]. Current research and reviews have established that polymer molecules can be preferentially accumulated in tumors due to passive or receptor-mediated active targeting [6]. However, the use of polymers can lead to uncontrollable drug release. To overcome this problem, stimuli-responsible polymers can provide a controlled release of drugs [7, 8]. However, for development of stimuli-responsive polymers for biomedical application, it needs a deep knowledge on polymer-nanocarrier biocompatibility, drug loading capacity, nanocarrier stability, and toxicity and etc.

It is known that poly(N-isopropylacrylamide) (PNIPAM) can undergo a tunable phase transition in the region of physiological temperature [9]. Several reviews devoted to the prospects of the biomedical use of PNIPAM have been published in the last decade [10, 11]. It was emphasized that the most promising fields for PNIPAM application are drug delivery, biosensors, and bioimaging. PNIPAM attracts attention due to its water solubility and existence of phase transition in the region of physiological temperatures. The lower critical solution temperature (LCST) for PNIPAM is close to human body temperature [12]. The reversibility of the hydrophilic/hydrophobic state occurs by varying the temperature below or above LCST. Below LCST the polymer is soluble in water, above the LCST, this polymer undergoes a phase transition; then collapses and forms aggregates. LCST value for linear PNIPAM is approximately 32 °C. PNIPAM is hydrophilic at room temperature, but after phase transition becomes partially hydrophobic, therefore can be used for controlled drug release [13].

We have synthesized star-shaped temperature sensitive copolymers with dextran core and grafted poly-N-isopropylacrylamide arms (D-PNIPAM) [14]. It was shown the possibility to shift the temperature of the conformational transition (phase transition) for the PNIPAM-containing polymer by playing with its molecular architecture. The LCST for D-PNIPAM is for 3–4 °C higher in comparison with the linear PNIPAM. Thus, LCST becomes closer to human body temperature.

In our recent research it was reported that D-PNIPAM copolymer has demonstrated the high in vitro efficiency as nanocarriers for anticancer drug Doxorubicin [15]. The nanosystems consisting of gold nanoparticles (AuNPs) into D-PNIPAM water solution were synthesized and studied [16–18]. It is known, that nanosized gold is practically non-toxic and has an affinity with biological factors that support vascular growth. Gold nanoparticles (AuNPs) enhance the therapeutic effect of the drugs for photodynamic and photothermal therapy [19]. The ability of AuNPs to penetrate the endothelium of tumor vessels provides higher selectivity for the accumulation of anticancer drugs in the malignant cells [20]. The size characteristics of AuNPs and their aggregative stability are very important factors for biomedical application [20]. Polymers can affect the processes of formation of metal particles, control their size, shape and can prevent the nanoparticle aggregation [21].

Recently, we have reported on the high efficiency of nanocomposites for photodynamic therapy, consisting of AuNPs, a photosensitizer of Chlorine e6 (Ce6) incorporated into the anionic star-shape copolymer dextran-polyacrylamide-(co)-polyacrylic acid [22, 23].

In the present research the behavior of the multicomponent nanosystem D-PNIPAM/AuNPs/Ce6 in water and in a Hank's buffer solution in the region of physiological temperatures were studied. Hank's balanced salt solution is composed of inorganic salts and supplemented with glucose. The solution may be used to wash cells and tissue and to maintain cells in a viable state. The solution is buffered with phosphate and maintains a physiological pH and osmotic pressure.

The presence of Hank's solution can drastically affect the nanocomposite properties, partially the aggregation process. Finally the prepared nanocomposites were tested for its photodynamic efficacy on the MT-4 cell line (human T-cell leukemia).

## 14.2 Polymer Nanocarrier

Dextran with molecular weights $M_w = 7 \times 10^5$ g/mol from Fluka (Switzerland), N-isopropylacrylamide (NIPAM) from Sigma-Aldrich (USA) were used for synthesis of Dextran-graft-PNIPAM copolymer. The sample was designated as D-PNIPAM. The synthesis and characterization of D-PNIPAM copolymer with 15 PNIPAM grafts were reported in [14]. The molecular characteristics of D-PNIPAM were obtained using Multidetection size exclusion chromatography (SEC) allowed determining the molecular weight distribution and average molecular weights of the branched PNIPAM sample. This SEC line involves a usual HPLC part (Shimadzu, Japan) and Malvern's triple detection TDA 302 (Viscotek, UK) with refractometer, viscosimeter, and two-angle (7 and 90°) light scattering apparatus. The molecular parameters of D-PNIPAM are given in the Table 14.1 (Mw -the weight average molecular weight of copolymer, $M_w/M_n$ - polydispersity of copolymer, $D_h$ - hydrodynamic diameter of macromolecule at 25 °C.

The behavior of D-PNIPAM in the region of phase transition was studied by Dynamic light scattering using Zetasizer Nano ZS90 (Malvern Instruments Ltd., UK). Figure 14.2 demonstrates decreasing of $D_h$ for D-PNIPAM macromolecule from 31 to 22 nm in the temperature range of 32.6–33.4 °C and sharp aggregation process at 33.7–34.1 °C. The drastic change in the shape of sizes distributions curves and scattering intensity increasing at 33.7–34.1 °C was observed [14]. Further heating to 36 °C doesn't cause additional aggregation.

**Table 14.1** Molecular parameters of D-PNIPAM copolymer

| Sample | $M_w \times 10^{-6}$, g/mol | $M_w/M_n$ | $D_h$ $^{25C}$, nm |
|---|---|---|---|
| D-PNIPAM | 1.03 | 1.52 | 40 |

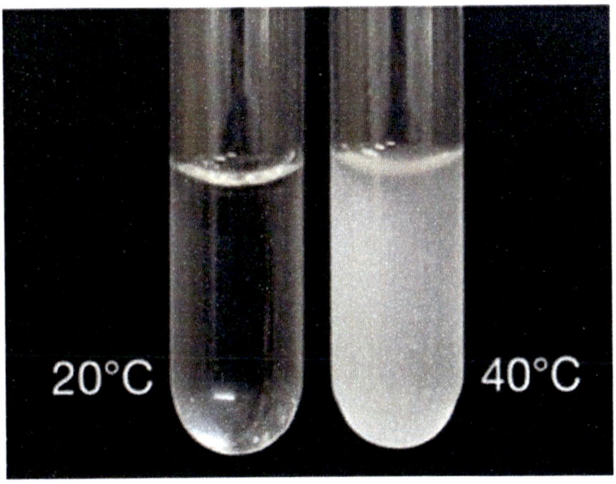


**Fig. 14.1** The image of the D-PNIPAM solution before and above LCST [14]

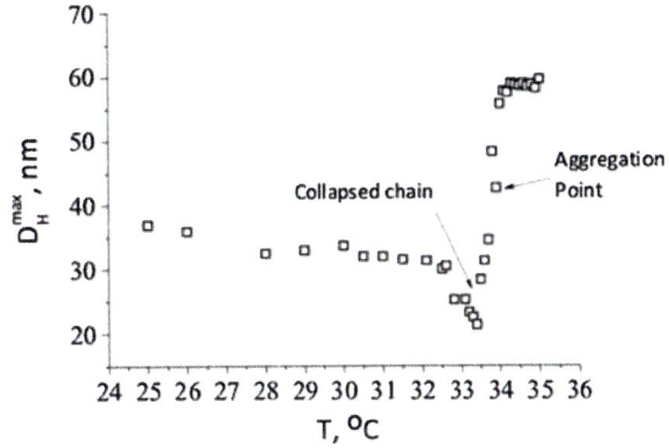


**Fig. 14.2** Dependence of hydrodynamic diameter value versus temperature for D-PNIPAM solution

## 14.3 Nanosystems for PDT

***Reagents Used for Nanosystems Synthesis.*** Photosensitizer Chlorin e6 (Ce6) was obtained from Santa Cruz Biotechnology (USA); $NaBH_4$, $HAuCl_4$ were purchase from Sigma-Aldrich (USA). Hank's balanced salt solution was purchased from Sigma-Aldrich (USA). Hank's solution was phenol red free and consisted of two components: glucose (1.0 g/L) and $NaHCO_3$ (0.35 g/L); Dimethyl sulfoxide (DMSO) was obtained from Serva (Germany).

***Polymer/AuNPs.*** Tetrachloroauric acid (Au precursor) and $NaBH_4$ (reductant) have been used for in situ AuNPs synthesis into polymer solution. Reduction of Au ions was performed in aqueous solutions of the D-PNIPAM as it was reported in [14–16]. 0.1 M $HAuCl_4$ aqueous solution was added to 1 ml of polymer solution (C = 1 g/L) and stirred during 20 min at t = 25 °C. Then, 0,1 M of $NaBH_4$ solution was added drop by drop at stirring. The obtained Au sols were stored in cold dark. The nanosystems were prepared in polymer solution below the concentration of crossover (Guiner regime) for D-PNIPAM.

***D-PNIPAM/AuNPs/Ce6 in Water and Hank's Solution.*** A stock solution of Ce6 (C = 0.182 mg/mL) was prepared in DMSO. Then, 0.55 mL of this solution was added to 0.30 mL of distilled water. This mixture was added to 1.15 mL of D-PNIPAM/AuNPs sol under constant stirring. Then, 8 mL of distilled water or Hank's buffer solution was added to 2 mL of obtained three-component D-PNIPAM/AuNPs/Ce6 system. The reaction mixture was stirred. The prepared three-component D-PNIPAM/AuNPs/Ce6 was centrifuged. The absorption spectra of the supernatant showed the absence of Ce6 in solution, thus indicating its total incorporation into polymer nanocarrier.

## 14.4 Methods Used for Nanosystems Study

***UV-Vis Spectroscopy.*** The UV absorption spectra of D-PNIPAM copolymer, D-PNIPAM/AuNPs and D-PNIPAM/AuNPs/Ce6 in water and Hank's solution were obtained using a Varian Cary 50 scan UV spectrophotometer (Palo Alto, CA, USA) in the range of 200–800 nm in the temperature region of 25–40 °C with step 1 °C. The studied systems were kept for 2 min at each temperature before measurement.

***Transmission Electron Microscopy (TEM).*** For the sample preparation 400 mesh Cu grids with plain carbon film (Elmo, Cordouan Technologies Bordeaux France). A 5 μl drop was deposited and let adsorbed for 1 min, then the excess of solution was removed with a piece of filter paper. TEM measurements were carried on two TEMs, Tecnai G2 or CM12 (FEI, Eindhoven Netherlands) and the images were acquired with a ssCCD Eagle camera on the Tecnai and a Megaview SIS Camera on the CM12.

***Dynamic Light Scattering (DLS).*** DLS measurements were carried out using Zetasizer Nano ZS90 (Malvern Instruments Ltd., UK). The apparatus contains a 4 mW He-Ne laser with a wavelength 632.8 nm and the scattered light is detected at an angle 173° (back scattering). For accurate transition study, correlogramms of nanosystems were collected in the temperature of physiological temperatures, namely at 25, 37 and 40 °C. Each temperature point was held for 5 min before measurements to equilibrate the sample. At least 10 correlation curves for each temperature points were treated by CONTIN algorithm [24] that is known to be reliable for complicate systems to get hydrodynamic diameter ($D_h$) distributions.

## 14.5 *In Vitro* Experiments with Malignant Cells

Cells of the MT-4 line (human T-cell leukemia) were used from the cell culture bank of the R.E. Kavetsky Institute of Experimental Pathology, Oncology and Radiobiology,

NAS of Ukraine. The cells were maintained in RPMI-1640 medium (Sigma-Aldrich, USA) containing 10% fetal bovine serum at a temperature of 37 °C in an incubator with 95% humidity and 5% $CO_2$ in air. For PDT cell suspensions in the logarithmic growth phase ($10^6$ cells/mL) were prepared in Hank's balanced salt solution. After 1.5 h of incubation of the cells at 37 °C with Ce6 e or with its nanocomposite, the samples were washed twice with a 10× volume of Hank's solution and exposed to a red laser (wavelength 658 nm, specific power—1.1 mW/cm$^2$, doses—1 J/cm$^2$). After irradiation, the cells were transferred to the culture medium and incubated at 37 °C for 18 h to complete the photodynamically induced apoptosis process. The cell viability was then determined by trypan blue dye exclusion test. Each experiment was repeated at least 5 times.

A semiconductor laser (PMNP Photonik Plus, Cherkasy) with a wavelength of 658 nm, which coincides with one of the Ce6 absorption maxima, was used as a light source for photodynamic damage to cells.

## 14.6 Results and Discussion

The typical TEM image of D-g-PNIPAM/AuNPs nanosystem obtained at temperature of 25 °C is shown in the Fig. 14.3. It is seen that AuNPs have size of $d = (8 \pm 3)$ nm and are spherical in shape. AuNPs were synthesized in dilute

**Fig. 14.3** TEM image of D-PNIPAM/AuNPs hybrid nanosystem

aqueous solution of D-PNIPAM (polymer concentration was below the concentration of crossover). Thus, the distance between the neighboring macromolecules is larger than the macromolecule size by several times.

The influence of AuNPs on the behavior of D-PNIPAM in water solution was studied by DLS [17]. It was shown that in the temperature region 21 to 32.5°C the objects which scatter light in the nanosystem D-PNIPAM/AuNPs decreases in size from 32 to 26 nm; then at 32.5–33.5 °C the sharp decrease in size from 26 to 14 nm was registered. Further, in the temperature range of 33.5–34.5 °C the sharp increase of the size of the scattering objects from 14 to 86 nm occurs that testifies of the aggregation process in the nanosystem D-PNIPAM. It should be noted that the size of macromolecule D-PNIPAM with incorporated AuNPs at the temperatures which were lower than LCST point was equal to 37 nm. It was larger than the size of D-PNIPAM without AuNPs (32 nm). Also, it was found that the size of individual AuNPs does not change during conformation transition of polymer that indicates the absence their direct aggregation [18]. Also, it was registered that the change in the size of the D-PNIPAM/AuNP nanosystems during the conformational transition is significantly less than for linear PNIPAM. That is result of the presence of AuNPs, which interact with the hydrophilic groups of the polymer and the presence of AuNPs into polymer macromolecule reduces a number of degrees of freedom for the polymer chains [18]. However, all our recent studies were performed in water solution. For biomedical application usually used the Hank's balance solution. That is why it is important to investigate the behavior of nanosystems for PDT in Hank's solution in comparison with water solutions.

Figure 14.4(a) demonstrates the UV spectra of the D-PNIPAM copolymer solutions prepared in water. The optical absorption spectra of PNIPAM exhibits maximum absorption ($\lambda_{max}$) at 228 nm corresponding to π and n electron transition localized on C = O group of PNIPAM [25].

The small peak at 250–275 nm obviously corresponds to some internal molecular structures formed by C = O groups caused by hydrogen bonding. Upon heating, absorption band intensity at 250–275 nm slight increases. It testifies the fact of some

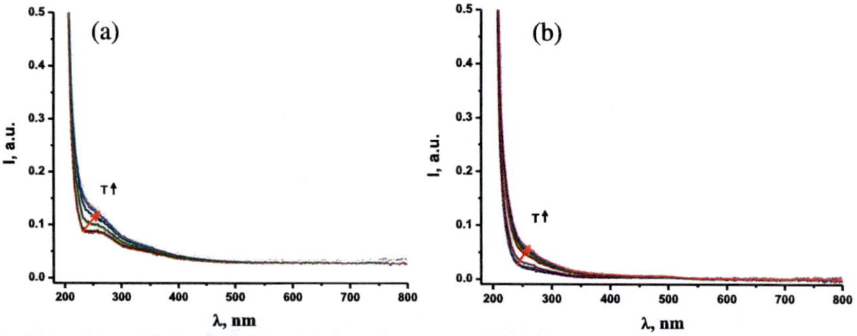


**Fig. 14.4** Absorption spectra of **a** D-PNIPAM water solution and of **b** of D-PNIPAM copolymer in Hank's solution within temperature range of 25–40 °C

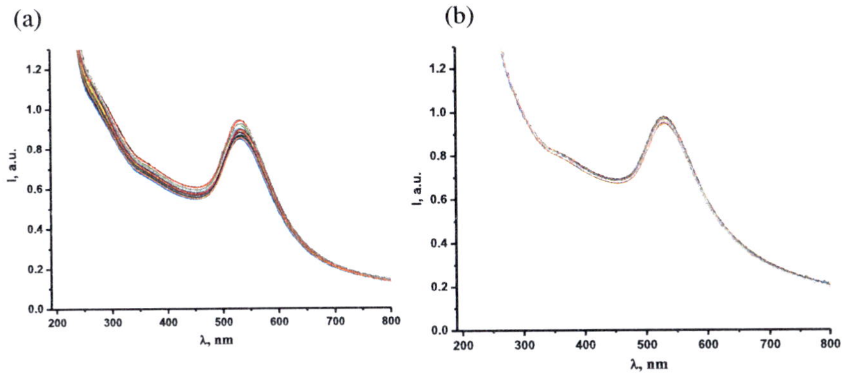


**Fig. 14.5** Absorption spectra of **a** Au sol into polymer water solution and of **b** Au sol into Hank's solution within temperature range of 25–40 °C

changes in studied system in the region of LCST. It is known, the macromolecule PNIPAM becomes partially hydrophobic after LCST, that leads to macrocoil conformational transition. The similar changes were observed for D-PNIPAM in Hank's balanced salt solution (Fig. 14.4(b)).

However, these changes are less pronounced. In is not surprising as Hank's solution consists of addition of salts which can affect the formation if internal hydrogen bond in system. Absorption spectra of Au sols synthesized in situ in the D-PNIPAM, in water and with addition of Hanks' solution are shown in Fig. 14.5(a) and (b). In both systems, the surface plasmon resonance (SPR) peak of Au nanoparticles at 530 nm is registered. This peak does not shift when heated. However, a slight change in the intensity of the SPR band is observed. That may also be caused by the conformational changes of the macromolecule within the LCST region leading to the change in the distance between AuNPs. In the Hank's solution, the change in the intensity of the SPR under heating is less pronounced than in the aqueous solution.

It is known that the Ce6 absorption spectrum has two characteristic bands: in the region of 470–670 nm (first electronic transition), in the region 350–440 nm (second electronic transition), the so-called Sore band [26].

As can be seen, for the nanosystem D-PNIPAM/AuNPs/Ce6 (Fig. 14.6), two components appear in the absorption spectrum: a photosensitizer (Sore band at 400 nm and one Q-band 671 nm) and gold nanoparticles (530 nm). As it is noted above, the polymer matrix does not absorb in this wavelength range.

The nanocomposite D-PNIPAM/AuNPs/Ce6 in Hank's solution was studied by DLS to estimate the size of nanoobject in nanosystems which can be used for in vitro PDT examination.

The experiment was performed at 25 and 37 °C, as the nanocomposites are preparing at room temperature, but biological testing is carried out at 37 °C. We have reported [16–18], that for D-PNIPAM/AuNPs in water solution few scattering objects are registered: the AuNPs of 10 nm in size, D-PNIPAM macromolecules

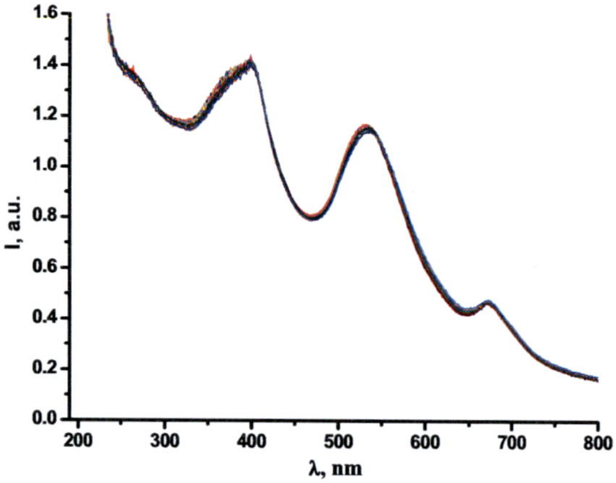


**Fig. 14.6** Absorption spectra of nanosystems D-PNIPAM/AuNPs/Ce6 in Hank's solution within temperature range of 25–40 °C

with incorporated AuNPs of 40 nm in size, and some aggregates of macromolecules. Size characteristics of D-PNIPAM/AuNPs/Ce6 in Hank's solution at 25 and 37 °C are shown in the Fig. 14.7.

It is seen that AuNPs of 10 nm in size in registered, and bimodal distribution of scattering object of 100 and 300 nm are present in nanosytem at 25 °C. Increasing temperature to 37 °C caused the decreasing of size of aggregates to 100 nm. Moreover, the individual macromolecules of 20–30 nm in size is registered.

Thus, it has been proved that no drastic changes accompanied by aggregation process in the nanosystems in Hank's saline solution. Nanosystem remains stable with heating to physiological temperature. Thus, this ternary nanosystems were recommended to test for PDT activity.

The in vitro examination the photodynamic activity of the D-PNIPAM/AuNPs/Ce6 nanosystem on the culture of MT-4 human malignant lymphocytes was performed (Fig. 14.8) in comparison to Ce14. In vitro study of toxicity of hybrid

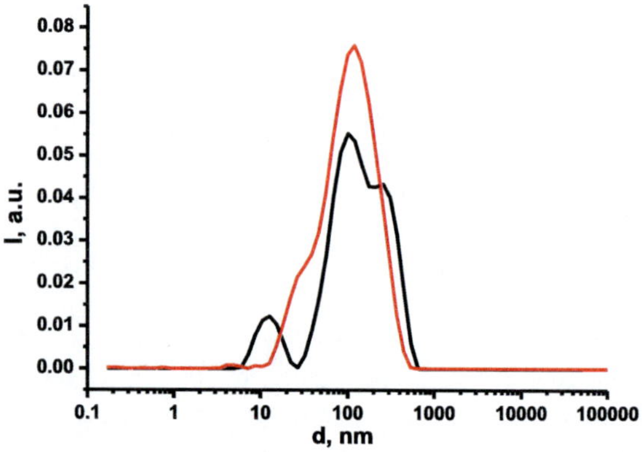


**Fig. 14.7** Hydrodynamic diameter distribution for D-PNIPAM/AuNPs/Ce6 nanocomposites in Hank's solution at 25 (black) and 37 (red) °C

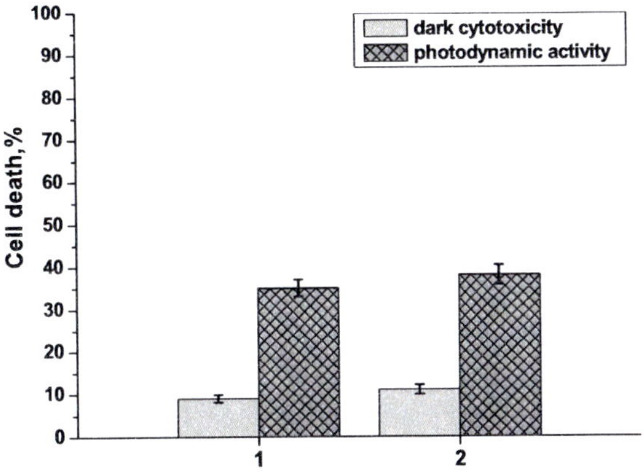


**Fig. 14.8** Dark toxicity and PDT efficacy: 1—Ce6; 2—D-PNIPAM/AuNPs/Ce6 nanocomposite

gold-polymer composites was reported in [27] and proved low toxicity of examined compounds even in high doses.

Figure 14.8 demonstrates the D-PNIPAM/AuNPs/Ce6 does not show cytotoxicity in the absence of red irradiation. At the same time, the mortality of lymphocytes pre-incubated with nanocomposite after laser irradiation was 40%. This result is comparable with efficacy of Ce14. However, the nanocomposite can prevent the aggregation of photosensitizer in vivo experiments as well as can provide target delivery of drug to tumor cell. Moreover, for nanosystem polymer/AuNPs under illumination with light with $\lambda = 650$ nm the local heating in the small volume near the laser beam was registered [28]. It means, that nanosystems with AuNPS during PDT can provide also hyperthermia effect.

## 14.7 Conclusions

Nowadays, there is a great interest to the synthesis and application of hybrid nanosystems containing gold nanoparticles incorporated into various polymer matrices. Such nanosystems can be used for modern cancer therapy. The absorption spectroscopy and DLS showed that the comparison of D-PNIPAM/AuNPs/Ce6 nanosystem in water and Hank's solution didn't reveale the increasing of aggregation process in the region of conformation transition of polymer nanocarrier. Thus, the nanosystem remained stable in the region of physiological temperature. The in vitro study demonstrated that the photodynamic activity of the D-PNIPAM/AuNPs/Ce6 nanosystem is approximately equal in efficiency to Ce14. However, the nanocomposite can prevent the aggregation of photosensitizer in vivo experiments and can provide target delivery of drug to tumor cells. Thus, this ternary nanosystems can be recommended for further in vivo experiments.

**Acknowledgements** The authors would like to thank I. Shton and E. Shishko from R.E. Kavetsky Institute of Experimental Pathology, Oncology and Radiobiology for in vitro experiment. The publication is based on the research provided partially by the grant support of the Ministry of the Education and Science of Ukraine, the Belarusian Republican Foundation for basic research—joint Ukrainian-Belarusian research and development projects "Design and physico-chemical properties of novel multicomponent nanosystems for the treatment and diagnostics of solid tumors" (2019–2020), and by National Research Foundation of Ukraine. Project 2020.02/0022 "Plasmon hybrid nanosystems "metal-polymer-fluorophore" with enhanced optical response for photonics and biomedical applications".

## References

1. Paszko E, Ehrhardt C, Senge MO et al (2011) Nanodrug applications in photodynamic therapy. Photodiagn Photodyn Ther 8:14–29. https://doi.org/10.1016/j.pdpdt.2010.12.001
2. Dolmans D, Fukumura D, Jain R (2003) Photodynamic therapy for cancer. Nat Rev Cancer 3:380–387. https://doi.org/10.1038/nrc1071
3. Wang H, Xu Y, Shi J et al (2015) Photodynamic therapy in the treatment of basal cell carcinoma: a systematic review and meta-analysis. Photodermatol Photoimmunol Photomed 31:44–53. https://doi.org/10.1111/phpp.12148
4. Kwiatkowski S, Knap B, Przystupski D et al (2018) Photodynamic therapy – mechanisms, photosensitizers and combinations. Biomed Pharmacother 106:1098–1107. https://doi.org/10.1016/j.biopha.2018.07.049
5. Feldman D (2016) Polymer nanocomposites in medicine. J Macromol Sci Part A Pure Appl Chem 53(1):55–62. https://doi.org/10.1080/10601325.20114.1110459
6. Asiri IAM, Mohammad A (eds) (2018) Applications of nanocomposite materials in drug delivery. In: Woodhead Punlishing Series in Biomaterils. https://doi.org/10.1016/C2016-0-05075-1
7. Wei M, Gao Y, Li X (2017) Stimuli-responsive polymers and their applications. Polym Chem 8:127–143.https://doi.org/10.1039/C6PY01585A
8. Kutsevol N, Harahuts L, Nadtoka O et al (2018) Hybrid nanocomposites synthesized into stimuli responsible polymer matrices: synthesis and application prospects, chap 12. In: Fesenko O, Yatsenko L (eds) Nanophotonics, nanooptics, nanobiotechnology, and their applications. Springer proceedings in physics, vol 222, pp 167–185. https://doi.org/10.1007/978-3-030-17759-1_12
9. Jain K, Vedarajan R, Watanabe M, Ishikiriyama M, Matsumi N (2015) Tunable LCST behavior of poly(*N*-isopropylacrylamide/ionic liquid) copolymers. Polym Chem 6(38):6819–6825. https://doi.org/10.1039/C5PY00998G
10. Lanzalaco S, Armelin E 2017 Poly(N-isopropylacrylamide) and copolymers: a review on recent progresses in biomedical applications. Gels 3:314.https://doi.org/10.3390/gels3040036
11. Najafi M, Hebels E, Hennink WE et al (2020) Poly(N-isopropylacrylamide): physicochemical properties and biomedical applications. In: Khutoryanskiy VV, Georgiou TK (eds) Temperature - responsive polymers. Wiley, pp 1–34. https://doi.org/10.1002/9781119157830.ch1
12. Guo Z, Li S, Wang C et al (2016) Biocompatibility and cellular uptake mechanisms of poly(N-isopropylacrylamide) in different cells. J Bioactive Compatible pol 32(1):17–31. https://doi.org/10.1177/0883911516648969
13. Narang P, Venkatesu P (2018) Unravelling the role of polyols with increasing carbon chain length and OH groups on the phase transition behavior of PNIPAM. New J Chem 42:13708–13717. https://doi.org/10.1039/C8NJ02510J
14. Chumachenko V, Kutsevol N, Harahuts Y et al (2017) Star-like dextran-graft-PNIPAM copolymers. Effect of internal molecular structure on the phase transition. J Mol Liq 235:77–82. https://doi.org/10.1016/j.molliq.2017.02.098

15. Matvienko T, Sokolova V, Prylutska S et al (2019) In vitro study of the anticancer activity of various Doxorubicin-containing dispersions. Bioimpacts 9(1):57–63. https://doi.org/10.15171/bi.2019.07
16. Yeshchenko O, Naumenko A, Kutsevol N et al (2018) Laser driven structural transformation in dextran-graft-PNIPAM copolymer/Au nanoparticles hybrid nanosystem: the role of plasmon heating and attractive optical forces. RSC Adv 8:38400–38409. https://doi.org/10.1039/C8RA07768A
17. Yeshchenko OA, Naumenko AP, Kutsevol NV et al (2018) Anomalous inverse hysteresis of phase transition in thermosensitive dextran-graft-PNIPAM copolymer/Au nanoparticles hybrid nanosystem. J Phys Chem C 122(14):8003–8010. https://doi.org/10.1021/acs.jpcc.8b01111
18. Kutsevol N, Glamazda A, Chumachenko V et al (2018) Behavior of hybrid thermosensitive nanosystem dextran-graft-PNIPAM/gold nanoparticles: characterization within LCTS. J Nanopart Res 20:2314. https://doi.org/10.1007/s11051-018-4331-2
19. Kim HS, Lee DY (2018) Near-infrared-responsive cancer photothermal and photodynamic therapy using gold nanoparticles. Polymers 10(9):961. https://doi.org/10.3390/polym10090961
20. Jain S, Hirst DG, O'Sullivan JM (2012) Gold nanoparticles as novel agents for cancer therapy. Br J Radiol 85(1010):101–113. https://doi.org/10.1259/bjr/59448833
21. Corbierre MK, Cameron NS, Sutton M et al (2005) Nanocomposites: dispersion of nanoparticles as a function of capping agent molecular weight and grafting density. Langmuir 21(13):6063–6072. https://doi.org/10.1021/la047193e
22. Chumachenko VA, Shton IO, Shishko ED et al (2016) Branched copolymers dextran-graft-polyacrylamide as nanocarriers for delivery of gold nanoparticles and photosensitizers to tumor cells. In: Fesenko O, Yatsenko L (eds) Nanophysics, nanophotonics, surface studies, and applications. Springer proceedings in physics, vol 183, pp 379–390. https://doi.org/10.1007/978-3-319-30737-4_32
23. Kutsevol N, Naumenko A, Harahuts Y et al (2019) New hybrid composites for photodynamic therapy: synthesis, characterization and biological study. Appl Nanosci 9:881. https://doi.org/10.1007/s13204-018-0768-y
24. Scotti A, Liu W, Hyatt JS et al (2015) The CONTIN algorithm and its application to determine the size distribution of microgel suspensions. J Chem Phys 142:234905. https://doi.org/10.1063/1.4921686
25. Makharza S, Auisa J, Sharkh SA et al (2010) Structural and thermal analysis of copperdoped Poly(N-isopropylacrylamide) films. Int J Pol Anal Charact 15:1–12.https://doi.org/10.1080/10236661003747031
26. Gattuso H, Monariab A, Marazzi M (2017) Photophysics of chlorin e6: from one- and two photon absorption to fluorescence and phosphorescence. RSC Adv 7:10992. https://doi.org/10.1039/c6ra28616j
27. Kutsevol N, Harahuts Y, Shton I et al (2018) In vitro study of toxicity of hybrid gold-polymer composites. Mol Cryst Liq Cryst 671(1):1–8. https://doi.org/10.1080/154214014.2018.1542078
28. Harahuts Y, Pavlov V, Mokrinskaya E (2018) Anomalous change of refractive index for Au sols under laser illumination, chap 3. In: Fesenko O, Yatsenko L (eds) Nanophotonics, nanooptics, nanobiotechnology, and their applications. Springer proceedings in physics, vol 222, pp 53–72. https://doi.org/10.1007/978-3-030-17755-3_3

# Chapter 15
# Biomedical Applications of Laponite®-Based Nanomaterials and Formulations

**Olena Samoylenko, Olena Korotych, Maryna Manilo, Yurii Samchenko, Volodymyr Shlyakhovenko, and Nikolai Lebovka**

**Abstract** Laponite® (Lap)-based materials and their biomedical applications are critically discussed. The properties of Lap grades and their modified species are presented. The current state of biomedical applications, including of Lap-based drug delivery systems for different pharmaceutical molecules is reviewed. Special attention is paid to medical hydrogels with incorporated Lap, and cytotoxicity and antimicrobial activity assessments for different Lap-based systems. The detailed reference data of Lap-based materials include Tables with decryptions the grades of Laps, popular silanation agents for edge modification, Lap-based therapeutic platforms, examples of biomedical applications hydrogels with incorporated Lap, hydrogel inks for 3D printing, popular assays for toxicity assessment, effects of Lap on cellular metabolism, and examples of bioactivity assessments for Lap-based materials.

| List of Abbreviations and Chemical Substances | |
|---|---|
| **Abbreviations** | |
| AFM | Atomic force microscopy |
| AMF | Alternating magnetic field |
| DSC | Differential scanning calorimetry |
| DLS | Dynamic light scattering |

(continued)

O. Samoylenko (✉) · V. Shlyakhovenko
Kavetsky Institute of Experimental Pathology, Oncology and Radiobiology, National Academy of Sciences of Ukraine, 45, Vasylkivska Street, Kyiv 03022, Ukraine
e-mail: a-samoilenko@ukr.net

O. Korotych (✉)
Chemical Engineering Department, The University of Florida, 1030 Center Dr, Gainesville, FL 32611, USA
e-mail: o.i.korotych@gmail.com

Biochemistry and Cellular and Molecular Biology Department, The University of Tennessee, 1311 Cumberland Avenue, Knoxville, TN 37916, USA

M. Manilo · Y. Samchenko · N. Lebovka
F. D. Ovcharenko Institute of Biocolloidal Chemistry, National Academy of Sciences of Ukraine, 42, Vernadsky Avenue, Kyiv 03142, Ukraine



L. Bulavin and N. Lebovka (eds.), *Soft Matter Systems for Biomedical Applications*, Springer Proceedings in Physics 266, https://doi.org/10.1007/978-3-030-80924-9_15

(continued)

| List of Abbreviations and Chemical Substances | |
|---|---|
| DTA | Differential thermal analysis |
| FTIR | Fourier-transform infrared |
| hBMSCs | Human bone marrow stromal cells |
| hMSC | Human marrow stem cells |
| LCST | Lower critical solution temperature |
| MIC | Minimum inhibitory concentration |
| MRI | Magnetic resonance imaging |
| SAXS | Small-angle x-ray scattering |
| SEM | Scanning electron microscopy |
| SMF | Steady magnetic field |
| TEM | Transmission electron microscopy |
| UCST | Upper critical solution temperature |
| UV | Ultraviolet |
| XRD | X-ray diffraction |
| wt% | Percentage weight concentration |
| **Chemical substances** | |
| (See also Table 15.2 for silane coupling agents and Table 15.3 for pharmaceutical molecules) | |
| AA | Acrylic acid, $C_3H_4O_2$ |
| AAm | Acrylamide, $C_3H_5NO$ |
| AMPS | 2-acrylamide-2-methylpropylsulphate, $C_7H_{13}NO_4S$ |
| CTAB | Hexadecyltrimethylammonium bromide, $C_{19}H_{42}N^+Br^-$ |
| DODAB | Dimethyldioctadecylammonium bromide, $C_{38}H_{80}BrN$ |
| DTA | Diatrizoic acid, $C_{11}H_9I_3N_2O_4$ |
| DTPA | Diethylenetriaminepentaacetic acid, $C_{14}H_{23}N_3O_{10}$ |
| Lap | Laponite® |
| LapA | Acid-activated Lap |
| LapM | Magnetically modified Lap |
| LapO | Organo-modified Lap |
| LapSi | Silanated Lap |
| NIPAAm | *N*-isopropylacrylamide, $(C_6H_{11}NO)_n$ |
| PAA | Poly(acrylic acid), $(C_3H_4O_2)_n$ |
| PAAm | Poly(acrylamide), $(C_3H_5NO)_n$ |
| PAH | Poly(allylamine) hydrochloride, $[CH_2CH(CH_2NH_2{\cdot}HCl)]_n$ |
| PAMA | Poly(2-(*N,N*-dimethylamino) ethyl methacrylate), $(C_8H_{15}NO_2)_n$ |
| Pc | Phthalocyanine, $C_{32}H_{18}N_8$ |
| PCL | Polycaprolactone, $(C_6H_{10}O_2)_n$ |
| PEG | Poly(ethylene glycol), $H(OCH_2CH_2)_nOH$ |

(continued)

(continued)

| List of Abbreviations and Chemical Substances | |
|---|---|
| PEO | Poly(ethylene oxide), $H(OCH_2CH_2)_nOH$ |
| PLA | Polylactic acid, or polylactide, $(C_3H_4O_2)_n$ |
| PMMA | Poly(methyl methacrylate), $(C_5O_2H_8)_n$ |
| polyNIPAAm | Poly(*N*-isopropylacrylamide), $C_4H_8O_2(C_6H_{11}NO)_n$ |
| PSS | Poly(sodium styrene sulfonate), $(C_8H_7NaO_3S)_n$ |
| PTZ | phenothiazine, $C_{12}H_9NS$ |
| PVA | Poly(vinyl alcohol), $(CH_2CH(OH))_n$ |
| PVP | Poly(*N*-vinylpyrrolidone), $(C_6H_9NO)_n$ |
| TEMED | *N,N,N′,N′*-tetramethylethylenediamine, $C_6H_{16}N_2$ |

## 15.1 Introduction

The Laponite® (Lap) was first synthesized in the 1960s for application in the paint industry as a material with excellent colloidal properties [1, 2]. Lap has a plate-like structure and belongs to the family of silicates. It is composed of salts of silicic acids with highly heterogeneous charge distribution and patchy interactions between platelets [3]. The Lap platelets in deionized water exist as 1 nm thick discs with diameters of 25 nm and their faces carry a net negative surface charge. Colloidal suspensions of the highly anisometric Lap can exhibit many unusual properties including interesting phase states, "house of cards" assembly, unique structural ordering, and dynamical arrest [4–6]. Pristine and modified Lap and Lap-based materials have already demonstrated excellent transport, chemical and biomedical properties [7–9]. Lap is widely used in many practical applications as the components of household items, agricultural and horticultural products, personal care and cosmetic products (toothpastes, skincare and sunscreen emulsions, cosmetics, shampoos, creams), and additives to the multifunctional composites, adsorbents, catalysts, glass, ceramics, enamels, barriers and polymer films, surface coatings, etc. [10]. Nowadays, these two-dimensional nanomaterials attract great attention for production of biomaterials for regenerative medicine, thickeners in dentifrices for hypersensitive teeth, biohydrogels, materials with antimicrobial potency towards various types of pathogens and materials for controlled antitumor drug delivery [11–20].

This chapter discusses grades of Lap, main properties of Lap-based materials, and their modified species, and current state of biomedical applications, including Lap-based drug delivery systems for different pharmaceutical molecules, medical hydrogels with incorporated Lap, cytotoxicity and antimicrobial activity assessments for different Lap systems.

## 15.2 Structure and Properties of Lap

Lap is a synthetic material with 2:1 layered structure similar to the natural trioctahedral clay mineral hectorite [21]. Lap synthesis was patented in June 1962 by Barbara Neumann [22, 23], and its production was started in May 1965 by former Laporte Industries Ltd., now BYK Additives, Inc. (Cheshire, UK). Nowadays, among producers/suppliers of Lap products, we can also refer to the companies Hemmelrath Lackfabrik (www.hemmelrath.de), Eckart Effect Pigments GmbH (www.eckart.net), Huizhi Jiangsu Fine Chemical Co., Ltd. (huizhichemical.lookchem.com), and Kremer Pigmente (www.kremer-pigmente.com).

In the process of hydrothermal synthesis of Lap, aqueous mixtures of LiF and/or LiCl, $Mg(OH)_2$, and $SiO_2$ are combined at specific molar ratio of Si:Mg:Li and specific reaction conditions (temperature, pH, etc.) [21, 24]. After the synthesis, the products are filtered, washed, dried, and milled to fine powders. The composition and crystallinity of Lap can be controlled during the hydrothermal process. The details on the synthesis and modification of synthetic hectorites have been recently reviewed [21]. According to the detailed structural studies [25], the term "Laponite" does not describe a specific mineral type of hectorite-like composition, but rather characterize a mixture of different minerals (depending on the synthetic conditions). Thus, the authors [25] suggested to use the term "synthetic Mg-Li phyllosilicate mixture" instead of Lap.

The produced Lap is free of crystalline silica and it has low content of transition metals and other impurities that might be found in natural clays. Moreover, compared to natural clays, Lap demonstrates high optical transparency, high dispersibility in water, and an excellent colloidal stability in suspensions [30]. Additional advantages of Lap include narrow size distribution and rather uniform shape of its individual platelets. As claimed by the manufacturer, the Lap platelets are ≈0.92 nm thick with diameter ranging from 20 to 30 nm [26, 27].

Figure 15.1 presents schematic drawing of the structure of Lap. It is composed of one Mg-O-Li central sheet sandwiched between two Si-O-Si tetrahedral sheets. The sheets are laterally continuous, but non-covalent bonds between them are weak, allowing water and other molecules to enter between layers causing their expansion. In the central sheet some divalent $Mg^{2+}$ ions are randomly substituted by monovalent ions of $Li^+$. An ideal chemical formula of Lap is

$$Na^{+\alpha}[Mg_{6-\alpha}Li_\alpha Si_8O_{20}(OH)_4], \tag{15.1}$$

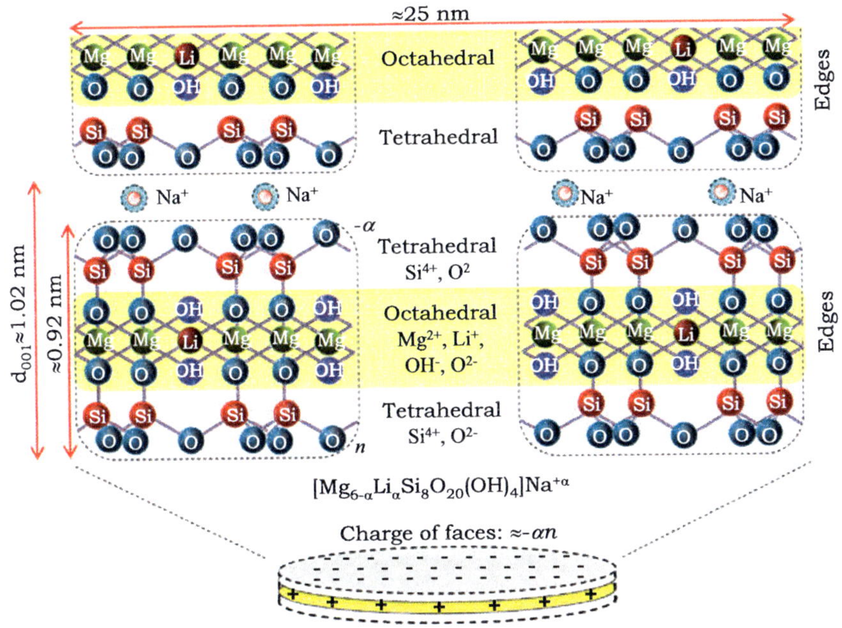


**Fig. 15.1** Schematic drawing of the structure of Lap particle with the formula of $Na^{+\alpha}[Mg_{6-\alpha}Li_{\alpha}Si_8O_{20}(OH)_4]$. The platelet has the thickness of ≈0.92 nm and diameter ranging from 20 to 30 nm [26, 27]. The distance between particles, $d_{001}$, depends upon the water contents and in dry state $d_{001} \approx 1.02$ nm [28]. Here $\alpha$ is the degree of the isomorphous substitution of $Li^+$ for $Mg^{2+}$, $n$ is the number of unit cells in the single Lap platelet, $-\alpha n$ is the total negative charge of the both faces. Typically, one Lap platelet includes approximately $n = 1100 \pm 100$ unit cells [29]

where $\alpha$ is the degree of the isomorphous substitution of $Li^+$ for $Mg^{2+}$.

This substitution results in appearance of the net negative charges on the basal faces that are compensated by cations in the interlayer space (e.g., the sodium ions, $Na^+$). In practice, some positions may be empty and in general case the empirical formula is

$$Na^{+\alpha}[Mg_{\beta}Li_{\gamma}Si_8O_{20}(OH)_4], \tag{15.2}$$

where $\alpha = 0.8$, $\beta = 5.4$ and $\gamma = 0.4$ [31] or $\alpha = 0.7$, $\beta = 5.5$ and $\gamma = 0.3$ [32]. For ideal structure without empty places $\beta + \gamma = 6$.

In the aqueous media, interlayer cations are released and Lap platelets have a strongly negatively charged faces. The edges of Lap platelets carry amphoteric hydroxyl groups such as Mg–OH (predominantly), Si–OH, and Li–OH. The edge charge depends on the acid–base behavior of these groups and changes from positive to negative at isoelectric point, associated with the hydrous oxides [33]. For Mg–OH groups an isoelectric point is localized around a pH of 12 [33, 34].

### 15.2.1 Grades of Lap

The different types of Lap products (grades) are commercially available [35]. Table 15.1 presents some popular grades of Lap. All the grades can be readily dispersed in water under intensive agitation. The grades can be classified according to their gel-forming or sol-forming (temporal and permanent) properties. Lap grades display different colloidal stability and rheological properties. The benefits/limitations of different Lap grades have been recently discussed [10, 14]. For example, the gel-forming grades $Lap_{RD}$ and $Lap_{XLG}$ produce thixotropic gels even at small concentration of Lap in water starting from concentration of $C \approx 2$ wt%. The $Lap_{RD}$ grade is the most popular and frequently used in different applications for rheology control in surface coatings, household products, and other industrial fields. The $Lap_{XLG}$ grade is used in personal care and cosmetic applications, particularly for preparation of emulsions, lotions and creams. The special purposes high-purity personal care grades $Lap_D$ and $Lap_{XL21}$ were developed for cosmetic and medical applications.

The sol-forming grades (with characters "S" in the grade names) include peptizing agents that delay the formation of gels. The produced sol-forming grades can be of temporary type (short-term stability) and permanent type (long-term stability).

**Table 15.1** Popular grades of Lap. For more information on the products refer to the technical data sheet presented at https://www.byk.com

| Grade | Description, the degree of the isomorphous substitution of $Li^+$ for $Mg^{2+}$, $\alpha$, chemical composition, *CC* (dry basis), and cation-exchange capacity, *CEC* (measured in moles of electric charge) |
|---|---|
| | **Gel-forming** |
| $Lap_{RD}$<br>$Lap_{XLG}$ | The RD (Readily dispersible) and XLG grades are chemically identical. The chemical composition of the $Lap_{RD}$ grade can include insignificant amount of Fe, Al, K, and Ca as chemicals impurities [36]. A personal care grade $Lap_{XLG}$ is a purer grade with certified low heavy metal and low microbiological content<br>$\alpha = 0.66$, $CC = [SiO_2$: 59.5%, MgO: 215.5%, $Li_2O$: 0.8%, $Na_2O$: 2.8%],<br>$CEC \approx 0.79$ mol/kg at pH 8.9 ($Lap_{RD}$) [36] |
| $Lap_D$<br>$Lap_{XL21}$ | High purity personal care grades for cosmetic and medical purposes. The $Lap_D$ grade is optimized for rapid dispersion in sorbitol solution, and $Lap_{XL21}$ (fluorohectorite) grade is a sodium magnesium fluorosilicate for application in skincare formulations |
| | **Sol-forming: temporal** |
| $Lap_{RDS}$<br>$Lap_{XLS}$ | The grades incorporating an inorganic polyphosphate peptiser (tetrasodium pyrophosphate) $Na_4P_2O_{15}$<br>$\alpha = 0.7$, $CC = [SiO_2$: 54.5%, MgO: 26.0%, $Li_2O$: 0.8%, $Na_2O$: 5.6%, $P_2O_5$: 4.4%],<br>$CEC = 1.04$ mol/kg. The grades are optimized for enhancing the dispersion of platelets in highly polar solvents including aqueous solutions |
| | **Sol-forming: permanent** |
| $Lap_{S482}$ | The grade incorporating a patented dispersing agent (1-hydroxyethylidene) bisphosphonate, $C_2H_4Na_4O_7P_2$. At concentration of 25% it remains free flowing for up to 1 year<br>$\alpha = 0.7$, $CC = [SiO_2$: 54.5%, MgO: 26.0%, $Li_2O$: 0.8%] |

The temporal sol-forming gels $Lap_{RDS}$ and $Lap_{XLS}$ grades include inorganic peptiser. Dissociation of peptiser produces tetravalent anions ($P_2O_7^{4-}$) that neutralize the rim positive charges and attenuate attraction between the Lap edges and faces [29, 34]. The peptizing agent can prevent gel formation. For these grades, at concentration of 10 wt%, highly-stable and low-viscous suspensions are formed in 24 h. Upon electrolyte addition, the highly thixotropic gels are formed rapidly. The grades are recommended for use in formulations with low contents of free water. The $Lap_{RDS}$ is the general-purpose grade and the $Lap_{XLS}$ is the similar high purity-personal care grade certified low heavy metal and low microbiological content.

The permanent sol-forming grade $Lap_{S482}$ are based on the special dispersing agent tetrasodium (1-hydroxyethylidene) bisphosphonate ($C_2H_4Na_4O_7P_2$) that act by the same mechanism as sodium pyrophosphate in $Lap_{RDS}$ and $Lap_{XLS}$ grades. This grade allows forming highly stable and colorless colloidal suspensions (sols) that can be used for preparation of electrically conductive, antistatic and barrier films.

Note that all sol-forming grades can be rapidly transformed into highly thixotropic gels after addition of small quantities of electrolyte.

### *15.2.2 Properties of Lap*

The most investigated variety in different studies is a general-purpose grade $Lap_{RD}$. In a dry state, the bulk density of $Lap_{RD}$ powder is 1 g/cm$^3$, while the particle density is 2.53 g/cm$^3$ [35]. The specific surface area, $S$, determined in the dry state by nitrogen adsorption was 330 m$^2$/g and the total value of $S$ of the completely exfoliated $Lap_{RD}$ particles was estimated as $S$ = 930–970 m$^2$/g [26]. In powder state, Lap platelets are grouped in parallel crystallite microdomains. Scanning electron micrographs of $Lap_{RD}$ powders revealed presence of large micrometer agglomerates (with size up to 500 μm) with irregular size and shape [37].

The Lap particles are highly hygroscopic and they can easily adsorb atmospheric moisture. The distance between Lap particles, $d_{001}$, in powders stored in atmospheres at different relative humidity, $RH$, was studied using X-ray diffraction (XRD) method [38].

Lap powders show rather broad XRD patterns due to the small size and low crystallinity of the Lap platelets (see inset in Fig. 15.2). The reflection at ≈5–12° was attributed to (001) crystal plane or the basal interlayer spacing filled with water ($d_{001}$) and the peak at 60.9° ($d_{060}$ = 0.152 nm) confirms the trioctahedral character of the Lap [39]. For completely dehydrated Lap the value $d_{001}$ equal to 1.02 ± 0.02 nm was estimated.

The size and morphology of $Lap_{RD}$ particles were studied using sedimentation and light scattering techniques [26], atomic force microscopy (AFM) [27], and transmission electron microscopy (TEM) [40]. The AFM study revealed an elliptical shape of $Lap_{RD}$ particle morphology with mean lateral dimensions of 24.0 ± 6.9 nm and 16.8 ± 4.9 nm [27]. The photon correlation spectroscopy was used to estimate a particle

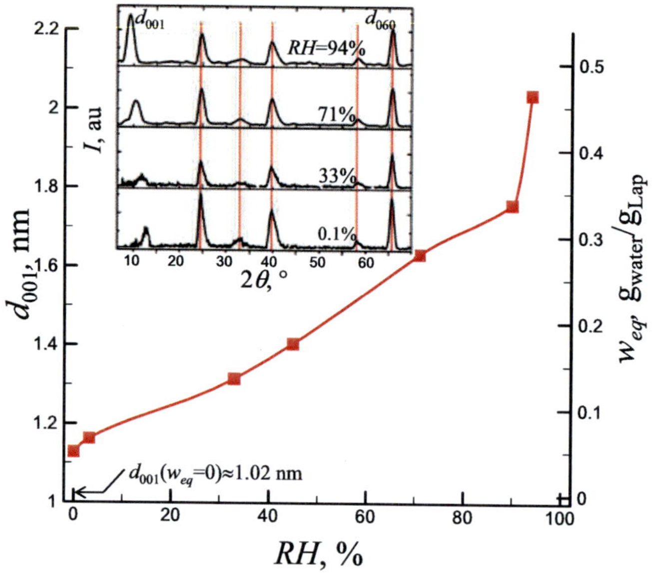


**Fig. 15.2** Interlayer spacing, $d_{001}$, and equilibrium moisture content, $w_{eq}$, versus relative humidity, *RH* (adapted from the data presented in [38]). Inset shows X-ray diffraction patterns of $Lap_{RD}$ conditioned under different *RH* (reprinted with permission from [38])

diameter, which was approximately 18.4 nm and an aspect ratio which was in the range of 20–21 [26].

TEM images of $Lap_{RD}$ evidenced of plate-like particle morphology, 20–30 nm in diameter [40]. The process of exfoliation Lap in water includes the swelling and dispersion of Lap tactoids with formation of platelet aggregates [42]. In aqueous suspensions the size distributions depend on pH and concentration of particles, *C*, and time of exfoliation process [43–45]. The exfoliation is a long-term process and even in diluted suspensions (0.1–0.2 wt%) it takes more than 5 h [46]. Figure 15.3 presents an example of the particle size distribution, measured by dynamic light scattering (DLS), for 1 wt% Lap (the pH is about 9.8) aqueous suspension [41]. The results evidenced the presence of individual platelets and small aggregates with sizes ranging between 5 and 100 nm. At relatively high concentration (above 1.5 wt%) a substantial fraction of the platelet aggregates was observed [47].

The Lap, dispersed in water, acts as a base, and the resulting pH and electrical conductivity, $\sigma$, increase with the concentration of Lap, *C* (Fig. 15.4a). The electrokinetic potential, $\zeta$, (absolute value) increases with pH and depends on *C* (Fig. 15.4b) [48, 49]. In alkaline media the estimated values of $\zeta$-potential were $\approx-52$ mV [50, 51], $\approx-40$–45 mV [52], $\approx-58$ mV [53], $\approx-45$ mV [54].

The estimated charges of one Lap platelet in deionized water (0.05–0.2 wt%) are $\approx-500e$ and $\approx+50e$ (here, *e* is elementary charge) for faces and edges, respectively

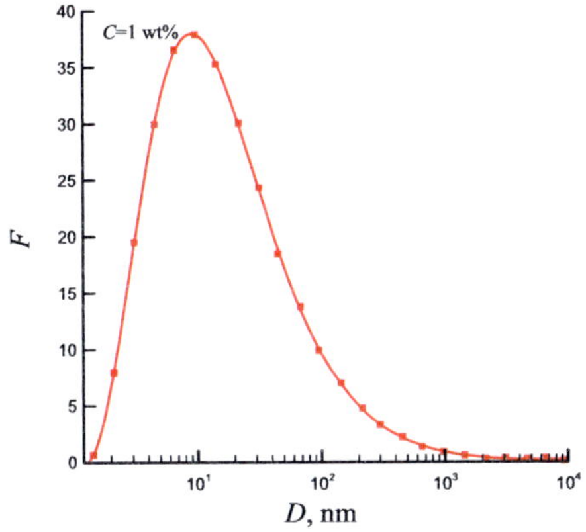


**Fig. 15.3** Particle size distribution for 1 wt% Lap (the pH is about 9.8) aqueous suspension (adapted from the data presented in [40])

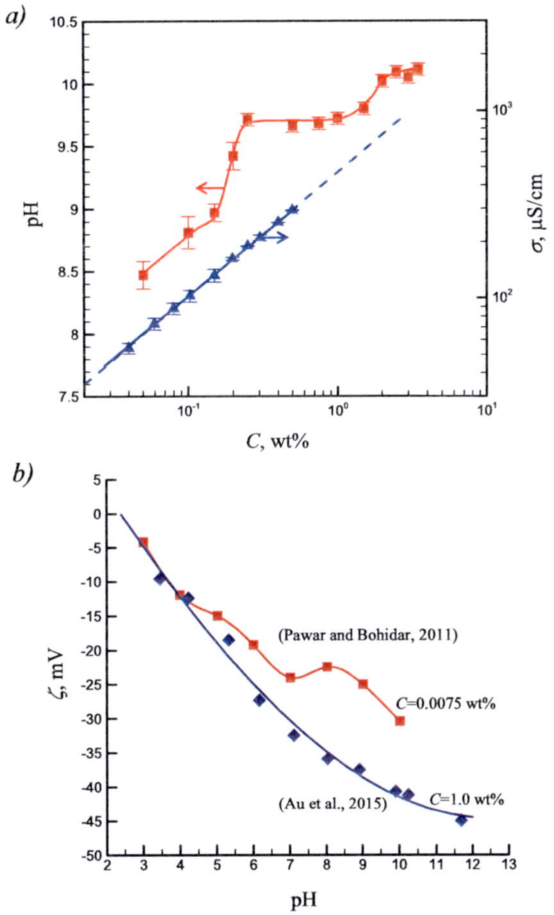


**Fig. 15.4** **a** pH and electrical conductivity, $\sigma$, versus $Lap_{RD}$ concentration, $C$, in water (adapted from the data presented in [55]). **b** $\zeta$-potential versus pH at two different concentrations of Lap (adapted from the data presented in [48] (0.0075 wt%) and in [49] (1 wt%))

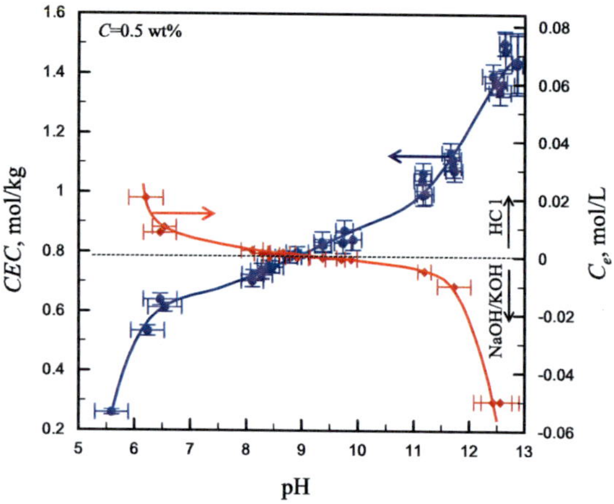


**Fig. 15.5** Cation-exchange capacity, *CEC*, of $Lap_{RD}$ as a function of pH obtained using the copper triethylenetetramine (Cu-trien) method (left) and pH of the suspensions as a function of the concentration of added HCl or NaOH/KOH, $C_e$, (right axis). Concentration of $Lap_{RD}$ was 0.5 wt%. Without pH adjustment the *CEC* value was estimated to be equal to $0.79 \pm 0.01$ mol/kg (adapted from the data presented in [36])

[56]. Negative charge on the surface of Lap can be characterized by the cation-exchange capacity (*CEC*) defined as the amount of positive charge that can be exchanged per mass of particles. Typically, *CEC* is measured in moles of electric charge (e.g., moles of $Na^+$ cations) per kilogram of Lap (mol/kg). The reported *CEC* values for $Lap_{RD}$ varied between 0.47 and 0.92 mol/kg depending on the applied method and pH of medium [31, 57, 58]. Recently, the *CEC* values of $Lap_{RD}$ were evaluated at different pH using the Cu-trien method [36]. The concentration of $Lap_{RD}$ was fixed at 1 wt%. For the usual Cu-trien procedure, without any addition of acid or base for pH adjustment, the *CEC* was measured to be equal to $0.79 \pm 0.01$ mol/kg. The value of pH was adjusted by addition of HCl or NaOH/KOH. The increase in pH from 6 to 13 resulted in increase the *CEC* value from 0.25 to 1.5 mol/kg (Fig. 15.5). The significant *CEC* increase at pH > 10 was explained by the deprotonation of the silanol edge sites.

In aqueous solutions the $Lap_{RD}$ is chemically unstable due to dissolution of magnesium (below pH 9) or silica (above pH 10) [31, 59]. For example, in acidic conditions at low pH, the following scheme was proposed for dissolution of $Lap_{RD}$ and release of $Mg^{2+}$

$$Na^{+0.7}[Mg_{5.45}Li_{0.4}Si_8O_{24}H_4] + 12H^+ + 8H_20 \rightarrow 0.7Na^+ + 8Si(OH)_4 + 5.45Mg^{2+} + 0.4Li^+ \quad (15.3)$$

The significant dissolution in inert conditions (in nitrogen atmosphere) was explained by absence of acidic environment stimulated by adsorption of the atmospheric $CO_2$ [59]. At low concentration of Lap (below 1.7 wt%) the leaching of magnesium ions was observed even at pH above 10, but it was absent at 2.8 wt% Lap [33]. The stabilizing effects against degradation were explained by the importance of the concentration of sodium ions in suspensions, and the similar effects were also observed with externally added salt (NaCl) [32].

### *15.2.3 Modified Lap*

Modification of Lap structure allows fine regulation of different properties of Lap platelets, their dispersibility, surface charge, colloidal stability and affinity toward biomolecules. Lap includes different types of active sites such as the inter-layer, surface, edge and inter-particle sites [21]. Existing methods of modification include acid activation, intercalation, modification with organic molecules, covalent functionalization, grafting, and pillaring.

#### 15.2.3.1 Acid Activation

Acid activation in hydrochloric (HCl) or sulphuric ($H_2SO_4$) acids is frequently used to prepare acid-activated Lap (LapA) [60]. During acid activation, the protons can penetrate into the Lap layers and attack the structural OH-groups.

The properties of the final products depend on the extent of acid activation (particularly, on type and concentration of acid, time of activation, temperature etc.) [63]. The mild acid treatment results in protonated products [64, 65], whereas the strong acid treatment, with severe attacks of the oxygen sites results in connection of the octahedral and tetrahedral sheets on the edge of the faces and partial damage of the structure [60, 66]. Acid treatment can affect the overall negative charge, specific surface area, and average pore volume [60, 63, 67, 68]. Moreover, acid activation can be used as a pretreatment method for further modification of composites with organic or inorganic cations, and macromolecules [69].

The transformations of $Lap_{XLS}$ structure under acid activation for 12 h have been studied using XRD method under mild [61] and strong [62] acid treatments with hydrochloric and sulfuric acids, respectively (Fig. 15.6). Under the mild treatment in hydrochloric acid (≈0.29 wt% suspension of Lap in 0.57 wt% HCl solution) the platelet crystallinity (observed at higher angular range, $2\theta > 15°$) was mainly retained (Fig. 15.6a) [61]. The position of reflection $d_{001}$ related to the basal interlayer spacing filled with water for pristine Lap and LapA were approximately the same. The higher sharpness of the peak for LapA was attributed to the increased gallery gap and ordering in the Lap upon acid activation [70–72]. Under the strong acid treatment for 12 h in sulfuric acid (4 wt% suspension of Lap in 6–18 wt% $H_2SO_4$ solution) the Lap crystallinity observed in XRD patterns was lost and the transformation into the amorphous products was observed (Fig. 15.6b). The same conclusion was made based on analysis of the FTIR spectra [62]. Samples of LapA have been tested for preparation of cross-linked hydrogels [62] and LapA-Nafion composite membranes for fuel cells [61]. The hydrogels cross-linked with LapA have demonstrated better swelling properties and higher mechanical strength [62]. Incorporation of LapA to Nafion improved the proton conductivity of the composites compared to the unmodified Nafion, and improved the mechanical properties, thermal stability, and water uptake [61].

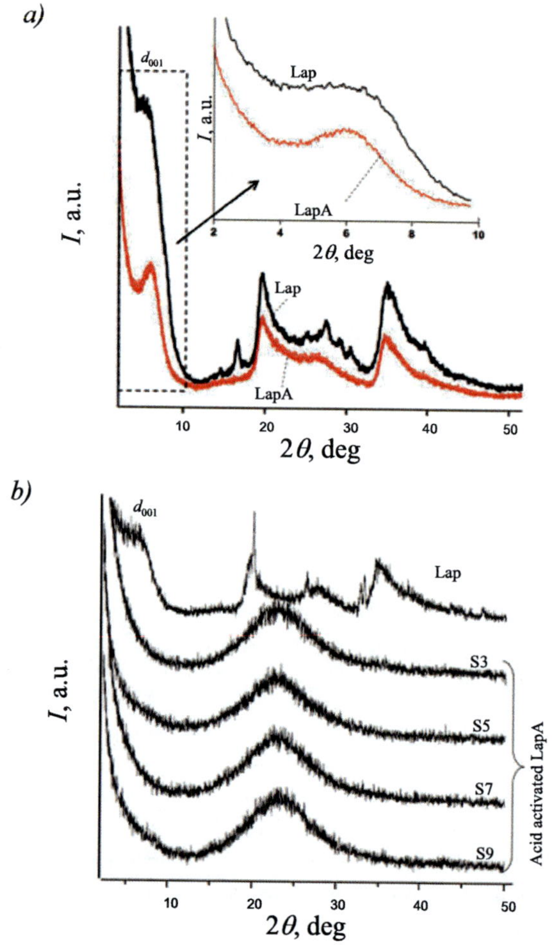


**Fig. 15.6** Examples of XRD patterns of Lap (grade XLS containing pyrophosphate based peptizer, Table 15.1) and acid-activated LapA. **a** Mild treatment in hydrochloric acid (≈0.29 wt% suspension of Lap in 0.57 wt% HCl solution for 10 h) [61]. **b** Strong treatment in sulfuric acid (4% suspension of Lap in 6 wt% (S3), 10 wt% (S5), 14 wt% (S7) and 18 wt% (S9) solutions of $H_2SO_4$ for 10 h) [62] (reprinted with permission from [61] **(a)** and [62] **(b)**)

#### 15.2.3.2 Organo-Modification

In aqueous systems, the positively charged polar parts of cationic surfactants can strongly interact with negatively charged faces of Lap platelets. The adsorption of surfactant molecules on Lap platelet leads to the coverage of its surface by the hydrophobic shells and formation of so-called organo-modified LapO products. The different types of cationic surfactants such as hexadecyltrimethylammonium bromide (CTAB) [53, 73], hexadecyltrimethylammonium chloride [74], dimethyldioctadecylammonium bromide [75], dioctadecyldimethylammonium chloride [76], Brij-35 [77], and many others have been used to study their interactions with Lap and preparation of LapO. At relatively small concentrations of surfactant the Lap suspensions were stable [74, 78] and Lap-surfactant complex retained platelets morphology [73].

However, at high concentrations of surfactant and strong neutralization of the anionic charge of Lap, the aggregation of particles and overcharging of the Lap surface were observed [53, 74, 78]. For example, in 2.0 wt% suspension the $\zeta$-potential of Lap platelets changed the sign from negative to positive at CTAB concentration of $\approx$1 *CEC* and after it was increasing with further increase in the CTAB content [53]. Decoration of Lap by CTAB increased hydrophobicity [79] and noticeably influenced the surface charge of Lap platelets [80]. The organo-modification also helps to increase the interlayer spacing in Lap products. In the presence of CTAB, the faces of Lap became patchy and covered by hydrophobic spots. These spots strongly interact with each other and form hydrophobic contacts (H bonds). In aqueous systems, realizations of perpendicular edge-to-face contacts (T bonds), parallel face-to-face contacts between hydrophobic spots (H bonds), and mixed T & H bonds were demonstrated [79].

Typically, the preparation procedures of LapO [76, 81, 82] include temperature incubation of a mixture of Lap and ionic surfactant in water [83, 84] or in water/propanol solution [75, 85] for several hours. The quantity of surfactant corresponds to the required surface coverage accounting for the *CEC* value of Lap. Then the treated Lap platelets are purified by repeated washing and dried (lyophilized). The incubation is done at room [75, 85–87] or at elevated (233–253 K) [76, 83, 84, 88–90] temperatures. The incubation temperature can noticeably affect the properties of LapO–doped composite. For example, the better exfoliation in liquid crystalline composites for LapO prepared at room temperature as compared with LapO platelets prepared at elevated temperatures was observed [91]. More details on preparation of LapO can be found elsewhere [89, 91, 92].

In some works the modifications of the edges of Lap with anionic surfactants, like sodium dodecyl sulfate (SDS) [55] or sodium polystyrene sulfonate (PSS) [93] were reported. The addition of SDS or PSS resulted in significant acceleration of aging dynamics [55, 93]. Moreover, the addition of PSS induced the transition from a homogeneous Wigner glass state to a gel network with a fractal structure [93]. Formation of cationic/anionic mixed surfactant bilayers on Lap was also reported [94]. The Lap particles were covered by cationic surfactant (cetylpyridinium bromide or dioctadecyldimethylammonium chloride) and then anionic surfactant SDS was added.

Hybrid Lap-phenothiazine (PTZ) material has been synthesized [95]. The aromatic amine PTZ and its derivatives could be used in pharmacology and biomedicine as inhibitors, antioxidants, antifungal, and antibacterial agents [96]. Formation of different charge-separated states was observed in Lap-PTZ hybrids depending on the procedure of synthesis [95]. PTZ can penetrate into the Lap interlayer space and stabilized as a dication. The effects of confinement on the structure and mobility of adsorbed and intercalated PTZ species were demonstrated.

Organic–inorganic hybrids based on $Lap_{RD}$ and phthalocyanine (Pc) dyes have been synthesized [97, 98]. Pc dyes are photosensitizers, capable of generating highly toxic singlet oxygen. However, the high hydrophobicity and poor solubility in water reduce their effectiveness. For example, prepared organo-modified hybrid of Lap with aluminum hydroxide Pc ($C_{32}H_{17}AlN_8O$) showed remarkable solubility and photophysical performance in water [97]. The hybrids of Lap with copper (II) Pc are also formed stable aqueous suspensions. These materials can be used in photodynamic therapy for the treatment of cancer and infectious diseases.

Promising Lap-dye hybrid materials with high fluorescence have been synthesized [99]. Ten different neutral and cationic fluorophores from perylene, coumarin, merocyanine, cyanine, and xanthene classes were tested. The prepared Lap-dye hybrids can be promising as nanoscale fluorescent reporters and optical probes for the bioimaging of pathogenic changes or disease-related biomarkers.

#### 15.2.3.3 Covalent Functionalization

Covalent functionalization of Lap allows preparation of multifunctional Lap-based materials with improved dispersing properties in water, organic solvents, and in nonpolar polymers. Lap has many reactive hydroxyl groups in form of Si–OH and Mg (Li)–OH on the edges and in the form of Mg–OH or Li–OH inside the layer that can chemically react with alkylsilane generating new Si–O bonds [36]. Such covalent functionalization or grafting by silylation was used to prepare the edge modified LapSi [72]. Table 15.2 shows popular silane coupling agents (with their abbreviations (Abbr) and formulas) used for preparation of LapSi.

As an example, a condensation reaction with mono- and trifunctional alkoxy silanes was used to prepare covalently functionalized LapSi [72]. APTMS, APMES, and TMPS were attached to Lap via condensation reactions, and were further reacted to yield attached methacrylate, benzophenone, and tertiary bromine groups. Binding mechanisms of silanes to Lap were explained by formation of covalent and ionic

**Table 15.2** Popular silane coupling agents used for preparation of edge modified LapSi

| Abbr | Name | Formula | Ref. |
|---|---|---|---|
| APTMS | (3-aminopropyl) trimethoxysilane | $C_6H_{17}NO_3Si$ | [72] |
| APMES | (3-aminopropyl) dimethylethoxysilane | $C_7H_{19}NOSi$ | [72, 100] |
| APTES | (3-aminopropyl) triethoxylsilane | $C_9H_{23}NO_3Si$ | [101] |
| MPMDS | (3-mercaptopropyl) methyldimethoxysilane | $C_6H_{16}O_2SSi$ | [102] |
| DMES | Dimethylethoxysilane | $C_4H_{12}OSi$ | [103, 104] |
| DMMOS | Dimethylmethoxyoctylsilane | $C_{11}H_{26}OSi$ | [105] |
| TMMS | Trimethoxymethylsilane | $C_4H_{12}OSi$ | [104] |
| TMES | Trimethylethoxysilane | $C_5H_{14}OSi$ | [106] |
| TMPS | Trimethoxypropylsilane | $C_6H_{16}O_3Si$ | [72] |
| TMOS | Trimethoxyoctylsilane | $C_{11}H_{26}O_3Si$ | [107] |
| γ-MPDES | γ-methacryloxypropyldimethyl-methoxysilane | $C_{10}H_{22}O_3Si$ | [108] |
| γ-MPTMS | γ-methacryloxypropyl-trimethoxysilane | $C_9H_{20}O_5Si$ | [108] |

bonds, as well as physical adsorption. LapSi materials demonstrated improved properties and can be potentially used in polymer composites. LapSi-based composites demonstrated improved mechanical properties and decreased flammability at relatively low loadings.

MPMDS was grafted as a linker to the rims of Lap for the subsequent covalent functionalization with of a rare earth β-diketonate complexes, co-coordinated with epoxy-functional phenanthroline (Eu $(ttfa)_3$ epoxiphen) that is an efficient luminescent material [102]. The Lap-based hybrid materials can be used as biomarkers in bioassays and they have interesting applicatory potential, e.g. in biophotonics.

The mechanisms of edge modification of Lap with the DMES [103], DMMOS [105], TMES [106], and TMOS [107] were discussed. Particularly, it was demonstrated that silylation of Lap can greatly enhance the intercalation rate of bulky and hydrophobic drug ITA (Table 15.3) without significant effects on the crystal structure and uptake capacity of LapSi [107]. XRD measurements proved that the silane is bonded onto the surface of the silicate with penetration into the gallery space [106]. LapSi (DMES) particles uniquely combined the Pickering effect with amphiphilic properties similar to the surfactant [103]. These amphiphilic platelets were used for preparation of very stable Pickering emulsions stabilized by solid particles.

Combined modification of Lap by ammonium surfactant and/or covalently bound poly(methyl methacrylate) (PMMA) was tested [109]. The polymer attachment was performed via reaction of a methacrylate compound with the Lap silanol group. The PMMA-modified Lap were highly dispersible in organic solvents and prepared composites were optically transparent and homogeneous. Effect of edge-modified LapSi (TMMS) and LapSi (DMMOS) on gelation behavior in suspensions was studied [104]. Gelation of LapSi (TMMS) was very similar to that of pristine Lap, but in contrast, LapSi (DMMOS) gelled more rapidly compared to Lap, and formed weak, turbid gels. The effects were explained by the differences in stacking of Lap and LapSi platelets. Lap materials functionalized by γ-MPDES and γ-MPTMS were also used for applications in emulsion polymerizations [108, 110]. LapSi platelets were used as seeds for synthesis of polymer composite latex particles [108].

In some works APTES has been used as grafting agent to promote interfacial behavior of Lap [101, 111, 112]. Such amino functionalization results in the formation of Si–O–Si covalent bonds between the hydrolysable alkoxy groups of APTES and hydroxyl groups located on the Lap edges. LapSi (APTES) particles can provide durable immobilization of organic molecules of biological interest via hydrogen bonds with organosilane functional groups, or hydrophobic and electrostatic interactions with surface of Lap [101]. LapSi (APTES) was used for preparation of alginate and alginic acid polysaccharide composites [111]. Covalent functionalization increased thermal resistance of composites. Some examples of biological and environmental applications of amino functionalization has been recently reviewed [112].

**Table 15.3** Therapeutic effects and other characteristics of some pharmaceutical molecules

| Drug | Therapeutic effects and other comments | Molecular weight, $M_w$, and solubility in water at ambient temperature, $C_w$ |
|---|---|---|
| DOX | Doxorubicin is used in therapeutics as a hydrochloride salt. It is composed of hydrophobic (adriamycinone) and hydrophilic (daunosamine) groups [118] and has a high chemotherapeutic activity against different types of solid and liquid tumors [119] | $M_w \approx$ 580 g/mol<br>$C_w \approx$ 50 mM ($\approx$30 g/L) |
| AMX | Amoxicillin is a penicillin-type antibiotic for treatment a variety of bacterial infections, such as ear infections, bladder infections etc. In solution, it exists in different ionic forms [120] depending on the pH of the medium [121] | $M_w \approx$ 365.4 g/mol<br>$C_w \approx$ 2.4 g/L (pH 5.6)<br>$C_w \approx$ 8.5 g/L (pH 15.6) |
| DEX | Dexamethasone is a corticosteroid anti-inflammatory medication that used for treatment of inflammation (swelling), blood/hormone/immune system disorders, eye disease, arthritis, allergic reactions, skin conditions, COVID-19, certain cancers etc. [122] | $M_w \approx$ 392.5 g/mol,<br>$C_w \approx$ 0.1 g/L |
| TTC | Tetracycline is a broad-spectrum naphthacene antibiotic used in the treatment and prevention of bacterial infections, antimicrobial agent, an antiprotozoal drug, and a protein synthesis inhibitor in *Escherichia coli* metabolism [123] | $M_w \approx$ 444.4 g/mol<br>$C_w \approx$ 0.311 g/L (base)<br>$C_w \approx$ 115.8 g/L (hydrochloride) |
| ITA | Itraconazole is an antifungal medication that is used to treat infections caused by fungus such as blastomycosis, histoplasmosis, and aspergillosis etc. [124]. Itraconazole leads the fungal membrane abnormalities or changes membrane-bound enzyme activity. Itraconazole poses low bioavailability due to its low solubility in water. The aqueous solution of itraconazole at concentration of 5 g/L requires the addition of 5% cyclodextrin solution | $M_w \approx$ 705.6 g/mol<br>Practically insoluble |
| SRA | Strontium ranelate (a strontium (II) salt of ranelic acid) is a medication for osteoporosis [125]. It acts as a "dual action bone agent" and demonstrates increasing the deposition of new bone by osteoblasts and decreasing the resorption of bone by osteoclasts | $M_w \approx$ 513.5 g/mol<br>$C_w \approx$ 3.7 g/L |

(continued)

**Table 15.3** (continued)

| Drug | Therapeutic effects and other comments | Molecular weight, $M_w$, and solubility in water at ambient temperature, $C_w$ |
|---|---|---|
| MTX | Methotrexate is a chemotherapy, an antimetabolite and antifolate agent with antitumor and immunosuppressant activities. It is a useful treatment for many dermatologic conditions [126]. Base form of methotrexate is practically insoluble in water, but its disodium salt is very soluble in water ($C_w > 25$ g/L) | $M_w \approx$ 454.5 g/mol<br>Practically insoluble |
| OFX | Ofloxacin is an antibiotic agent useful for the treatment of a bacterial infections of skin, soft tissues, urinary tract, and lower respiratory tract including multidrug resistant tuberculosis [127] | $M_w \approx$ 361.4 g/mol<br>$C_w \approx$ 36.02 g/L at pH 1.2<br>$C_w \approx$ 1.79 g/L at pH 15.4 |
| VCM | Vancomycinis is an antibiotic agent useful for treatment of complicated skin infections, endocarditis, and bone and joint infections [128] | $M_w \approx$ 1449.3 g/mol<br>$C_w \approx$ 100 g/L |
| LDC | Lidocaine is a local anesthetic. It is used to reduce pain or discomfort caused by skin irritations (sunburn, insect bites, scratches, burns etc.) [129] | $M_w \approx$ 234.3 g/mol<br>$C_w \approx$ 0.59 g/L (hydrochloride) |
| GCT | Glucantime (meglumine antimoniate) is a medicine agent useful for treatment of leishmaniasis, including visceral, mucocutaneous, and cutaneous leishmaniasis [130] | $M_w \approx$ 366 g/mol (varies)<br>$C_w \approx$ 240 g/L |
| DCF | Diclofenac (voltaren) is a nonsteroidal anti-inflammatory drug available in both a sodium and a potassium salts [131] | $M_w \approx$ 296.2 g/mol<br>$C_w \approx$ 4.82 mg/L (sodium salt) |

#### 15.2.3.4 Lap-Based Contrast Agents

Nowadays different computed tomography imaging techniques are widely used for disease diagnosis [113]. These techniques are based on using molecular contrast agents. For example, in conventional X-ray imaging technique iodine-based contrast agents (e.g., diatrizoic acid (DTA), omnipaque, or iohexol) have been used. In magnetic resonance imaging (MRI) technique the molecular contrast agents can be divided into longitudinal $T_1$-positive agents (mainly gadolinium (Gd) or manganese (Mn) agents) and transverse $T_2$-negative agents (superparamagnetic iron oxide agents).

To overcome the drawbacks of existing molecular contrast agents (poor colloidal stability, short imaging time, insufficient contrast effect, renal toxicity, nonspecificity, etc.) different functionalization strategies for preparation of effective Lap-based contrast agents have been developed.

For example, Lap modified by DTA has been tested as the iodinated contrast agent for X-ray tomography [100]. Lap platelets were first silanized with APMES (Table 15.2) to render the particles with amine groups. Then LapSi (APMES) particles were conjugated with DTA. The cell viability assay revealed that the LapSi (APMES) + DTA particles were cytocompatible (in the given concentration range of 0.1–1.0 g/L) and showed an improved imaging performance of the major organs (heart, liver, and bladder) and a tumor model.

Magnetized LapM has been used as a contrast agent for *in vivo* MRI detection of tumors [114]. LapM platelets were prepared using a facile controlled co-precipitation route of $Fe_3O_4$ in Lap aqueous suspensions. Detailed analysis of XRD patterns, TEM-images, hydrodynamic sizes and zeta potential values, and FTIR spectra evidenced that $Fe_3O_4$ is immobilized directly on Lap platelets. The formed magnetized LapM platelets displayed the effective uptake by tumor cells, good biocompatibility, and shown about two-fold increase of $T_2$ relaxivity compared to $Fe_3O_4$. It was concluded that LapM system has the potential for application as a versatile theranostic platform.

Dendrimer-functionalized Lap loaded with gadolinium (Gd) have been synthesized for MRI applications [115]. Lap was sequentially silanized with APMES (Table 15.2), carboxylated, and modified with G2 dendrimers. Then, the dendrimer-modified Lap was conjugated with Gd chelator diethylenetriaminepentaacetic acid (DTPA) and finally, LapSi (APMES)-G2-DTPA(Gd) complexes were prepared. The designed complexes were non-cytotoxic. They have shown promising contrast properties for $T_1$-weighted MRI of cancer cells *in vitro* and animal organ/tumor model *in vivo*. Moreover, high loading capacity of the Lap complexes is promising for tumor theranostic applications.

## 15.3 Drug Delivery Systems

Lap-based materials have excellent ability to entrap or release organic moieties and they have shown high efficacy for delivery of different therapeutics [116, 117]. These materials demonstrated excellent biocompatibility, stimuli responsiveness, and sustained drug release [9]. Moreover, Lap-based materials showed outstanding hydrolytic stability and antimicrobial properties. Selective surface modification (covalent or non-covalent) of Lap platelets [7] allows improving degree of their exfoliation and drug loading efficiency. The various types of Lap-based hybrids with improved biocompatibility and prolonged bioactivity periods for applications in drug delivery systems and regenerative medicine were developed.

### 15.3.1 Examples of Pharmaceutical Molecules

The pharmaceutical medium-sized molecules typically composed of several inter-linked aromatic cores and multiple hetero-atoms such as N, P, O, S, and F or Cl. Various drug molecules, such as doxorubicin (DOX) and many others presented in Fig. 15.7 and Table 15.3 have been tested in drug delivery systems based on Lap.

These drug molecules can display different therapeutic actions and bioactivities, and their efficiency can be significantly improved using the Lap@drug platforms. For example, DOX molecules can be transported inside the cells via channel proteins and their interactions with intracellular compartment result in blocking of DNA (deoxyribonucleic acid) replication and transcription.

The DOX initiated toxicity includes formation of reactive oxygen species and lipid peroxidation (oxidative degradation of lipids in cell membranes). However, DOX has rather low therapeutic index and the main obstacles for its applications are:

i) development of cancer cell adaptation and resistance to DOX action;
ii) decreased drug uptake;
iii) increased drug efflux;
iv) strong binding tendency of DOX with cellular macromolecules.

To overcome these problems the increased DOX doses are required. However, the fast diffusion of DOX and its accumulation in normal cells can have numerous cytotoxic effects on normal cells and healthy tissues during the treatment. Commonly, DOX treatment can result in different post-treatment debacles (cardiomyopathy, heart failure, weakening of the immune system, and other health disorders) [132]. Moreover, the reversal of tumor and formation of secondary (metastatic) tumors are possible. The similar problems exist with applications of other pharmaceutical molecules [133, 134]. To overcome these problems different Lap-based platforms have been fabricated and their therapeutic efficacies have been tested. The existing

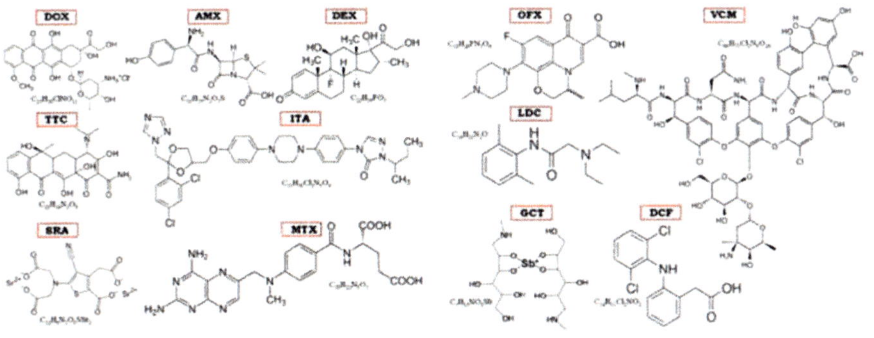


**Fig. 15.7** Molecular structures of some pharmaceutical molecules tested in drug delivery systems based on Lap platelets: DOX (doxorubicin), AMX (amoxicillin), DEX (dexamethasone), TTC (tetracycline), ITA (itraconazole), SRA (strontium ranelate), MTX (methotrexate), OFX (ofloxacin), VCM (vancomycinis), LDC (lidocaine), GCT (glucantime), and DCF (diclofenac or voltaren)

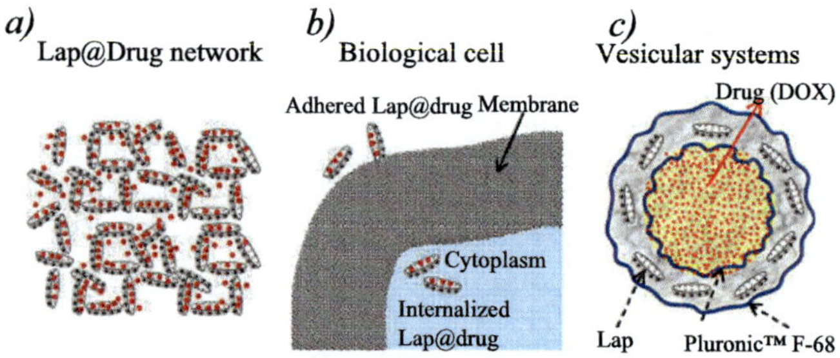


**Fig. 15.8** Examples of drug delivery systems on Lap-based platforms: **a** restricted transport in Lap barrier hydrogel networks [135]; **b** Lap@drug platforms adhered to the cellular membrane or internalized inside cell [40]; **c** DOX-encapsulated vesicle with Lap particles inside vesicular walls for controlled release of drug [137]

platforms account for the aqueous solubility of pharmaceutical molecules, mechanisms of their therapeutic actions, possibility of fine regulation of the aggregation of drug-bearing particles, their localized delivery to the target sites, regulation of drug release rate (sustained drug release) and their temperature and pH-dependence, possibility of targeted transport of multiple drugs, cytocompatibility, and physiological stability.

The drug molecules can be trapped in Lap-based materials in different ways, including the cooperation between Lap and drug at the nanoscale (e.g., by physical absorption or covalent attachment to the Lap surface, intercalation between Lap lamellae etc.), or incorporation of drug into three-dimensional hydrogel networks or networks formed by Lap at the microscale [116].

The different ways of drug delivery can be achieved using Lap-based platforms. Some of them are presented in Fig. 15.8. Inside the Lap barrier hydrogel networks, the interlayer diffusion and delivery properties may be significantly affected by distribution of Lap particles in the networks (Fig. 15.8a) [135]. The interaction of individual drug-loaded Lap particles or their aggregates with biological cells can be achieved in different ways [40]. The Lap-based platforms can be adhered on cell surface (membrane) or internalized inside biological cells (Fig. 15.8b). The strong electrostatic interaction of the positively charged edges of Lap platelets with negatively charged cell membranes was suggested [136]. Through these cationic edges, the platelets can pass through the membranes and destabilize the entire membrane structure. In vesicular system controlled release of DOX can be limited by Lap particles incorporated inside the walls with thickness of 30 nm [137] (Fig. 15.8c).

## 15.3.2 Nanoscale Drug Delivery Systems

The nanoscale Lap platforms are based on ability of Lap particles to accumulate drug molecules on their surface. For example, the DOX hydrochloride salt (DOX·HCl) is highly soluble in water. In aqueous suspension the DOX molecules are attracted to Lap by strong electrostatic interactions, and at high DOX concentration, the hydrophobic parts of DOX cover the Lap surface, i.e., hydrophobization occurs. The DOX has been most frequently used for testing efficacy of Lap-based drug delivery platforms (Table 15.4). Different types of drug delivery platforms were fabricated (Fig. 15.9).

In the pioneering experimental works [138, 139], the simple single drug Lap@DOX platforms (Fig. 15.9a) have been studied. It was demonstrated that hydrophobization induces the aggregation of Lap@DOX particles. Estimated hydrodynamic size for pristine Lap particles was ≈31 nm as compared with size of ≈98 nm for Lap@DOX hybrid particles (at DOX/Lap weight ratio of 1/6). The observed

**Table 15.4** Lap-based DOX-loaded therapeutic platforms fabricated by different methods

| Type of platforms and therapeutics | Comments | Ref. |
|---|---|---|
| Lap@DOX<br>Single drug | The drug release was faster at acidic pH condition (pH = 5.4) than at physiological pH condition (pH 15.4). The enhanced antitumor efficacy was explained by higher cellular uptake of hybrides as compared to free DOX | [138] |
| Lap@DOX<br>Single drug | Platform demonstrated an exceptionally high DOX loading, biocompatibility, and ability to be cleared out of body at 45 days post treatment | [139] |
| ($Lap_{XLG}$@DOX) @MTX<br>Dual drugs | A combinative way for drug delivery was tested. The encapsulation efficiencies were ≈85% for DOX and ≈80% for MTX. Platforms allowed a sequential release of the loaded drugs with a rapid initial release of MTX which was followed by a slow release of DOX for 120 h. The acceleration of drug release under acidic pH and upon heating was observed. The human liver cell line (HepG2) was used in intracellular drug accumulation tests | [140] |
| $Lap_{RDS}$@DOX +<br>(PEG-PLA)<br>Single drug<br>One-layer shell | Platforms demonstrated improved cytocompatibility, stability, sustainability under physiological conditions, and pH-responsive release | [141] |
| $Lap_{XLG}$@DOX +<br>PVP<br>Single drug<br>One-layer shell | The functionalization with poly(*N*-vinylpyrrolidone) (PVP) improved colloidal stability and drug loading capacity of the platforms. Enhanced pH- and thermosensitive drug release as compared to the PEG-modified Lap@DOX platforms was also observed | [148] |
| $Lap_{RDS}$@DOX +<br>(oHA-APBA)<br>Single drug<br>One-layer shell | Lap@DOX was assembled with oligomeric hyaluronic acid-aminophenylboronic acid (oHA-APBA) copolymer. The satisfactory drug encapsulation efficiency (>80%) was observed. The *in vitro* studies demonstrated the sustained and acid-dependent DOX release from the platforms. The fabricated platforms can be used as effective targeted delivery systems for breast cancer therapy | [143] |
| ($Lap_{XLG}$@DOX + PVP)@MTX<br>Dual drugs<br>One-layer shell | Lap@DOX was decorated with a hydrophilic poly(*N*-vinylpyrrolidone) (PVP) and then outer layer was loaded with methotrexate (MXT). Proposed combinative approach can be useful for co-delivery of various drugs to cancer cells | [144] |
| $Lap_{XLG}$@DOX +<br>(PAH/PSS)<br>Single drug<br>Multilayer shell | The electrostatic layer-by-layer self-assembly approach was used to cover the Lap@DOX particles by *n* layers (*n* = 1–6) of cationic poly(allylamine) hydrochloride and anionic poly(sodium styrene sulfonate) (PAH/PSS) polyelectrolytes. The platforms allowed a fine-tuning of the extension of drug release at neutral and acidic pH | [146] |

(continued)

**Table 15.4** (continued)

| Type of platforms and therapeutics | Comments | Ref. |
|---|---|---|
| Lap + (PEG-PLA) + PEI($Au^0$) + HA@DOX DOX-$Au^0$ combination Multi-layer shell | Multilayer theranostic platforms for computerized tomography imaging and chemotherapy have been fabricated. Lap particles were covered by amphiphilic poly (ethylene glycol)–poly(lactic acid) diblock copolymer (PEG-PLA) and then conjugated with polyethylenimine (PEI). These templates were embedded by zerovalent gold nanoparticles ($Au^0$) and modified by hyaluronic acid (HA). Finally, DOX was loaded. Formed platforms demonstrated decreased level of negative side effects of DOX and inhibited cancer cells over-expressing CD44 receptors. The platforms can be also used as a targeted contrast agent for computerized tomography imaging of tumors | [147] |

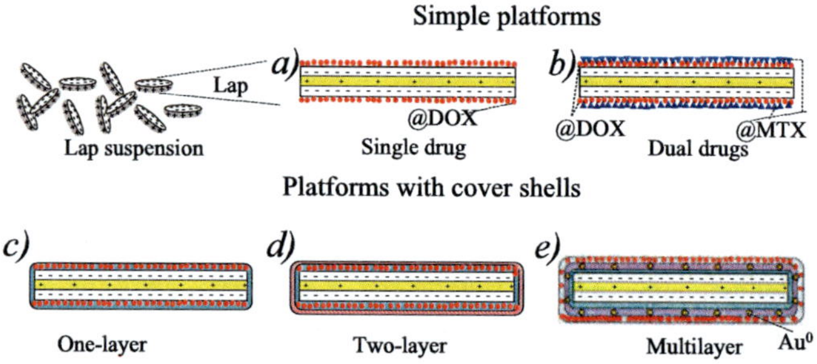


**Fig. 15.9** Schematic presentation of different drug delivery platforms fabricated in aqueous suspensions of Lap. Platforms without cover shells: **a** single drug Lap@DOX platform [138, 139] and **b** dual-drug (Lap@DOX)@MTX platform [140]. Platforms with cover shells: **c** one-layer, **d** two-layer, and **e** multilayer

DOX loading efficiency was rather high (≈98%). The obtained Lap@DOX platforms demonstrated pH-dependent release profiles of DOX. For example, the DOX was released at increased rate under acidic pH microenvironments (mimicking solid tumor and endo-lysosomal compartments). The dependence on pH was attributed to the different hydrophilicity of the DOX drug under different pH conditions. It was demonstrated that Lap@DOX platforms enable an enhanced cellular internalization of DOX. They have enhanced *in vivo* antitumor efficacy, demonstrated a low systemic toxicity, and were considered as a promising systems for tumor therapy applications [139].

Dual-drug platforms with hydrophobic core Lap@DOX covered by supplementary hydrophobic drug MTX have been also developed (Fig. 15.9b) [140]. The dual platforms (Lap@DOX)@MTX allowed a rapid initial release of MTX, followed by a slow release of DOX for up to 120 h. These platforms can be useful to improve therapeutic accumulation of drugs in cancer cells.

However, the low colloidal stability of simple hydrophobic hybrids can limit their clinical applications. The stability can be increased by using the hydrophilic cover or protecting shells. For example, in simple one-layer shell platform (Fig. 15.9c) the colloidal stability of Lap@DOX platforms was enhanced by functionalization of their surface with an amphiphile poly(ethylene glycol)–poly(lactic acid) diblock copolymer (PEG-PLA) [141]. In this platform, the hydrophobic block of copolymer acts as an anchor, which binds to the surface of drug-loaded Lap@DOX particles. The hydrophilic PEG block forms the protective layers that improve the colloidal stability. Application of cover shell also allows preventing the burst release of drug. One-layer shell platforms, based on alginate [142] or oligomeric hyaluronic acid-aminophenylboronic acid copolymer [143], were also synthesized. Fabricated dual-drug (DOX + MTX) platforms with one-layer shell of hydrophilic poly (*N*-vinylpyrrolidone) demonstrated sequential therapeutic delivery and good colloidal stability [144].

In recent years layer-by-layer platforms for drug delivery became very popular [145]. For example, the single drug two-layer shell platforms were produced by sequential covering of Lap@DOX particles by cationic poly(allylamine) hydrochloride and anionic poly(sodium styrene sulfonate) polyelectrolytes [146] (Fig. 15.9d). These platforms can be effectively internalized by cells. In another approach, three-layer shell platforms doped by DOX with imbedded gold nanoparticles were fabricated (Fig. 15.9e) [147]. The produced platforms displayed a high drug loading efficiency, good colloidal stability, and pH-sensitive profiles.

Table 15.5 presents examples of drug platforms based on silanized LapSi platelets. In these platforms, the LapSi (APMES) platelets (Table 15.2) have been chemically modified (conjugated) with different bioactive substances (e.g., folic acid, hyaluronic acid, D-α-tocopheryl poly(ethylene glycol) succinate, etc.) and loaded with drugs [101, 149–153]. The platforms were pH-responsible and demonstrated the improved drug loading efficiency (≈85–98%), good biocompatibility, good colloidal stability, and exhibited targeted anti-tumor effects. In some cases, applied approach allowed overcoming the multidrug resistance of tumor cells.

Examples of others (non-DOX) Lap-based therapeutic platforms are presented in Table 15.6.

## 15.4 Medical Lap-Based Hydrogels

### *15.4.1 Types and General Properties of Hydrogels*

Polymers in addition to be widely used in our daily lives, also found their application in science, medicine, and engineering. Biopolymers such as polysaccharides, proteins, and nucleic acids are the basis of all living organisms. Synthetic polymers such as rubber, polyethylene, polypropylene, polystyrene, polyamides, polyesters, polyurethanes, and plastics are produced in millions of tons and are widely used in industry. The properties of polymers vary in a wide range depending on polymer application. For biomedical applications, polymers should have the inherent specific set of properties – a combination of polymer biocompatibility, three-dimensional hydrophilic structure, ability to retain shape, and mechanical properties similar to native tissues. The most suitable polymer-based systems for biomedical application, which meet those requirements, are hydrogels – three-dimensional spatially cross-linked hydrophilic polymers [162]. In addition, hydrogels are characterized by the

**Table 15.5** Therapeutic platforms based on silanized LapSi platelets. Lap was preliminary silanized using APMES (Table 15.2) to endow them with abundant surface amine groups (–$NH_2$). The resulting LapSi (APMES) particles were further conjugated with different substances and loaded with drugs

| Type of platforms and therapeutics | Comments | Ref. |
|---|---|---|
| LapSi + FA@DOX | LapSi was conjugated with folic acid (FA). Platforms were able to specifically target cancer cells overexpressing high-affinity FA receptors | [149] |
| LapSi + HA@DOX | LapSi was conjugated with hyaluronic acid (HA). Platforms demonstrated good biocompatibility and exhibited targeted anti-tumor effects against human epithelial (HeLa) cell over-expressing CD44 receptors | [150] |
| LapSi + TPGS@DOX | LapSi was conjugated with D-α-tocopheryl polyethylene glycol 1000 succinate (TPGS) (P-glycoprotein inhibitor). Platforms displayed good colloidal stability and a high DOX encapsulation efficiency. *In vitro* experiments evidenced the efficient drug delivery into DOX-resistant MCF-7/ADR cells (the multidrug-resistant breast cancer cell models) and the inhibition of proliferation of cancer cells | [151] |
| LapSi + EG-LA@DOX<br>LapSi + mPEG-COOH@DOX | LapSi was conjugated with poly(ethylene glycol)-linked lactobionic acid (PEG-LA) or PEG monomethyl ether with one carboxyl terminal group (mPEG-COOH). The LapSi + PEG-LA@DOX platforms demonstrated high therapeutic efficacy, good biocompatibility, and improved colloidal stability. The platforms may serve as efficient targeted delivery carriers to liver cancer cells | [152] |
| LapSi + SAH + G2@DOX | LapSi was modified with amine responsive succinic anhydride (SAH) and conjugated with amine-terminated G2 PAMAM dendrimers. Platforms allowed prolonging the circulation time, and targeting the tumor tissues through enhanced permeability and retention effect. The high efficiency for inhibition the proliferation of KB cells (a human epithelial carcinoma cell line) was demonstrated | [153] |

(continued)

**Table 15.5** (continued)

| Type of platforms and therapeutics | Comments | Ref. |
|---|---|---|
| $Lap_{RD}Si@CHXDG$ | LapSi was loaded with chlorhexidine digluconate (CHXDG) antiseptic and disinfectant against a wide range of Gram negative and Gram-positive bacteria, yeasts, dermatophyte fungi, and lipophilic viruses. The percentage of drug loading was 99%. Adsorption of drug onto LapSi surface involved formation of hydrogen bonds between the CHX NH groups and the LapSi $NH_2$ and OH-groups. These strong interactions prevented fast CHXDG release into the medium | [101] |

similarity of their structure to the human body tissues [163], specifically to extracellular matrix, while the available functional groups can be modified yielding hydrogel systems with novel properties.

The term "hydrogel" has been used in the scientific literature since 1894 and it was applied to "colloidal gel of inorganic salt" [164]. Hydrogels in the modern sense of the term (spatially cross-linked polymers) were first synthesized based on hydroxyethyl methacrylate by O. Wichterle in 1960 [165]. Since then hydrogels have been widely used for manufacturing soft contact lenses. Nowadays hydrogels are also used for producing hygiene products, wound dressings, drug delivery systems (the material for capsules), and in agriculture (e.g., soil moistener).

Three-dimensional polymer network gives hydrogels the ability to swell in "good" solvents (especially in water and aqueous solutions) with an increase in their mass and volume (sometimes tens and hundreds of times) without being dissolved. The cross-linking of hydrogels can be achieved via chemical (covalent bonds) or physical (electrostatic or hydrophobic interactions, hydrogen bonding, or hydrophobic interactions) factors [166].

Covalently cross-linked hydrogels can be synthesized using free-radical, cationic, anionic, addition, or condensation polymerization [166]. To initiate gelation, the chemical initiators (oxidizing agents at elevated temperatures or redox systems at room temperatures) can be added to the mixture or monomers can be polymerized under ultraviolet, gamma-, or electron radiation [167, 168]. As initiators, enzymes such as chrome peroxidase, transglutaminase, and tyrosinase were also used [169]. The advantage of hydrogels cross-linked by enzymes is the lack of toxicity compared to chemically cross-linked systems, but their application is limited due to the low stability of cross-linking enzymes.

Hydrogels can be also physically cross-linked or cross-linked by using simultaneously both covalent and non-covalent bonds. For example, both physically and chemically cross-linked hydrogels were synthesized by photopolymerization of ethylene glycol diacrylate in the presence of $Lap_{RD}$ platelets [170]. Covalent cross-linking of polymer chains resulted in the formation of an elastic network, whereas physical cross-linking between Lap platelets and polymer chains induced viscoelastic properties [170].

Hydrogels can be classified based on:

**Table 15.6** Lap-based therapeutic platforms loaded by different pharmaceutical molecules (Table 15.3)

| Type of platforms and therapeutics | Comments | Ref. |
|---|---|---|
| Lap@DEX | Platforms were prepared in ethanol. The interactions between Lap and dexamethasone (DEX) were explained by hydrogen bonding involving hydroxyl and carbonyl groups. The slow release of DEX into saline solutions (used as models for vitreous tumor) was demonstrated | [154] |
| LapRDS@DEX | Platforms were prepared in water. The loading efficiency by dexamethasone (DEX) was ≈95% at pH 3 and it was significantly higher compared to loading at natural and alkaline pH. The platforms demonstrated excellent stability under physiological conditions and sustained pH-responsive drug release | [155] |
| LapRD@DEX | Platforms were prepared by mixing of aqueous Lap suspensions with ethanol solution of dexamethasone (DEX) and tested for intravitreal and suprachoroidal administration in rabbit eyes. The ocular tissues parameters (lens, vitreous humor, choroid-retina unit, and sclera) were studied. The intraocular retention time for formulations noticeably increased when compared with conventional solutions of DEX | [122] |
| LapXLG@ITA | Platforms were prepared by mixing of aqueous Lap suspension (1 wt%, pH 4) with itraconazole (ITA) dissolved in methylene chloride | [156, 157] |
| LapXLG@TTC | Platforms at high loading efficiencies demonstrated a sustained release of tetracycline (TTC) into model saline solution over a 72 h period. The strong interaction between Lap and TCC was supported by the data of FTIR spectroscopy and the DTA analyses. The acidic environment increased the extent of release of TTC from Lap carrier | [158] |

(continued)

**Table 15.6** (continued)

| Type of platforms and therapeutics | Comments | Ref. |
|---|---|---|
| Lap@AMX + PLGA, nanofibers | Platforms Lap@AMX (amoxicillin) were incorporated within poly(lactic-*co*-glycolic acid) (PLGA) nanofibers through electrospinning. Prepared systems displayed good cytocompatibility and a sustained drug release (much slower than for Lap@AMX or PLGA@AMX platforms) | [159] |
| LapRD@LL-37 | Lap@LL-37 (human cathelicidin peptide) hybrid particles have been prepared. LL-37 has demonstrated wound-healing properties, antimicrobial, and anti-biofilm activity against multiple Gram-positive and Gram-negative human pathogens. The loading of net positively charged LL-37 increased with increasing pH. The peptide was localized primarily on the outer surface of Lap in a helical conformation. The combined Lap@LL-37 hybrids demonstrated both bacterial killing through membrane disruption and bacterial flocculation effects | [160] |
| LapRD@SRA + PCL | Lap@SRA (strontium ranelate) platforms were prepared in aqueous phase, blended with polycaprolactone (PCL), and lyophilized to obtain composite scaffold. Release profile of SRA exhibited an inverse relation on the Lap-SRA complex content. *In vitro* evaluation of the scaffolds using human osteosarcoma cells revealed an optimum composition at 3 wt% loading of PCL for bone tissue engineering applications | [161] |

i) type of cross-linking agent (chemically and physically cross-linked);
ii) origin of monomers and polymers (natural, synthetic, or hybrid);
iii) presence of inorganic components (hydrogel systems or hydrogel composites);
iv) cross-linking procedure (polymerization of monomers with simultaneous cross-linking, polymer cross-linking, synthesis of interpenetrating networks);
v) polymerization conditions (bulk polymerization or dispersed polymerization suspension or emulsion polymerization for synthesis of hydrogel micro- and nanoparticles) [171];
vi) physical structure (amorphous, semi-crystalline, or supra-molecular) [172];
vii) polymer concentration distribution (gradient or slab gels);

viii) responsiveness to the changes in the environment (conventional or stimuli-responsive) [173];

ix) charge (neutral, ionic (anionic, cationic, or amphoteric [containing simultaneously the acidic and basic groups]), or zwitterionic (having a net charge equal to 0 at a specific pH) [163].

Synthetic hydrogels are preferably synthesized using a wide range of vinyl monomers and macromers, specifically acrylamide, acrylic acid, acrylonitrile, *N*-isopropylacrylamide, hydroxyethyl methacrylate, poly(ethylene glycol), and polyvinyl alcohol [174]. Unlike natural hydrogels (e.g., based on proteins or polysaccharides: collagen [175], chitosan, alginate [176], hyaluronic acid [177], dextran, starch, etc.), synthetic hydrogels are characterized by higher reproducibility of their physicochemical and exploitational characteristics [178]. However, commonly they have high levels of toxicity due to the presence of harmful impurities (unreacted monomers and cross-linking agents). Washing of hydrogels, which is usually controlled by UV spectroscopy or chromatography, can improve biocompatibility of synthetic hydrogels. Combining different polymers, either synthetic or natural, the hybrid copolymer hydrogels with a much wider range of properties (sometimes even with their synergistic improvement) compared to homo-polymer hydrogels (derived from a single species of monomers) can be synthesized [179].

Among different hydrogel types special role belongs to stimuli-responsive hydrogels, also known as stimuli-sensitive or "smart" hydrogels. These hydrogel systems can be designed to respond to changes in one or a few factors and will be discussed in detail in the following sections.

Hydrogel composites represent three-dimensional organic–inorganic hybrids formed by physical or chemical cross-linking of polymer networks in aqueous solution in the presence of functional inorganic nanomaterials [180]. Hydrogel composites, besides combining properties of polymer hydrogels and inorganic nanoparticles, also can exhibit new or enhanced properties (e.g., mechanical, electrical, optical, magnetic, and biological properties) due to the synergistic effects [181].

Haraguchi's pioneering work [182] started the intensive research of synthesis and application of hydrogel composites with incorporated Lap. Application of Lap as a physical cross-linker instead of traditionally used chemical cross-linking agents yields hydrogels with improved properties: optical clarity, mechanical characteristics, swelling/deswelling rates, and homogeneity [183]. Using thermoresponsive hydrogels based on Lap and N-isopropylacrylamide (NIPAAm) the improvement in mechanical properties was demonstrated. Particularly, the modulus of elasticity and tensile strength increased proportionally to the Lap concentration. The effect of Lap, NIPAAm, initiator, inorganic salt concentrations, and type of initiator on viscosity, mechanical and optical properties of hydrogels were discussed [184]. The mechanism of gelation of hydrogel composites with incorporated Lap was investigated and the properties of physically cross-linked hydrogels with incorporated Lap were compared to the properties of chemically cross-linked hydrogels [185]. Polymer hydrogels with incorporated Lap improves the mechanical properties of thermosensitive hydrogels based on NIPAAm.

Composites based on Lap and acrylamide with improved sorptivity for cationic dyes such as Basic Blue 9, Basic Blue 12, and Basic Violet 1 were also synthesized [186]. The increase in Lap concentration resulted in decrease the swelling degree of hydrogels. The effect was explained by the increase in the cross-linking frequency of polymer chains with Lap. At the same time, the rate of sorption of dyes by the composite and its sorption capacity increased significantly. The selective sorption of cationic dyes on Lap platelets was explained by electrostatic interactions. Sorption can also be enhanced by the formation of hydrogen bonds between the amino groups of the dye and the carbonyl groups of the hydrogel. Stimuli-responsive hydrogels with improved mechanical (tensile strength and elasticity was about 50 times higher than for chemically cross-linked gels), optical, and sorption properties were also synthesized and studied [187]. Improvement of mechanical properties by incorporation of Lap in hydrogels on acrylamide, poly (ethylene oxide) [188], poly (vinyl alcohol), and chitosan [189] was demonstrated.

Biocompatible hydrogel composites based on poly (acrylic acid) and Lap have been successfully synthesized by in situ free radical polymerization of acrylic acid (AA) in aqueous suspensions of Lap [190]. The resulting hydrogel composites were homogeneous and transparent. Their viscosity, storage modulus, and loss modulus significantly increased with increasing Lap and AA concentration, showing a maximum with increasing the degree of neutralization of AA. It has been shown that adhesive properties can be easily adjusted by varying the degree of AA neutralization, and concentration of both Lap and monomers. Hydrogels with adjustable adhesive properties seem to be the ideal candidates for the manufacture of dental adhesives, dressings, and for tissue engineering.

Hydrogels with excellent rheological properties (increased strength and elasticity) based on hydrolyzed poly (acrylamide) physically cross-linked with $Lap_{XLG}$ were synthesized using electrostatic interactions between negatively charged acrylate groups and positively charged edged of platelets [191]. The synthesized hydrogel composites were proposed to be used for sealing aquifers to increase the efficiency of oil production. The authors also demonstrated that the mechanical properties increase proportionally to the clay concentration.

Transparent and strong ionic hydrogels are extremely promising as a material for the manufacturing soft contact lenses or ophthalmic implants. In situ copolymerization of 2-acrylamide-2-methylpropylsulphate (AMPS) and acrylamide in Lap suspension was performed [192]. Due to homogeneous distribution of Lap platelets within the systems, the ionic hydrogels with high optical clarity and improved mechanical properties were obtained. The synergetic interactions between amide and sulphate groups in AMPS played a key role in stabilizing the Lap suspensions.

The process of redox synthesis of hydrogels based on NIPAAm with incorporated Lap has been studied by SAXS, FTIR and UV techniques [193]. The effect of NIPAAm and the components of redox initiating system (based on ammonium peroxide and *N,N,N′,N′*-tetramethylethylenediamine (TEMED) on gelation and hydrogel structure was studied. Both ammonium peroxide and TEMED initiators promote aggregation of Lap due to the formation of strong ionic interactions between Lap and persulfate or amino groups [193]. The structure of the hydrogel was affected by the mixing time during preparation and the degree of Lap exfoliation, especially at high Lap concentrations. The better results for pyrophosphate-modified $Lap_{XLS}$ as compared to $Lap_{XLG}$ were explained by better exfoliation of the $Lap_{XLS}$ [193].

### 15.4.2 Gradient Hydrogels

Gradient hydrogels represent materials with a gradual spatiotemporal change in physico-chemical properties [194]. Gradient hydrogels are considered as promising materials for robotics and artificial muscle manufacturing.

The electrophoresis technique to move the charged components of the gel-forming composition during polymerization was used for creation of a gradient structure [195]. Electrically active (negatively charged) Lap platelets were dispersed in a monomer solution containing hydroxyethyl acrylate and NIPAAm. Gelation was carried out between two electrodes at electric field strength of 1 V/cm using direct electric current. The electrophoretic movement of the Lap platelets towards the anode resulted in formation of a gradient gel structure. SEM images of lyophilized samples showed a porosity gradient along the direction of the electric field due to the concentration effect of the cross-linking action of Lap. An increase in the average mesh parameter from 0.35 μm on the anode side to 4.83 μm on the cathode side was observed. The designed materials present the temperature-responsive hydrogel actuators that can be utilized in sensors, artificial muscles, soft robotics and temperature-controlled microfluidic switches [196]. Based on Lap-polyNIPAAm gradient hydrogels, actuators with fast, large, and reversible bending were obtained [197]. A similar technique was successfully used to create hydrogels based on Lap incorporated into NIPAAm and thermosensitive *N,N*-diethyl acrylamide and (2-dimethylamino)ethyl methacrylate [198]. Gradient hydrogels based on NIPAAm and $Lap_{XLG}$ were synthesized by polymerization in the presence of a constant electric field using spherical graphite electrodes at electric field strength of 400 V/m [199]. The synthesized gradient hydrogels were characterized by a slower release of incorporated drugs (rhodamine B and sodium salicylate).

Synthesis of thermosensitive self-assembled two-layer hydrogels based on Lap and poly(*N*-isopropylacrylamide-*co*-acrylamide) was reported [200]. The hydrogels were able quickly, reversibly, and repeatedly change their shape under the change in temperature. Upon heating in the temperature range from 24 to 42 °C, the hydrogel strip got bent, and when the temperature was decreased back to 24 °C – the strip got unfolded.

### 15.4.3 Stimuli-Responsive Hydrogels

Stimuli-responsive hydrogels, also known as stimuli-sensitive or “smart” hydrogels, can be designed to respond to the changes in one or a few stimuli. They are capable of changing their shape and properties abruptly and in a controllable way under the slight changes in external factors [201]. Classification of "smart" hydrogels is based on the type of stimulus they respond to. "Smart" hydrogels can be classified as thermosensitive, pH-sensitive, electro-sensitive, as well as hydrogels sensitive to ionic strength, magnetic field, pressure, different types of ions/molecules, for example, enzymes, etc.

#### 15.4.3.1 Thermoresponsive Hydrogels

Among stimuli-responsive hydrogel systems, thermosensitive hydrogels have been widely studied. Hydrogels which respond to temperature changes can be divided into two groups: negatively and positively thermosensitive hydrogels. Thermosensitive polymers or hydrogels can be characterized either by upper critical solution temperature (UCST) or lower critical solution temperature (LCST) [202]. For given system, the components of a mixture are miscible in all proportions above UCST or below LCST.

Positively temperature sensitive polymers and hydrogels are characterized by UCST. They increase their water solubility and swelling degree, respectively, as the temperature increases. At temperatures below UCST they are in a folded (collapsed) state, and above UCST, they are in unfolded (swollen) state [203]. In contrast, negatively temperature sensitive polymers and hydrogels are characterized by LCST. At temperatures below LCST they are in the expanded or swollen state, and above LCST they are in the folded or collapsed state. Therefore, hydrogels with LCST collapse as the temperature increases. In such systems at low temperatures the hydrogen bond between the hydrophilic groups of the polymer chain and water molecules prevails. These favorable interactions lead to the dissolution of linear polymers in water and to the swelling of cross-linked polymers. However, as the temperature increases, the hydrophobic interactions between the hydrophobic groups of polymer increase. The hydrogen bonds becomes weaker causing gelation of linear polymers and collapse of cross-linked systems [204].

The most natural polymers such as gelatin, agarose, amylose, amylopectin, and some cellulose derivatives are positively thermosensitive (UCST) [204]. In addition to the aforementioned natural polymers, positively thermosensitive polymers include poly (acrylic acid), polyacrylamide, and poly (acrylamide-co-butyl methacrylate). The most studied representative of negatively thermosensitive polymers and hydrogels (LCST) are based on poly (*N*-isopropylacrylamide) (polyNIPAAm), poly (ethylene glycol), poly (propylene glycol), poly (methacrylic acid), poly (vinyl alcohol), poly (*N*-vinylpyrrolidone), poly [2-(dimethylamino) ethylmethacrylate], hydroxypropylcellulose etc. [204–206].

Particularly, coil-to-globule transition for polyNIPAAm-based systems was observed at LCST of about 32 °C that is very close to the temperature of the human body [180]. The side-group of polyNIPAAm contains both hydrophilic (amide) and hydrophobic (isopropyl) groups in each monomer unit. In other words, the polyNIPAAm molecule is hydrophilic with expanded polymer chains below LCST, but it is hydrophobic at temperatures above LCST with the polymer chains being in folded globular conformation.

#### 15.4.3.2 pH-Responsive Hydrogel Composites

pH-responsive hydrogels can swell or collapse in response to changes in pH. In the human body the values of pH vary greatly, ranging from strongly acidic in the stomach (pH 1.7) to weakly acidic in the vagina (pH 4.5–5.0) and approximately neutral in blood and colon (pH ~ 15.3) [201]. The pH values also vary between healthy and diseased tissues. For example, the pH in the tumor and its microenvironment is ~ 5.5–6.0 that is lower than in normal tissues due to the high metabolic rate and production of lactic acid [207]. Therefore, given the specific physiological environment, hydrogels that respond to pH are very relevant for oral, subcutaneous, and targeted organ therapy [208].

Anionic and cationic hydrogels are the two main classes of pH-sensitive hydrogels. The high degrees of swelling of such hydrogel systems can reflect the osmotic swelling of deprotonated anionic hydrogels and protonated cationic hydrogels. Anionic hydrogels have functional groups that can be ionized at a pH higher than their acid dissociation constant, p$K_a$, and show an increased equilibrium swelling degree at a higher, preferably alkaline, pH. Their polymer chains remain collapsed at a pH lower than their p$K_a$ due to the presence of physical interactions, specifically, hydrogen bonding, between the polymer chains. Instead, cationic hydrogels swell at a pH below p$K_a$ and collapse when pH exceeds the p$K_a$ value. Anionic hydrogels are based on monomers with carboxyl groups, in particular acrylic acid, methacrylic acid or carboxymethyl agarose and their copolymers. Cationic hydrogels are obtained by (co) polymerization of monomers with an amine or amide group, for example, acrylamide, dimethyl aminoethyl methacrylate, and 2-(diethylamino) ethyl methacrylate [201]. Hydrogels based on natural polymers such as chitosan, alginate [209], albumin [210], gelatin [211] also show a pronounced pH sensitivity, combined with biocompatibility and biodegradability.

#### 15.4.3.3 Electro-Responsive Hydrogels

Electro-responsive or electrically-active hydrogels can contract (collapse) or expand (swell) upon applying electric field. Due to fast, precise, and reproducible response to even minor changes in voltage and current, these hydrogels are extremely promising for applications in biomedical engineering [212]. Electro-sensitive hydrogels are polyelectrolytes, containing ionizable groups. Characteristic representatives of this class are hydrogels based on poly (acrylic acid), polyvinylpyrrolidone, polyvinyl alcohol, and sodium acrylate. The natural polyelectrolytes such as polysaccharides and proteins also belong to electro-responsive systems [201]. Incorporation of Lap in hydrogels can also result in improving of hydrogel electroconductivity.

For example, polyampholytic hydrogels based on Lap, polyacrylamide, and poly (3-acrylamidopropyl) trimethylammonium chloride cross-linked with ethylene glycol dimethacrylate showed controllable release of paracetamol by changing the pH or applying electric field [213].

#### 15.4.3.4 Magneto-Responsive Hydrogels

Magneto-responsive hydrogels, also known as ferrogels, are a class of hydrogel systems which response to changes in magnetic field [214]. Application of external magnetic field allow spatial and temporal control of the shape and elasticity of hydrogels [201]. This creates prerequisites for non-invasive remote manipulations with potential application in different pharmaceutical and biomedical applications, and development of soft actuators, microfluidic and microelectronic devices [215]. The magnetic hydrogels have also potential for construction of magnetically driven platform for targeted delivery and drug release, as well as for 3D and 4D printing [216].

In different studies the interactions in the mixtures of Lap with magnetic particles [217–220] and procedures for preparation of magnetic LapM particles have been discussed [30, 221–224].

Magnetic spherical particles of ferric oxide of maghemite ($\gamma$-$Fe_2O_3$) with a mean diameter of 8 nm were incorporated as probes for the determination of the structure of Lap suspensions. SANS and microrheological studies repealed the regions of homogeneous and microscopically phase segregated states in mixtures of Lap particles and maghemite [217], citrate-coated maghemite [218], and silica-coated maghemite [219]. Studies by SANS, or at intermediate scales by optical microscopy, revealed existence of densely connected domains of Lap platelets surrounding magnetic liquid pockets of maghemite nanoparticles [219]. Microscopic phase separation was explained by competition between orientational entropy and long-range electrostatic interactions.

Covered by maghemite ($\gamma$-$Fe_2O_3$) Lap particles have been prepared by a chemical route in 0.1 wt% aqueous suspension of Lap and characterized by TEM, XRD, magnetization measurements, Mossbauer spectroscopy, DLS, and MRI and $\zeta$-potential measurements [30]. Magnetized Lap particles exhibited superparamagnetic behavior with perfect ferrimagnetic ordering and negligible spin frustration, and strong $T_2$-weighted MRI contrast. These particles had a 200 nm hydrodynamic diameter, and formed rather stable hydrosols and/or hydrogels. Covered by maghemite Lap platelets have been used in magnetic hyperthermia experiments [225]. Hyperthermia is induced by irradiation of the magnetic fluids with an alternating electromagnetic field and it has promising application in the field of nanomedicine. The iron oxide/Lap particles had magnetic loading based on $Fe_2O_3$ phase. The specific absorption rate (SAR) studies revealed the influence of the initial susceptibility to the SAR values and the effect of its field dependence.

Gravitational and magnetic separation in aqueous colloidal mixtures of Lap with magnetic citrated cobalt ferrite nanoparticles have been studied [220]. The SAXS was mainly due to the magnetic nanoparticles and on a local scale, the mixtures behaved like a perfect gas of isolated magnetic particles. It was concluded that interactions between magnetic particles are mediated by the presence of Lap platelets.

In preparation procedures of LapM, the magnetic content is usually co-precipitated on Lap platelets from the solutions containing Fe (II) or Fe (III) salts, or both, at alkaline pH. LapM modified with magnetite ($Fe_3O_4$), hematite $\gamma$-$Fe_2O_3$, or cobalt ferrite ($CoFe_2O_4$) were synthesized [30, 221]. Then LapM can be incorporated into hydrogel composition yielding magnetic hydrogels. Compared to unmodified Lap, cross-linking of hydrogels with LapM can improve both the physicochemical properties of resulting hydrogels and provide magnetic sensitivity necessary for the development of targeted drug delivery systems and controlled drug release. Moreover, such modification could improve biocompatibility of Lap platelets [226].

One-step co-precipitation method was used to synthesize LapM with different weight ratios $X = Fe_3O_4/Lap$ (=0–2) [222–224]. This method allowed fine regulation of LapM particles magnetization. For characterization of LapM platelets the particle size distributions and sedimentation behavior in an external magnetic field were studied [222]. An increased aggregation of LapM particles in aqueous suspensions has been revealed, but all the systems demonstrated high sedimentation stability. Significant effects of value of $X$ on rate of sedimentation LapM particles in magnetic fields were also revealed. The temperature sensible hydrogels on the base of NIPAAm cross-linked by LapM were also synthesized [222–224]. Obtained magnetic thermosensitive hydrogels demonstrated a distinct volume phase transition within the physiologically acceptable temperature range of 33–36 °C and with controllable temperature-induced drug release. The DSC analysis [223, 224] evidenced that phase transition between the swollen and collapsed states depends on the polymer cross-linking method (covalent or physical). For magnetic hydrogels physically cross-linked by LapM, the phase transition temperature shifts toward higher temperatures compared to that for the chemically cross-linked magnetic hydrogels. Moreover, magnetization of Lap particles affected the hydrogel swelling and incorporation of LapM can also increase the sorption capacity. For example at $X = 2$ the increase in swelling by $\approx$2.7 was observed as compared with swelling for hydrogels based on pure Lap [222]. The effects of composition and structure of hydrogel composites on its thermo-responsive properties and hydrogel sorption and release of a model anticancer drug (5-fluorouracil) were reported [223]. With increase in magnetite concentration the magnetization of LapM nanoparticles increased, while fluorouracil absorptivity decreased.

Synthesized dual (magnetically-controlled and thermosensitive) hydrogels were mostly based on *N*-isopropylacrylamide (NIPAAm) with incorporated LapM [227], but there are numerous examples of the use of other monomers as well. For example, the highly tough, biocompatible, and magneto-responsive hydrogels based on *N,N*-dimethylacrylamide with incorporated Lap and magnetite were studied [228]. Due to the physical cross-linking of the hydrogel, its strength and elasticity increased significantly. Incorporation of magnetite, $Fe_3O_4$, allowed preparation of magneto-responsive hydrogels that comprise the magnetic nanoparticles dispersed in polymeric networks.

The hydrogels based on triblock copolymers (acrylamide, acrylic acid, 2-acrylamido methylpropane sulfonic acid, p(AM-*co*-AA-*co*-AMPS)) and Lap have been prepared in presence and absence of $Fe_3O_4$ nanoparticles by the *in situ* polymerization method [229]. The observed good mechanical properties and high sensitivity to magnetic field was explained by formation of unique multiple hierarchical structures composed by cross-linked polymer network and micro magnetic fluid units.

Chitosan-based magnetic beads with incorporated magnetic $Lap_{RD}M$ particles were prepared by solution-mixing method [230]. Magnetite, $Fe_3O_4$, was *in situ* immobilized on $Lap_{RD}$. Then $Lap_{RD}M$ platelets were dispersed in PVA and mixed with chitosan solution. The manufactured magnetic beads were tested for adsorption study of a bovine serum albumin (BSA). Introduction of $Lap_{RD}M$ noticeably improved the adsorption capacity. The modified magnetic LapM particles, prepared by co-precipitation the iron ions in the presence of $Lap_{RD}$, have been used for synthesis of PVA-based hydrogels by freeze–thaw cycles and subsequently, applying freeze-drying method [231]. The hydrogels were then characterized by different experimental techniques and the adsorption of BSA on magnetic hydrogels was investigated. Introduction of LapM led to an increase in adsorption capacity of hydrogels. The mechanism of BSA adsorption was attributed to the electrostatic interactions and physical hydrogen bonding. It was concluded that LapM-PVA hydrogels have attractive potential for biomedical applications such as drug delivery and enzyme immobilization.

Note, that LapM platelets or LapM-based hydrogels can be easily separated or manipulated in magnetic field. The effect of alternating and steady magnetic field (AMF and SMF) on diffusion of a model drug (dye methylene blue) was investigated using the hydrogel composite based on carboxymethyl cellulose and $CoFe_2O_4$ [232]. Application of AMF resulted in the increased release of methylene blue, especially at low and high frequencies, while under SMF the release of methylene blue was decreased.

In several works, the magnetic hydrogels have been tested as adsorbent for removal of contaminants and dyes from aqueous solutions [233–235]. The hydrogels based on hydroxypropyl methylcellulose—polyacrylamide, $Lap_{RD}$, and magnetite showed sensitivity to changes in both pH and external magnetic field [233]. Magnetically modified $Lap_{RD}M$ was incorporated into the hydrogels based on poly(vinyl alcohol) (PVA) which was cross-linked using the repeating freeze–thaw cycles [236]. The magnetite weight fraction was equal to 20, 40, and 60 wt%. The results indicated the suitability of the obtained composites as inexpensive adsorbents for removal of $Cd^{2+}$ from aqueous solutions. Similarly, the magnetic hydrogels based on sodium alginate and PVA were physically cross-linked with $Ca^{2+}$ using the repeating freeze–thaw cycles with incorporation of magnetically modified $Lap_{RD}M$. The incorporated LapM improved the strength and magnetosensitive properties of hydrogel. The maximal sorptivity of hydrogel granules towards bovine serum albumin was observed at pH 4.5 [221].

The magnetic hydrogels based on PVA with incorporated magnetic $Lap_{RD}M$ particles were prepared through dispersing $Lap_{RD}M$ in PVA solution, followed by the freeze–thaw technique for cross-linking [234]. The prepared materials were evaluated for removal of methylene blue dye from aqueous solutions. Introduction of magnetic nanoparticles resulted in decrease of the degree of swelling of hydrogels and increase of the dye adsorption capacity. It was concluded that such materials have a great potential application in the wastewater treatment.

The hydrogels based on $Lap_{RD}M$ and κ-carrageenan have been also prepared [235]. The $Lap_{RD}M$ nanoparticles were prepared through *in situ* method by introduction of iron salts into aqueous Lap suspension. The structures of pristine $Lap_{RD}$, $Lap_{RD}M$ and magnetic hydrogels were characterized using XRD, SEM, TEM, and vibrating sample magnetometer techniques. The decrease in water absorbency was found by introducing $Lap_{RD}M$. Introduction of $Lap_{RD}M$ allowed increasing adsorption capacity for dye (crystal violet) from aqueous solutions compared to the neat κ-carrageenan hydrogel.

#### 15.4.3.5 Photo-Responsive Hydrogels

The properties of photo-responsive hydrogels can be changed under illumination. Such hydrogels respond to illumination by transformations at molecular level caused by breaking or formation of bonds with corresponding changes in physicochemical properties, such as viscosity, elasticity, shape, and swelling degree [201]. Photo-responsive hydrogels are mainly functionalized by photo-sensitizers which are characterized by energy transitions, molecular conformational changes, or induced photochemical reactions due to light absorption in UV range [237].

#### 15.4.3.6 Chemically-Responsive Hydrogels

Chemoresponsive gels offer opportunities for biomimicry and the development of adaptive structures. Also, due to the enormous diversity facilitated by chemical synthesis, chemical compounds can be a highly specific and programmable stimulus for such hydrogels [238].

Hydrogels sensitive to biomolecules, specifically to enzymes, are an important class of chemically-responsive biomaterials [239]. Biochemical interactions, such as ligand-receptor, substrate-enzyme, and Watson–Crick DNA base pairing, are extremely highly specific and work in mild physiologically acceptable conditions, enabling the creation of autonomous, programmable, and smart hydrogels.

### *15.4.4 Medical Applications of Lap-Based Hydrogels*

Existing examples of medical applications of Lap-based hydrogels include different drug delivery systems, matrixes for cell culturing, and materials for regenerative medicine, specifically for tissue engineering.

#### 15.4.4.1 Drug Delivery

Stimuli-responsive hydrogel composites with incorporated Lap can be extremely promising for developing targeted and controlled drug delivery systems. The activation and control of drug release can be achieved by the application of external stimuli

**Table 15.7** Drug delivery systems based on hydrogels physically cross-linked with Lap for different incorporated drug molecules (Table 15.3)

| Polymer | Lap grade and drug | Application | Ref. |
|---|---|---|---|
| Alginate | $Lap_{RDS}$@DOX | Anticancer drug delivery system: transmembrane transport of DOX (high load) | [142] |
| Alginate | $Lap_{RDS}$@DOX | Transmembrane transport of DOX and controlled release | [241] |
| Alginate | $Lap_{XLG}$@proteins | Localized and controlled delivery of recombinant proteins from porous cryogels | [242] |
| Alginate | $Lap_{XLG}$@MAF (mafenide) | Wound healing | [243] |
| Alginate and PVA | $Lap_{RDS}$, bovine serum albumin (BSA) adsorbed by swelled gel | Wound healing | [244] |
| Chitosan | $Lap_{RD}$, VCM (vancomycin) | Prolonged drug release | [245] |
| Chitosan cross-linked with sodium tripolyphosphate | Lap, OFX (ofloxacin) | Hydrogel beads for pH-induced drug release | [246] |
| Poly(*N*-vinylpyrrolidone) - chitosan cross-linked by γ irradiation | $Lap_{RD}$, GCT (glucantime) | For treatment of injuries caused by Leishmaniasis disease | [247] |
| Gellan gum methacrylate (GG-MA) cross-linked by UV irradiation | $Lap_{XLG}$, OFX (ofloxacin) | Innovative wound dressing material for the treatment of burn wounds, which are subjected to chronic infections | [135] |
| Hydroxypropyl methylcellulose-g-poly(acrylamide) (HPMC-g-PAAm) | $Lap_{RD}$, DCF (sodium diclofenac) | pH- and magneto-induced drug release | [233] |
| (2-methoxy-ethylacrylate-co–*N*,*N*-dimethyl-acrylamide) (MEA-DMAAm) | $Lap_{XLG}$, LDC (lidocaine) hydrochloride | Thermoreversible drugs or chemicals release | [248] |

such as temperature, pH, magnetic field, light, or electric field [9, 172, 240] (Table 15.7).

When developing drug delivery systems, it is essential to control the drug release profile. If too much drug will get released at once, it can be harmful to the body, if too little, it can limit the drug efficacy. Delivery of drugs in the optimal dose over the optimal period of time can noticeably improve the therapy efficacy [249]. The excellent ability of Lap to interact with other molecules makes it an ideal carrier for transport and sustained release of a wide range of drugs, fluorescent molecules, biomolecules, and nanoparticles [12, 13]. Lap platelets can also protect drugs from degradation in the physiological environment [13].

Different Lap-based therapeutic platforms have been already discussed in Table 15.4 (unmodified Lap loaded with DOX), in Table 15.5 (silanized LapSi conjugated with different substances and loaded with different drugs), and in Table 15.6 (unmodified Lap loaded with variety of drugs).

Drug delivery systems based on hydrogels physically cross-linked with Lap and loaded with some pharmaceutical molecules (Table 15.3) are presented in Table 15.15.

Different strategies were applied using Lap@Drug platforms and alginate (a biocompatible natural polymer). For example, covering of Lap@DOX platforms by alginate shells improved gel stability and removed the initial burst release of the DOX [142]. A similar approach using a biodegradable alginate hydrogel cross-linked with $Ca^{2+}$ to coat the Lap@DOX platforms was proposed [241]. These hybrid hydrogels can serve as an effective system of transporting DOX across the cell membrane and preventing ion-trapping of DOX by the acidic extracellular environment of tumors and endolysosomes. In addition, a significant, 17 day prolongation of DOX release was achieved. Lap particles with pre-adsorbed protein cargo were used for preparation of injectable cryogels for versatile protein drug delivery [242]. Porous cryogels were formed by bio-orthogonal cross-linking of alginate using tetrazine-norbornene coupling. Incorporation of Lap platelets within the walls of the cryogels substantially hindered the release of proteins and the kinetics of release could be precisely tuned by modifying the Lap content.

Lap@MAF (mafenide) platforms loaded with antimicrobial agent mafenide (Table 15.3) were introduced in alginate solutions, and finally Lap@MAF-alginate gel films were prepared [243]. Obtained systems showed suitable characteristics in application for the delivery of MAF to wound site. Hybrid PVA-alginate-PVA hydrogels with incorporated Lap platelets have been prepared [244]. Incorporation of Lap significantly reduced swelling of hydrogel and degradation rate, and improved hydrogel mechanical properties. The hydrogels were nontoxic toward human fibroblast skin. It was assumed that this hydrogels could be the ideal materials for wound healing applications.

The hydrogels based on the chitosan with incorporated Lap platelets have been also prepared [245–247]. Chitosan is a natural polysaccharide with excellent biocompatibility and non-toxic characteristics, and it has been widely used for preparation of hydrogels [250]. These hydrogels can be potentially used in many practical applications like wastewater treatment, drug delivery, and tissue engineering [251,

252]. By electrophoretic deposition method a uniform chitosan films containing 25–55 wt% $Lap_{RD}$ and loaded with vancomycin (VCM) (Table 15.3) were prepared [245]. The drug-eluting Lap-chitosan coatings demonstrated promising potential for orthopaedic applications.*In vitro* drug-release experiments showed an initial burst release followed by a long-term diffusion-controlled release. Hydrogel beads based on Lap and chitosan were prepared by ionic cross-linking reaction using sodium tripolyphosphate as the cross-linker [246]. It was suggested that incorporated Lap particles could act as a physical cross-linker to facilitate the formation of network structure. Incorporation of Lap improved the swelling ability, enhanced the drug entrapment efficiency, and slowed down the drug release behavior.

Hydrogels of poly(*N*-vinylpyrrolidone) (PVP) containing chitosan and $Lap_{RD}$ were cross-linked by gamma irradiation and characterized for glucantime (GCT) drug delivery [246]. Incorporation of Lap decreased the hydrogel swelling and drug release rate.

Gellan gum methacrylate (GG-MA) combined with $Lap_{XLG}$ (0.1–1 wt%) was cross-linked by UV irradiation for the fabrication of injectable hydrogels [135]. The highest value of storage modulus was found for the hydrogel containing 1 wt% Lap. The effects of Lap on the swelling and release capability of the model drug ofloxacin (OFX) (Table 15.3) were evaluated. Presence of Lap not affected the diffusion coefficient of OFX, but significantly decreased the amount of antibiotic released over the first 8 h.

pH-responsive magnetic hydrogels based on hydroxypropyl methylcellulose-g-poly(acrylamide) (HPMC-g-PAAm) with incorporated $Lap_{RD}$ were prepared [233]. The magnetized LapM particles were synthesized inside HPMC-g-PAA matrix through an *in situ* method. The TEM micrographs evidenced the immobilization of magnetite onto the Lap surface. The higher swelling capacity of the magnetic hydrogels was explained by the formation of carboxylate pendants on the LapM particles. Due to theses carboxylate groups, the magnetic hydrogels demonstrated a pH dependent swelling behavior. The drug release from hydrogels was tested using sodium diclofenac (DCF) (Table 15.3). The content of the released drug was dependent upon pH and could be regulated by the strength of the external magnetic field. In acidic environment (at pH 1.2) less than 5% of the absorbed amount of the drug was released. However, at pH 7.4, approximately 30–50% release was achieved. In alternating magnetic field the percentage of released drugs increased proportionally to magnetic field strength up to 80%.

Thermoreversible hydrogels based on (2-methoxyethylacrylate-*co*–*N*,*N*-dimethylacrylamide) (MEA-DMAAm) copolymer with incorporated $Lap_{XLG}$ were synthesized by in situ free-radical polymerization of two monomers (MEA and DMAA) in the presence of exfoliated Lap in aqueous media [248]. The mechanical and swelling properties were largely dependent on the DMAA/MEA mole ratio and Lap content. The thermoresponsive release of lidocaine (LDC) hydrochloride drug (Table 15.3) or glycerin from hydrogels was observed.

Note, that in addition to the ability to sustain controlled and prolonged release of immobilized drugs, hybrid carriers must maintain their colloidal stability in ion-rich media such as blood and extracellular fluid [13]. This can cause some complication in practical application of the developed systems given that Lap is prone to aggregation in the presence of polyelectrolytes. To prevent the aggregation of Lap, it can be modified with PEG [253]. A several orders of magnitude increase in the circulation of modified Lap platelets in the bloodstream was achieved. To prevent flocculation of $Lap_{XLG}$ nanoparticles, its modification (α-acetalpoly) (ethylene glycol)-block-[poly(2-(*N*, *N*-dimethylamino) ethyl methacrylate)] (acetal-PEG-b-PAMA) was used [254]. Authors concluded that the negative charge on the Lap surface was completely neutralized by alkali resistant cationic flocculant PAMA, and PEG formed an outer peripheral stabilizing layer around the Lap nanoparticles.

#### 15.4.4.2 Cell Culturing

Hydrogels and hydrogel composites can be used for culturing of cells, biomedical engineering, and regenerative medicine. A wide variety of cells including mesenchymal stem cells, osteoblasts, fibroblasts, and epithelial cells have been effectively cultured on the surface of hydrogel composites with incorporated Lap and copolymer hydrogels based on NIPAAm/polysilocsan and 2-methoxyethyl acrylate/*N*, *N*-dimethylacrylamide [255]. When the temperature was decreased, the cells were quantitatively separated from the hydrogel surface without the use of chemical agents such as trypsin. Incorporation of Lap into hydrogels may improve cytocompatibility of hydrogel composites, including cell adhesion, proliferation, and differentiation [18].

In many works, hydrogels based on Lap and poly(ethylene glycol)/poly(ethylene oxide) (PEG/PEO) have been tested. The PEG/PEO are hydrophilic and non-fouling polymers with very low adhesion of proteins and cells [256]. For example, introduction of Lap to the composition of PEG resulted in concentration-dependent increase in cell adhesion and proliferation of murine fibroblasts NIH3T3 [257], murine prosteoblasts MC3T3-E1 [258], and human bone marrow stromal cells (hBMSCs) [188]. The *in vitro* cytocompatibility of Lap cross-linked PEO (Lap + PEO) composite films using MC3T3-E1 mouse preosteoblast cells was tested and cell adhesion, spreading, proliferation, and mineralization were investigated as a function of film composition [259]. The ability of using of Lap + PEO composites to create bioactive scaffolds was demonstrated.

Hydrogels based on Lap and PEO with improved mechanical properties and suitable for stem cell cultivation have been described [188]. With increase of content of Lap, the mechanical properties and the attachment and propagation of stem cells were improved. Incorporation of Lap into hydrogels based on PEG enhanced the adhesion and spreading of cells on the surface of hydrogels [170, 256, 260]. It was also demonstrated that Lap-PEG diacrylate hydrogels can support cell cultures [261, 262]. Lap-polyNIPAAm hydrogels were suitable for culturing a wide variety of cells [263]. The cell adhesion and proliferation were depended on the Lap concentration and the highest number of cultured cells was observed at $\approx$4.6 wt% Lap. In contrast, the culturing of cells on conventional polyNIPAAm hydrogels (chemically cross-linked) was practically impossible regardless of the concentration of the cross-linking agent [263].

#### 15.4.4.3 Tissue Engineering

Tissue engineering is an important field of regenerative medicine that combines transplantation of biological scaffolds and the formation or regeneration of tissues, or both [264]. Due to the similarity of three-dimensional structure of hydrogels with extracellular matrix, high water content, and biocompatibility, they are widely used to mimic the structure and properties of native tissues.

Many hydrogels have high biocompatibility and structural similarity with native tissues. However, their low mechanical strength may hinder their biomedical applications. For example, scaffolds for tissue regeneration in plastic surgery require high strength and significant elongation [265]. Development of nanoreinforced hydrogels with an improved set of properties, including enhanced structural characteristics, contributed to the rapid development of regenerative medicine [266].

In recent years Lap-based hydrogels have found many attractive medical applications for bone regeneration, wound dressing with minimal inflammatory response, cardiac tissue repair, and as injectable hemostatic agents (Table 15.8).

Incorporation of Lap improved osteogenic differentiation of human mesenchymal cells in the absence of growth factors [270, 272–274]. Lap is an attractive rheological enhancer and has been extensively studied for the inherent [275] and indirect [276, 277] ability to stimulate the osteogenic differentiation of hBMSCs *in vitro*, but with

**Table 15.8** Examples of medical applications of hydrogels physically cross-linked with Lap [14]

| Polymer | Cell type | Application | Ref. |
|---|---|---|---|
| Poly(2-methoxyethyl acrylate-co–*N*,*N*-dimethyl-acrylamide) | Human mesenchymal stem cells, neonatal human dermal fibroblasts, and human osteosarcoma cells | Bone regeneration | [255] |
| Regenerated silk fibroin of Bombyx mori | Primary osteoblasts from skull of newborn Sprague–Dawley rats | Bone regeneration | [267] |
| Dopamine grafted four-armed poly(ethylene glycol) | L929 mouse fibroblasts | Wound dressing with minimal inflammatory response | [268] |
| Gelatin methacrylate | Human bone marrow-derived mesenchymal stem cells and human umbilical vein endothelial cells | Cardiac tissue repair and regeneration | [269] |
| Gelatin methacryloyl | Bone marrow-derived human mesenchymal stem cells | Bone regeneration | [270] |
| Gelatin (type A) | Mouse monocyte/macrophage RAW 264.7 cells | Injectable hemostatic agent | [271] |

limited exploration of its *in vivo* potential. The cross-linked by Lap hydrogels with shear-thinning properties were also synthesized [278].

*In vivo* experiments have confirmed that physical cross-linking by Lap can promote capillary formation, reduce scarring, and improve cardiac function. Hydrogels based on Lap and dopamine-modified PEG were used to make bio-adhesive substrates with improved bioactivity for muscle tissue regeneration [268]. The hydrogels based on gelatin (type A, from pig skin) and Lap demonstrated the increased hemocompatibility, hemostasis, increased blood clotting rate, and could be a promising hemostatic agent [271].

The injectable system with high mechanical stiffness was developed based on natural polysaccharide (κ-carrageenan) and $Lap_{XLG}$ [279]. Addition of Lap increased protein adsorption on hydrogels. It enhanced cell adhesion and spreading, increased platelets binding and reduced blood clotting time. The developed multifunctional composite facilitated *in vitro* tissue regeneration and wound healing and can be used as an injectable hemostat or as a vehicle for therapeutic delivery to facilitate tissue regeneration.

Numerous studies have demonstrated the direct bioactive effect of Lap on osteogenic differentiation of skeletal populations. An improved osteogenic response of MC3T3 preosteoblast cell line and bone marrow stromal cells with incorporated Lap into PEO-based polymer [259], poly(glycerol sebacate) [280], PEG [170], polycaprolactone, and gelatin methacryloyl [259, 270] have been reported. Addition of Lap suspension to cell culture medium resulted in dose-dependent specific regulation of osteogenic gene expression (RUNX2, BGLAP, and SPARC), increased alkaline phosphatase activity, synthesis of collagen type I, and calcium phosphate formation [40, 281]. A significant osteogenic effect was observed even in the absence of standard additives for osteogenic differentiation (dexamethasone, ascorbate-2-phosphate, and $\beta$-glycerophosphate), indicating a direct effect of Lap on the activation of osteogenic pathways.

Incorporation of Lap not only enhanced the mechanical properties and affected drug release characteristics of hydrogels, but also improved their biological properties as Lap contain biominerals such as Si, Mg, and Ca. Lap incorporated to the composition of thiol-modified hyaluronic acid and alginate (in amounts up to 0.5%) can be used for effective bone regeneration [282]. A significant increase in mechanical strength of composite material, increased viability of human marrow stem cells (hMSC), and improved cell proliferation and differentiation compared to the original hydrogel mixture was observed. Hydrogels were characterized by a 36-fold increase in hMSC activity and simultaneous increase in biomineralization. This improvement in a wide range of parameters was attributed to the improved affinity of the polymer composite, additional spatial cross-linking of polymer, and partially intercalated/exfoliated morphology of the hydrogels.

Excellent osteogenic differentiation of hMSC and material biocompatibility even in the absence of growth factors has led to the prospect of using for bone regeneration the composite based on Lap and modified gelatin [270]. DNA- and Lap-based hydrogels [283] improved osteogenic stem cell differentiation, calcium deposition,

**Table 15.9** Lap-based hydrogel inks for 3D printing. Incorporation of Lap provided physical cross-linking

| Composition | Type of cross-linking | Ref. |
|---|---|---|
| 2,2,6,6-tetramethylpiperidinyl-1-oxyl -oxidized bacterial cellulose + sodium alginate + $Lap_{XLS}$ | Ionic-covalent entanglement | [287] |
| Alginate + methylcellulose + $Lap_{XLG}$ | Ionic–covalent entanglement | [288] |
| Gelatin methacryloyl + methylcellulose + Lap | Induced by visible light illumination | [289] |
| Gelatin methacryloyl + kappa-carrageenan + $Lap_{XLG}$ | Ionic-covalent entanglement | [285] |

long-term biodegradation, resulted in stable drug release, and significant bone growth in cranial defects.

The growing demand on customized implants [284, 285] is accompanied by both the development of advanced bioactive materials and processing strategies for *in vitro* and *in vivo* fabrication of complex 3D structures. The 3D printing allows producing systems that resemble and mimic the structure and functions of native tissues and organs [284]. Nowadays, this technique attracts great attention in tissue engineering and regenerative medicine [285]. 4D printing involves the fourth dimension of time and allows production of materials with ability to change their shape/function over time. The hydrogels can be used as ink in 3D and 4D printing [286]. Table 15.9 presents examples of hydrogel inks based on polysaccharide materials physically cross-linked with Lap. To introduce bioactive properties to printed materials, a variety of biomolecules and cells can be added to the formulations.

Among different grades of Lap, for biomedical application $Lap_{XLG}$ was studied the most due to its lower heavy metal content [284]. The bio-ink formulations should be further improved to facilitate cell spreading, proliferation, and cell growth within printed scaffold. Lap can be used to improve hydrogel printability [285, 287, 288], specifically shear thinning properties [285, 289], and to decrease the swelling [287, 289] and degradation of the scaffolds [289]. Incorporation of anisotropic Lap platelets can provide *in situ* cross-linking and also add extra functionality to polymer matrix [285, 288, 289]. Moreover, integration of Lap in printed products allows development of smart medical systems that can be used as biosensors, bioactuators, and biorobots.

## 15.5 Cytotoxicity and Antimicrobial Activity Assessments

The screening of cytotoxicity is important for different biomedical applications of Lap-based materials. These materials are applied in drug delivery systems, polymer hydrogels, bioactive, bioimaging, and biosensor systems, different cosmetic pharmaceutical products and formulations. The existing tests include *in vitro* cytotoxicity and cell viability assays, and *in vivo* acute and chronic toxicity assays [290, 291]. *In vitro* assays evolve relatively fast, convenient and less expensive experiments performed with isolated tissues, organs or cells. These tests are aimed to mimic approximately the cellular environments. In contrast, *in vivo* assays performed with living organisms (e.g., with model animals, mice, rats, and rabbits) are time consuming, more expensive, and can include concerns of ethical conflicts. The toxicity tests are also characterized by exposure duration. Acute toxicity tests usually include effects from a single or from multiple exposures in a short period (usually 24–96 h). This helps in calculating the dosage of the exposure to assess the lethal concentration. Chronic toxicity tests usually include effects of repeated exposures at lower levels conducted for longer periods (months or years).

### *15.5.1 Toxicity Tests*

*In vitro* assays for toxicity assessment are based on evaluation of cell membrane integrity by level of penetration of dyes into the cells and staining intracellular components [136, 292]. Many authors have tested cytotoxic effects of Lap and its modifications on various mammalian cell lines. Some popular gold standards designed to measure cell viability or proliferation are shown in Table 15.10.

These assays are based on selective reducing ability of different tetrazolium (MTT, MTS, XTT, or WST), aeurhodin (neutral red), and phenoxazine (resazurin) dyes. For characterization of antagonist drug potency *in vitro* pharmacological tests the value of $IC_{50}$ (half maximal inhibitory concentration required for 50% mortality in test population) is commonly used [296].

### *15.5.2 Antimicrobial Tests*

Existing *in vitro* antimicrobial tests are based on the broth dilution, gradient, and agar disk diffusion methods [297] In the most popular agar diffusion test the inhibition zones of bacterial growth around each of the antibiotic platelets are measured. In this method, the Petri dishes containing Muller-Hinton agar plates are inoculated either with the Gram-negative or Gram-positive bacteria [19]. The method is simple, accurate, reproducible, relatively inexpensive, and the results are "qualitative". For characterization of drug antimicrobial efficiency the minimum inhibitory concentration, *MIC*, is commonly used [19, 298]. This concentration corresponds to the lowest concentration that inhibits the growth of microbial pathogen after overnight incubation.

**Table 15.10** Popular colorimetric assays for toxicity assessment [293–295]

| Assay | Chemical structure of dye, absorbance maximum λm and other comments |
|---|---|
| MTT | 3-(4,5-Dimethyl-2-thiazolyl)-2,5-diphenyl-2H tetrazolium bromide dye. It is bioreduced by viable cells into the insoluble in water purple colored formazan crystals. Obvious disadvantage is that MTT assay requires additional step necessary for dissolving of the formazan crystals in appropriated solvent (dimethyl sulfoxide or in acidified ethanol solution) for further absorbance analysis (at λm≈550 nm). The main uncertainties are related with dependence of the conversion efficiency on metabolic rate and number of mitochondria |
| MTS | 3-(4,5-dimethylthiazol-2-yl)-5-(3-carboxymethoxyphenyl)-2-(4-sulfophenyl)-2H-tetrazolium. A "one-step" assay. MTS, in the presence of phenazine methosulfate (PMS), is bioreduced by viable cells into a formazan product (λm = 490–500 nm) |
| XTT | 2,3-bis-(2-methoxy-4-nitro-5-sulfophenyl)-2H-tetrazolium-5-carboxanilide. A 'one-step' assay yielding higher sensitivity and a higher dynamic range as compared with MTT. Water soluble XTT is bioreduced by actively respiring cells into the water-soluble, orange colored formazan, avoiding a final solubilization step (λm = 475 nm) |
| WST-8<br>CCK-8 | 2-(2-methoxy-4-nitrophenyl)-3-(4-nitrophenyl)-5-(2,4-disulfophenyl)-2H-tetrazolium, or mono sodium. In living cells the dye is bioreduced by dehydrogenase activities to give a yellow-color formazan dye (λm = 460 nm). The detection sensitivity is higher than the other tetrazolium salts such as MTT, MTS or XTT. Cell Counting Kit-8 (CCK-8) allows convenient assays using WST-8 |
| NR | 3-Amino-7-dimethylamino-2-methylphenazine hydrochloride. Neutral red is aeurhodin dye. It is act as pH indicator, changing its color from red (pH 6.8) to yellow (pH 8.0). NR is bioincorporated by viable cells into the lysosomes. The loss of NR uptake corresponds to loss of cell viability. It provides a quantitative measurement the cell viability (λm = 540 nm) |
| Resazurin<br>Alamar Blue® | 7-Hydroxy-3H-phenoxazin-3-one 10-oxide. Resazurin is a phenoxazine dye. Resazurin is bioreduced into the pink-colored and highly fluorescent resorufin by aerobic respiration of viable cells. The assays can be monitored by absorbance or fluorescence measurements |

### *15.5.3 Examples of Bioactivity Assessments*

The Lap particles do not contain different impurities typical for natural clays and it allows avoiding side effects caused by possible toxic impurities. Moreover, significant degradation (dissolution) of Lap particles in aqueous systems (especially at pH lower than 9 [31]) can result in releasing of non-toxic products like $Na^+$, $Si(OH)_4$, $Mg^{2+}$, $Li^+$, etc. These degradation products can be used in many attractive biomedical applications [13], e.g., they can induce osteogenic differentiation of mesenchymal stem cells [299], promote bone formation by regulating cell adhesion [300], and activate the Wnt/β-catenin signaling pathway [301].

#### 15.5.3.1 Pristine Lap Particles

Effects of pristine Lap platelets on cellular metabolism were investigated in series of works (Table 15.11) [40, 136, 302]. The cytotoxicity tests with relatively low concen-

**Table 15.11** Effects of loading of pristine Lap particles on the cellular metabolism determined by colorimetric assays (Table 15.10)

| Lap grade | Type of assays and cells | Comments | Ref. |
|---|---|---|---|
| $Lap_{RDS}$ | Resazurin: mesenchymal stem cells (MSCs) | With increasing of Lap concentration (from 1 to 7 mg/L), the cell viability decreased. After 1 day for 7 mg/L Lap ($7 \times 10^{-4}$ wt%) the viability was ≈83%, which indicated that Lap is biocompatible | [302] |
| $Lap_{RD}$ | MTT: human embryonic kidney 293 cell line (HEK-293) and Cervical cancer cell line (SiHa)<br>Antimicrobial test: *Escherichia coli* Gram-negative bacterial cells | For Lap concentration of 0.0125 mg/L within 24 h of exposure on HEK and SiHa cell lines the cytotoxicity was 21.2% and 8.6%, respectively<br>Lap exhibited maximum efficacy as an antimicrobial agent against *Escherichia coli* | [136] |
| $Lap_{XLG}$ | MTT: human Mesenchymal Stem Cells (hMSCs) | The metabolic activity was unaffected at small concentration of Lap (<1 g/L) but at high concentration (>5 g/L) the metabolic activity dropped drastically. The estimated value of $IC_{50}$ was ≈4 g/L (≈0.4 wt%) | [40] |
| $Lap_{XLG}$ | MTS: human adipose stromal cells (hASC) | Lap was non-toxic only up to 10 μg/L ($10^{-6}$ wt%). From 10 μg/L and onwards, the toxicity and Lap internalization inside the cells increased | [306] |

trations of Lap in water ($C_{max} = 7$ mg/L for $Lap_{RDS}$ [302] and $C_{max} = 0.0125$ mg/L for $Lap_{RD}$ [136]) indicated the biocompatibility of Lap.

Experiments with higher concentration of $Lap_{XLG}$ (up to 2 wt%) revealed the half maximal inhibitory concentration $IC_{50}$ of ≈4 g/L (0.4 wt%) [136]. It was concluded that Lap platelets show nearly no cytotoxicity at concentrations lower than 1 g/L (0.1 wt%) and these concentrations were recommended for using in biomedical applications. Note that estimated value of $IC_{50}$ for $Lap_{XLG}$ was significantly higher compared to that for other nanoparticles such as hydroxyapatite ($IC_{50}\approx$ 0.25 g/L) [303], silica ($IC_{50}\approx$ 0.4–0.5 g/L) [304], and graphene ($IC_{50}\approx$ 0.1 g/L) [305]. However, other estimations for the same $Lap_{XLG}$ gave significantly smaller cytotoxicity threshold of order of 10 μg/L [306]. The Lap induced cytotoxicity effects were be explained by the restriction of cellular functionality related with adhesion of Lap particles on membrane surface and internalization of Lap inside cells (Fig. 15.8b) or interaction of charged Lap particles with the media proteins [136]. At relatively large concentration of Lap ($C > 1$ g/L), enhanced generations of intracellular reactive oxygen species and reactive nitrogen species were observed.

The cytotoxicity of Lap can also reflect the destabilization of the entire membrane structure through the strong electrostatic interactions between the positively charged edges of Lap platelets with negatively charged cell membranes [136]. In general, Lap demonstrates high biocompatibility towards the human epithelial carcinoma cells KB at high concentrations ($C = 1$ g/L) [138]. Note that cytotoxicity of nano-sized clay particles may depend on chemical composition and nanoparticle shape, presence of impurities and other factors). For example, the cytotoxicity of montmorillonite was lower than for Lap [136], and cytotoxicity of halloysite was even lower compared to montmorillonite [308].

#### 15.5.3.2 Drug Delivery Platforms and Hydrogel Networks

In drug delivery systems, the bioactivity can reflect presence of Lap particles, drug, supplementary materials (polymers, bioactive molecules), and different effects related with their interactions. Table 15.12 presents some examples of cytotoxicity and antimicrobial evaluations for different systems used in drug delivery platforms and hydrogel networks.

## 15.6 Final Remarks and Perspectives

In this review work, we have discussed recent advances in biomedical applications of Lap-based nanomaterials and formulations. The topic is really “hot” at the moment and each year considerable numbers of research papers, reviews, and book are issued on biomedical applications of Lap-based materials. Lap is a unique two-dimensional material with a list of amazing properties. The aqueous suspensions of Lap represent a testing ground for studies of the fundamental phenomena of colloidal science

**Table 15.12** Examples of bioactivity assessments for Lap-based materials

| Drug delivery nanoplatforms | | | |
|---|---|---|---|
| System | Type of assays and cells | Comments | Ref. |
| Lap@DOX | *In vivo* animal experiments: Mouse model, intra-tumor injection of human epidermoid carcinoma (KB) cells | Platforms inhibited tumor growth more significantly than free DOX. Enhanced antitumor efficacy was attributed to platforms accumulation in the tumor region via the known enhanced permeability and retention effect. The blood and histology examinations evidenced Lap@DOX is a promising platform for anti-tumor therapy applications | [139] |
| Lap + (PEG-PLA) + PEI($Au^0$) + HA@ DOX $Au^0$-DOX combination | CCK-8 (WST-8): human cervical cancer cell line (HeLa cells). Mouse fibroblast cell line (L929 cells). *In vivo* targeted cancer cell inhibition (mice animals) | *In vitro* viability tests with HeLa cells evidenced that the antitumor efficacy of combined DOX-$Au^0$ platforms is solely associated with the encapsulated DOX. The value of inhibitory concentration $IC_{50}$ of platforms (5.8 mg/L) was higher than that of free DOX (1.8 mg/L). The specific targeting effects of the combined DOX-$Au^0$ platforms were also demonstrated in the viability tests with L929 cells. *In vivo* experiments illustrated significant efficiency of platform for inhibition the growth of tumors. Moreover, the platform can be used as a promising theranostic platform for targeted chemotherapy and imaging of tumor overexpressing CD44 receptors | [147] |
| $Lap_{RDS}$@DOX) + (PEG-PLA) | Resazurin: human osteosarcoma cell line (CAL-72) | DOX-loaded platforms were effectively internalized by onside CAL-72 cells and exhibited a remarkable higher anticancer cytotoxicity than free DOX | [141] |
| $Lap_{XLG}$@DOX + PVP | MTT: human epithelial carcinoma cell line (KB cells) | The blank Lap + poly(*N*-vinylpyrrolidone) (PVP) platforms have not demonstrated distinct cytotoxicity and observed cytotoxicity reflected only the encapsulation of DOX. The intensive internalization ability (tracked by fluorescence microscopy) and high therapeutic accumulation of platforms inside cells was observed. The DOX-loaded platforms exhibited higher anticancer cytotoxicity as compared to free DOX | [148] |

(continued)

**Table 15.12** (continued)

| Drug delivery nanoplatforms | | | |
|---|---|---|---|
| System | Type of assays and cells | Comments | Ref. |
| $Lap_{RDS}$@DEX | MTT: human osteosarcoma cell line (MG63) | Fabricated platforms combined the respectable cytocompatibility together with sustained DEX release. DEX release from platforms did not reveal any cytotoxic effect. The increased viability of MG63 cells (24 and 48 h periods of culture) compared to Lap-free samples (DEX enriched medium) was observed. It was explained by positive role of Lap in control of burst release of DEX. Platforms may be used for local delivery of DEX in bone tissue engineering application | [155] |
| $Lap_{RD}S$@CHXDG | XTT: normal human lung fibroblasts (GM07492A) | Silanized organophilic particles LapSi (APMES) were loaded with disinfectant chlorhexidine digluconate (CHXDG). The following $IC_{50}$ values were observed when the cells were treated with different materials: >5 g/L (Lap), >5 g/L (APMES), ≈0.455 g/L (sLap), ≈0.070 g/L (sLap@CHXDG) and ≈0.024 g/L (CHXDG). The antibacterial tests by the agar diffusion technique revealed that sLap@CHXDG in the powder form was not effective against *Staphylococcus aureus*, but it is effective against *Streptococcus pyogenes* | [101] |
| $Lap_{XLG}$@TTC | Antimicrobial tests: *Escherichia coli* (NCTC 9001), (Gram-negative). *Staphylococcus aureus* (Oxford NCTC 6571), (Gram-positive) | Lap@TTC platform for local treatment of periodontal disease was fabricated. TCC released from Lap was biologically active and inhibited cell growth with similar *MIC* as for TCC alone. For following value were estimated: *MIC* ≈ 1.875 mg/L (*Escherichia coli*) and *MIC* ≈ 0.468 mg/L (*Staphylococcus aureus*) | [243] |

(continued)

**Table 15.12** (continued)

| Drug delivery nanoplatforms | | | |
|---|---|---|---|
| System | Type of assays and cells | Comments | Ref. |
| $Lap_{RDS}$@DOX + (oHA-APBA) | MTT: Human breast cancer cell line (MCF-7) | Lap@DOX was assembled with oligomeric hyaluronic acid-aminophenylboronic acid (oHA-APBA) copolymer. DOX loaded platforms revealed preferable cytotoxicity. The values of half-maximal inhibitory concentration were found to be $IC_{50} \approx 0.66$ g/L as compared to $IC_{50} \approx 12.85$ g/L for free DOX The enhanced cytotoxicity of these platforms for MCF-7 cells was attributed to the increased intracellular concentration of DOX owing to the active uptake mediated by oHA-APBA | [143] |
| $Lap_{XLG}$@(DOX + PVP)@MTX Dual drugs | MTT: Human epithelial carcinoma cell line (KB cells) | Lap@DOX was covered by poly(*N*-vinylpyrrolidone) (PVP) and then outer layer was loaded by MXT. The enhanced anticancer bioactivity of platforms was associated with enhanced therapeutic intracellular accumulation of two drugs in cancer cells | [144] |
| $Lap_{XLG}$@DOX + PAH/PSS Single drug Two-layer shell | Resazurin: Human breast cancer cell line (MCF-7) | The core of Lap@DOX was covered by different numbers of layers (1–6) of cationic poly(allylamine) hydrochloride and anionic poly(sodium styrene sulfonate) (PAH/PSS). The numerous aggregated platforms were observed both around the cells and over the cell surfaces Moreover, the cell viability tests showed effective internalization of DOX (both in its free form and associated to drug platform) by MCF-7 cancer cells | [146] |
| **Composites and hydrogels** | | | |
| System | Type of assays and cells | Comments | Ref. |
| $Lap_{XLG}$ + alginate | MTS: Human osteosarcoma cell line (MG63) | Alginate is a natural anionic polysaccharide (obtained from brown seaweed) with good biocompatibility and low toxicity. Lap-alginate hydrogels were non-toxic and the higher cell viability observed in the extracts can reflect the release of silica from the Lap platelets that improves the cellular growth | [292] |

(continued)

**Table 15.12** (continued)

| Drug delivery nanoplatforms | | | |
|---|---|---|---|
| System | Type of assays and cells | Comments | Ref. |
| $Lap_{XLG}$@MAF + alginate | MTT: Human skin fibroblast cells Antimicrobial tests: *Escherichia coli, Pseudomonas aeruginosa* (Gram-negative). *Staphylococcus aureus* (Gram-positive) | The dressing compositions Lap@mafenide (MAF) (gel) and (Lap@MAF)-alginate (film) were tested for the delivery of mafenide to wound site. Here, MAF is an antimicrobial agent with broad bacteriostatic action against bacteria. It commonly used for treating burn wounds. The cytotoxic effects of eluted MAF extract showed on fibroblasts were significantly less than at equivalent concentration of MAF in culture medium. Moreover, the prepared dressings have the potential for wound healing application. Antimicrobial tests evidenced that *MIC* values of the extracts from these compositions were similar to that of control MAF solution with the equivalent concentration (*MIC* = 6.25 mg/L for *Escherichia coli*, *MIC* = 1.56 mg/L for *Pseudomonas aeruginosa*, and *MIC* = 1.56 mg/L for *Staphylococcus aureus*. The control pristine Lap did not demonstrate any bactericidal effect | [243] |
| $Lap_{RDS}$ + alginate<br>$Lap_{RDS}$ + alginate @DOX<br>alginate @DOX | Resazurin: Osteosarcoma cells (CAL-72) | The antitumor efficacy was solely related to the loaded DOX within the hydrogels and unloaded hydrogels did not display any cytotoxicity. The (Lap - alginate)@DOX platforms were effectively internalized by cancer cells and exhibited higher anticancer efficacy than the corresponding alginate@DOX ones | [241] |
| $LAP_{RDS}$ + PVA- alginate | MTT: Human dermal fibroblasts cells. Human osteosarcoma cells (MG63) | Interpenetrating hydrogels composed of LAP + polyvinyl alcohol (PVA) and alginate revealed improved biocompatibility. Moreover, hemolysis and clotting tests indicated the hydrogel promoted hemostasis which could be helpful for the wound dressing | [244] |
| $Lap_{XLG}$ + GG-MA | NR: Human fibroblast cells (WI-38) | Injectable and photocrosslinkable Lap + polymeric gellan gum methacrylate (GG-MA). The hydrogel slightly affected the cell viability (the reduction was less than 30%) and according to the international guidelines on biological evaluation of cytotoxicity (ISO 10993–5:2009), the material can be considered biocompatible | [135] |

(continued)

**Table 15.12** (continued)

| Drug delivery nanoplatforms | | | |
|---|---|---|---|
| System | Type of assays and cells | Comments | Ref. |
| $Lap_{XLG}$ + Si-HPMC | MTS: Human adipose stromal cells (hASC) | The Lap + silated hydroxypropylmethyl cellulose (Si-HPMC) hydrogels for cartilage tissue engineering applications were fabricated. The hydrogels in culture medium were biocompatible and cells appeared to remain viable | [306] |
| $Lap_B$ + chitosan | MTT (ISO10993-5, 2009): mouse fibroblast cells (BALB/C 3 T3 clone A31). Antimicrobial tests: *Staphylococus aureus* (Gram-positive) and *Escherichia coli* (Gram-negative) | Non-cytotoxic Lap + chitosan scaffolds for wound dressing applications were fabricated. The antimicrobial tests evidenced that the scaffolds have a bactericidal potential. *MIC* values for different bacteria and all composite were nearly the same (≈125 mg/L) | [307] |

related to the formation of "house of cards" arrangements, Wigner glasses, and aging. Lap platelets can be modified and functionalized with fine regulation of their surface charge, colloidal stability and affinity to other species like macromolecules, biomolecules and drugs. Existing methods of Lap modification (acid activation, organic modification, and covalent attachment) allow formation of different Lap-drug platforms with many attractive applications in pharmacology, biomedicine, and drug delivery systems. Lap platelets have shown a high potential for significant improvement of properties of medical hydrogels. The unique gradient and stimuli-responsive hydrogels with incorporated Lap platelets have already been synthesized and studied. Exiting examples of medical applications of Lap-based hydrogels include drug delivery, cell culturing, and tissue engineering. Moreover, these materials have great potential for diverse applications in regenerative medicines, drug delivery and implantable devices, bioprinting, bioactuation, biosensing, and biorobotics with improved performance and biocompatibility [11, 14, 266, 286]. However, the more detailed studies with the focus on the cytotoxicity and antimicrobial activity assessments of Lap-based materials are crucial and necessary in future.

**Acknowledgements** This work was supported by the funding of the National research foundation of Ukraine, Project #2020.02/0138 "Electrokinetic phenomena in natural nano/micro-fluidic and disperse systems: characterizing, treatment, modelling" (MM), complex interdisciplinary research program of NAS of Ukraine "Molecular-Biological Factors of Heterogeneity of Malignant Cells and Variability of Clinical Course of Hormone-Dependent Tumors", Projects # 0117U002034 (OS,VS), and by the funding from the National Academy of Sciences of Ukraine, Projects 7/9/3-f-4-1230-2020 #0120U100226 and # 0120U102372/20-N (YuS,NL).

## References


1. Neumann BS (1965) Behaviour of a synthetic clay in pigment dispersions. Rheol Acta 4:250–255. https://doi.org/10.1007/BF01973660
2. Shafran K, Jeans C, Kemp SJ, Murphy K (2020) Dr Barbara S. Neumann: Clay scientist and industrial pioneer; creator of Laponite®. Clay Miner 1–9. https://doi.org/10.1180/clm.2020.35
3. Ghadiri M, Chrzanowski W, Rohanizadeh R (2015) Biomedical applications of cationic clay minerals. RSC Adv 5:29467–29481. https://doi.org/10.1039/C4RA16945J
4. Lebovka N, Lisetski L, Bulavin LA (2018) Organization of nano-disks of laponite® in soft colloidal systems. In: Bulavin L, Xu L (eds) Modern Problems of the Physics of Liquid Systems, pp 137–164
5. Pujala RK (2014) Dispersion stability, microstructure and phase transition of anisotropic nanodiscs. Springer International Publishing, ISBN 978-3-319-04555-9, Switzerland
6. Becher TB et al (2019) The structure--property relation Laponite® materials: from Wigner glasses to strong self-healing hydrogels formed by non-covalent interactions. Soft Matter 15:1278–1289. https://doi.org/10.1039/C8SM01965G
7. Massaro M, Cavallaro G, Lazzara G, Riela S (2020) Covalently modified nanoclays: synthesis, properties and applications. In: Cavallaro G, Fakhrullin R, Pasbakhsh P (eds) Clay Nanoparticles. Elsevier, pp 305–333
8. Peña-Parás L, Sánchez-Fernández JA, Vidaltamayo R (2018) Nanoclays for biomedical applications. In: Torres-Martinez LM, Kharissova OV, Kharisov BI (eds) Handbook of Ecomaterials. Springer International Publishing, pp 1–19
9. Murugesan S, Scheibel T (2020) Copolymer/Clay nanocomposites for biomedical applications. Adv Funct Mater 30:1908101. https://doi.org/10.1002/adfm.201908101
10. Gholamipour-Shirazi A, Carvalho MS, Huila MFG, Araki K, Dommersnes P, Fossum JO (2016) Transition from glass-to gel-like states in clay at a liquid interface. Sci Rep 6:37239. https://doi.org/10.1038/srep37239
11. Morariu S, Teodorescu M (2020) Laponite–a versatile component in hybrid materials for biomedical applications. Mem Sci Sect Rom Acad 43:1–25
12. Chimene D, Alge DL, Gaharwar AK (2015) Two-dimensional nanomaterials for biomedical applications: emerging trends and future prospects. Adv Mater 27:7261–7284. https://doi.org/10.1002/adma.201502422
13. Tomás H, Alves CS, Rodrigues J (2018) Laponite®: A key nanoplatform for biomedical applications? Nanomedicine Nanotechnology. Biol Med 14:2407–2420. https://doi.org/10.1016/j.nano.2017.04.016
14. Das SS, Hussain K, Singh S, Hussain A, Faruk A, Tebyetekerwa M et al (2019) Laponite-based nanomaterials for biomedical applications: a review. Curr Pharm Des 25:424–443. https://doi.org/10.2174/1381612825666190402165845
15. De Melo BR, Ferreira MA, Meirelles LMA, Zorato N, Raffin FN (2020) Nanoclays in drug delivery systems. In: Cavallaro G, Fakhrullin R, Pasbakhsh P (eds) Clay Nanoparticles. Elsevier, Properties and Applications. Micro and Nano Technologies, pp 185–202
16. Ianchis R et al (2020) Hydrogel-clay nanocomposites as carriers for controlled release. Curr Med Chem 27:919–954. https://doi.org/10.2174/0929867325666180831151055
17. Jayakumar A, Surendranath A, Mohanan PV (2018) 2D materials for next generation healthcare applications. Int J Pharm 551:309–321. https://doi.org/10.1016/j.ijpharm.2018.09.041
18. Mousa M, Evans ND, Oreffo ROC, Dawson JI (2018) Clay nanoparticles for regenerative medicine and biomaterial design: a review of clay bioactivity. Biomaterials 159:204–214. https://doi.org/10.1016/j.biomaterials.2017.12.024
19. Ogunsona EO, Muthuraj R, Ojogbo E, Valerio O, Mekonnen TH (2020) Engineered nanomaterials for antimicrobial applications: a review. Appl Mater Today 18:100473. https://doi.org/10.1016/j.apmt.2019.100473

20. Pramanik S, Das DS (2020) Chapter 9 - Future prospects and commercial viability of two-dimensional nanostructures for biomedical technology. In: Khan R, Barua S (eds) Two-Dimensional Nanostructures for Biomedical Technology. Elsevier, pp 281–302
21. Zhang J, Zhou CH, Petit S, Zhang H (2019) Hectorite: synthesis, modification, assembly and applications. Appl Clay Sci 177:114–138. https://doi.org/10.1016/j.clay.2019.05.001
22. Neumann BS (1970) Synthetic hectorite-type clay minerals. United States Patent No 3586478. Claims priority, applicatign Great Britain Ser. No 298401, June 20, 1963
23. Jeans CV (2009) Contrasting books on clay mineral science–how should they be judged? (shortened title two books on clay mineral science). Acta Geodyn Geromaterialia 6:45–59
24. Wang S et al (2019) Synthesis and biocompatibility of two-dimensional biomaterials. Colloids Surfaces A Physicochem Eng Asp 583:124004. https://doi.org/10.1016/j.colsurfa.2019.124004
25. Christidis GE, Aldana C, Chryssikos GD, Gionis V, Kalo H, Stöter M, Breu J, Robert J-L (2018) The nature of Laponite: pure hectorite or a mixture of different trioctahedral phases? Minerals 8:314. https://doi.org/10.3390/min8080314
26. Gantenbein D, Schoelkopf J, Matthews GP, Gane PAC (2011) Determining the size distribution-defined aspect ratio of platy particles. Appl Clay Sci 53:544–552. https://doi.org/10.1016/j.clay.2011.04.020
27. Balnois E, Durand-Vidal S, Levitz P (2003) Probing the morphology of laponite clay colloids by atomic force microscopy. Langmuir 19:6633–6637. https://doi.org/10.1021/la0340908
28. López-Angulo D et al (2020) Effect of Laponite® on the structure, thermal stability and barrier properties of nanocomposite gelatin films. Food Biosci 35:100596. https://doi.org/10.1016/j.fbio.2020.100596
29. Suman K, Joshi YM (2018) Microstructure and soft glassy dynamics of an aqueous laponite dispersion. Langmuir 34:13079–13103. https://doi.org/10.1021/acs.langmuir.8b01830
30. Tzitzios V et al (2010) Immobilization of magnetic iron oxide nanoparticles on laponite discs – an easy way to biocompatible ferrofluids and ferrogels. J Mater Chem 20:5418. https://doi.org/10.1039/c0jm00061b
31. Thompson DW, Butterworth JT (1992) The nature of laponite and its aqueous dispersions. J Colloid Interface Sci 151:236–243. https://doi.org/10.1016/0021-9797(92)90254-J
32. Mohanty RP, Joshi YM (2016) Chemical stability phase diagram of aqueous Laponite dispersions. Appl Clay Sci 119:243–248. https://doi.org/10.1016/j.clay.2015.10.021
33. Jatav S, Joshi YM (2014) Chemical stability of Laponite in aqueous media. Appl Clay Sci 97:72–77. https://doi.org/10.1016/j.clay.2014.06.004
34. Suman K, Mittal M, Joshi YM (2020) Effect of sodium pyrophosphate and understanding microstructure of aqueous LAPONITE® dispersion using dissolution study. J Phys Condens Matter 32:224002. https://doi.org/10.1088/1361-648X/ab724d
35. Anonymous (2018) Laponite. Performance Additives. BYK. Technical Information B-RI 21
36. Delavernhe L, Pilavtepe M, Emmerich K (2018) Cation exchange capacity of natural and synthetic hectorite. Appl Clay Sci 151:175–180. https://doi.org/10.1016/j.clay.2017.10.007
37. Capello C, Leandro GC, Campos CEM, Hotza D, Carciofi BAM, Valencia GA (2019) Adsorption and desorption of eggplant peel anthocyanins on a synthetic layered silicate. J Food Eng 262:162–169. https://doi.org/10.1016/j.jfoodeng.2019.06.010
38. Valencia GA, Djabourov M, Carn F, Sobral PJA (2018) Novel insights on swelling and dehydration of laponite. Colloid Interface Sci Commun 23:1–5. https://doi.org/10.1016/j.colcom.2018.01.001
39. Bippus L, Jaber M, Lebeau B (2009) Laponite and hybrid surfactant/laponite particles processed as spheres by spray-drying. New J Chem 33:1116–1126. https://doi.org/10.1039/B820429B
40. Gaharwar AK et al (2013) Bioactive silicate nanoplatelets for osteogenic differentiation of human mesenchymal stem cells. Adv Mater 25:3329–3336. https://doi.org/10.1002/adma.201300584
41. Li C, Liu Q, Mei Z, Wang J, Xu J, Sun D (2009) Pickering emulsions stabilized by paraffin wax and Laponite clay particles. J Colloid Interface Sci 336:314–321. https://doi.org/10.1016/j.jcis.2009.03.080

42. Ramsay JDF, Swanton SW, Bunce J (1990) Swelling and dispersion of smectite clay colloids: determination of structure by neutron diffraction and small-angle neutron scattering. J Chem Soc Faraday Trans 86:3919–3926. https://doi.org/10.1039/FT9908603919
43. Rosta L, Von Gunten HR (1990) Light scattering characterization of laporite sols. J Colloid Interface Sci 134:397–406. https://doi.org/10.1016/0021-9797(90)90149-I
44. Karpovich A, Vlasova M, Sapronova N, Sukharev V, Ivanov V (2016) Exfoliation dynamics of laponite clay in aqueous suspensions studied by NMR relaxometry. Orient J Chem 32:1679–1683. https://doi.org/10.13005/ojc/320346
45. Bakk A, Fossum JO, da Silva GJ, Adland HM, Mikkelsen A, Elgsaeter A (2002) Viscosity and transient electric birefringence study of clay colloidal aggregation. Phys Rev E 65:21407. https://doi.org/10.1103/PhysRevE.65.021407
46. Nicolai T, Cocard S (2000) Light scattering study of the dispersion of laponite. Langmuir 16:8189–8193. https://doi.org/10.1021/la9915623
47. Ali S, Bandyopadhyay R (2013) Use of ultrasound attenuation spectroscopy to determine the size distribution of clay tactoids in aqueous suspensions. Langmuir 29:12663–12669. https://doi.org/10.1021/la402478h
48. Pawar N, Bohidar HB (2011) Anisotropic domain growth and complex coacervation in nanoclay-polyelectrolyte solutions. Adv Colloid Interface Sci 167:12–23. https://doi.org/10.1016/j.cis.2011.06.007
49. Au P-I, Hassan S, Liu J, Leong Y-K (2015) Behaviour of LAPONITE® gels: rheology, ageing, pH effect and phase state in the presence of dispersant. Chem Eng Res Des 101:65–73. https://doi.org/10.1016/j.cherd.2015.07.023
50. Labanda J, Llorens J (2005) Influence of sodium polyacrylate on the rheology of aqueous Laponite dispersions. J Colloid Interface Sci 289:86–93. https://doi.org/10.1016/j.jcis.2005.03.055
51. Labanda J, Sabaté J, Llorens J (2007) Rheology changes of Laponite aqueous dispersions due to the addition of sodium polyacrylates of different molecular weights. Colloids Surfaces A Physicochem Eng Asp 301:8–15. https://doi.org/10.1016/j.colsurfa.2007.01.011
52. Huang AY, Berg JC (2006) High-salt stabilization of Laponite clay particles. J Colloid Interface Sci 296:159–164. https://doi.org/10.1016/j.jcis.2005.08.068
53. Zhang S, Lan Q, Liu Q, Xu J, Sun D (2008) Aqueous foams stabilized by Laponite and CTAB. Colloids Surfaces A Physicochem Eng Asp 317:406–413. https://doi.org/10.1016/j.colsurfa.2007.11.010
54. Manilo M, Lebovka N, Barany S (2014) Characterization of the electric double layers of multi-walled carbon nanotubes, laponite and nanotube+ laponite hybrids in aqueous suspensions. Colloids Surfaces A Physicochem Eng Asp 462:211–216. https://doi.org/10.1016/j.colsurfa.2014.09.006
55. Savenko VS (2014) Aging of Laponite aqueous suspensions in presence of anionic surfactant. Bull Taras Shevchenko Natl Univ Kyiv Ser Phys Math 2:277–282
56. Tawari SL, Koch DL, Cohen C (2001) Electrical double-layer effects on the Brownian diffusivity and aggregation rate of Laponite clay particles. J Colloid Interface Sci 240:54–66. https://doi.org/10.1006/jcis.2001.7646
57. Bergaya F, Vayer M (1997) CEC of clays: measurement by adsorption of a copper ethylenediamine complex. Appl Clay Sci 12:275–280. https://doi.org/10.1016/S0169-1317(97)00012-4
58. Borden D, Giese RF (2001) Baseline studies of the clay minerals society source clays: cation exchange capacity measurements by the ammonia-electrode method. Clays Clay Miner 49:444–445. https://doi.org/10.1346/CCMN.2001.0490510
59. Mourchid A, Levitz P (1998) Long-term gelation of laponite aqueous dispersions. Phys Rev E 57:R4887. https://doi.org/10.1103/PhysRevE.57.R4887
60. Komadel P (2016) Acid activated clays: materials in continuous demand. Appl Clay Sci 131:84–99. https://doi.org/10.1016/j.clay.2016.05.001
61. Mishra AK, Kuila T, Kim NH, Lee JH (2012) Effect of peptizer on the properties of Nafion-Laponite clay nanocomposite membranes for polymer electrolyte membrane fuel cells. J Memb Sci 389:316–323. https://doi.org/10.1016/j.memsci.2011.10.043

62. Li P, Kim NH, Hui D, Rhee KY, Lee JH (2009) Improved mechanical and swelling behavior of the composite hydrogels prepared by ionic monomer and acid-activated Laponite. Appl Clay Sci 46:414–417. https://doi.org/10.1016/j.clay.2009.10.007
63. Komadel P, Madejová J (2006) Acid activation of clay minerals. In: Bergaya F, Theng BKG, Lagaly G (eds) Developments in Clay Science. Elsevier, pp 263–287
64. Tkáč I, Komadel P, Müller D (1994) Acid-treated montmorillonites—a study by 29 Si and 27 Al MAS NMR. Clay Miner 29:11–19. https://doi.org/10.1180/claymin.1994.029.1.02
65. Breen C, Madejová J, Komadel P (1995) Characterisation of moderately acid-treated, size-fractionated montmorillonites using IR and MAS NMR spectroscopy and thermal analysis. J Mater Chem 5:469–474. https://doi.org/10.1039/JM9950500469
66. Bickmore BR, Bosbach D, Hochella MF Jr, Charlet L, Rufe E (2001) In situ atomic force microscopy study of hectorite and nontronite dissolution: Implications for phyllosilicate edge surface structures and dissolution mechanisms. Am Mineral 86:411–423. https://doi.org/10.2138/am-2001-0404
67. Van Rompaey K, Van Ranst E, De Coninck F, Vindevogel N (2002) Dissolution characteristics of hectorite in inorganic acids. Appl Clay Sci 21:241–256. https://doi.org/10.1016/S0169-1317(02)00086-8
68. Franco F, Pozo M, Cecilia JA, Benítez-Guerrero M, Lorente M (2016) Effectiveness of microwave assisted acid treatment on dioctahedral and trioctahedral smectites. The influence of octahedral composition. Appl Clay Sci 120:70–80. https://doi.org/10.1016/j.clay.2015.11.021
69. Kotal M, Bhowmick AK (2015) Polymer nanocomposites from modified clays: recent advances and challenges. Prog Polym Sci 51:127–187. https://doi.org/10.1016/j.progpolymsci.2015.10.001
70. Mishra AK, Chattopadhyay S, Nando GB (2010) Effect of modifiers on morphology and thermal properties of novel thermoplastic polyurethane-peptized laponite nanocomposite. J Appl Polym Sci 115:558–569. https://doi.org/10.1002/app.30975
71. Mishra AK, Rajamohanan PR, Nando GB, Chattopadhyay S (2011) Structure–property of thermoplastic polyurethane–clay nanocomposite based on covalent and dual-modified Laponite. Adv Sci Lett 4:65–73. https://doi.org/10.1166/asl.2011.1174
72. Wheeler PA, Wang J, Baker J, Mathias LJ (2005) Synthesis and characterization of covalently functionalized laponite clay. Chem Mater 17:3012–3018. https://doi.org/10.1021/cm050306a
73. Hanley HJM, Muzny CD, Butler BD (1997) Surfactant adsorption on a clay mineral: application of radiation scattering. Langmuir 13:5276–5282. https://doi.org/10.1021/la962048p
74. Nakamura T, Thomas JK (1987) Formation of surfactant double layers on laponite clay colloids. Langmuir 3:234–239. https://doi.org/10.1021/la00074a016
75. Pizzey C, Klein S, Leach E, van Duijneveldt JS, Richardson RM (2004) Suspensions of colloidal plates in a nematic liquid crystal: a small angle x-ray scattering study. J Phys Condens Matter 16:2479. https://doi.org/10.1088/0953-8984/16/15/002
76. Borsacchi S, Geppi M, Ricci L, Ruggeri G, Veracini CA (2007) Interactions at the surface of organophilic-modified laponites: a multinuclear solid-state NMR study. Langmuir 23:3953–3960. https://doi.org/10.1021/la063040a
77. Li C-F, Zhang S-Y, Wang J, Feng X-S, Sun D-J, Xu J (2008) Interactions between Brij surfactants and Laponite nanoparticles and emulsions stabilized by their mixtures. Acta Chim Sin -Chinese Ed 66(21):2313–2320
78. Liu Q, Zhang S, Sun D, Xu J (2009) Aqueous foams stabilized by hexylamine-modified Laponite particles. Colloids Surfaces A Physicochem Eng Asp 338:40–46. https://doi.org/10.1016/j.colsurfa.2008.12.035
79. Savenko V, Bulavin L, Rawiso M, Loginov M, Vorobiev E, Lebovka NI (2013) Sedimentation stability and aging of aqueous dispersions of Laponite in the presence of cetyltrimethylammonium bromide. Phys Rev E 88:52301. https://doi.org/10.1103/PhysRevE.88.052301

80. Manilo MV, Lebovka N, Barany S (2017) Combined effect of cetyltrimethylammonium bromide and laponite platelets on colloidal stability of carbon nanotubes in aqueous suspensions. J Mol Liq 235:104–110. https://doi.org/10.1016/j.molliq.2017.01.090
81. Connolly J, van Duijneveldt JS, Klein S, Pizzey C, Richardson RM (2006) Effect of surfactant and solvent properties on the stacking behavior of non-aqueous suspensions of organically modified clays. Langmuir 22:6531–6538. https://doi.org/10.1021/la0609219
82. Lambert Y et al (2006) Second-harmonic generation imaging of $LiIO_3$/laponite nanocomposite waveguides. Jpn J Appl Phys 45:7525. https://doi.org/10.1143/JJAP.45.7525
83. Yaroshchuk O, Tomylko S, Kovalchuk O, Lebovka N (2014) Liquid crystal suspensions of carbon nanotubes assisted by organically modified Laponite nanoplatelets. Carbon N Y 68:389–398. https://doi.org/10.1016/j.carbon.2013.11.015
84. Lysenkov EA, Lebovka NI, Yakovlev YV, Klepko VV, Pivovarova NS (2012) Percolation behaviour of polypropylene glycol filled with multiwalled carbon nanotubes and Laponite. Compos Sci Technol 72:1191–1195. https://doi.org/10.1016/j.compscitech.2012.04.002
85. Pizzey C, Van Duijneveldt J, Klein S (2004) Liquid crystal clay composites. Mol Cryst Liq Cryst 409:51–57. https://doi.org/10.1080/15421400490435657
86. Li W, Yu L, Liu G, Tan J, Liu S, Sun D (2012) Oil-in-water emulsions stabilized by Laponite particles modified with short-chain aliphatic amines. Colloids Surfaces A Physicochem Eng Asp 400:44–51. https://doi.org/10.1016/j.colsurfa.2012.02.044
87. Bruno TJ, Lewandowska A, Tsvetkov F, Miller KE, Hanley HJM (2002) Wall-coated open-tubular column chromatography on an organo–clay stationary phase. J Chromatogr A 973:143–149. https://doi.org/10.1016/S0021-9673(02)01124-X
88. Mirau PA, Serres JL, Jacobs D, Garrett PH, Vaia RA (2008) Structure and dynamics of surfactant interfaces in organically modified clays. J Phys Chem B 112:10544–10551. https://doi.org/10.1021/jp801479h
89. Wang B, Zhou M, Rozynek Z, Fossum JO (2009) Electrorheological properties of organically modified nanolayered laponite: influence of intercalation, adsorption and wettability. J Mater Chem 19:1816–1828. https://doi.org/10.1039/B818502F
90. Leach ESH, Hopkinson A, Franklin K, van Duijneveldt JS (2005) Nonaqueous suspensions of laponite and montmorillonite. Langmuir 21:3821–3830. https://doi.org/10.1021/la0503909
91. Bulavin LA et al (2018) Microstructure and optical properties of nematic and cholesteric liquid crystals doped with organo-modified platelets. J Mol Liq 267:279–285. https://doi.org/10.1016/j.molliq.2017.12.078
92. Loyens W, Jannasch P, Maurer FHJ (2005) Poly (ethylene oxide)/Laponite nanocomposites via melt-compounding: effect of clay modification and matrix molar mass. Polymer (Guildf) 46:915–928. https://doi.org/10.1016/j.polymer.2004.11.076
93. Savenko V, Bulavin L, Rawiso M, Lebovka N (2014) Aging of aqueous Laponite dispersions in the presence of sodium polystyrene sulfonate. Ukr J Phys 59(6):589–595. https://doi.org/10.15407/ujpe59.06.0589
94. Capovilla L, Labbe P, Reverdy G (1991) Formation of cationic/anionic mixed surfactant bilayers on laponite clay suspensions. Langmuir 7:2000–2003. https://doi.org/10.1021/la00058a004
95. Shaydyuk Y, Turrell S, Moissette A, Hureau M, Gomza Y, Klepko V, Lebovka N (2014) New phenothiazine–laponite hybrid systems: adsorption and ionization. J Mol Struct 1056:1–6. https://doi.org/10.1016/j.molstruc.2013.10.022
96. Ohlow MJ, Moosmann B (2011) Phenothiazine: the seven lives of pharmacology's first lead structure. Drug Discov Today 16:119–131. https://doi.org/10.1016/j.drudis.2011.01.001
97. Staniford MC et al (2015) Photophysical efficiency-boost of aqueous aluminium phthalocyanine by hybrid formation with nano-clays. Chem Commun 51:13534–13537. https://doi.org/10.1039/C5CC05352H
98. Staniford MC, Lezhnina MM, Kynast UH (2015) Phthalocyanine blue in aqueous solutions. RSC Adv 5:3974–3977. https://doi.org/10.1039/C4RA11139G
99. Grabolle M, Starke M, Resch-Genger U (2016) Highly fluorescent dye--nanoclay hybrid materials made from different dye classes. Langmuir 32:3506–3513. https://doi.org/10.1021/acs.langmuir.5b04297

100. Mustafa R et al (2016) Synthesis of diatrizoic acid-modified LAPONITE® nanodisks for CT imaging applications. RSC Adv 6:57490–57496. https://doi.org/10.1039/C6RA11755D
101. Peraro GR et al (2020) Aminofunctionalized LAPONITE® as a versatile hybrid material for chlorhexidine digluconate incorporation: Cytotoxicity and antimicrobial activities. Appl Clay Sci 195:105733. https://doi.org/10.1016/j.clay.2020.105733
102. Kaup G, Felbeck T, Staniford M, Kynast U (2016) Towards the rare earth functionalization of nano-clays with luminescent reporters for biophotonics. J Lumin 169:581–586. https://doi.org/10.1016/j.jlumin.2015.03.009
103. Yang Y, Liu Z, Wu D, Wu M, Tian Y, Niu Z, Huang Y (2013) Edge-modified amphiphilic Laponite nano-discs for stabilizing Pickering emulsions. J Colloid Interface Sci 410:27–32. https://doi.org/10.1016/j.jcis.2013.07.060
104. Patil SP, Mathew R, Ajithkumar TG, Rajamohanan PR, Mahesh TS, Kumaraswamy G (2008) Gelation of covalently edge-modified laponites in aqueous media. 1. Rheology and nuclear magnetic resonance. J Phys Chem B 112:4536–4544. https://doi.org/10.1021/jp710489n
105. Daniel LM, Frost RL, Zhu HY (2008) Edge-modification of laponite with dimethyl-octylmethoxysilane. J Colloid Interface Sci 321:302–309. https://doi.org/10.1016/j.jcis.2008.01.032
106. Hegyesi N, Simon N, Pukánszky B (2019) Silane modification of layered silicates and the mechanism of network formation from exfoliated layers. Appl Clay Sci 171:74–81. https://doi.org/10.1016/j.clay.2019.01.023
107. Park M, Shim I-K, Jung E-Y, Choy J-H (2004) Modification of external surface of laponite by silane grafting. J Phys Chem Solids 65:499–501. https://doi.org/10.1016/j.jpcs.2003.10.031
108. Herrera NN, Letoffe J-M, Reymond J-P, Bourgeat-Lami E (2005) Silylation of laponite clay particles with monofunctional and trifunctional vinyl alkoxysilanes. J Mater Chem 15:863–871. https://doi.org/10.1039/B415618H
109. Wheeler PA, Wang J, Mathias LJ (2006) Poly (methyl methacrylate)/laponite nanocomposites: exploring covalent and ionic clay modifications. Chem Mater 18:3937–3945. https://doi.org/10.1021/cm0526361
110. Herrera NN, Letoffe J-M, Putaux J-L, David L, Bourgeat-Lami E (2004) Aqueous dispersions of silane-functionalized laponite clay platelets. A first step toward the elaboration of water-based polymer/clay nanocomposites. Langmuir 20:1564–1571. https://doi.org/10.1021/la0349267
111. Bandeira LC et al (2012) Preparation of composites of laponite with alginate and alginic acid polysaccharides. Polym Int 61:1170–1176. https://doi.org/10.1002/pi.4196
112. Bui VKH, Park D, Lee Y-C (2018) Aminoclays for biological and environmental applications: an updated review. Chem Eng J 336:757–772. https://doi.org/10.1016/j.cej.2017.12.052
113. Samei E, Pelc NJ (2020) Computed Tomography: Approaches, Applications, and Operations. Springer Nature, Switzerland AG
114. Ding L et al (2016) LAPONITE® -stabilized iron oxide nanoparticles for in vivo MR imaging of tumors. Biomater Sci 4:474–482. https://doi.org/10.1039/C5BM00508F
115. Mustafa R, Zhou B, Yang J, Zheng L, Zhang G, Shi X (2016) Dendrimer-functionalized LAPONITE® nanodisks loaded with gadolinium for T 1-weighted MR imaging applications. RSC Adv 6:95112–95119. https://doi.org/10.1039/C6RA18718H
116. De Melo Barbosa R, Ferreira MA, Meirelles LMA, NicoleZorato, Raffin FN (2020) 8 - Nanoclays in drug delivery systems. In: Cavallaro G, Fakhrullin R, Pasbakhsh P (eds) Clay Nanoparticles. Elsevier, pp 185–202
117. Snigdha S, Kalarikkal N, Thomas S, Radhakrishnan EK (2020) Engineered phyllosilicate clay-based antimicrobial surfaces. In: Engineered Antimicrobial Surfaces. Springer, pp 95–108
118. Gonçalves M, Mignani S, Rodrigues J, Tomás H (2020) A glance over doxorubicin based-nanotherapeutics: from proof-of-concept studies to solutions in the market. J Control Release 317:347–374. https://doi.org/10.1016/j.jconrel.2019.11.016
119. Roychoudhury S, Kumar A, Bhatkar D, Sharma NK (2020) Molecular avenues in targeted doxorubicin cancer therapy. Futur Oncol 16:687–700. https://doi.org/10.2217/fon-2019-0458

120. Barraud L et al (2005) Increase of doxorubicin sensitivity by doxorubicin-loading into nanoparticles for hepatocellular carcinoma cells in vitro and in vivo. J Hepatol 42:736–743. https://doi.org/10.1016/j.jhep.2004.12.035
121. Bezerra IM, Chiavone-Filho O, Mattedi S (2013) Solid-liquid equilibrium data of amoxicillin and hydroxyphenylglycine in aqueous media. Brazilian J Chem Eng 30:45–54. https://doi.org/10.1590/S0104-66322013000100006
122. Prieto E et al (2020) Dexamethasone delivery to the ocular posterior segment by sustained-release Laponite formulation. Biomed Mater. https://doi.org/10.1088/1748-605X/aba445
123. Varanda F, de Melo MJ, Caco AI, Dohrn R, Makrydaki FA, Voutsas E, Tassios D, Marrucho IM (2006) Solubility of antibiotics in different solvents. 1. Hydrochloride forms of tetracycline, moxifloxacin, and ciprofloxacin. Ind Eng Chem Res 45:6368–6374. https://doi.org/10.1021/ie060055v
124. Tsai Y-C, Tsai T-F (2019) Itraconazole in the treatment of nonfungal cutaneous diseases: a review. Dermatol Ther (Heidelb) 9:271–280. https://doi.org/10.6084/m9.figshare.8010563
125. Shin YH, Shin WC, Kim JW (2020) Effect of osteoporosis medication on fracture healing: an evidence based review. J Bone Metab 27:15–26. https://doi.org/10.11005/jbm.2020.27.1.15
126. Aickara D, Bashyam AM, Pichardo RO, Feldman SR (2020) Topical methotrexate in dermatology: a review of the literature. J Dermatolog Treat 1–21. https://doi.org/10.1080/09546634.2020.1770170
127. Sanders WE Jr (1992) Oral ofloxacin: a critical review of the new drug application. Clin Infect Dis 14:539–554. https://doi.org/10.1093/clinids/14.2.539
128. Marsot A, Boulamery A, Bruguerolle B, Simon N (2012) Vancomycin. Clin Pharmacokinet 51:1–13. https://doi.org/10.2165/11596390-000000000-00000
129. Caracas HCPM, Maciel JVB, de Souza MMG, Maia LC et al (2009) The use of lidocaine as an anti-inflammatory substance: a systematic review. J Dent 37:93–97. https://doi.org/10.1016/j.jdent.2008.10.005
130. Jaffary F, Abdellahi L, Nilforoushzaheh MA (2017) Review of the prevalence and causes of antimony compounds resistance in different societies: review article. Tehran Univ Med J TUMS Publ 75:399–407
131. Lonappan L, Brar SK, Das RK, Verma M, Surampalli RY (2016) Diclofenac and its transformation products: environmental occurrence and toxicity-a review. Environ Int 96:127–138. https://doi.org/10.1016/j.envint.2016.09.014
132. Prathumsap N, Shinlapawittayatorn K, Chattipakorn SC, Chattipakorn N (2020) Effects of doxorubicin on the heart: from molecular mechanisms to intervention strategies. Eur J Pharmacol 866:172818. https://doi.org/10.1016/j.ejphar.2019.172818
133. Amalina ND, Nurhayati IP, Meiyanto E (2017) Doxorubicin induces lamellipodia formation and cell migration. Indones J Cancer Chemoprevention 8:61–67. https://doi.org/10.14499/indonesianjcanchemoprev8iss2pp61-67
134. Martins-Neves SR, Cleton-Jansen A-M, Gomes CMF (2018) Therapy-induced enrichment of cancer stem-like cells in solid human tumors: where do we stand? Pharmacol Res 137:193–204. https://doi.org/10.1016/j.phrs.2018.10.011
135. Pacelli S, Paolicelli P, Moretti G, Petralito S, Di Giacomo S, Vitalone A, Casadei MA (2016) Gellan gum methacrylate and laponite as an innovative nanocomposite hydrogel for biomedical applications. Eur Polym J 77:114–123. https://doi.org/10.1016/j.eurpolymj.2016.02.007
136. Rawat K, Agarwal S, Tyagi A, Verma AK, Bohidar HB (2014) Aspect ratio dependent cytotoxicity and antimicrobial properties of nanoclay. Appl Biochem Biotechnol 174:936–944. https://doi.org/10.1007/s12010-014-0983-2
137. Nair BP, Sharma CP (2012) Poly (lactide-co-glycolide)-laponite-F68 nanocomposite vesicles through a single-step double-emulsion method for the controlled release of doxorubicin. Langmuir 28:4559–4564. https://doi.org/10.1021/la300005c
138. Wang S et al (2013) Laponite® nanodisks as an efficient platform for doxorubicin delivery to cancer cells. Langmuir 29:5030–5036. https://doi.org/10.1021/la4001363

139. Li K et al (2014) Enhanced in vivo antitumor efficacy of doxorubicin encapsulated within laponite nanodisks. ACS Appl Mater Interfaces 6:12328–12334. https://doi.org/10.1021/am502094a
140. Zheng L et al (2019) Direct assembly of anticancer drugs to form Laponite-based nanocomplexes for therapeutic co-delivery. Mater Sci Eng C 99:1407–1414. https://doi.org/10.1016/j.msec.2019.02.083
141. Wang G et al (2014) Amphiphilic polymer-mediated formation of laponite-based nanohybrids with robust stability and pH sensitivity for anticancer drug delivery. ACS Appl Mater Interfaces 6:16687–16695. https://doi.org/10.1021/am5032874
142. Gonçalves M et al (2014) pH-sensitive Laponite®/doxorubicin/alginate nanohybrids with improved anticancer efficacy. Acta Biomater 10:300–307. https://doi.org/10.1016/j.actbio.2013.09.013
143. Yang Y, Li J, Chen F, Qiao S, Li Y, Pan W (2020) Synthesis, formulation, and characterization of doxorubicin-loaded laponite/oligomeric hyaluronic acid-aminophenylboronic acid nanohybrids and cytological evaluation against MCF-7 breast cancer cells. AAPS PharmSciTech 21:5. https://doi.org/10.1208/s12249-019-1533-6
144. Zhou B et al (2018) Drug-mediation formation of nanohybrids for sequential therapeutic delivery in cancer cells. Colloids Surfaces B Biointerfaces 163:284–290. https://doi.org/10.1016/j.colsurfb.2017.12.046
145. Alkekhia D, Hammond PT, Shukla A (2020) Layer-by-layer biomaterials for drug delivery. Annu Rev Biomed Eng 22:1–24. https://doi.org/10.1146/annurev-bioeng-060418-052350
146. Xiao S et al (2016) Fine tuning of the pH-sensitivity of laponite-doxorubicin nanohybrids by polyelectrolyte multilayer coating. Mater Sci Eng C 60:348–356. https://doi.org/10.1016/j.msec.2015.11.051
147. Zhuang Y et al (2017) Laponite-polyethylenimine based theranostic nanoplatform for tumor-targeting CT imaging and chemotherapy. ACS Biomater Sci Eng 3:431–442. https://doi.org/10.1021/acsbiomaterials.6b00528
148. Wang G et al (2016) In Situ formation of pH-/thermo-sensitive nanohybrids via friendly-assembly of poly (N-vinylpyrrolidone) onto LAPONITE®. RSC Adv 6:31816–31823. https://doi.org/10.1039/C5RA25628C
149. Wu Y et al (2014) Folic acid-modified laponite nanodisks for targeted anticancer drug delivery. J Mater Chem B 2:7410–7418. https://doi.org/10.1039/C4TB01162G
150. Jiang T, Chen G, Shi X, Guo R (2019) Hyaluronic acid-decorated laponite® nanocomposites for targeted anticancer drug delivery. Polymers (Basel) 11:137. https://doi.org/10.3390/polym11010137
151. Jiang T et al (2020) Doxorubicin encapsulated in P-glycoprotein-modified 2D-nanodisks overcomes multidrug resistance. Chem Eur J. https://doi.org/10.1002/chem.201905097
152. Chen G et al (2015) Targeted doxorubicin delivery to hepatocarcinoma cells by lactobionic acid-modified laponite® nanodisks. New J Chem 39:2847–2855. https://doi.org/10.1039/C4NJ01916D
153. Mustafa R, Luo Y, Wu Y, Guo R, Shi X (2015) Dendrimer-functionalized laponite nanodisks as a platform for anticancer drug delivery. Nanomaterials 5:1716–1731. https://doi.org/10.3390/nano5041716
154. Fraile JM et al (2016) Laponite as carrier for controlled in vitro delivery of dexamethasone in vitreous humor models. Eur J Pharm Biopharm 108:83–90. https://doi.org/10.1016/j.ejpb.2016.08.015
155. Roozbahani M, Kharaziha M, Emadi R (2017) pH sensitive dexamethasone encapsulated laponite nanoplatelets: Release mechanism and cytotoxicity. Int J Pharm 518:312–319. https://doi.org/10.1016/j.ijpharm.2017.01.001
156. Jung H, Kim H-M, Bin CY, Hwang S-J, Choy J-H (2008) Itraconazole-Laponite: Kinetics and mechanism of drug release. Appl Clay Sci 40:99–107. https://doi.org/10.1016/j.clay.2007.09.002
157. Jung H, Kim H-M, Bin CY, Hwang S-J, Choy J-H (2008) Laponite-based nanohybrid for enhanced solubility and controlled release of itraconazole. Int J Pharm 349:283–290. https://doi.org/10.1016/j.ijpharm.2007.08.008

158. Ghadiri M, Hau H, Chrzanowski W, Agus H, Rohanizadeh R (2013) Laponite® clay as a carrier for in situ delivery of tetracycline. RSC Adv 3:20193–20201. https://doi.org/10.1039/C3RA43217C
159. Wang S et al (2012) Encapsulation of amoxicillin within laponite-doped poly (lactic-co-glycolic acid) nanofibers: preparation, characterization, and antibacterial activity. ACS Appl Mater Interfaces 4:6393–6401. https://doi.org/10.1021/am302130b
160. Häffner SM et al (2019) Interaction of laponite with membrane components-consequences for bacterial aggregation and infection confinement. ACS Appl Mater Interfaces 11:15389–15400. https://doi.org/10.1021/acsami.9b03527
161. Nair BP, Sindhu M, Nair PD (2016) Polycaprolactone-laponite composite scaffold releasing strontium ranelate for bone tissue engineering applications. Colloids Surfaces B Biointerfaces 143:423–430. https://doi.org/10.1016/j.colsurfb.2016.03.033
162. Reddy NS, Rao KK (2016) Polymeric hydrogels: recent advances in toxic metal ion removal and anticancer drug delivery applications. Indian J Adv Chem Sci 4(2):214–234
163. Ahmed EM (2015) Hydrogel: preparation, characterization, and applications: a review. J Adv Res 6:105–121. https://doi.org/10.1016/j.jare.2013.07.006
164. Van Bemmelen JM (1894) Das hydrogel und das krystallinische hydrat des kupferoxyds. Zeitschrift für Anorg Chemie 5:466–483. https://doi.org/10.1002/zaac.18940050156
165. Wichterle O, Lim D (1960) Hydrophilic gels for biological use. Nature 185:117–118. https://doi.org/10.1038/185117a0
166. Maitra J, Shukla VK (2014) Cross-linking in hydrogels—a review. Am J Polym Sci 4:2531. https://doi.org/10.5923/j.ajps.20140402.01
167. Devi L, Gaba P (2019) Hydrogel: an updated primer. J Crit Rev 6:1–10. https://doi.org/10.22159/jcr.2019v6i4.33266
168. Baroli B (2006) Photopolymerization of biomaterials: issues and potentialities in drug delivery, tissue engineering, and cell encapsulation applications. J Chem Technol Biotechnol Int Res Process Environ Clean Technol 81:491–499. https://doi.org/10.1002/jctb.1468
169. Hu B-H, Messersmith PB (2005) Enzymatically cross-linked hydrogels and their adhesive strength to biosurfaces. Orthod Craniofacial Res 8:145–149. https://doi.org/10.1111/j.1601-6343.2005.00330.x
170. Gaharwar AK, Rivera CP, Wu C-J, Schmidt G (2011) Transparent, elastomeric and tough hydrogels from poly (ethylene glycol) and silicate nanoparticles. Acta Biomater 7:4139–4148. https://doi.org/10.1016/j.actbio.2011.07.023
171. Liu H, Wang C, Gao Q, Liu X, Tong Z (2010) Magnetic hydrogels with supracolloidal structures prepared by suspension polymerization stabilized by $Fe_2O_3$ nanoparticles. Acta Biomater 6:275–281. https://doi.org/10.1016/j.actbio.2009.06.018
172. Sharma G et al (2018) Applications of nanocomposite hydrogels for biomedical engineering and environmental protection. Environ Chem Lett 16:113–146. https://doi.org/10.1007/s10311-017-0671-x
173. Li J, Wu C, Chu PK, Gelinsky M (2020) 3D printing of hydrogels: rational design strategies and emerging biomedical applications. Mater Sci Eng R Rep 140:100543. https://doi.org/10.1016/j.mser.2020.100543
174. Buwalda SJ, Boere KWM, Dijkstra PJ, Feijen J, Vermonden T, Hennink WE (2014) Hydrogels in a historical perspective: From simple networks to smart materials. J Control Release 190:254–273. https://doi.org/10.1016/j.jconrel.2014.03.052
175. Antoine EE, Vlachos PP, Rylander MN (2014) Review of collagen I hydrogels for bioengineered tissue microenvironments: characterization of mechanics, structure, and transport. Tissue Eng Part B Rev 20:683–696. https://doi.org/10.1089/ten.teb.2014.0086
176. Bidarra SJ, Barrias CC, Granja PL (2014) Injectable alginate hydrogels for cell delivery in tissue engineering. Acta Biomater 10:1646–1662. https://doi.org/10.1016/j.actbio.2013.12.006
177. Collins MN, Birkinshaw C (2013) Hyaluronic acid based scaffolds for tissue engineering—a review. Carbohydr Polym 92:1262–1279. https://doi.org/10.1016/j.carbpol.2012.10.028

178. Bae KH, Wang L-S, Kurisawa M (2013) Injectable biodegradable hydrogels: progress and challenges. J Mater Chem B 1:5371–5388. https://doi.org/10.1039/c3tb20940g
179. Khan S, Ullah A, Ullah K, Rehman N (2016) Insight into hydrogels. Des Monomers Polym 19:456–478. https://doi.org/10.1080/15685551.2016.1169380
180. Kehr NS, Atay S, Ergün B (2015) Self-assembled monolayers and nanocomposite hydrogels of functional nanomaterials for tissue engineering applications. Macromol Biosci 15:445–463. https://doi.org/10.1002/mabi.201400363
181. Taylor DL, In Het Panhuis M (2016) Self-healing hydrogels. Adv Mater 28:9060–9093. https://doi.org/10.1002/adma.201601613
182. Haraguchi K, Takehisa T (2002) Nanocomposite hydrogels: a unique organic–inorganic network structure with extraordinary mechanical, optical, and swelling/deswelling properties. Adv Mater 14:1120. https://doi.org/10.1002/1521-4095(20020816)14:16%3c1120::AID-ADMA1120%3e3.0.CO;2-9
183. Haraguchi K, Takehisa T, Fan S (2002) Effects of clay content on the properties of nanocomposite hydrogels composed of poly (N -isopropylacrylamide) and clay. Macromolecules 35:10162–10171. https://doi.org/10.1021/ma021301r
184. Haraguchi K, Li H-J, Matsuda K, Takehisa T, Elliott E (2005) Mechanism of forming organic/inorganic network structures during *in situ* free-radical polymerization in PNIPA−clay nanocomposite hydrogels. Macromolecules 38:3482–3490. https://doi.org/10.1021/ma047431c
185. Miyazaki S, Endo H, Karino T, Haraguchi K, Shibayama M (2007) Gelation mechanism of poly (N-isopropylacrylamide)−clay nanocomposite gels. Macromolecules 40:4287–4295. https://doi.org/10.1021/ma070104v
186. Li P, Siddaramaiah KNH, Yoo G-H, Lee J-H (2009) Poly(acrylamide/Laponite) nanocomposite hydrogels: swelling and cationic dye adsorption properties. J Appl Polym Sci 111:1786–1798. https://doi.org/10.1002/app.29061
187. Haraguchi K (2011) Stimuli-responsive nanocomposite gels. Colloid Polym Sci 289:455–473. https://doi.org/10.1007/s00396-010-2373-9
188. Gaharwar AK, Kishore V, Rivera C, Bullock W, Wu C-J, Akkus O, Schmidt G (2012) Physically crosslinked nanocomposites from silicate-crosslinked PEO: mechanical properties and osteogenic differentiation of human mesenchymal stem cells. Macromol Biosci 12:779–793. https://doi.org/10.1002/mabi.201100508
189. Manjula B et al (2017) Hydrogels and its nanocomposites from renewable resources: biotechnological and biomedical applications. In: Thakur VK, Thakur MK, Kessler MR (eds) Handbook of Composites from Renewable Materials. John Wiley & Sons, Inc., Hoboken, pp 67–95
190. Shen M, Li L, Sun Y, Xu J, Guo X, Prud'homme RK (2014) Rheology and adhesion of poly (acrylic acid)/Laponite nanocomposite hydrogels as biocompatible adhesives. Langmuir 30:1636–1642. https://doi.org/10.1021/la4045623
191. Tongwa P, Nygaard R, Bai B (2013) Evaluation of a nanocomposite hydrogel for water shut-off in enhanced oil recovery applications: design, synthesis, and characterization. J Appl Polym Sci 128:787–794. https://doi.org/10.1002/app.38258
192. Chen P, Xu S, Wu R, Wang J, Gu R, Du J (2013) A transparent Laponite polymer nanocomposite hydrogel synthesis via in-situ copolymerization of two ionic monomers. Appl Clay Sci 72:196–200. https://doi.org/10.1016/j.clay.2013.01.012
193. Strachota B et al (2015) Poly(N-isopropylacrylamide)–clay based hydrogels controlled by the initiating conditions: evolution of structure and gel formation. Soft Matter 11:9291–9306. https://doi.org/10.1039/C5SM01996F
194. Zinkovska N, Smilek J, Pekar M (2020) Gradient hydrogels - the state of the art in preparation methods. Polymers (Basel) 12:966. https://doi.org/10.3390/polym12040966
195. Tan Y et al (2018) Rapid recovery hydrogel actuators in air with bionic large-ranged gradient structure. ACS Appl Mater Interfaces 10:40125–40131. https://doi.org/10.1021/acsami.8b13235

196. Ionov L (2014) Hydrogel-based actuators: possibilities and limitations. Mater Today 17:494–503. https://doi.org/10.1016/j.mattod.2014.07.002
197. Zhang Y et al (2019) Thermo-responsive and shape-adaptive hydrogel actuators from fundamentals to applications. Eng Sci 6:1–11. https://doi.org/10.30919/es8d788
198. Tan Y et al (2018) A fast, reversible, and robust gradient nanocomposite hydrogel actuator with water-promoted thermal response. Macromol Rapid Commun 39:1700863. https://doi.org/10.1002/marc.201700863
199. Xu P et al (2020) Multidimensional gradient hydrogel and its application in sustained release. Colloid Polym Sci 298:1187–1195. https://doi.org/10.1007/s00396-020-04688-3
200. Yao C et al (2016) Smart hydrogels with inhomogeneous structures assembled using nanoclay-cross-linked hydrogel subunits as building blocks. ACS Appl Mater Interfaces 8:21721–21730. https://doi.org/10.1021/acsami.6b07713
201. Erol O, Pantula A, Liu W, Gracias DH (2019) Transformer hydrogels: a review. Adv Mater Technol 4:1900043. https://doi.org/10.1002/admt.201900043
202. Augé A, Zhao Y (2016) What determines the volume transition temperature of UCST acrylamide–acrylonitrile hydrogels? RSC Adv 6:70616–70623. https://doi.org/10.1039/C6RA12720G
203. Seuring J, Agarwal S (2012) Polymers with upper critical solution temperature in aqueous solution. Macromol Rapid Commun 33:1898–1920. https://doi.org/10.1002/marc.201200433
204. Huang H, Qi X, Chen Y, Wu Z (2019) Thermo-sensitive hydrogels for delivering biotherapeutic molecules: a review. Saudi Pharm J 27:990–999. https://doi.org/10.1016/j.jsps.2019.08.001
205. Teotia AK, Sami H, Kumar A (2015) Thermo-responsive polymers. In: Zhang J (ed) Switchable and Responsive Surfaces and Materials for Biomedical Applications. Elsevier, pp 3–43
206. Parmar V, Patel G, Abu-Thabit NY (2018) Responsive cyclodextrins as polymeric carriers for drug delivery applications. In: Makhlouf ASH, Abu-Thabit NY (eds) Stimuli Responsive Polymeric Nanocarriers for Drug Delivery Applications, vol. 1. Elsevier, pp 555–580
207. Song CW, Griffin R, Park HJ (2007) Influence of Tumor pH on Therapeutic Response. Cancer Drug Resistance. Humana Press, Totowa, pp 21–42
208. Li H, Go G, Ko SY, Park J-O, Park S (2016) Magnetic actuated pH-responsive hydrogel-based soft micro-robot for targeted drug delivery. Smart Mater Struct 25:027001, p 9. https://doi.org/10.1088/0964-1726/25/2/027001
209. Thakur S, Arotiba OA (2018) Synthesis, swelling and adsorption studies of a pH-responsive sodium alginate–poly (acrylic acid) superabsorbent hydrogel. Polym Bull 75:4587–4606. https://doi.org/10.1007/s00289-018-2287-0
210. Raja STK, Thiruselvi T, Mandal AB, Gnanamani A (2015) pH and redox sensitive albumin hydrogel: a self-derived biomaterial. Sci Rep 5:15977. https://doi.org/10.1038/srep15977
211. Yoon S, Chen B (2018) Elastomeric and pH-responsive hydrogels based on direct crosslinking of the poly (glycerol sebacate) pre-polymer and gelatin. Polym Chem 9:3727–3740. https://doi.org/10.1039/C8PY00544C
212. Wang Q, Wang Q, Teng W (2016) Injectable, degradable, electroactive nanocomposite hydrogels containing conductive polymer nanoparticles for biomedical applications. Int J Nanomedicine 11:131. https://doi.org/10.2147/IJN.S94777
213. Ekici S, Tetik A (2015) Development of polyampholyte hydrogels based on Laponite for electrically stimulated drug release. Polym Int 64:335–343. https://doi.org/10.1002/pi.4816
214. Weeber R, Hermes M, Schmidt AM, Holm C (2018) Polymer architecture of magnetic gels: a review. J Phys Condens Matter 30:63002. https://doi.org/10.1088/1361-648X/aaa344
215. Thévenot J, Oliveira H, Sandre O, Lecommandoux S (2013) Magnetic responsive polymer composite materials. Chem Soc Rev 42:7099. https://doi.org/10.1039/c3cs60058k
216. Frachini ECG, Petri DFS (2019) Magneto-responsive hydrogels: preparation, characterization, biotechnological and environmental applications. J Braz Chem Soc 30:2010–2028. https://doi.org/10.21577/0103-5053.20190074

217. Cousin F, Cabuil V, Levitz P (2002) Magnetic colloidal particles as probes for the determination of the structure of laponite suspensions. Langmuir 18:1466–1473. https://doi.org/10.1021/la010947u
218. Galicia JA, Sandre O, Cousin F, Guemghar D, Ménager C, Cabuil V (2003) Designing magnetic composite materials using aqueous magnetic fluids. J Phys Condens Matter 15:S1379. https://doi.org/10.1088/0953-8984/15/15/306
219. Cousin F, Cabuil V, Grillo I, Levitz P (2008) Competition between entropy and electrostatic interactions in a binary colloidal mixture of spheres and platelets. Langmuir 24:11422–11430. https://doi.org/10.1021/la8015595
220. Paula FL de O et al (2009) Gravitational and magnetic separation in self-assembled clay-ferrofluid nanocomposites. Brazilian J Phys 39:163–170. https://doi.org/10.1590/S0103-97332009000200007
221. Mahdavinia GR, Mousanezhad S, Hosseinzadeh H, Darvishi F, Sabzi M (2016) Magnetic hydrogel beads based on PVA/sodium alginate/Laponite RD and studying their BSA adsorption. Carbohydr Polym 147:379–391. https://doi.org/10.1016/j.carbpol.2016.04.024
222. Lebovka NI et al (2020) Temperature sensitive hydrogels cross-linked by magnetic Laponite RD: Effects of particle magnetization. Mater Adv 1:2994–2999. https://doi.org/10.1039/d0ma00687d
223. Goncharuk O et al (2020) Thermoresponsive hydrogels physically crosslinked with magnetically modified LAPONITE® nanoparticles. Soft Matter 16:5689–5701. https://doi.org/10.1039/D0SM00929F
224. Goncharuk O et al (2020) Thermosensitive hydrogel nanocomposites with magnetic laponite nanoparticles. Appl Nanosci 10:4559–4569. https://doi.org/10.1007/s13204-020-01388-w
225. Diamantopoulos G et al (2013) Magnetic hyperthermia of laponite based ferrofluid. J Magn Magn Mater 336:71–74. https://doi.org/10.1016/j.jmmm.2013.02.032
226. Aguiar AS et al (2020) The use of a laponite dispersion to increase the hydrophilicity of cobalt-ferrite magnetic nanoparticles. Appl Clay Sci 193:105663. https://doi.org/10.1016/j.clay.2020.105663
227. Jalili NA, Muscarello M, Gaharwar AK (2016) Nanoengineered thermoresponsive magnetic hydrogels for biomedical applications. Bioeng Transl Med 1:297–305. https://doi.org/10.1002/btm2.10034
228. Lee JH, Han WJ, Jang HS, Choi HJ (2019) Highly tough, biocompatible, and magneto-responsive $Fe_3O_4$/Laponite/PDMAAm nanocomposite hydrogels. Sci Rep 9:15024. https://doi.org/10.1038/s41598-019-51555-5
229. Sun Y, Wang Y, Yao J, Gao L, Li D, Liu Y (2017) Highly magnetic sensitivity of polymer nanocomposite hydrogels based on magnetic nanoparticles. Compos Sci Technol 141:40–47. https://doi.org/10.1016/j.compscitech.2017.01.006
230. Mahdavinia GR, Soleymani M, Etemadi H, Sabzi M, Atlasi Z (2018) Model protein BSA adsorption onto novel magnetic chitosan/PVA/laponite RD hydrogel nanocomposite beads. Int J Biol Macromol 107:719–729. https://doi.org/10.1016/j.ijbiomac.2017.09.042
231. Soleymani M, Akbari A, Mahdavinia GR (2019) Magnetic PVA/laponite RD hydrogel nanocomposites for adsorption of model protein BSA. Polym Bull 76:2321–2340. https://doi.org/10.1007/s00289-018-2480-1
232. Uva M, Pasqui D, Mencuccini L, Fedi S, Barbucci R (2014) Influence of alternating and static magnetic fields on drug release from hybrid hydrogels containing magnetic nanoparticles. J Biomater Nanobiotechnol 05:116–127. https://doi.org/10.4236/jbnb.2014.52014
233. Mahdavinia GR, Ettehadi S, Amini M, Sabzi M (2015) Synthesis and characterization of hydroxypropyl methylcellulose-g-poly(acrylamide)/LAPONITE RD nanocomposites as novel magnetic- and pH-sensitive carriers for controlled drug release. RSC Adv 5:44516–44523. https://doi.org/10.1039/C5RA03731J
234. Mahdavinia GR, Soleymani M, Sabzi M, Azimi H, Atlasi Z (2017) Novel magnetic polyvinyl alcohol/laponite RD nanocomposite hydrogels for efficient removal of methylene blue. J Environ Chem Eng 5:2617–2630. https://doi.org/10.1016/j.jece.2017.05.017

235. Mahdavinia GR, Rahmani Z, Mosallanezhad A, Karami S, Shahriari M (2016) Effect of magnetic laponite RD on swelling and dye adsorption behaviors of κ-carrageenan-based nanocomposite hydrogels. Desalin Water Treat 57:20582–20596. https://doi.org/10.1080/19443994.2015.1111808
236. Mola-ali-abasiyan S, Mahdavinia GR (2018) Polyvinyl alcohol-based nanocomposite hydrogels containing magnetic Laponite RD to remove cadmium. Environ Sci Pollut Res 25:14977–14988. https://doi.org/10.1007/s11356-018-1485-5
237. Liu Q, Liu L (2019) Novel light-responsive hydrogels with antimicrobial and antifouling capabilities. Langmuir 35:1450–1457. https://doi.org/10.1021/acs.langmuir.8b01663
238. Kuksenok O, Yashin VV, Dayal P, Balazs AC (2010) Copying from nature: designing adaptive, chemoresponsive gels. J Polym Sci Part B Polym Phys 48:2533–2541. https://doi.org/10.1002/polb.22113
239. Chandrawati R (2016) Enzyme-responsive polymer hydrogels for therapeutic delivery. Exp Biol Med 241:972–979. https://doi.org/10.1177/1535370216647186
240. Raghavendra GM, Jayaramudu T, Varaprasad K, Mohan Reddy GS, Raju KM (2015) Antibacterial nanocomposite hydrogels for superior biomedical applications: a facile eco-friendly approach. RSC Adv 5:14351–14358. https://doi.org/10.1039/C4RA15995K
241. Gonçalves M et al (2014) Antitumor efficacy of doxorubicin-loaded laponite/alginate hybrid hydrogels. Macromol Biosci 14:110–120. https://doi.org/10.1002/mabi.201300241
242. Koshy ST, Zhang DKY, Grolman JM, Stafford AG, Mooney DJ (2018) Injectable nanocomposite cryogels for versatile protein drug delivery. Acta Biomater 65:36–43. https://doi.org/10.1016/j.actbio.2017.11.024
243. Ghadiri M, Chrzanowski W, Rohanizadeh R (2014) Antibiotic eluting clay mineral (Laponite®) for wound healing application: an in vitro study. J Mater Sci Mater Med 25:2513–2526. https://doi.org/10.1007/s10856-014-5272-7
244. Golafshan N, Rezahasani R, Esfahani MT, Kharaziha M, Khorasani SN (2017) Nanohybrid hydrogels of laponite: PVA-Alginate as a potential wound healing material. Carbohydr Polym 176:392–401. https://doi.org/10.1016/j.carbpol.2017.08.070
245. Ordikhani F, Dehghani M, Simchi A (2015) Antibiotic-loaded chitosan–laponite films for local drug delivery by titanium implants: cell proliferation and drug release studies. J Mater Sci Mater Med 26:269. https://doi.org/10.1007/s10856-015-5606-0
246. Yang H, Hua S, Wang W, Wang A (2011) Composite hydrogel beads based on chitosan and laponite: preparation, swelling, and drug release behaviour. Iran Polym J 20(6):479–490
247. Oliveira MJA et al (2014) Influence of chitosan/clay in drug delivery of glucantime from PVP membranes. Radiat Phys Chem 94:194–198. https://doi.org/10.1016/j.radphyschem.2013.05.050
248. Haraguchi K, Murata K, Takehisa T (2013) Stimuli-responsive properties of nanocomposite gels comprising (2-methoxyethylacrylate-co-N, N-dimethylacrylamide) copolymer-clay networks. Macromol Symp 329:150–161. https://doi.org/10.1002/masy.201300026
249. Jafarbeglou M, Abdouss M, Shoushtari AM, Jafarbeglou M (2016) Clay nanocomposites as engineered drug delivery systems. RSC Adv 6:50002–50016. https://doi.org/10.1039/C6RA03942A
250. Pellá MCG et al (2018) Chitosan-based hydrogels: From preparation to biomedical applications. Carbohydr Polym 196:233–245. https://doi.org/10.1016/j.carbpol.2018.05.033
251. Pakdel PM, Peighambardoust SJ (2018) Review on recent progress in chitosan-based hydrogels for wastewater treatment application. Carbohydr Polym 201:264–279. https://doi.org/10.1016/j.carbpol.2018.08.070
252. Qu B, Luo Y (2020) Chitosan-based hydrogel beads: preparations, modifications and applications in food and agriculture sectors–a review. Int J Biol Macromol. https://doi.org/10.1016/j.ijbiomac.2020.02.240
253. Owens D III, Peppas N (2006) Opsonization, biodistribution, and pharmacokinetics of polymeric nanoparticles. Int J Pharm 307:93–102. https://doi.org/10.1016/j.ijpharm.2005.10.010

254. Takahashi T, Yamada Y, Kataoka K, Nagasaki Y (2005) Preparation of a novel PEG–clay hybrid as a DDS material: dispersion stability and sustained release profiles. J Control Release 107:408–416. https://doi.org/10.1016/j.jconrel.2005.03.031
255. Kotobuki N, Murata K, Haraguchi K (2013) Proliferation and harvest of human mesenchymal stem cells using new thermoresponsive nanocomposite gels. J Biomed Mater Res Part A 101A:537–546. https://doi.org/10.1002/jbm.a.34355
256. Gaharwar AK, Schexnailder PJ, Jin Q, Wu C-J, Schmidt G (2010) Addition of chitosan to silicate cross-linked PEO for tuning osteoblast cell adhesion and mineralization. ACS Appl Mater Interfaces 2:3119–3127. https://doi.org/10.1021/am100609t
257. Gaharwar AK et al (2010) Highly extensible bio-nanocomposite films with direction-dependent properties. Adv Funct Mater 20:429–436. https://doi.org/10.1002/adfm.200901606
258. Peak CW, Carrow JK, Thakur A, Singh A, Gaharwar AK (2015) Elastomeric cell-laden nanocomposite microfibers for engineering complex tissues. Cell Mol Bioeng 8:404–415. https://doi.org/10.1007/s12195-015-0406-7
259. Gaharwar AK, Schexnailder PJ, Kline BP, Schmidt G (2011) Assessment of using Laponite cross-linked poly (ethylene oxide) for controlled cell adhesion and mineralization. Acta Biomater 7:568–577. https://doi.org/10.1016/j.actbio.2010.09.015
260. Surdu I, Vătuiu D, Jurcoane Ş, Olteanu M, Vătuiu I (2018) The antimicrobial activity of neutral electrolyzed water against germs and fungi from feedstuffs, eggshells and laying henhouse. Rom Biotechnol Lett 3:13607–13614. https://www.e-repository.org/rbl/vol.23/iss.3/7.pdf
261. Chang C-W, van Spreeuwel A, Zhang C, Varghese S (2010) PEG/clay nanocomposite hydrogel: a mechanically robust tissue engineering scaffold. Soft Matter 6:5157–5164. https://doi.org/10.1039/c0sm00067a
262. Shibayama M, Suda J, Karino T, Okabe S, Takehisa T, Haraguchi K (2004) Structure and dynamics of poly (N-isopropylacrylamide)−clay nanocomposite gels. Macromolecules 37:9606–9612. https://doi.org/10.1021/ma048464v
263. Haraguchi K, Takehisa T, Ebato M (2006) Control of cell cultivation and cell sheet detachment on the surface of polymer/clay nanocomposite hydrogels. Biomacromol 7:3267–3275. https://doi.org/10.1021/bm060549b
264. Zhao H, Liu M, Zhang Y, Yin J, Pei R (2020) Nanocomposite hydrogels for tissue engineering applications. Nanoscale 12:14976–14995. https://doi.org/10.1039/D0NR03785K
265. Jaiswal MK, Xavier JR, Carrow JK, Desai P, Alge D, Gaharwar AK (2016) Mechanically Stiff nanocomposite hydrogels at ultralow nanoparticle content. ACS Nano 10:246–256. https://doi.org/10.1021/acsnano.5b03918
266. Mehrali M et al (2017) Nanoreinforced hydrogels for tissue engineering: Biomaterials that are compatible with load-bearing and electroactive tissues. Adv Mater 29:1603612. https://doi.org/10.1002/adma.201603612
267. Su D, Jiang L, Chen X, Dong J, Shao Z (2016) Enhancing the gelation and bioactivity of injectable silk fibroin hydrogel with Laponite nanoplatelets. ACS Appl Mater Interfaces 8:9619–9628. https://doi.org/10.1021/acsami.6b00891
268. Liu Y, Meng H, Konst S, Sarmiento R, Rajachar R, Lee BP (2014) Injectable dopamine-modified poly (ethylene glycol) nanocomposite hydrogel with enhanced adhesive property and bioactivity. ACS Appl Mater Interfaces 6:16982–16992. https://doi.org/10.1021/am504566v
269. Waters R, Pacelli S, Maloney R, Medhi I, Ahmed RPH, Paul A (2016) Stem cell secretome-rich nanoclay hydrogel: a dual action therapy for cardiovascular regeneration. Nanoscale 8:7371–7376. https://doi.org/10.1039/C5NR07806G
270. Paul A et al (2016) Nanoengineered biomimetic hydrogels for guiding human stem cell osteogenesis in three dimensional microenvironments. J Mater Chem B 4:3544–3554. https://doi.org/10.1039/C5TB02745D
271. Gaharwar AK et al (2014) Shear-thinning nanocomposite hydrogels for the treatment of hemorrhage. ACS Nano 8:9833–9842. https://doi.org/10.1021/nn503719n
272. Gaharwar AK, Peppas NA, Khademhosseini A (2014) Nanocomposite hydrogels for biomedical applications. Biotechnol Bioeng 111:441–453. https://doi.org/10.1002/bit.25160

273. Xavier JR et al (2015) Bioactive nanoengineered hydrogels for bone tissue engineering: a growth-factor-free approach. ACS Nano 9:3109–3118. https://doi.org/10.1021/nn507488s
274. Dawson JI, Oreffo ROC (2013) Clay: new opportunities for tissue regeneration and biomaterial design. Adv Mater 25:4069–4086. https://doi.org/10.1002/adma.201301034
275. Carrow JK et al (2018) Widespread changes in transcriptome profile of human mesenchymal stem cells induced by two-dimensional nanosilicates. Proc Natl Acad Sci 115:E3905–E3913. https://doi.org/10.1073/pnas.1716164115
276. Dawson JI, Kanczler JM, Yang XB, Attard GS, Oreffo ROC (2011) Clay gels for the delivery of regenerative microenvironments. Adv Mater 23:3304–3308. https://doi.org/10.1002/adma.201100968
277. Shi P, Kim Y-H, Mousa M, Sanchez RR, Oreffo ROC, Dawson JI (2018) Self-assembling nanoclay diffusion gels for bioactive osteogenic microenvironments. Adv Healthc Mater 7:1800331. https://doi.org/10.1002/adhm.201800331
278. Waters R, Alam P, Pacelli S, Chakravarti AR, Ahmed RPH, Paul A (2018) Stem cell-inspired secretome-rich injectable hydrogel to repair injured cardiac tissue. Acta Biomater 69:95–106. https://doi.org/10.1016/j.actbio.2017.12.025
279. Lokhande G et al (2018) Nanoengineered injectable hydrogels for wound healing application. Acta Biomater 70:35–47. https://doi.org/10.1016/j.actbio.2018.01.045
280. Kerativitayanan P, Tatullo M, Khariton M, Joshi P, Perniconi B, Gaharwar AK (2017) Nanoengineered osteoinductive and elastomeric scaffolds for bone tissue engineering. ACS Biomater Sci Eng 3:590–600. https://doi.org/10.1021/acsbiomaterials.7b00029
281. Mihaila SM, Gaharwar AK, Reis RL, Khademhosseini A, Marques AP, Gomes ME (2014) The osteogenic differentiation of SSEA-4 sub-population of human adipose derived stem cells using silicate nanoplatelets. Biomaterials 35:9087–9099. https://doi.org/10.1016/j.biomaterials.2014.07.052
282. Hasany M et al (2018) Combinatorial screening of nanoclay-reinforced hydrogels: a glimpse of the "Holy Grail" in orthopedic stem cell therapy? ACS Appl Mater Interfaces 10:34924–34941. https://doi.org/10.1021/acsami.8b11436
283. Basu S, Pacelli S, Feng Y, Lu Q, Wang J, Paul A (2018) Harnessing the noncovalent interactions of DNA backbone with 2D silicate nanodisks to fabricate injectable therapeutic hydrogels. ACS Nano 12:9866–9880. https://doi.org/10.1021/acsnano.8b02434
284. Heid S, Boccaccini AR (2020) Advancing bioinks for 3D bioprinting using reactive fillers: a review. Acta Biomater 113:1–22. https://doi.org/10.1016/j.actbio.2020.06.040
285. Sears C et al (2020) Conditioning of 3D printed nanoengineered ionic–covalent entanglement scaffolds with iP-hMSCs derived matrix. Adv Healthc Mater 9:1901580. https://doi.org/10.1002/adhm.201901580
286. Zhu W, Webster TJ, Zhang LG (2019) 4D printing smart biosystems for nanomedicine. Nanomedicine (Lond) 14(13):1643–1645. https://doi.org/10.2217/nnm-2019-0134
287. Wei J et al (2020) A 3D-printable TEMPO-oxidized bacterial cellulose/alginate hydrogel with enhanced stability via nanoclay incorporation. Carbohydr Polym 238:116207. https://doi.org/10.1016/j.carbpol.2020.116207
288. Cidonio G et al (2020) Nanoclay-based 3D printed scaffolds promote vascular ingrowth ex vivo and generate bone mineral tissue in vitro and in vivo. Biofabrication 12:35010. https://doi.org/10.1088/1758-5090/ab8753
289. Adib AA et al (2020) Direct-write 3D printing and characterization of a GelMA-based biomaterial for intracorporeal tissue engineering. Biofabrication 12:045006. https://doi.org/10.1088/1758-5090/ab97a1
290. Bidoia ED, Montagnolli RN (2018) Toxicity and Biodegradation Testing. Springer Science+Business Media LLC, New York, USA
291. Pillai SC, Lang Y (2019) Toxicity of Nanomaterials: Environmental and Healthcare Applications. CRC Press, Taylor & Francis Group, Boca Raton, USA
292. Ghadiri M, Chrzanowski W, Lee WH, Fathi A, Dehghani F, Rohanizadeh R (2013) Physico-chemical, mechanical and cytotoxicity characterizations of Laponite®/alginate nanocomposite. Appl Clay Sci 85:64–73. https://doi.org/10.1016/j.clay.2013.08.049

293. Repetto G, Del Peso A, Zurita JL (2008) Neutral red uptake assay for the estimation of cell viability/cytotoxicity. Nat Protoc 3:1125. https://doi.org/10.1038/nprot.2008.75
294. Präbst K, Engelhardt H, Ringgeler S, Hübner H (2017) Basic colorimetric proliferation assays: MTT, WST, and resazurin. In: Gilbert D, Friedrich O (eds) Cell Viability Assays. Methods in Molecular Biology, vol 1601. Humana Press, New York, NY, pp 1–17. https://doi.org/10.1007/978-1-4939-6960-9_1
295. Van Meerloo J, Kaspers GJL, Cloos J (2011) Cell sensitivity assays: the MTT assay. In: Cree I (ed) Cancer Cell Culture. Humana Press, Methods in Molecular Biology (Methods and Protocols), pp 237–245. https://doi.org/10.1007/978-1-61779-080-5_20
296. Voigt M, Bartels I, Nickisch-Hartfiel A, Jaeger M (2019) Determination of minimum inhibitory concentration and half maximal inhibitory concentration of antibiotics and their degradation products to assess the eco-toxicological potential. Toxicol Environ Chem 101:315–338. https://doi.org/10.1080/02772248.2019.1687706
297. Reller LB, Weinstein M, Jorgensen JH, Ferraro MJ (2009) Antimicrobial susceptibility testing: a review of general principles and contemporary practices. Clin Infect Dis 49:1749–1755. https://doi.org/10.1086/647952
298. Andrews JM (2001) Determination of minimum inhibitory concentrations. J Antimicrob Chemother 48:5–16. https://doi.org/10.1093/jac/48.suppl_1.5
299. Tan L et al (2020) Osteogenic differentiation of mesenchymal stem cells by silica/calcium micro-galvanic effects on the titanium surface. J Mater Chem B 8:2286–2295. https://doi.org/10.1039/D0TB00054J
300. Venkatraman SK, Swamiappan S (2020) Review on calcium-and magnesium-based silicates for bone tissue engineering applications. J Biomed Mater Res Part A 108:1546–1562. https://doi.org/10.1002/jbm.a.36925
301. Kurgan N et al (2019) Low dose lithium supplementation activates Wnt/β-catenin signalling and increases bone OPG/RANKL ratio in mice. Biochem Biophys Res Commun 511:394–397. https://doi.org/10.1016/j.bbrc.2019.02.066
302. Li Y, MacIel D, Tomás H, Rodrigues J, Ma H, Shi X (2011) Ph sensitive Laponite/alginate hybrid hydrogels: Swelling behaviour and release mechanism. Soft Matter 7:6231–6238. https://doi.org/10.1039/c1sm05345k
303. Motskin M et al (2009) Hydroxyapatite nano and microparticles: correlation of particle properties with cytotoxicity and biostability. Biomaterials 30:3307–3317. https://doi.org/10.1016/j.biomaterials.2009.02.044
304. Napierska D et al (2009) Size-dependent cytotoxicity of monodisperse silica nanoparticles in human endothelial cells. Small 5:846–853. https://doi.org/10.1002/smll.200800461
305. Akhavan O, Ghaderi E, Akhavan A (2012) Size-dependent genotoxicity of graphene nanoplatelets in human stem cells. Biomaterials 33:8017–8025. https://doi.org/10.1016/j.biomaterials.2012.07.040
306. Boyer C et al (2018) Laponite nanoparticle-associated silated hydroxypropylmethyl cellulose as an injectable reinforced interpenetrating network hydrogel for cartilage tissue engineering. Acta Biomater 65:112–122. https://doi.org/10.1016/j.actbio.2017.11.027
307. Gonzaga V de AM et al (2020) Chitosan-laponite nanocomposite scaffolds for wound dressing application. J Biomed Mater Res Part B Appl Biomater 108:1388–1397. https://doi.org/10.1002/jbm.b.34487
308. Vergaro V et al (2010) Cytocompatibility and uptake of halloysite clay nanotubes. Biomacromolecules 11:820–826. https://doi.org/10.1021/bm9014446